Technological Applications of Nano-Hybrid Composites

Virat Khanna
Maharaja Agrasen University, India

Prianka Sharma
Maharaja Agrasen University, India

Priyanka Mahajan
Maharaja Agrasen University, India

A volume in the Advances in Chemical and Materials Engineering (ACME) Book Series

IGI Global
PUBLISHER of TIMELY KNOWLEDGE

Published in the United States of America by
IGI Global
Engineering Science Reference (an imprint of IGI Global)
701 E. Chocolate Avenue
Hershey PA, USA 17033
Tel: 717-533-8845
Fax: 717-533-8661
E-mail: cust@igi-global.com
Web site: http://www.igi-global.com

Library of Congress Cataloging-in-Publication Data

Names: Khanna, Virat, editor. | Sharma, Prianka, 1978- editor. | Mahajan, Priyanka, 1986- editor.
Title: Technological applications of nano-hybrid composites / edited by: Virat Khanna, Prianka Sharma, Priyanka Mahajan.
Description: Hershey PA : Engineering Science Reference, [2024] | Includes bibliographical references. | Summary: "The proposed book focuses on the investigation and comprehension of multifunctional materials and their many applications in numerous technical domains"-- Provided by publisher.
Identifiers: LCCN 2023047362 (print) | LCCN 2023047363 (ebook) | ISBN 9798369312612 (hardcover) | ISBN 9798369312629 (ebook)
Subjects: LCSH: Nanocomposites (Materials)
Classification: LCC TA418.9.N35 T42 2024 (print) | LCC TA418.9.N35 (ebook) | DDC 620.1/18--dc23/eng/20240402
LC record available at https://lccn.loc.gov/2023047362
LC ebook record available at https://lccn.loc.gov/2023047363

This book is published in the IGI Global book series Advances in Chemical and Materials Engineering (ACME) (ISSN: 2327-5448; eISSN: 2327-5456)

British Cataloguing in Publication Data
A Cataloguing in Publication record for this book is available from the British Library.

For electronic access to this publication, please contact: eresources@igi-global.com.

Advances in Chemical and Materials Engineering (ACME) Book Series

ISSN:2327-5448
EISSN:2327-5456

Editor-in-Chief: J. Paulo Davim, University of Aveiro, Portugal

MISSION

The cross disciplinary approach of chemical and materials engineering is rapidly growing as it applies to the study of educational, scientific and industrial research activities by solving complex chemical problems using computational techniques and statistical methods.

The **Advances in Chemical and Materials Engineering (ACME) Book Series** provides research on the recent advances throughout computational and statistical methods of analysis and modeling. This series brings together collaboration between chemists, engineers, statisticians, and computer scientists and offers a wealth of knowledge and useful tools to academics, practitioners, and professionals through high quality publications.

COVERAGE

- Fracture Mechanics
- Toxicity of Materials
- Sustainable Materials
- Nanomaterials
- Industrial Chemistry
- Chemical engineering
- Heat Treatments
- Composites
- Multifuncional and Smart Materials
- Biomaterials

IGI Global is currently accepting manuscripts for publication within this series. To submit a proposal for a volume in this series, please contact our Acquisition Editors at Acquisitions@igi-global.com or visit: http://www.igi-global.com/publish/.

The Advances in Chemical and Materials Engineering (ACME) Book Series (ISSN 2327-5448) is published by IGI Global, 701 E. Chocolate Avenue, Hershey, PA 17033-1240, USA, www.igi-global.com. This series is composed of titles available for purchase individually; each title is edited to be contextually exclusive from any other title within the series. For pricing and ordering information please visit http://www.igi-global.com/book-series/advances-chemical-materials-engineering/73687. Postmaster: Send all address changes to above address.

Titles in this Series

For a list of additional titles in this series, please visit:
http://www.igi-global.com/book-series/advances-chemical-materials-engineering/73687

Machining Polymer Matrix Composites Tools, Techniques, and Sustainability
Francisco Mata Cabrera (Universidad de Castilla-La Mancha, Spain) and Issam Hanafi (Université Abdelmalek Essaâdi, Morocco)
Engineering Science Reference • copyright 2024 • 320pp • H/C (ISBN: 9781668499276) • US $300.00 (our price)

Building a Low-Carbon Society Through Applied Environmental Materials Science
Takaomi Kobayashi (Nagaoka University of Technology, Japan)
Engineering Science Reference • copyright 2024 • 300pp • H/C (ISBN: 9798369300039) • US $255.00 (our price)

FlexPDE and Finite Element Method Applications in Thermal Energy Storage and Cavities
Khashayar Hosseinzadeh (Sharif University of Technology, Iran) Mehdi Mahboobtosi (Babol Noshirvani University of Technology, Iran) and Davood Domairry Ganji (Babol Noshirvani University of Technology, Iran)
Engineering Science Reference • copyright 2024 • 300pp • H/C (ISBN: 9798369309322) • US $225.00 (our price)

Advancements in Powder Metallurgy Processing, Applications, and Properties
Shashanka Rajendrachari (Bartin University, Turkey) and Rayappa Shrinivas Mahale (Jain College of Engineering and Research, India)
Engineering Science Reference • copyright 2024 • 300pp • H/C (ISBN: 9781668493854) • US $275.00 (our price)

Innovative Multifunctional Nanomaterial for Photocatalysis, Sensing, and Imaging
Azad Kumar (Siddharth University, India)
Engineering Science Reference • copyright 2023 • 283pp • H/C (ISBN: 9781668487433) • US $270.00 (our price)

Innovations in Materials Chemistry, Physics, and Engineering Research

For an entire list of titles in this series, please visit:
http://www.igi-global.com/book-series/advances-chemical-materials-engineering/73687

IGI Global
PUBLISHER of TIMELY KNOWLEDGE

701 East Chocolate Avenue, Hershey, PA 17033, USA
Tel: 717-533-8845 x100 • Fax: 717-533-8661
E-Mail: cust@igi-global.com • www.igi-global.com

Table of Contents

Preface xiv

Chapter 1
A Futuristic Approach on the Multifunctionality of Nanomaterials: Relevance of Nanoparticles 1
Archna Talwar, Isabella Thoburn College, Lucknow, India
Avni Nayyar, Era University, Lucknow, India
Shruti Anand, Era University, Lucknow, India
Manaal Zahera, Era University, Lucknow, India
Faria Fatima, Integral University, India

Chapter 2
Characterization Methods for Assessing Hybrid Nanomaterial Performance 37
Prem Sagar, The Technological Institute of Textile Sciences, Bhiwani, India
Sushma Sangwan, Sumit Dhariwal Memorial Group of institutions, Hisar, India

Chapter 3
Hybrid Nanomaterials: A Sustainable Tool to Detect Environmental Problems. 63
Archna Talwar, Isabella Thoburn College, Lucknow, India
Shruti Anand, Isabella Thoburn College, Lucknow, India
Avni Nayyar, Isabella Thoburn College, Lucknow, India
Faria Fatima, Integral University, India
Manaal Zahera, Era University, Lucknow, India

Chapter 4
Integration of Hybrid Nanomaterials and Artificial Intelligence for Sustainable Agriculture 97
Madhu Bala, Maharaja Agrasen University, India
Ritika Sharma, Maharaja Agrasen University, India
Shilpa Gupta, Maharaja Agrasen University, India

Chapter 5
Energy Harvesting and Storage Applications for Hybrid Nanomaterials..........119
Jagadeesh Ramadoss, Bharathidasan University, India
Arumugam Sonachalam, Bharathidasan University, India

Chapter 6
Unleashing the Potential: Exploring Hybrid Nanomaterials for Advanced Energy Harvesting and Storage Systems 138
Pooja Kapoor, School of Basic and Applied Sciences, Maharaja Agrasen University, Baddi, India
Yogyata Pathania, DAV Post Graduate College, Chandigarh, India

Chapter 7
Biomedical Applications of Hybrid Nanomaterials..170
Bancha Yingngam, Faculty of Pharmaceutical Sciences, Ubon Ratchathani University, Thailand
Benjabhorn Sethabouppha, Faculty of Pharmaceutical Sciences, Ubon Ratchathani University, Thailand
Krit Yingngam, Faculty of Pharmaceutical Sciences, Ubon Ratchathani University, Thailand
Rojjares Netthong, Faculty of Pharmaceutical Sciences, Ubon Ratchathani University, Thailand
Jeerisuda Khumsikiew, Faculty of Pharmaceutical Sciences, Ubon Ratchathani University, Thailand
Sisira Donsamak, Faculty of Pharmaceutical Sciences, Ubon Ratchathani University, Thailand
Abhiruj Navabhatra, College of Pharmacy, Rangsit University, Thailand

Chapter 8
Exploring the Transformative Potential of Hybrid Nanoparticles in Biomedical Applications: Relevance of Hybrid Nanoparticles 226
Arshya Hashim, Dr. D.Y. Patil Arts, Commerce, and Science College, India
Faria Fatima, Integral University, India

Chapter 9
Hybrid Nanomaterial Employment in Clinical and Therapeutic Applications..247
Marriam Zargham, University of Health Sciences, Lahore, Pakistan
Muhammad Farhan Khan, University of Health Sciences, Lahore, Pakistan
Anam Munawar, University of Health Sciences, Lahore, Pakistan

Chapter 10
A Review on the Chemical Vapor Deposition Synthesis of 2D Materials and Their Applications 270
Sumit Kumar, Department of Physics, Patna University, India

Compilation of References 295

About the Contributors 370

Index 374

Detailed Table of Contents

Preface...xiv

Chapter 1
A Futuristic Approach on the Multifunctionality of Nanomaterials: Relevance of Nanoparticles..1
Archna Talwar, Isabella Thoburn College, Lucknow, India
Avni Nayyar, Era University, Lucknow, India
Shruti Anand, Era University, Lucknow, India
Manaal Zahera, Era University, Lucknow, India
Faria Fatima, Integral University, India

The realm of analytical chemistry has been invigorated by the advent of multifunctional nanoparticles. Nanomaterials, with their distinct properties stemming from quantum effects and high surface-to-volume ratios, are poised to reshape industries ranging from electronics to medicine and environmental sustainability as they deliver unprecedented performance by integrating semiconducting, plasmonic, and piezoelectronic properties. Furthermore, multifunctional nano species play a pivotal role in personalized medicine and targeted therapies. Magnetic nanoparticles respond to magnetic fields and are employed in hyperthermia therapy and targeted drug delivery. The utilization of nanostructures for promoting environmental sustainability is highly commendable. They have the remarkable ability to pinpoint pollutants and their degradation. Therefore, as the research progresses, there are transformative breakthroughs to harness the multifunctionality of nanomaterials across scientific domains, driving society towards a future characterized by technological marvels and sustainable progress.

Chapter 2
Characterization Methods for Assessing Hybrid Nanomaterial Performance.....37
Prem Sagar, The Technological Institute of Textile Sciences, Bhiwani, India
Sushma Sangwan, Sumit Dhariwal Memorial Group of institutions, Hisar, India

The rapid development of nanotechnology has led to the emergence of hybrid nanomaterials with diverse applications across various fields. These materials combine different types of nanoparticles, organic molecules, or polymers to achieve enhanced properties and multifunctionality. To fully exploit the potential of hybrid nanomaterials, accurate and comprehensive characterization methods are crucial for assessing their performance and understanding their structure-function relationships. Characterization techniques such as electron microscopy, spectroscopy, and atomic force microscopy play a pivotal role in unravelling hybrid nanomaterials' intricate structures and properties, facilitating their integration into cutting-edge technologies and innovative solutions. This chapter aims to provide an in-depth exploration of the characterization techniques employed to evaluate the performance of hybrid nanomaterials, showcasing their significance in guiding material design and application development.

Chapter 3
Hybrid Nanomaterials: A Sustainable Tool to Detect Environmental Problems. 63
Archna Talwar, Isabella Thoburn College, Lucknow, India
Shruti Anand, Isabella Thoburn College, Lucknow, India
Avni Nayyar, Isabella Thoburn College, Lucknow, India
Faria Fatima, Integral University, India
Manaal Zahera, Era University, Lucknow, India

Hybrid nanomaterials, a novel class formed by integrating organic and inorganic components at the nanoscale, have gained attention in nanotechnology and materials science. This synergy creates structures with enhanced functionalities. In environmental detection, they play a crucial role due to their sustainable and cutting-edge attributes. This chapter examines their pivotal role in addressing environmental issues, leveraging diverse nanomaterial properties. With tunable properties, high surface area, and sensitivity, hybrid nanomaterials are ideal for sensing environmental parameters. Their sustainability aligns with the global push for eco-friendly solutions, extending to applications in renewable energy and waste reduction. The chapter explores synthesis methods, sensor design intricacies, and applications, highlighting their potential to revolutionize environmental monitoring and contribute to a sustainable future through advancements in detection technologies.

Chapter 4
Integration of Hybrid Nanomaterials and Artificial Intelligence for Sustainable Agriculture 97
Madhu Bala, Maharaja Agrasen University, India
Ritika Sharma, Maharaja Agrasen University, India
Shilpa Gupta, Maharaja Agrasen University, India

The agricultural industry is a primary source of livelihood for the world's population. The importance of agriculture in ensuring a sustainable future is becoming even more evident as we deal with the problems generated by a growing global population. To address the challenges of food security, environmental preservation, and resource efficiency, the integration of hybrid nanomaterials with artificial intelligence (AI) represents a transformative approach to attain sustainable agriculture. This chapter will concentrate on the applications of hybrid nanomaterials and AI in agriculture, emphasising their functions in sustainable and smart farming. Further, researchers working in this domain will get an insight that how these two cutting edge techniques can be used for characterizing and quantifying the interactions between nanomaterials, plants, soil, and pests thereby paving the way for a more sustainable and productive future for agriculture.

Chapter 5
Energy Harvesting and Storage Applications for Hybrid Nanomaterials..........119
Jagadeesh Ramadoss, Bharathidasan University, India
Arumugam Sonachalam, Bharathidasan University, India

Hybrid nanomaterials (HNs) have developed as promising candidates for advancing energy harvesting and storage technologies due to their unique combination of properties. This chapter focuses on developments in the design, synthesis, and applications of HNs for energy storage/harvesting applications. These materials, formed by integrating distinct nanoscale components such as nanoparticles, nanowires, and nanosheets, exhibit synergistic effects that enhance energy conversion, storage capacity, and overall device performance. These strategies encompass bottom-up and top-down approaches, self-assembly methods, and surface functionalization techniques. It aims to understand how these combinations affect charge transport, catalytic activity, and electrochemical behavior. These applications span a wide range, including solar cells, thermoelectric generators, electrochemical energy storage, and conversion.

Chapter 6
Unleashing the Potential: Exploring Hybrid Nanomaterials for Advanced Energy Harvesting and Storage Systems 138
Pooja Kapoor, School of Basic and Applied Sciences, Maharaja Agrasen University, Baddi, India
Yogyata Pathania, DAV Post Graduate College, Chandigarh, India

The fast-rising market for portable electronic devices and the development of hybrid electric vehicles require the use of clean energy sources at a much higher level than presently in use. Nanomaterials have the potential to revolutionize energy harvesting and storage, providing a sustainable solution for our ever-growing energy needs. These materials combine the unique properties of different nanomaterials, leading to the development of new hybrid nanomaterials, to create synergistic effects, enhancing their overall performance and efficiency. Owing to their tunability and versatility, scientists can tailor the composition, morphology, and structure of these materials to optimize their performance. This technology has led to the development of advanced supercapacitors, fuel cells, solar cells, supercapacitors, rechargeable batteries, etc. with high energy density, fast charging capabilities, and long cycle life. This chapter is focused on specially designed hybrid nanomaterials for energy harvesting and storage applications.

Chapter 7
Biomedical Applications of Hybrid Nanomaterials..170
Bancha Yingngam, Faculty of Pharmaceutical Sciences, Ubon Ratchathani University, Thailand
Benjabhorn Sethabouppha, Faculty of Pharmaceutical Sciences, Ubon Ratchathani University, Thailand
Krit Yingngam, Faculty of Pharmaceutical Sciences, Ubon Ratchathani University, Thailand
Rojjares Netthong, Faculty of Pharmaceutical Sciences, Ubon Ratchathani University, Thailand
Jeerisuda Khumsikiew, Faculty of Pharmaceutical Sciences, Ubon Ratchathani University, Thailand
Sisira Donsamak, Faculty of Pharmaceutical Sciences, Ubon Ratchathani University, Thailand
Abhiruj Navabhatra, College of Pharmacy, Rangsit University, Thailand

The ever-evolving field of nanotechnology has led to groundbreaking advancements in biomedical science, particularly through the development of hybrid nanomaterials. This chapter explores the synthesis, characterization, and diverse biomedical applications of these materials, which combine the advantageous properties of both organic and inorganic constituents for optimized functionality. The text delves into the use of hybrid nanomaterials in areas such as drug delivery, diagnostics, tissue engineering, and biosensors. A key case study demonstrating the wound-healing applications of silver nanoparticle-loaded nanofibrous scaffolds is included. The chapter also addresses the challenges and ethical considerations associated with the clinical translation of these nanomaterials. Therefore, this chapter serves as a comprehensive guide for readers interested in understanding and harnessing the immense potential of hybrid nanomaterials in biomedical applications.

Chapter 8
Exploring the Transformative Potential of Hybrid Nanoparticles in Biomedical Applications: Relevance of Hybrid Nanoparticles 226
Arshya Hashim, Dr. D.Y. Patil Arts, Commerce, and Science College, India
Faria Fatima, Integral University, India

Hybrid nanoparticles are nanoscale particles that are composed by interaction between the different components, resulting in enhanced properties that can be harnessed for wide range of application across fields like medicine, electronics energy, and more. These nanoparticles are typically in the size range from 1-100 nanometers, which is advantageous. Because at this scale, they often exhibit novel behavior due to their quantum and surface effects. The choice of material and the way they are combined can be tailored to achieve specific goals. For example, in biomedical applications, hybrid nanoparticles can be engineered to have specific targeting abilities such as targeted drug delivery, theranostics, gene therapy, phototherapy, tissue regeneration, vaccines, antibacterial, biomolecules detection, imaging probes, tissue engineering, biosensing, and cancer treatment. They have enhanced qualities with increased target specificity and sensitivity, extended circulation times, and resistance to biological barriers.

Chapter 9
Hybrid Nanomaterial Employment in Clinical and Therapeutic Applications..247
Marriam Zargham, University of Health Sciences, Lahore, Pakistan
Muhammad Farhan Khan, University of Health Sciences, Lahore, Pakistan
Anam Munawar, University of Health Sciences, Lahore, Pakistan

Hybrid nanomaterials are based on a combination of materials resulting in amalgamation of unique features that can address various challenges in medicine and healthcare. Such materials offer a wide range of functionalities that can revolutionize diagnostics, therapeutics, and medical imaging. In drug delivery, tunable properties of hybrid nanomaterials enable precise targeting of specific cells or tissues, enhancing the therapeutic efficacy while minimizing side effects while their utility in diagnostic techniques facilitate early disease detection with improved sensitivity and selectivity.

Chapter 10
A Review on the Chemical Vapor Deposition Synthesis of 2D Materials and Their Applications 270
Sumit Kumar, Department of Physics, Patna University, India

This chapter explains the principle and synthesis method involved in the chemical vapor deposition (CVD) process. A smart method for making layered 2D atomic crystal nanomaterials on a compatible substrate through bottom-up approach is explored. CVD is a vacuum deposition method used to produce high-quality, high-performance solid materials. CVD is a scalable method to fabricate one layer to multi-layer thin films to bulk layer hybrid materials and their hetero-structure composites by changing the precursor compound with combination of an appropriate element having affinity toward the precursor complex cations. The process stacks two or more atomic layers of different compounds in the rich variety of 2D library to explore novel and collective quantum phenomena at the interfaces. Recently, 2D-SnXSY semiconductor materials have been synthesized by using CVD. These hybrid materials find various applications in sensors, photovoltaics, and quantum phenomena. Latest trends on fabricating hetero-structures of SnS and SnS2 with other hybrid composites by CVD have also been discussed.

Compilation of References 295

About the Contributors 370

Index 374

Preface

In the captivating landscape of nanotechnology, the amalgamation of diverse materials into hybrid composites has ignited a paradigm shift, opening up a plethora of possibilities at the intersection of innovation and application. As editors of *Technological Applications of Nano-Hybrid Composites*, it is with great enthusiasm that we present this meticulously curated and comprehensive edited reference book. This anthology endeavors to unravel the intricate tapestry of nanomaterials endowed with hybrid reinforcements, and their profound impact across an array of technological domains.

This collaborative effort is a testament to the collective wisdom of our esteemed contributors, each an authority in their respective fields. The chapters within this volume transcend disciplinary boundaries to provide a holistic exploration of the fundamental principles, state-of-the-art fabrication techniques, precise characterization methods, and real-world applications of nano-hybrid composites.

The book unfolds systematically, guiding readers through the fundamental aspects of nanomaterials. It commences with a lucid exposition of the principles underpinning the design and development of these materials, illuminating their unique properties and the scientific principles governing their functionality. The subsequent sections delve into the latest and most innovative fabrication techniques, placing a spotlight on methodologies that empower researchers to engineer materials with tailored functionalities.

A pivotal focus of the book is the importance of characterization methods. The chapters herein shed light on the methods and techniques used to discern the performance and behavior of nanohybrid composite materials. This emphasis on characterization not only enhances our understanding of these materials but also serves as a guide for researchers and practitioners in optimizing their applications.

The scope of the book extends beyond theoretical discussions, venturing into diverse technological applications such as electronics, energy, aerospace, biomedical engineering, construction, and environmental sensing. Each chapter unfolds as a detailed exploration, offering insights into how multifunctional materials are reshaping and revolutionizing their respective industries. Through a nuanced examination of their advantages, challenges, and potential future developments, readers gain a comprehensive understanding of the transformative power of nano-hybrid composite materials.

ORGANIZATION OF THE BOOK

Chapter 1: This chapter explores the transformative role of multifunctional nanoparticles in analytical chemistry. Emphasizing the quantum effects and high surface-to-volume ratios of nanomaterials, it illustrates their potential to reshape industries, from electronics to medicine and environmental sustainability. The chapter particularly highlights their integration of semiconducting, plasmonic, and piezoelectronic properties, showcasing breakthroughs in personalized medicine, targeted therapies, hyperthermia therapy, and targeted drug delivery. The discussion extends to nanomaterials' commendable role in environmental sustainability, pinpointing pollutants and contributing to a future characterized by technological marvels and sustainable progress.

Chapter 2: This chapter delves into the realm of hybrid nanomaterials, born out of the rapid development of nanotechnology. Focusing on diverse applications across fields, it highlights the significance of accurate characterization methods for assessing performance and understanding structure-function relationships. The chapter explores techniques such as electron microscopy, spectroscopy, and atomic force microscopy, showcasing their pivotal role in unraveling intricate structures. By providing an in-depth exploration of these techniques, the chapter serves as a guide for material design and application development in cutting-edge technologies.

Chapter 3: This chapter explores the novel class of hybrid nanomaterials, combining organic and inorganic components at the nanoscale, with a specific focus on their pivotal role in addressing environmental issues. It examines their sustainable and cutting-edge attributes, demonstrating their suitability for sensing environmental parameters. The chapter covers synthesis methods, sensor design intricacies, and applications, emphasizing their potential to revolutionize environmental monitoring and contribute to a sustainable future through advancements in detection technologies.

Chapter 4: Focusing on the intersection of hybrid nanomaterials and artificial intelligence (AI), this chapter addresses the challenges of food security, environmental preservation, and resource efficiency in agriculture. It emphasizes the transformative approach of integrating hybrid nanomaterials and AI for sustainable and smart farming. The chapter explores applications in characterizing interactions between nanomaterials, plants, soil, and pests, paving the way for a more sustainable and productive future in agriculture.

Chapter 5: This chapter delves into the developments in the design, synthesis, and applications of hybrid nanomaterials for energy storage and harvesting. By integrating distinct nanoscale components, these materials exhibit synergistic effects enhancing energy conversion, storage capacity, and overall device performance. The discussion encompasses bottom-up and top-down approaches, self-assembly methods, and surface functionalization techniques, providing insights into their applications in solar cells, thermoelectric generators, and electrochemical energy storage and conversion.

Chapter 6: Focusing on the ever-growing demand for clean energy in portable electronic devices and hybrid electric vehicles, this chapter explores the role of nanomaterials in energy harvesting and storage. The discussion centers on specially designed hybrid nanomaterials tailored for applications such as advanced supercapacitors, fuel cells, solar cells, and rechargeable batteries. The chapter highlights their tunability, versatility, and potential to revolutionize energy solutions for our increasing energy needs.

Chapter 7: Groundbreaking advancements in biomedical science are explored in this chapter, focusing on the synthesis, characterization, and diverse biomedical applications of hybrid nanomaterials. These materials combine the properties of both organic and inorganic constituents for optimized functionality, with applications in drug delivery, diagnostics, tissue engineering, and biosensors. A case study on silver nanoparticle-loaded nanofibrous scaffolds showcases the potential for wound-healing applications, while addressing challenges and ethical considerations associated with clinical translation.

Chapter 8: This chapter delves into the realm of hybrid nanoparticles, nanoscale particles composed through the interaction of different components. It discusses their enhanced properties applicable across diverse fields, such as medicine, electronics, energy, and more. Specifically, it explores how these nanoparticles can be engineered for targeted drug delivery, theranostics, gene therapy, tissue regeneration, vaccines, antibacterial applications, biomolecule detection, imaging probes, and cancer treatment. Their enhanced qualities, including increased target specificity and sensitivity, are highlighted.

Chapter 9: This chapter explores the diverse functionalities offered by hybrid nanomaterials in medicine and healthcare. The tunable properties of these materials play a crucial role in revolutionizing diagnostics, therapeutics, and medical imaging. The discussion extends to drug delivery, emphasizing the precision of targeting specific cells or tissues for enhanced therapeutic efficacy while minimizing side effects. The utility of hybrid nanomaterials in diagnostic techniques is also highlighted for early disease detection with improved sensitivity and selectivity.

Chapter 10: This chapter provides an in-depth exploration of the principles and synthesis methods involved in the chemical vapor deposition (CVD) process for creating layered 2D atomic crystal nano-materials. It outlines CVD as a scalable method for fabricating thin films, bulk layer hybrid materials, and hetero-structure composites. The focus is on 2D-SnXSY semiconductor materials synthesized by CVD and their applications in sensors, photovoltaics, and quantum phenomena. The chapter concludes with discussions on the latest trends in fabricating hetero structures of SnS and SnS2 with other hybrid composites using CVD.

This anthology is not just a collection of chapters; it is an invitation to embark on a journey through the fascinating world of nano-hybrid composites. We encourage readers to immerse themselves in the narratives presented within these pages, as we collectively explore the transformative potential that nano-hybrid composite materials hold for technological advancements across a myriad of industries.

We extend our gratitude to our contributors, whose expertise and dedication have made this volume possible, and we hope this book serves as an indispensable resource for scholars, researchers, and industry professionals alike, seeking to unlock the boundless potential of nano-hybrid composites in shaping the future of technology.

Virat Khanna
Maharaja Agrasen University, India

Prianka Sharma
Maraja Agrasen University, India

Priyanka Mahajan
Maharaja Agrasen University, India

Chapter 1
A Futuristic Approach on the Multifunctionality of Nanomaterials:
Relevance of Nanoparticles

Archna Talwar
Isabella Thoburn College, Lucknow, India

Avni Nayyar
Era University, Lucknow, India

Shruti Anand
https://orcid.org/0009-0005-8930-0652
Era University, Lucknow, India

Manaal Zahera
Era University, Lucknow, India

Faria Fatima
Integral University, India

ABSTRACT

The realm of analytical chemistry has been invigorated by the advent of multifunctional nanoparticles. Nanomaterials, with their distinct properties stemming from quantum effects and high surface-to-volume ratios, are poised to reshape industries ranging from electronics to medicine and environmental sustainability as they deliver unprecedented performance by integrating semiconducting, plasmonic, and piezoelectronic properties. Furthermore, multifunctional nano species play a pivotal role in personalized medicine and targeted therapies. Magnetic nanoparticles respond to magnetic fields and are employed in hyperthermia therapy and targeted drug

DOI: 10.4018/979-8-3693-1261-2.ch001

delivery. The utilization of nanostructures for promoting environmental sustainability is highly commendable. They have the remarkable ability to pinpoint pollutants and their degradation. Therefore, as the research progresses, there are transformative breakthroughs to harness the multifunctionality of nanomaterials across scientific domains, driving society towards a future characterized by technological marvels and sustainable progress.

1. INTRODUCTION

Nanobiotechnology has made remarkable strides in the development of nanoparticle-based solutions for both diagnostics and therapeutics. The creation of multifunctional nanoparticles is a prominent trend in the emerging field of nanomedicine. A significant category of nanoparticles is composed of inorganic materials such as metal and its oxides, rare earth minerals, silica, and semiconductors characterized by exceptional electric, optical, magnetic, and plasmonic properties at the nanometer scale (Alivisatos, 2004). Within this realm of inorganic nanoparticles, several types naturally possess multiple functions. For instance, gold nanoparticles serve as remarkable agents for computed tomography, optical imaging, and photoacoustic imaging (Popovtzer et al., 2008; Gobin et al., 2007). It is also possible to harness both therapeutic and imaging capabilities from the gold nanoparticles. Furthermore, these inorganic entities exhibit a wide array of established surface chemistry and surface modification techniques that allow the conjugation of small molecules such as targeting ligands, therapeutic agents, and dyes to nanoparticles through precisely controlled chemical reactions. The coating strategies of nanoparticles involving both physical and chemical attachment of small molecules hold great promise for various applications in nanomedicine and beyond. Another promising category of nanostructures that has garnered significant attention for therapeutic and imaging applications is organic nanoparticles that are crafted from biodegradable polymers like polyglycolide and polycaprolactones (Panyam and Labhasetwar, 2003; Vassileva and Koseva, 2010). The encapsulation of these nanoparticles with a wide range of therapeutic agents like nucleic acids, proteins, and peptides allows for precisely targeted delivery of these therapeutic payloads (Hans and Lowman, 2002; Byrne et al., 2008). Solid lipid nanoparticles offer advantages such as prolonged drug release, large surface area, and favorable zeta potential. One notable example is DOXIL®, a liposomal system containing doxorubicin, which is used for treating AIDS-associated Kaposi's sarcoma (Northfelt et al., 1998). Knowingly, the COVID-19 pandemic presents a significant challenge, urging researchers to seek rapid solutions. Studies indicate that SARS-CoV-2, the virus responsible for COVID-19, can be detected in various settings like specimens, blood, feces, and solid waste from infected individuals (Wibowo et al., 2023).

On the other hand, agricultural production worldwide has been significantly affected by crop diseases and insect pests, resulting in substantial economic losses. These agricultural challenges have been prevented and managed by the use of traditional pesticides (Kumar et al., 2019; Lykogianni et al., 2021).

Moreover, supporting the economies of developing nations, agriculture plays a vital role by supplying food, textiles, timber, and raw materials for various industries. To tackle the food crisis caused by a growing global population, the Food and Agricultural Organization (FAO) projected a necessary increase in agricultural production by 25–70% by 2050. Emerging nanotechnology research is significantly influencing agricultural advancements and the practice of precision farming. Recent research indicates that various nanoparticles (NPs) like carbon nanotubes (CNTs), silicon dioxide (SiO2), zinc oxide (ZnO), titanium dioxide (TiO2), and gold (Au) NPs have been instrumental in enhancing seed germination in crops such as wheat, pearl millet, tomatoes, soybeans, barley, rice, and maize (Tripathi et al., 2023).

Additionally, the integration of multifunctionality into environmentally sustainable nanoparticles marks a significant advancement in the field of nanotechnology. These sustainable nanoparticles offer a promising avenue to address several challenges promising a future where nanotechnology becomes a cornerstone in fostering a cleaner, healthier, and more sustainable planet.

The focal point of this chapter revolves around exploring diverse nanoparticles and their applications, particularly in diagnostics and medicine. These nanoparticles, known for their magnetic, photothermal, and photochemical properties, hold substantial promise in medical applications and beyond. Among these, nanomicelles, possessing an amphiphilic nature, encapsulate drugs and imaging agents in a manner that eliminates the need for harsh chemicals, presenting a safer avenue for drug delivery. Addressing challenges in water treatment, especially concerning microcontaminants like endocrine-disrupting compounds (EDCs), traditional methods have fallen short. Nanotechnology emerges as an advanced approach, offering nanoscale materials with heightened reactivity and sorption capabilities owing to their substantial surface area-to-volume ratio. This unique trait renders nanomaterials indispensable in the realms of sensing, treatment, remediation, and pollution prevention within water systems.

2. ROLE OF NANOCARRIERS IN MEDICINE

The exciting developments in nanotechnology have sparked enthusiasm in the minds of researchers for using these tiny materials in the field of medicine to address various problems. The purpose of combining these nanostructures with therapeutic agents is to minimize toxicity, increase solubility, improving their distribution, enhance their circulation half-life, and reduce their immunogenicity. The nanoparticles of nearly 100 nm in size are responsible for modifying drug accumulation, its absorption, and its effectiveness in treating diseases. The physical and chemical morphology of these particles can be altered and allows scientists to fine-tune the biological as well as physical traits of these particles to achieve the desired outcome (Xia, 2014). The binding of drug molecules to nanocarriers followed by remodeling their surfaces is crucial for effective drug delivery. The drug entities can either stick to the carrier's surfaces or attach covalently to them. Moreover, the surface of nanocarriers can be coated by specific targeting molecules which may be hydrophilic or hydrophobic, or by molecules like aptamers, antibodies, peptides, or small compounds. On the site of diseased tissue, these drug-loaded molecules release the therapeutic agents in response to changes in the body's natural conditions such as pH level, temperature, osmolarity, and enzymatic activity (Panzarini et al., 2013). There is a widespread use of nanoparticle-based delivery systems in various medical fields as they can be easily customized so these tiny structures possess remarkable ability to precisely target specific cells.

2.1 Physical Properties of Nanosized Drug Molecules

The size, shape, and surface properties of nanoparticles have a vital impact on how they can enter tissues and achieve desired therapeutic results.

2.1.1. Size and Shape

The interaction of nanoparticles with the carbohydrates, proteins, and lipids of cell membranes activates the cell's internal uptake mechanisms. The small-sized nanoparticles have a larger surface area to adhere to and interact with the cell membrane, consequently leading to increased mobility. The reduced size of the nanoparticle increases its mobility, which, in turn, enhances its ability to be taken up by the cells (Ha et al., 2015). Additionally, these nanostructures typically don't get recognized and engulfed by the macrophages and are not processed through the body's disposal system meant for larger particles (Choi et al., 2009). A study conducted by Li and coworkers investigated the influence of shape on cellular uptake and found that spherical structures required the least amount of energy to overcome the membrane-bending barrier, making them the most easily internalized by cells followed by cubic-shaped, rod-shaped, and disk-shaped nanoparticles (Li et al., 2015).

2.1.2. Surface Chemistry

The chemical makeup of the surface of nanoparticles is another important factor that determines the interaction of nanoparticles with the biological environment. The hydrophilic nature of nanoparticles is ideal to avoid their capture by immune cells like macrophages (Moghimi and Szebeni, 2003). This is achieved by coating the surface of nanostructures with hydrophilic polymer such as polyethylene glycol (PEG). The PEG-coated nanoparticles stay in circulation for longer periods because they readily don't bind to blood proteins, which can lead to blood aggregation. Moreover, the tumor-targeting ability of the nanoparticles can be modified by customizing the PEG chains with amines, alcohols, carboxylic acids, and thiols.

2.1.3. Targeting Ligands

To achieve precise targeting of the cells both in vitro and in vivo the surface of tiny structures is customized with several targeting agents such as antibody fragments, aptamers, carbohydrates, proteins, and vitamins (Zhang, Gao, and Bao, 2015; Narang and Mahato, 2010). Monoclonal antibodies are preferable choices as targeting agents as they are highly specific towards their molecular targets and are readily available. For example, monoclonal antibodies like trastuzumab used for breast cancer, and bevacizumab used for colorectal cancer have made their way into clinical use for cancer therapy (Milano et al., 2007). However, these antibodies are complex and provoke immune responses.

2.2 Nanomaterials Utilized in the Fabrication of Nanoparticles

2.2.1. Inorganic Nanoparticles

Inorganic nanoparticles possess diagnostic applications due to their magnetic, photothermal, and photochemical properties. The versatile nature of these nanostructures can be refined by complexing their sturdy outer layer and robust core with a variety of drug agents. Lanthanide-doped nanoparticles, a class of inorganic nanoparticles are a type of fluorescent materials that can transform low energy light, like near-infrared into high-energy visible or UV light. This property of long-lasting luminescence has been utilized in cancer diagnostic trials to improve imaging (Coll, 2011; Naccache et al., 2012) as well as to target tumors (Cheng et al., 2012). The superparamagnetic iron oxide particles can be used to improve the clarity of images in techniques like magnetic resonance imaging (MRI) to detect tumors. The coating of surfactants and the formation of connections between specific chemical moieties like amine and carboxyl groups on the surface of tiny vehicles make them better with specific targeting elements (Kievit and Zhang, 2011). Additionally, gold nanoparticles are tiny and shiny with electrical and optical properties that can be useful for imaging tumors (Kim and Jon, 2012). Quantum dots are minuscule semiconductors that serve as tiny beacons for creating imaging tools inside the body (Huang et al., 2011). The hybrid system of gold nanoparticles and quantum dots with organic polymers such as plastics serves a dual purpose of carrying drugs to the target and monitoring the state of the tumor in real-time (Bao et al., 2014; Liang et al., 2013).

2.2.2. Organic Nanoparticles

The biodegradable organic nanoparticles are made of nanogels, micelles, dendrimers, and liposomes with flexible outer surfaces and cores that break down naturally in the body with high payload loading efficiency (Jia et al., 2013). Polymeric nanoparticles are like sponges of tiny strands capable of holding a large amount of medicine and simultaneously delivering multiple treatment modalities (Chacko et al., 2012). Nanomicelles are amphiphilic entities with inner hydrophobic nature (such as D, L- lactic acid, L-lysine, aspartic acid, propylene oxide, and others) and outer hydrophilic nature (such as polyethylene glycol, poly N- vinyl- 2- pyrrolidone, polyvinyl alcohol) and are responsible for holding drugs and imaging agents in their water- repellent core, thus overcoming the use of harmful chemicals (Felber, Dufresne, and Leroux, 2012) (Figure 1). Liposomes are small, spherical structures made up of phospholipids along with cholesterol and other lipids that encapsulate genetic material into their water-loving core and directly aim at tumor sites (Lin et al., 2013). Dendrimers are small tree-shaped molecules made up of polyamide, polyglycerol, triazine, polypropylene imine, and polyamidoamine that encapsulate drugs into the core and target molecules, imaging agents can be customized on its branches (Kojima et al., 2000).

3. ROLE OF MAGNETIC NANOCARRIERS

Superparamagnetic nanoparticles are highly promising in the field of nanotechnology due to their unique characteristics. These nanoparticles are often composed of various types of iron oxides like maghemite (γ-Fe_2O_3) and magnetite (Fe_3O_4) offer significant advantages in analytical applications due to their easy separation from a complex matrix in the presence of a magnetic field. Additionally, there is no such issue of agglomeration of these superparamagnetic nanoparticles as they lose their magnetic properties once the magnetic field is removed. The property of biocompatibility of these nanostructures has been utilized in bio-labeling and bio-separation (Teja and Koh, 2009). Iron oxide nanoparticles or magnetic nanobeads

Figure 1. a) Nanomicelles (amphiphilic blocks) formed through the self assembly of hydrophobhic head and hydrophilic tail for encapsulating therapeutic and imaging agents. b) Lipid bilayers of liposome nanoparticles encompassing drug hyrophobhic core.

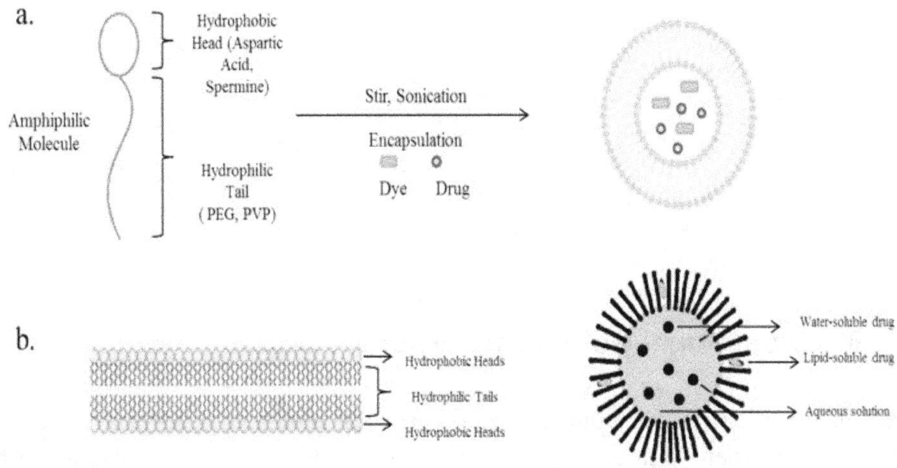

hold significant promise in the development of electrochemical biosensors due to their exceptionally large surface area that allows for a higher density of receptors to be immobilized on their surface. Magnetic nanobeads are intentionally designed with magnetic properties that enable their efficient manipulation, concentration, and separation using external magnetic fields. The magnetic tiny structures are well suited for integration into microfluidic devices because they are highly efficient and involved in sample processing making them valuable tools in diagnostics and point-of-care testing (Lin and Samia, 2006; Qadri, Ganoe, and Haik, 2009; Yuan and Williams, 2007).

Uncoated magnetic nanoparticles face challenges like aggregation due to their large surface area. Iron oxide nanoparticles lose magnetic properties through oxidation and are susceptible to biodegradation. To address these issues, effective surface coatings, such as organic molecules (surfactants, polymers) or inorganic layers (silica, metal sulphides, metal oxides), are crucial. Coatings act as shields, preventing exposure to reactive substances and reducing surface energy, thus maintaining stability and discouraging aggregation. Safeguarding the stability of magnetic nanoparticles is essential for their successful application. (Figure 2) (Wu, He, and Jiang, 2008).

Figure 2. Silane-based surface modification of magnetic nanoparticle

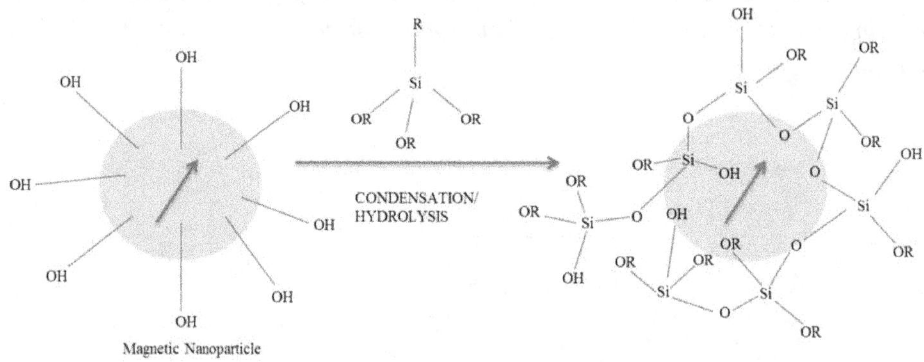

3.1 Multifunctional Gold Nanocarriers

Gold nanoparticles exhibit unique optical properties, making them ideal for analytical applications. Their light absorption stems from coherent oscillation induced by electromagnetic field interactions (El-Sayed, 2001). The high surface-to-volume ratio makes their plasmon frequency sensitive to interface changes, resulting in observable color shifts for colorimetric analyses (El-Sayed, 2001; Yguerabide and Yguerabide, 1998). Immobilizing biomolecules involves sulfur-containing ligands, forming strong covalent bonds. Monolayer-protected gold nanoparticles serve as design elements for biomolecular binding (Goodman and Rotello, 2004; Liz-Marzan et al., 1996; Jana et al., 2007). Gold-silica hybrid structures, achieved through controlled silica coating, offer unique properties explored in diverse research fields. The bifunctional Au-Fe_3O_4 nanoparticles reported by Li et al. offered unique properties by chemically linking gold and iron-oxide nanoparticles conjugating their properties in a single entity. The researchers modified Au-Fe_3O_4 nanostructures with nitrilotriacetic acid through an Au-S interaction to be used in the separation of proteins with the assistance of a magnet. This approach was utilized in the separation of proteins as well as in the separation of other types of biomolecules (Bao et al., 2007).

Inspired by the bio barcode assay concept, Wong, Wu, and colleagues introduced a novel approach using gold core or iron oxide nanoparticles to enhance complexity and functionality in a scanometric array detection system. These unique structures, with distinct regions made of different materials within the same particle, leverage both iron oxide and gold shell nanoparticles in an ultrasensitive glycans array. Conjugated with antibodies or proteins, these nanoparticles allow specific binding to glycans, offering selective binding platforms and optical signal enhancement (Liang et al., 2009). The technique, applied in clinical settings, holds promise for early detection of diseases associated with low levels of specific carbohydrate antigens.

3.2 Multifunctional Quantum Dots

Quantum dots, tiny colloidal semiconductor nanoparticles, offer unique properties suitable for biological labeling, imaging, detection, and fluorescence resonance energy transfer (FRET) (William et al., 2006). Composed of materials like cadmium selenide (CdSe) and cadmium sulfide (CdS), or indium phosphide (InP) and indium arsenide (InAs), they emit light at distinct wavelengths, enabling multiplexed assays (Wang and Herron, 1991; Murray et al., 1993; Mićić and Nozik, 1996; Pötschke et al., 2004). Silica coating enhances safety, reducing toxicity by blocking the release of toxic ions (Bailey and Nie, 2003; Mazumder et al., 2009; Jamieson et al., 2007; Huo, 2007; Lovrić et al., 2005; Derfus et al., 2004). Silica-coated quantum dots, modified for hydrophilicity, find versatile applications in biotechnology. Magnetic quantum dots and core/shell hybrid nanoparticles, such as cobalt or cadmium selenide (Co/CdSe), offer bifunctional behavior for diverse applications (Green and Howman, 2005; Jin et al., 2007; Kirchner et al. 2005; Gerion et al., 2001). Conjugation with molecularly imprinted polymers (MIPs) enhances quantum dots' functionality in detecting analyte molecules, contributing to applications in biosensing and analytical chemistry (Schwartz et al., 2003; Kim et al., 2005; Diltemiz et al., 2008).

3.3 Multifunctional Carbon and Inorganic Nanotubes

Wang and Li's review explores functionalized carbon nanotubes (CNTs) in biosensing, highlighting advancements in electrochemical biosensors using CNTs and carbon nanofibers (Wang and Lin, 2008). Daniel et al's (2007) review emphasizes the significance of CNT functionalization, particularly attaching DNA to SWCNTs and MWCNTs for developing DNA-selective biosensors. The synergy of gold nanoparticles with functionalized CNTs is gaining interest in gas sensing and catalysis (Satishkumar et al., 1996; Jiang et al., 2003; Ellis et al., 2003; Jiang and Gao, 2003; Fási et al., 2003). In 2007, Gogotsi et al. incorporated either gold or iron oxide nanoparticles into the multifunctional carbon nanotubes for detecting glycine by using a non-catalytic carbon vapor deposition (CVD) method with a template-assisted approach (Mattia et al., 2007). The incorporation of gold nanoparticles into carbon nanotubes creates a phenomenon of plasmon resonance leading to surface-enhanced Raman spectroscopic (SERS) studies to detect single molecules. Kulesza et al. fabricated a specialized film by combining multi-walled carbon nanotubes (MWCNTs), cobalt porphyrin, and tungsten oxide to be deposited onto a glassy carbon electrode substrate. This system fulfilled the purpose of developing an electrocatalytic system capable of reducing oxygen in an acidic environment (0.5 mol L^{-1} H_2SO_4). Furthermore, the multi-component electrocatalytic film exhibited higher activity in reducing hydrogen-peroxide as well as demonstrated high ethanol tolerance, making it particularly valuable for applications requiring selective cathode electrocatalysts, such as fuel cells that utilize alcohol-based fuels (Dembinska and Kulesza, 2009).

Inorganic oxide nanotubes are functionalized using various molecular surface linkages such as siloxane (Ghosh and Spiro, 1980; Bookbinder and Wrighton, 1983; Ford and Rodgers, 1994), acetyl acetonate (Heimer et al., 1995), carboxylic acid (Duoss et al., 2007), catecholate (Tahir et al., 2006a, 2006b), or phosphonate (Péchy et al., 1995; Alberti et al., 1994; Nagtegaal et al., 1999) through covalent bonds that allows the researchers to tailor the properties of these nanotubes for specific applications. Tremel et al. reported a facile method for the functionalization of WS_2 nanotubes by using multifunctional polymeric ligands with different functional groups like NTA (nitrilotriacetic acid) linker, anchor group, and fluorescent dye (Tahir et al., 2007). Zhang and Zhao used a nanocrystal self-assembly method to create arrays of rare earth fluoride nanotubes (REF-NTs) and magnetite-doped rare earth fluoride nanotubes (Fe_3O_4 REF-NTs) by utilizing porous anodic aluminium oxide (AAO) as a hard template. During the preparation of REF-NTs, α-$NaYF_4$ nanocrystals doped with Yb^{3+} and Er^{3+} were impregnated into the pores of an AAO membrane. Similarly, Fe_3O_4 REF-NTs were formed by self-assembling Fe_3O_4 and Yb/Er-doped α-$NaYF_4$ nanocrystals within the AAO pore channels. These nanotube arrays demonstrated a robust magnetic response as well as exhibited strong up-conversion fluorescent emissions when excited in the near-infrared (NIR) regions (Zhang et al., 2009).

4. ROLE OF NANOCARRIERS IN AGRICULTURE

Nanotechnology holds immense promise in various fields, one particularly exciting area of nanotechnology research is its application in plant science and agriculture. These nano-sized structures deliver substances that protect plants from pests and serve important purposes of improving plant health, expanding pest control, and reducing the use of pesticides, thereby contributing to more sustainable and eco-friendly agricultural practices (Dasgupta et al., 2015; Mali et al., 2020). Furthermore, this technology facilitates the delivery of biological pesticides such as toxic *Bacillus thuringiensis* (Bt) proteins and double-stranded RNAs (Yan et al., 2020a, 2020b, 2021). Nanopesticides are created at the nanometer level using various methods involving physical chemistry and physics. Kah and Hofmann (2014) described nanopesticides as plant protection products with particle sizes smaller than 1000 nanometres. The advantage of using these small-sized structures is that they mix well with water contributing to their stability and increasing their efficiency (Müller and Peters, 1998; Yang et al., 2017). For instance, magnesium nanoparticles can be used to combat *Ralstonia solanacearum* which causes bacterial wilt in tomatoes and magnesium oxide nanoparticles reduce the impact of diseases on plants (Imada et al., 2016). Silver nanoparticles control early tomato blight as well as boost the growth of tomatoes along with increasing the content of chlorophyll by 23.52% (Kumari et al., 2017). The application of copper oxide and calcium nanoparticles inhibited the growth of *Spodoptera littoralis* (Ayoub et al., 2018).

5. NANOCARRIERS FOR REMOVAL OF POLLUTANTS

The world is facing a growing demand for clean water as our freshwater sources are dwindling due to prolonged droughts, a rising global population, threats posed by climate change, and increasingly stringent water quality standards (Weber, 2002). The presence of microcontaminants (endocrine-disrupting compounds, EDS) in polluted water and wastewater has made traditional water and wastewater treatment plans less effective in meeting environmental standards. Conventional treatment technologies such as oxidation methods, activated sludge, activated carbon, and reverse osmosis membranes have been proven insufficient in treating highly polluted water (pollutants). Among various commercial and non-commercial technological innovations, nanotechnology has emerged as a highly advanced approach to water and wastewater treatment. The materials at the nanoscale exhibit unique properties like high reactivity and strong sorption capabilities due to higher surface area to volume ratio and make these structures valuable for a wide range of applications including sensing and detection, treatment and remediation, and pollution prevention (Rickerby and Morrison, 2007; Vaseashta et al., 2007; Qu et al., 2013). Nanoparticles have played a significant role in advancing water filtration processes, particularly in the context of membrane technology. Nanoparticles are frequently integrated into the manufacturing of membranes and polymeric or inorganic membranes can be produced by incorporating nanoparticles into their structures (Bottino et al., 2002; Kim et al., 2003; Li et al., 2009). The development of advanced membranes facilitates water reuse and desalination processes.

The different types of nanomaterials including titanium, zinc, and silver have demonstrated the capability to disinfect waterborne microbes due to their charge-carrying capacity, and TiO_2 photocatalysts and metallic or metal-oxide nanoparticles are particularly promising due to their antimicrobial properties. The most widely used nanoparticle for disinfection in water is silver nanoparticle due to its low toxicity (Spadaro et al., 1974; Kumar et al., 2004). The antibacterial mechanism of silver nanoparticle is well documented (Feng et al., 2000; Yamanaka et al., 2005) and the mechanism involved in the bactericidal effects of silver nanoparticles include the generation of free radicals that damage bacterial membranes (Xiu et al., 2011, 2012), interactions with DNA, adhesion to the bacterial cell membrane, which alters membrane properties, and damage to bacterial enzymes (Kim et al., 2007; Liau et al., 1997; Danilczuk et al., 2006). A direct electrospinning method involving cellulose acetate fibres embedded with silver nanoparticles was found to be effective against both types of bacteria that are Gram-positive and Gram-negative. The incorporation of silver nanoparticles into different types of polymers to create antimicrobial nanofibres and nanocomposites has also been reported (Balogh et al., 2001; Chen et al., 2003; Botes and Eugene Cloete, 2010). Moreover, poly (ε- caprolactone) based polyurethane nanofibre mats containing silver nanostructures have been used as antimicrobial nanofilters. Another promising photocatalyst for water purification is TiO_2 nanoparticles. The fundamental mechanism of these semiconductor-based photocatalysts involves the generation of highly reactive oxidants like hydroxyl radicals (OH), which are effective for disinfecting a wide range of microorganisms including bacteria, fungi, algae, viruses, and others (Einaga et al., 1999; Fujishima et al., 2000; Shephard et al., 2002; Ibanez et al., 2003; Liu and yang, 2003; Cho et al., 2004, 2005; Hajkova et al., 2007; Zan et al., 2007). The synthesis of visible light-activated TiO_2 nanoparticles is significant because when they are irradiated with UV-visible light demonstrate strong bactericidal activity against *E.coli* (Ireland et al., 1993; Wei et al., 1994; Bekbölet and Araz; Maness et al., 1999). Furthermore, metal-doped TiO_2 nanoparticles, iron, and sulfur have shown potent antibacterial effects against *E.coli* (Yu et al., 2005; Egerton et al., 2006). The modification of these tiny structures with nanoparticles of transition metal oxides, and nanostructured TiO_2 photocatalysts has shown great potential for water disinfection (Yu et al., 2005; Egerton et al., 2006).

Additionally, carbon nanotubes (CNTs) are highly effective in removing bacterial pathogens and have gained significant attention for their exceptional capabilities in eliminating biological contaminants from water (Savage and Diallo, 2005). The antimicrobial effect of carbon nanotubes is due to their unique physical properties, cytotoxicity, and surface functionalization characteristics (Kabra et al., 2004). The fibrous shape of CNTs, the size and length of the tubes, and the number of layers (single-walled or multi-walled) also contribute to their antimicrobial activity. It has been observed that single-walled CNTs are more detrimental to microorganisms than multi-walled CNTs due to the exceptionally high adsorption rate of bacteria by single-walled CNTs (Li et al., 2008; Wick et al., 2007). The mechanism adopted by CNTs to kill bacteria includes the induction of oxidative stress and disturbances to the bacterial cell membrane (Vecitis et al., 2010).

The surface modifications of various nanomaterials such as nanoscale TiO_2 through the coupling with a second catalytic metal significantly enhance the quality of water. Surface modification improves the photocatalytic activity of the compounds due to the shorter lifetime of reactive oxygen species and increases the affinity of the modified nanomaterials for various emerging contaminants and bimetallic nanoparticles have also proven to be effective in remediating water contaminants. Additionally, the technique of electrospinning offers a way to modify the surface properties of nanomaterials and is used to create nanofibrous filters for antifouling water filtration membranes.

6. APPLICATIONS OF MULTIFUNCTIONAL NANOCARRIERS

6.1 Application in Therapeutics

The unique physical and chemical properties of nanoparticles make them incredibly valuable in the field of biomedicine. Inorganic nanoparticles exhibit special optical properties, making them valuable tools for various biomedical applications. Nanoparticles used in biomedical applications can be categorized into six categories based on their function that involves- (1) Signal Transmitters:Some nanoparticles behave as tiny signal broadcasters within the body and can change electromagnetic waves to send biological signals to monitoring devices (Lee et al., 2007; Lee et al., 2013; Kneipp et al., 2008). (2) Energy Convertors: These types of nanoparticles convert energy from passing waves into heat or chemical energy. This property is utilized in various therapies like magnetic hyperthermia, sonodynamic therapy (SDT), photothermal therapy (PTT), and high-intensity focussed ultrasound (HIFU) therapy (Huang et al., 2006; Devarakonda et al., 2017; Lucky et al., 2015; Deepagan et al., 2016). (3) Catalytic Nanoparticles: The catalytic properties of nanoparticles can scavenge reactive oxygen species (ROS) and promote specific chemical reactions, thereby enhancing disease therapy (Walkey et al., 2015). (4) Drug- Delivery Vehicles: Nanoparticles serve as carriers for drug molecules like nucleic acids and anticancer agents. The drug-loaded nanoparticles are directed to specific target sites, and the release of the drugs is controlled to improve therapeutic effectiveness and reduce side effects (Cho et al., 2008; Brigger et al., 2012). (5) Self-emitting Nanoparticles: The property of generating electromagnetic emissions without the need for external energy sources is valuable for imaging techniques like single-photon emission computed tomography (SPECT), positron emission tomography (PET), and persistence luminescence imaging (Sun et al., 2015), (6) Targeting Agents: Nanoparticles can act as targeting agents to enhance the precision of drug delivery or imaging. For instance, magnetic nanoparticles can be guided to specific locations within the body using a magnetic field.

Multifunctional nanoparticles address the limitations of multimodal imaging (Louie, 2010), serving as imaging probes and contrast agents simultaneously. For example, "trimodal" imaging, combining MRI, photoacoustic imaging, and surface-enhanced Raman Spectroscopy, allows for brain tumor detection with spatial resolution (Kircher et al., 2012). These nanoparticles enable therapies like MRI-guided sonodynamic therapy and facilitate combination treatments, such as chemotherapy, photodynamic therapy, and photothermal therapy, for enhanced cancer cell destruction with lower drug doses. The use of metal oxide nanoparticles with persistent catalytic activities benefits low-oxygen tumor tissues (Lee et al., 2015). The integration of diagnosis and therapy, known as "theranostics," is a promising approach in nanomedicine, offering precise drug delivery, therapeutic evaluation, and assistance in medical therapies (Miller et al., 2015).

Multifunctional nanoparticles for biomedical applications prioritize water dispersibility, biocompatibility, and structural stability. Synthesis in organic solvents with biocompatible surfactants maintains stability in physiological conditions (Kim et al., 2018; Huang and Juang, 2011). Three main groups include nanoparticles with multiple inorganic phases or dopants (core-shell, heterostructured, and doped nanostructures), polymer-nanoparticle composites, and porous/hollow structured nanoparticles, each tailored for specific functionalities in the biomedical sector.

6.2 Application of Magnetic Nanoparticles

Combining magnetic nanoparticles with other particles like iron oxide enhances their features, expanding their applications. These multifunctional structures facilitate protein purification, contamination removal, enzyme immobilization, hyperthermia for tumor treatment, and act as contrast agents in Magnetic Resonance Imaging (MRI). Additionally, their near-infrared light-absorbing property makes them valuable for diseases like cancer in Photothermal Therapy.

6.3 Application of Nanoparticles in Crop Management

The research towards the creation of smart nanocarriers that can precisely target, and effectively deliver various plant protection substances to crops has revolutionized the agricultural sector (Rajput et al., 2020; Baranowska-Wójcik et al., 2020; Staroń et al., 2020). The effects of the absorption of nanoparticles by crops in agricultural ecosystems may be positive, negative, or neutral which directly leaves an impact on agricultural productivity and human health. The positive effects of nanoparticles like cesium (Cs) and titanium (Ti) (Hernandez-Viezcas et al., 2013; Rico et al., 2013; Song et al., 2013; Zhao et al., 2013) has been observed when they are accumulated in plants such as tomato, cucumber, rice, and soybean. The growth and yield of peanut seedlings (Wu et al., 2015) and soybean (Sheykhbaglou et al., 2010) have been enhanced by iron (Fe) nanomaterials while copper (Cu) promoted the growth of mung bean (Pradhan et al., 2013), and increased the pod weight of peanuts (Prasad et al., 2012). Moreover, the yield of wheat was improved by the uptake of copper (Cu) and iron (Fe) nanoparticles (Yasmeen et al., 2017). Furthermore, the use of unmanned aerial vehicles (UAVs) has advanced significantly, which is a small, high-speed sprinkler to spray nanopesticides to protect plants. The extensive research and refinement efforts have led to the creation of organic and inorganic nanoparticles due to their biocompatible properties towards protection of crops from foreign entities (Table 1).

6.4 Application of Nanoparticles in Wastewater Treatment

Nanotechnology-based systems provide an affordable solution for wastewater treatment (Qu et al., 2013) because of their distinctive properties such as the ability to kill microbes (Durán et al., 2007) large surface area, and others. Nanotechnology is commonly used in water treatments to capture pollutants, use light to break down contaminants, disinfect, separate substances, and sense and monitor water quality (Zhang et al., 2016). A

Table 1. Nanopesticides prepared from inorganic and organic materials

S.No.	Material (Inorganic/Organic)	Pesticide	Target	Function	Reference
1.	Fe(III)-MOFs (Inorganic)	Azoxystrobin	Wheat Seed	Increases weight of the upper parts of wheat.	(Shan et al., 2020)
2.	AM@CuBTC (Inorganic)	Avermectin	Pine wilt disease	Improves pesticide efficacy, solubility, and photolysis performance.	(Liu et al., 2022)
3.	MSN (Inorganic)	Mildamine	Cucumber	Improves fungicidal activity against *Magnaporthe oryzae* with longer duration.	(Xu et al., 2021)
4.	PRCRC (Inorganic)	Chlorpyrifos	Tick	Better insecticidal performance and pH sensitivity.	(Xiang et al., 2018)
5.	Biochar (Inorganic)	Glyphosate	Weeds	Better control of weeds.	(Chen et al., 2018)
6.	SPc-dinotefuran (Organic)	Dinotefuran	Aphids	Improves distribution and uptake efficiency.	(Jiang et al., 2022)
7.	ABA@AL-CTAB (Organic)	Abscisic Acid	Rice	Provides resistance to photolysis and slow-release performance.	(Yin et al., 2020, 2022)
8.	SPc- calcium glycine (Organic)	Calcium glycine	Tomato Mosaic Virus	Improves photostability and control virus.	(Yan et al., 2022)
9.	SPc-chitosan (Organic)	Chitosan	*Phytophthora infestans*	Enhances control effect.	(Wang et al., 2021)
10.	AL/PEG- acetamiprid (Organic)	Acetamiprid	*Xanthogaleruca luteola*	Increases insecticidal performance.	(Ebadollahi et al., 2022)

nanocomposite involving the conjugation of TiO_2 / ZnO material on carbon nanofibres (CNF) (Pant et al., 2013) could efficiently get rid of methylene blue (chemical dye) and *Escherichia coli* (microbe) simultaneously and displayed excellent antimicrobial activities in the presence of ultraviolet light. Furthermore, a study by Tsydenova et al. reviewed the use of TiO_2 alone for degrading pollutants and disinfecting pathogens extensively. Nanocomposites like Er-Al co-doped ZnO photo anode, mesoporous Fe_3O_4@ Ag@TiO_2, Ag/AgCl- magnetically activated carbon composites, and macroporous Ag/ TiO_2 (Tseng et al., 2018) composite breaks down organic pollutants in the presence of light and disinfects microorganisms. A range of nanocomposites such as metal organic frameworks, carbon based adsorbants and others have been investigated for the removal of diverse pollutants depicted in Table 2 (Hlongwane et al., 2019).

7. CONCLUSION

The versatility of nanoparticles holds immense potential across scientific and industrial realms, impacting medicine, materials science, and environmental science. Nanoparticles play a pivotal role in personalized medicine, leveraging advancements

Table 2. Nanomaterials used to adsorb diverse range of pollutants

S.No.	Nanomaterial	Method of Nanomaterial Preparation	Pollutants Treated	Pollutants
1.	Magnetite Nanoparticles	Co-precipitation	Metal ions	Pb (II), Cr (III)
2.	Malachite Nanoparticles	Chemical Synthesis	Anion Compounds (Polyatomic ions)	Chromate, Arsenate
3.	Amino-functionalized silica magnetite	Sonochemical Synthesis	Organic Compounds	Reactive black 5, Sodium dodecylbenzenesulfonate
4.	Nickel sulphide nanoparticles loaded on activated carbon	Chemical Synthesis	Organic Compounds	Methylene Blue, Safranin-O
5.	Magnetic graphene oxide nanocomposite	Co-precipitation	Metal ion and Organic Compound	Cd (II), Methylene Blue, Orange G
6.	Nanoscale zerovalent iron (nZVI) and Au doped nZVI particles	Chemical Synthesis	Metal ion and polyatomic ion	Cd (II), Nitrate
7.	Magnetic multi-walled carbon nanotubes (MWCNTs)	Chemical Synthesis	Metal ion and Organic Compound	Cu (II), Atrazine
8.	Magnesium Oxide	Sol-Gel Method	Metal ions	Pb (II), Cd (II)
9.	Graphene oxide-hydrated zirconium oxide nanocomposites	Hydro-thermal co-precipitation	Metal ions	As (III), As (V)
10.	Nano-alumina modified with 2, 4-dinitrophenylhydrazine	Chemical Synthesis	Metal ions	Pb(II), Cd(II)

in genomics and proteomics for targeted cancer treatments. Challenges in drug delivery systems persist, but magnetic nanoparticles, coupled with ligands, show promise in biological applications and diagnostics like Magnetic Resonance Imaging (MRI). In wastewater treatment, innovative nanocomposites effectively degrade diverse contaminants. As nanomaterials accumulate in the environment, future research must delve into their interactions with biological systems, emphasizing the importance of understanding factors like size, shape, charge, and hydrophobicity for toxicological assessments. A balanced approach, synthesizing biocompatible and degradable nanomaterials, is crucial for responsible and informed nanotechnology applications, ensuring innovation aligns with environmental stewardship.

REFERENCES

Alberti, G., Marmottini, F., Murcia-Mascarós, S., & Vivani, R. (1994). Preparation and Preliminary Characterization of a Covalently Pillared Zirconium Phosphate-Diphosphonate with Interlayer Microporosity. *Angewandte Chemie International Edition in English*, *33*(15-16), 1594–1597. doi:10.1002/anie.199415941

Alivisatos, P. (2004). The use of nanocrystals in biological detection. *Nature Biotechnology*, *22*(1), 47–52. doi:10.1038/nbt927 PMID:14704706

Ayoub, H. A., Khairy, M., Elsaid, S., Rashwan, F. A., & Abdel-Hafez, H. F. (2018). Pesticidal activity of nanostructured metal oxides for generation of alternative pesticide formulations. *Journal of Agricultural and Food Chemistry*, *66*(22), 5491–5498. doi:10.1021/acs.jafc.8b01600 PMID:29757642

Bailey, R. E., & Nie, S. (2003). Alloyed semiconductor quantum dots: Tuning the optical properties without changing the particle size. *Journal of the American Chemical Society*, *125*(23), 7100–7106. doi:10.1021/ja035000o PMID:12783563

Balogh, L., Swanson, D. R., Tomalia, D. A., Hagnauer, G. L., & McManus, A. T. (2001). Dendrimer– silver complexes and nanocomposites as antimicrobial agents. *Nano Letters*, *1*(1), 18–21. doi:10.1021/nl005502p

Bao, J., Chen, W., Liu, T., Zhu, Y., Jin, P., Wang, L., Liu, J., Wei, Y., & Li, Y. (2007). Bifunctional Au-Fe3O4 nanoparticles for protein separation. *ACS Nano*, *1*(4), 293–298. doi:10.1021/nn700189h PMID:19206679

Bao, Q. Y., Zhang, N., Geng, D. D., Xue, J. W., Merritt, M., Zhang, C., & Ding, Y. (2014). The enhanced longevity and liver targetability of Paclitaxel by hybrid liposomes encapsulating Paclitaxel-conjugated gold nanoparticles. *International Journal of Pharmaceutics*, *477*(1-2), 408–415. doi:10.1016/j.ijpharm.2014.10.040 PMID:25455782

Baranowska-Wójcik, E., Szwajgier, D., Oleszczuk, P., & Winiarska-Mieczan, A. (2020). Effects of titanium dioxide nanoparticles exposure on human health—A review. *Biological Trace Element Research*, *193*(1), 118–129. doi:10.1007/s12011-019-01706-6 PMID:30982201

Bekbölet, M., & Araz, C. V. (1996). Inactivation of Escherichia coli by photocatalytic oxidation. *Chemosphere*, *32*(5), 959–965. doi:10.1016/0045-6535(95)00359-2 PMID:8867143

Bookbinder, D. C., & Wrighton, M. S. (1983). Electrochromic polymers covalently anchored to electrode surfaces. Optical and electrochemical properties of a viologen-based polymer. *Journal of the Electrochemical Society*, *130*(5), 1080–1087. doi:10.1149/1.2119888

Botes, M., & Eugene Cloete, T. (2010). The potential of nanofibers and nanobiocides in water purification. *Critical Reviews in Microbiology*, *36*(1), 68–81. doi:10.3109/10408410903397332 PMID:20088684

Bottino, A., Capannelli, G., & Comite, A. (2002). Preparation and characterization of novel porous PVDF-ZrO2 composite membranes. *Desalination*, *146*(1-3), 35–40. doi:10.1016/S0011-9164(02)00469-1

Brigger, I., Dubernet, C., & Couvreur, P. (2012). Nanoparticles in cancer therapy and diagnosis. *Advanced Drug Delivery Reviews*, *64*, 24–36. doi:10.1016/j.addr.2012.09.006 PMID:12204596

Byrne, J. D., Betancourt, T., & Brannon-Peppas, L. (2008). Active targeting schemes for nanoparticle systems in cancer therapeutics. *Advanced Drug Delivery Reviews*, *60*(15), 1615–1626. doi:10.1016/j.addr.2008.08.005 PMID:18840489

Chacko, R. T., Ventura, J., Zhuang, J., & Thayumanavan, S. (2012). Polymer nanogels: A versatile nanoscopic drug delivery platform. *Advanced Drug Delivery Reviews*, *64*(9), 836–851. doi:10.1016/j.addr.2012.02.002 PMID:22342438

Chen, C., Zhang, G., Dai, Z., Xiang, Y., Liu, B., Bian, P., Zheng, K., Wu, Z., & Cai, D. (2018). Fabrication of light-responsively controlled-release herbicide using a nanocomposite. *Chemical Engineering Journal*, *349*, 101–110. doi:10.1016/j.cej.2018.05.079

Chen, Y., Wang, L., Jiang, S., & Yu, H. J. (2003). Study on novel antibacterial polymer materials (I) preparation of zeolite antibacterial agents and antibacterial polymer composite and their antibacterial properties. *Journal of Polymer Materials*, *20*(3), 279–284.

Cheng, L., Yang, K., Li, Y., Zeng, X., Shao, M., Lee, S. T., & Liu, Z. (2012). Multifunctional nanoparticles for upconversion luminescence/MR multimodal imaging and magnetically targeted photothermal therapy. *Biomaterials*, *33*(7), 2215–2222. doi:10.1016/j.biomaterials.2011.11.069 PMID:22169825

Cho, K., Wang, X. U., Nie, S., Chen, Z., & Shin, D. M. (2008). Therapeutic nanoparticles for drug delivery in cancer. *Clinical Cancer Research*, *14*(5), 1310–1316. doi:10.1158/1078-0432.CCR-07-1441 PMID:18316549

Cho, M., Chung, H., Choi, W., & Yoon, J. (2004). Linear correlation between inactivation of E. coli and OH radical concentration in TiO2 photocatalytic disinfection. *Water Research*, *38*(4), 1069–1077. doi:10.1016/j.watres.2003.10.029 PMID:14769428

Cho, M., Chung, H., Choi, W., & Yoon, J. (2005). Different inactivation behaviors of MS-2 phage and Escherichia coli in TiO2 photocatalytic disinfection. *Applied and Environmental Microbiology*, *71*(1), 270–275. doi:10.1128/AEM.71.1.270-275.2005 PMID:15640197

Choi, J., Zhang, Q., Reipa, V., Wang, N. S., Stratmeyer, M. E., Hitchins, V. M., & Goering, P. L. (2009). Comparison of cytotoxic and inflammatory responses of photoluminescent silicon nanoparticles with silicon micron-sized particles in RAW 264.7 macrophages. *Journal of Applied Toxicology*, *29*(1), 52–60. doi:10.1002/jat.1382 PMID:18785685

Coll, J. L. (2011). Cancer optical imaging using fluorescent nanoparticles. *Nanomedicine (London)*, *6*(1), 7–10. doi:10.2217/nnm.10.144 PMID:21182412

Daniel, S., Rao, T. P., Rao, K. S., Rani, S. U., Naidu, G. R. K., Lee, H. Y., & Kawai, T. (2007). A review of DNA functionalized/grafted carbon nanotubes and their characterization. *Sensors and Actuators. B, Chemical*, *122*(2), 672–682. doi:10.1016/j.snb.2006.06.014

Danilczuk, M., Lund, A., Sadlo, J., Yamada, H., & Michalik, J. (2006). Conduction electron spin resonance of small silver particles. *Spectrochimica Acta. Part A: Molecular and Biomolecular Spectroscopy*, *63*(1), 189–191. doi:10.1016/j.saa.2005.05.002 PMID:15978868

Dasgupta, N., Ranjan, S., Mundekkad, D., Ramalingam, C., Shanker, R., & Kumar, A. (2015). Nanotechnology in agro-food: From field to plate. *Food Research International*, *69*, 381–400. doi:10.1016/j.foodres.2015.01.005

Deepagan, V. G., You, D. G., Um, W., Ko, H., Kwon, S., Choi, K. Y., Yi, G.-R., Lee, J. Y., Lee, D. S., Kim, K., Kwon, I. C., & Park, J. H. (2016). Long-circulating Au-TiO2 nanocomposite as a sonosensitizer for ROS-mediated eradication of cancer. *Nano Letters*, *16*(10), 6257–6264. doi:10.1021/acs.nanolett.6b02547 PMID:27643533

Dembinska, B., & Kulesza, P. J. (2009). Multi-walled carbon nanotube-supported tungsten oxide-containing multifunctional hybrid electrocatalytic system for oxygen reduction in acid medium. *Electrochimica Acta*, *54*(20), 4682–4687. doi:10.1016/j.electacta.2009.03.043

Derfus, A. M., Chan, W. C., & Bhatia, S. N. (2004). Probing the cytotoxicity of semiconductor quantum dots. *Nano Letters*, *4*(1), 11–18. doi:10.1021/nl0347334 PMID:28890669

Devarakonda, S. B., Myers, M. R., Lanier, M., Dumoulin, C., & Banerjee, R. K. (2017). Assessment of gold nanoparticle-mediated-enhanced hyperthermia using MR-guided high-intensity focused ultrasound ablation procedure. *Nano Letters*, *17*(4), 2532–2538. doi:10.1021/acs.nanolett.7b00272 PMID:28287747

Diltemiz, S. E., Say, R., Büyüktiryaki, S., Hür, D., Denizli, A., & Ersöz, A. (2008). Quantum dot nanocrystals having guanosine imprinted nanoshell for DNA recognition. *Talanta*, *75*(4), 890–896. doi:10.1016/j.talanta.2007.12.036 PMID:18585161

Duoss, E. B., Twardowski, M., & Lewis, J. A. (2007). Sol-gel inks for direct-write assembly of functional oxides. *Advanced Materials*, *19*(21), 3485–3489. doi:10.1002/adma.200701372

Durán, N., Marcato, P. D., De Souza, G. I., Alves, O. L., & Esposito, E. (2007). Antibacterial effect of silver nanoparticles produced by fungal process on textile fabrics and their effluent treatment. *Journal of Biomedical Nanotechnology*, *3*(2), 203–208. doi:10.1166/jbn.2007.022

Ebadollahi, A., Valizadeh, B., Panahandeh, S., Mirhosseini, H., Zolfaghari, M., & Changbunjong, T. (2022). Nanoencapsulation of Acetamiprid by Sodium Alginate and Polyethylene Glycol Enhanced Its Insecticidal Efficiency. *Nanomaterials (Basel, Switzerland)*, *12*(17), 2971. doi:10.3390/nano12172971 PMID:36080008

Egerton, T. A., Kosa, S. A., & Christensen, P. A. (2006). Photoelectrocatalytic disinfection of E. coli suspensions by iron doped TiO 2. *Physical Chemistry Chemical Physics*, *8*(3), 398–406. doi:10.1039/B507516E PMID:16482283

Einaga, H., Futamura, S., & Ibusuki, T. (1999). Photocatalytic decomposition of benzene over TiO2 in a humidified airstream. *Physical Chemistry Chemical Physics*, *1*(20), 4903–4908. doi:10.1039/a906214i

El-Sayed, M. A. (2001). Some interesting properties of metals confined in time and nanometer space of different shapes. *Accounts of Chemical Research*, *34*(4), 257–264. doi:10.1021/ar960016n PMID:11308299

Ellis, A. V., Vijayamohanan, K., Goswami, R., Chakrapani, N., Ramanathan, L. S., Ajayan, P. M., & Ramanath, G. (2003). Hydrophobic anchoring of monolayer-protected gold nanoclusters to carbon nanotubes. *Nano Letters*, *3*(3), 279–282. doi:10.1021/nl025824o

Fási, A., Pálinkó, I., Seo, J. W., Kónya, Z., Hernádi, K., & Kiricsi, I. (2003). Sonication assisted gold deposition on multiwall carbon nanotubes. *Chemical Physics Letters*, *372*(5-6), 848–852. doi:10.1016/S0009-2614(03)00527-X

Felber, A. E., Dufresne, M. H., & Leroux, J. C. (2012). pH-sensitive vesicles, polymeric micelles, and nanospheres prepared with polycarboxylates. *Advanced Drug Delivery Reviews*, *64*(11), 979–992. doi:10.1016/j.addr.2011.09.006 PMID:21996056

Feng, Q. L., Wu, J., Chen, G. Q., Cui, F. Z., Kim, T. N., & Kim, J. O. (2000). A mechanistic study of the antibacterial effect of silver ions on Escherichia coli and Staphylococcus aureus. *Journal of Biomedical Materials Research*, *52*(4), 662–668. doi:10.1002/1097-4636(20001215)52:4<662::AID-JBM10>3.0.CO;2-3 PMID:11033548

Ford, W. E., & Rodgers, M. A. (1994). Interfacial electron transfer in colloidal SnO2 hydrosols photosensitized by electrostatically and covalently attached ruthenium (II) polypyridine complexes. *Journal of Physical Chemistry*, *98*(14), 3822–3831. doi:10.1021/j100065a044

Fujishima, A., Rao, T. N., & Tryk, D. A. (2000). Titanium dioxide photocatalysis. *Journal of Photochemistry and Photobiology C, Photochemistry Reviews*, *1*(1), 1–21. doi:10.1016/S1389-5567(00)00002-2

Gerion, D., Pinaud, F., Williams, S. C., Parak, W. J., Zanchet, D., Weiss, S., & Alivisatos, A. P. (2001). Synthesis and properties of biocompatible water-soluble silica-coated CdSe/ZnS semiconductor quantum dots. *The Journal of Physical Chemistry B*, *105*(37), 8861–8871. doi:10.1021/jp0105488

Ghosh, P., & Spiro, T. G. (1980). Photoelectrochemistry of tris (bipyridyl) ruthenium (II) covalently attached to n-type tin (IV) oxide. *Journal of the American Chemical Society*, *102*(17), 5543–5549. doi:10.1021/ja00537a021

Gobin, A. M., Lee, M. H., Halas, N. J., James, W. D., Drezek, R. A., & West, J. L. (2007). Near-infrared resonant nanoshells for combined optical imaging and photothermal cancer therapy. *Nano Letters*, *7*(7), 1929–1934. doi:10.1021/nl070610y PMID:17550297

Goodman, C. M., & Rotello, V. M. (2004). Biomacromolecule surface recognition using nanoparticles. *Mini-Reviews in Organic Chemistry*, *1*(1), 103–114. doi:10.2174/1570193043488935

Green, M., & Howman, E. (2005). Semiconductor quantum dots and free radical induced DNA nicking. *Chemical Communications*, (1), 121–123. doi:10.1039/b413175d PMID:15614393

Ha, H. K., Kim, J. W., Lee, M. R., Jun, W., & Lee, W. J. (2015). Cellular uptake and cytotoxicity of β-lactoglobulin nanoparticles: The effects of particle size and surface charge. *Asian-Australasian Journal of Animal Sciences*, *28*(3), 420–427. doi:10.5713/ajas.14.0761 PMID:25656189

Hajkova, P., Spatenka, P., Horsky, J., Horska, I., & Kolouch, A. (2007). Photocatalytic effect of TiO2 films on viruses and bacteria. *Plasma Processes and Polymers*, *4*(S1), S397–S401. doi:10.1002/ppap.200731007

Hans, M. L., & Lowman, A. M. (2002). Biodegradable nanoparticles for drug delivery and targeting. *Current Opinion in Solid State and Materials Science*, *6*(4), 319–327. doi:10.1016/S1359-0286(02)00117-1

Heimer, T. A., D'Arcangelis, S. T., Farzad, F., Stipkala, J. M., & Meyer, G. J. (1996). An acetylacetonate-based semiconductor– sensitizer linkage. *Inorganic Chemistry*, *35*(18), 5319–5324. doi:10.1021/ic960419j

Hernandez-Viezcas, J. A., Castillo-Michel, H., Andrews, J. C., Cotte, M., Rico, C., Peralta-Videa, J. R., Ge, Y., Priester, J. H., Holden, P. A., & Gardea-Torresdey, J. L. (2013). In situ synchrotron X-ray fluorescence mapping and speciation of CeO2 and ZnO nanoparticles in soil cultivated soybean (Glycine max). *ACS Nano*, *7*(2), 1415–1423. doi:10.1021/nn305196q PMID:23320560

Hlongwane, G. N., Sekoai, P. T., Meyyappan, M., & Moothi, K. (2019). Simultaneous removal of pollutants from water using nanoparticles: A shift from single pollutant control to multiple pollutant control. *The Science of the Total Environment*, *656*, 808–833. doi:10.1016/j.scitotenv.2018.11.257 PMID:30530150

Huang, H. C., Barua, S., Sharma, G., Dey, S. K., & Rege, K. (2011). Inorganic nanoparticles for cancer imaging and therapy. *Journal of Controlled Release*, *155*(3), 344–357. doi:10.1016/j.jconrel.2011.06.004 PMID:21723891

Huang, S. H., & Juang, R. S. (2011). Biochemical and biomedical applications of multifunctional magnetic nanoparticles: A review. *Journal of Nanoparticle Research, 13*(10), 4411–4430. doi:10.1007/s11051-011-0551-4

Huang, X., El-Sayed, I. H., Qian, W., & El-Sayed, M. A. (2006). Cancer cell imaging and photothermal therapy in the near-infrared region by using gold nanorods. *Journal of the American Chemical Society, 128*(6), 2115–2120. doi:10.1021/ja057254a PMID:16464114

Huo, Q. (2007). A perspective on bioconjugated nanoparticles and quantum dots. *Colloids and Surfaces. B, Biointerfaces, 59*(1), 1–10. doi:10.1016/j.colsurfb.2007.04.019 PMID:17544637

Ibáñez, J. A., Litter, M. I., & Pizarro, R. A. (2003). Photocatalytic bactericidal effect of TiO2 on Enterobacter cloacae: Comparative study with other Gram (−) bacteria. *Journal of Photochemistry and Photobiology A Chemistry, 157*(1), 81–85. doi:10.1016/S1010-6030(03)00074-1

Imada, K., Sakai, S., Kajihara, H., Tanaka, S., & Ito, S. (2016). Magnesium oxide nanoparticles induce systemic resistance in tomato against bacterial wilt disease. *Plant Pathology, 65*(4), 551–560. doi:10.1111/ppa.12443

Ireland, J. C., Klostermann, P., Rice, E. W., & Clark, R. M. (1993). Inactivation of Escherichia coli by titanium dioxide photocatalytic oxidation. *Applied and Environmental Microbiology, 59*(5), 1668–1670. doi:10.1128/aem.59.5.1668-1670.1993 PMID:8390819

Jamieson, T., Bakhshi, R., Petrova, D., Pocock, R., Imani, M., & Seifalian, A. M. (2007). Biological applications of quantum dots. *Biomaterials, 28*(31), 4717–4732. doi:10.1016/j.biomaterials.2007.07.014 PMID:17686516

Jana, N. R., Earhart, C., & Ying, J. Y. (2007). Synthesis of water-soluble and functionalized nanoparticles by silica coating. *Chemistry of Materials, 19*(21), 5074–5082. doi:10.1021/cm071368z

Jia, F., Liu, X., Li, L., Mallapragada, S., Narasimhan, B., & Wang, Q. (2013). Multifunctional nanoparticles for targeted delivery of immune activating and cancer therapeutic agents. *Journal of Controlled Release, 172*(3), 1020–1034. doi:10.1016/j.jconrel.2013.10.012 PMID:24140748

Jiang, K., Eitan, A., Schadler, L. S., Ajayan, P. M., Siegel, R. W., Grobert, N., Mayne, M., Reyes-Reyes, M., Terrones, H., & Terrones, M. (2003). Selective attachment of gold nanoparticles to nitrogen-doped carbon nanotubes. *Nano Letters, 3*(3), 275–277. doi:10.1021/nl025914t

Jiang, L., & Gao, L. (2003). Modified carbon nanotubes: An effective way to selective attachment of gold nanoparticles. *Carbon*, *41*(15), 2923–2929. doi:10.1016/S0008-6223(03)00339-7

Jiang, Q., Xie, Y., Peng, M., Wang, Z., Li, T., Yin, M., Shen, J., & Yan, S. (2022). A nanocarrier pesticide delivery system with promising benefits in the case of dinotefuran: Strikingly enhanced bioactivity and reduced pesticide residue. *Environmental Science. Nano*, *9*(3), 988–999. doi:10.1039/D1EN00752A

Jin, Y., Kannan, S., Wu, M., & Zhao, J. X. (2007). Toxicity of luminescent silica nanoparticles to living cells. *Chemical Research in Toxicology*, *20*(8), 1126–1133. doi:10.1021/tx7001959 PMID:17630705

Kabra, K., Chaudhary, R., & Sawhney, R. L. (2004). Treatment of hazardous organic and inorganic compounds through aqueous-phase photocatalysis: A review. *Industrial & Engineering Chemistry Research*, *43*(24), 7683–7696. doi:10.1021/ie0498551

Kah, M., & Hofmann, T. (2014). Nanopesticide research: Current trends and future priorities. *Environment International*, *63*, 224–235. doi:10.1016/j.envint.2013.11.015 PMID:24333990

Kievit, F. M., & Zhang, M. (2011). Surface engineering of iron oxide nanoparticles for targeted cancer therapy. *Accounts of Chemical Research*, *44*(10), 853–862. doi:10.1021/ar2000277 PMID:21528865

Kim, D., & Jon, S. (2012). Gold nanoparticles in image-guided cancer therapy. *Inorganica Chimica Acta*, *393*, 154–164. doi:10.1016/j.ica.2012.07.001

Kim, D., Shin, K., Kwon, S. G., & Hyeon, T. (2018). Synthesis and biomedical applications of multifunctional nanoparticles. *Advanced Materials*, *30*(49), 1802309. doi:10.1002/adma.201802309 PMID:30133009

Kim, H., Achermann, M., Balet, L. P., Hollingsworth, J. A., & Klimov, V. I. (2005). Synthesis and characterization of Co/CdSe core/shell nanocomposites: Bifunctional magnetic-optical nanocrystals. *Journal of the American Chemical Society*, *127*(2), 544–546. doi:10.1021/ja047107x PMID:15643877

Kim, J. S., Kuk, E., Yu, K. N., Kim, J. H., Park, S. J., Lee, H. J., Kim, S. H., Park, Y. K., Park, Y. H., Hwang, C.-Y., Kim, Y.-K., Lee, Y.-S., Jeong, D. H., & Cho, M. H. (2007). Antimicrobial effects of silver nanoparticles. *Nanomedicine; Nanotechnology, Biology, and Medicine*, *3*(1), 95–101. doi:10.1016/j.nano.2006.12.001 PMID:17379174

Kim, S. H., Kwak, S. Y., Sohn, B. H., & Park, T. H. (2003). Design of TiO2 nanoparticle self-assembled aromatic polyamide thin-film-composite (TFC) membrane as an approach to solve biofouling problem. *Journal of Membrane Science*, *211*(1), 157–165. doi:10.1016/S0376-7388(02)00418-0

Kircher, M. F., De La Zerda, A., Jokerst, J. V., Zavaleta, C. L., Kempen, P. J., Mittra, E., Pitter, K., Huang, R., Campos, C., Habte, F., Sinclair, R., Brennan, C. W., Mellinghoff, I. K., Holland, E. C., & Gambhir, S. S. (2012). A brain tumor molecular imaging strategy using a new triple-modality MRI-photoacoustic-Raman nanoparticle. *Nature Medicine*, *18*(5), 829–834. doi:10.1038/nm.2721 PMID:22504484

Kirchner, C., Liedl, T., Kudera, S., Pellegrino, T., Muñoz Javier, A., Gaub, H. E., Stölzle, S., Fertig, N., & Parak, W. J. (2005). Cytotoxicity of colloidal CdSe and CdSe/ZnS nanoparticles. *Nano Letters*, *5*(2), 331–338. doi:10.1021/nl047996m PMID:15794621

Kneipp, J., Kneipp, H., & Kneipp, K. (2008). SERS—A single-molecule and nanoscale tool for bioanalytics. *Chemical Society Reviews*, *37*(5), 1052–1060. doi:10.1039/b708459p PMID:18443689

Kojima, C., Kono, K., Maruyama, K., & Takagishi, T. (2000). Synthesis of polyamidoamine dendrimers having poly (ethylene glycol) grafts and their ability to encapsulate anticancer drugs. *Bioconjugate Chemistry*, *11*(6), 910–917. doi:10.1021/bc0000583 PMID:11087341

Kumar, S., Nehra, M., Dilbaghi, N., Marrazza, G., Hassan, A. A., & Kim, K. H. (2019). Nano-based smart pesticide formulations: Emerging opportunities for agriculture. *Journal of Controlled Release*, *294*, 131–153. doi:10.1016/j.jconrel.2018.12.012 PMID:30552953

Kumar, V. S., Nagaraja, B. M., Shashikala, V., Padmasri, A. H., Madhavendra, S. S., Raju, B. D., & Rao, K. R. (2004). Highly efficient Ag/C catalyst prepared by electro-chemical deposition method in controlling microorganisms in water. *Journal of Molecular Catalysis A Chemical*, *223*(1-2), 313–319. doi:10.1016/j.molcata.2003.09.047

Kumari, M., Pandey, S., Bhattacharya, A., Mishra, A., & Nautiyal, C. S. (2017). Protective role of biosynthesized silver nanoparticles against early blight disease in Solanum lycopersicum. *Plant Physiology and Biochemistry*, *121*, 216–225. doi:10.1016/j.plaphy.2017.11.004 PMID:29149700

Lee, H., Lee, Y., Song, C., Cho, H. R., Ghaffari, R., Choi, T. K., Kim, K. H., Lee, Y. B., Ling, D., Lee, H., Yu, S. J., Choi, S. H., Hyeon, T., & Kim, D. H. (2015). An endoscope with integrated transparent bioelectronics and theranostic nanoparticles for colon cancer treatment. *Nature Communications*, *6*(1), 10059. doi:10.1038/ncomms10059 PMID:26616435

Lee, J. H., Huh, Y. M., Jun, Y. W., Seo, J. W., Jang, J. T., Song, H. T., Kim, S., Cho, E.-J., Yoon, H.-G., Suh, J.-S., & Cheon, J. (2007). Artificially engineered magnetic nanoparticles for ultra-sensitive molecular imaging. *Nature Medicine*, *13*(1), 95–99. doi:10.1038/nm1467 PMID:17187073

Lee, N., Choi, S. H., & Hyeon, T. (2013). Nano-sized CT contrast agents. *Advanced Materials*, *25*(19), 2641–2660. doi:10.1002/adma.201300081 PMID:23553799

Li, J. H., Xu, Y. Y., Zhu, L. P., Wang, J. H., & Du, C. H. (2009). Fabrication and characterization of a novel TiO2 nanoparticle self-assembly membrane with improved fouling resistance. *Journal of Membrane Science*, *326*(2), 659–666. doi:10.1016/j.memsci.2008.10.049

Li, Q., Mahendra, S., Lyon, D. Y., Brunet, L., Liga, M. V., Li, D., & Alvarez, P. J. (2008). Antimicrobial nanomaterials for water disinfection and microbial control: Potential applications and implications. *Water Research*, *42*(18), 4591–4602. doi:10.1016/j.watres.2008.08.015 PMID:18804836

Li, Y., Kröger, M., & Liu, W. K. (2015). Shape effect in cellular uptake of PEGylated nanoparticles: Comparison between sphere, rod, cube and disk. *Nanoscale*, *7*(40), 16631–16646. doi:10.1039/C5NR02970H PMID:26204104

Liang, C. H., Wang, C. C., Lin, Y. C., Chen, C. H., Wong, C. H., & Wu, C. Y. (2009). Iron oxide/gold core/shell nanoparticles for ultrasensitive detection of carbohydrate– protein interactions. *Analytical Chemistry*, *81*(18), 7750–7756. doi:10.1021/ac9012286 PMID:19689135

Liang, R., Wang, J., Wu, X., Dong, L., Deng, R., Wang, K., Sullivan, M., Liu, S., Wu, M., Tao, J., Yang, X., & Zhu, J. (2013). Multifunctional biodegradable polymer nanoparticles with uniform sizes: Generation and in vitro anti-melanoma activity. *Nanotechnology*, *24*(45), 455302. doi:10.1088/0957-4484/24/45/455302 PMID:24145641

Liau, S. Y., Read, D. C., Pugh, W. J., Furr, J. R., & Russell, A. D. (1997). Interaction of silver nitrate with readily identifiable groups: Relationship to the antibacterialaction of silver ions. *Letters in Applied Microbiology*, *25*(4), 279–283. doi:10.1046/j.1472-765X.1997.00219.x PMID:9351278

Lin, X. M., & Samia, A. C. (2006). Synthesis, assembly and physical properties of magnetic nanoparticles. *Journal of Magnetism and Magnetic Materials, 305*(1), 100–109. doi:10.1016/j.jmmm.2005.11.042

Lin, Y. Y., Kao, H. W., Li, J. J., Hwang, J. J., Tseng, Y. L., Lin, W. J., Lin, M.-H., Ting, G., & Wang, H. E. (2013). Tumor burden talks in cancer treatment with PEGylated liposomal drugs. *PLoS One*, *8*(5), e63078. doi:10.1371/journal.pone.0063078 PMID:23675454

Liu, H. L., & Yang, T. C. K. (2003). Photocatalytic inactivation of Escherichia coli and Lactobacillus helveticus by ZnO and TiO2 activated with ultraviolet light. *Process Biochemistry (Barking, London, England), 39*(4), 475–481. doi:10.1016/S0032-9592(03)00084-0

Liu, Y., Zhang, Y., Xin, X., Xu, X., Wang, G., Gao, S., Qiao, L., Yin, S., Liu, H., Jia, C., Shen, W., Xu, L., Ji, Y., & Zhou, C. (2022). Design and Preparation of Avermectin Nanopesticide for Control and Prevention of Pine Wilt Disease. *Nanomaterials (Basel, Switzerland), 12*(11), 1863. doi:10.3390/nano12111863 PMID:35683719

Liz-Marzán, L. M., Giersig, M., & Mulvaney, P. (1996). Synthesis of nanosized gold–silica core– shell particles. *Langmuir, 12*(18), 4329–4335. doi:10.1021/la9601871

Louie, A. (2010). Multimodality imaging probes: Design and challenges. *Chemical Reviews, 110*(5), 3146–3195. doi:10.1021/cr9003538 PMID:20225900

Lovrić, J., Bazzi, H. S., Cuie, Y., Fortin, G. R., Winnik, F. M., & Maysinger, D. (2005). Differences in subcellular distribution and toxicity of green and red emitting CdTe quantum dots. *Journal of Molecular Medicine, 83*(5), 377–385. doi:10.1007/s00109-004-0629-x PMID:15688234

Lucky, S. S., Soo, K. C., & Zhang, Y. (2015). Nanoparticles in photodynamic therapy. *Chemical Reviews, 115*(4), 1990–2042. doi:10.1021/cr5004198 PMID:25602130

Lykogianni, M., Bempelou, E., Karamaouna, F., & Aliferis, K. A. (2021). Do pesticides promote or hinder sustainability in agriculture? The challenge of sustainable use of pesticides in modern agriculture. *The Science of the Total Environment, 795*, 148625. doi:10.1016/j.scitotenv.2021.148625 PMID:34247073

Mali, S. C., Raj, S., & Trivedi, R. (2020). Nanotechnology a novel approach to enhance crop productivity. *Biochemistry and Biophysics Reports, 24*, 100821. doi:10.1016/j.bbrep.2020.100821 PMID:33015378

Maness, P. C., Smolinski, S., Blake, D. M., Huang, Z., Wolfrum, E. J., & Jacoby, W. A. (1999). Bactericidal activity of photocatalytic TiO2 reaction: Toward an understanding of its killing mechanism. *Applied and Environmental Microbiology, 65*(9), 4094–4098. doi:10.1128/AEM.65.9.4094-4098.1999 PMID:10473421

Mattia, D., Korneva, G., Sabur, A., Friedman, G., & Gogotsi, Y. (2007). Multifunctional carbon nanotubes with nanoparticles embedded in their walls. *Nanotechnology, 18*(15), 155305. doi:10.1088/0957-4484/18/15/155305

Mazumder, S., Dey, R., Mitra, M. K., Mukherjee, S., & Das, G. C. (2009). Biofunctionalized quantum dots in biology and medicine. *Journal of Nanomaterials, 2009*, 2009. doi:10.1155/2009/815734

Mićić, O. I., & Nozik, A. J. (1996). Synthesis and characterization of binary and ternary III–V quantum dots. *Journal of Luminescence, 70*(1-6), 95–107. doi:10.1016/0022-2313(96)00047-6

Milano, A., Nasti, G., Iaffaioli, R. V., & Caponigro, F. (2007). First line targeted therapies in breast cancer: Focus on bevacizumab. *Biologics, 1*(1), 3–10. PMID:19707343

. Miller, M. A., Gadde, S., Pfirschke, C., Engblom, C., Sprachman, M. M., Kohler, R. H., ... Weissleder, R. (2015). Predicting therapeutic nanomedicine efficacy using a companion magnetic resonance imaging nanoparticle. *Science Translational Medicine, 7*(314), 314ra183-314ra183.

Moghimi, S. M., & Szebeni, J. (2003). Stealth liposomes and long circulating nanoparticles: Critical issues in pharmacokinetics, opsonization and protein-binding properties. *Progress in Lipid Research, 42*(6), 463–478. doi:10.1016/S0163-7827(03)00033-X PMID:14559067

Müller, R. H., & Peters, K. (1998). Nanosuspensions for the formulation of poorly soluble drugs: I. Preparation by a size-reduction technique. *International Journal of Pharmaceutics, 160*(2), 229–237. doi:10.1016/S0378-5173(97)00311-6

Naccache, R., Rodríguez, E. M., Bogdan, N., Sanz-Rodríguez, F., de la Cruz, M. D. C. I., de la Fuente, Á. J., ... Capobianco, J. A. (2012). High resolution fluorescence imaging of cancers using lanthanide ion-doped upconverting nanocrystals. *Cancers (Basel), 4*(4), 1067–1105. doi:10.3390/cancers4041067 PMID:24213500

Nagtegaal, M., Küther, J., Ensling, J., Gütlich, P., & Tremel, W. (1999). Hydrothermal deposition of small α-Fe 2 O 3 (hematite) particles on ordered zirconium phosphonate multilayer SAMs on gold. *Journal of Materials Chemistry, 9*(5), 1115–1120. doi:10.1039/a809319i

Narang, A. S., & Mahato, R. I. (Eds.). (2010). *Targeted delivery of small and macromolecular drugs*. CRC Press. doi:10.1201/9781420087734

Northfelt, D. W., Dezube, B. J., Thommes, J. A., Miller, B. J., Fischl, M. A., Friedman-Kien, A., Kaplan, L. D., Du Mond, C., Mamelok, R. D., & Henry, D. H. (1998). Pegylated-liposomal doxorubicin versus doxorubicin, bleomycin, and vincristine in the treatment of AIDS-related Kaposi's sarcoma: Results of a randomized phase III clinical trial. *Journal of Clinical Oncology*, *16*(7), 2445–2451. doi:10.1200/JCO.1998.16.7.2445 PMID:9667262

Pant, B., Pant, H. R., Barakat, N. A., Park, M., Jeon, K., Choi, Y., & Kim, H. Y. (2013). Carbon nanofibers decorated with binary semiconductor (TiO2/ZnO) nanocomposites for the effective removal of organic pollutants and the enhancement of antibacterial activities. *Ceramics International*, *39*(6), 7029–7035. doi:10.1016/j.ceramint.2013.02.041

Panyam, J., & Labhasetwar, V. (2003). Biodegradable nanoparticles for drug and gene delivery to cells and tissue. *Advanced Drug Delivery Reviews*, *55*(3), 329–347. doi:10.1016/S0169-409X(02)00228-4 PMID:12628320

Panzarini, E., Inguscio, V., Tenuzzo, B. A., Carata, E., & Dini, L. (2013). Nanomaterials and autophagy: New insights in cancer treatment. *Cancers (Basel)*, *5*(1), 296–319. doi:10.3390/cancers5010296 PMID:24216709

Péchy, P., Rotzinger, F. P., Nazeeruddin, M. K., Kohle, O., Zakeeruddin, S. M., Humphry-Baker, R., & Grätzel, M. (1995). Preparation of phosphonated polypyridyl ligands to anchor transition-metal complexes on oxide surfaces: Application for the conversion of light to electricity with nanocrystalline TiO 2 films. *Journal of the Chemical Society. Chemical Communications*, *0*(1), 65–66. doi:10.1039/C39950000065

Popovtzer, R., Agrawal, A., Kotov, N. A., Popovtzer, A., Balter, J., Carey, T. E., & Kopelman, R. (2008). Targeted gold nanoparticles enable molecular CT imaging of cancer. *Nano Letters*, *8*(12), 4593–4596. doi:10.1021/nl8029114 PMID:19367807

Pötschke, K., Müller-Kirsch, L., Heitz, R., Sellin, R. L., Pohl, U. W., Bimberg, D., Zakharov, N., & Werner, P. (2004). Ripening of self-organized InAs quantum dots. *Physica E, Low-Dimensional Systems and Nanostructures*, *21*(2-4), 606–610. doi:10.1016/j.physe.2003.11.089

Pradhan, S., Patra, P., Das, S., Chandra, S., Mitra, S., Dey, K. K., Akbar, S., Palit, P., & Goswami, A. (2013). Photochemical modulation of biosafe manganese nanoparticles on Vigna radiata: A detailed molecular, biochemical, and biophysical study. *Environmental Science & Technology*, *47*(22), 13122–13131. doi:10.1021/es402659t PMID:24144189

Prasad, T. N. V. K. V., Sudhakar, P., Sreenivasulu, Y., Latha, P., Munaswamy, V., Reddy, K. R., Sreeprasad, T. S., Sajanlal, P. R., & Pradeep, T. (2012). Effect of nanoscale zinc oxide particles on the germination, growth and yield of peanut. *Journal of Plant Nutrition*, *35*(6), 905–927. doi:10.1080/01904167.2012.663443

Qadri, S., Ganoe, A., & Haik, Y. (2009). Removal and recovery of acridine orange from solutions by use of magnetic nanoparticles. *Journal of Hazardous Materials*, *169*(1-3), 318–323. doi:10.1016/j.jhazmat.2009.03.103 PMID:19406571

Qu, X., Alvarez, P. J., & Li, Q. (2013). Applications of nanotechnology in water and wastewater treatment. *Water Research*, *47*(12), 3931–3946. doi:10.1016/j.watres.2012.09.058 PMID:23571110

Qu, X., Alvarez, P. J., & Li, Q. (2013). Applications of nanotechnology in water and wastewater treatment. *Water Research*, *47*(12), 3931–3946. doi:10.1016/j.watres.2012.09.058 PMID:23571110

Rajput, V., Minkina, T., Mazarji, M., Shende, S., Sushkova, S., Mandzhieva, S., Burachevskaya, M., Chaplygin, V., Singh, A., & Jatav, H. (2020). Accumulation of nanoparticles in the soil-plant systems and their effects on human health. *Annals of Agricultural Science*, *65*(2), 137–143. doi:10.1016/j.aoas.2020.08.001

Rickerby, D. G., & Morrison, M. (2007). Nanotechnology and the environment: A European perspective. *Science and Technology of Advanced Materials*, *8*(1-2), 19–24. doi:10.1016/j.stam.2006.10.002

Rico, C. M., Morales, M. I., Barrios, A. C., McCreary, R., Hong, J., Lee, W. Y., Nunez, J., Peralta-Videa, J. R., & Gardea-Torresdey, J. L. (2013). Effect of cerium oxide nanoparticles on the quality of rice (Oryza sativa L.) grains. *Journal of Agricultural and Food Chemistry*, *61*(47), 11278–11285. doi:10.1021/jf404046v PMID:24188281

Satishkumar, B. C., Vogl, E. M., Govindaraj, A., & Rao, C. N. R. (1996). The decoration of carbon nanotubes by metal nanoparticles. *Journal of Physics. D, Applied Physics*, *29*(12), 3173–3176. doi:10.1088/0022-3727/29/12/037

Savage, N., & Diallo, M. S. (2005). Nanomaterials and water purification: Opportunities and challenges. *Journal of Nanoparticle Research*, *7*(4-5), 331–342. doi:10.1007/s11051-005-7523-5

Schwartz, D. A., Norberg, N. S., Nguyen, Q. P., Parker, J. M., & Gamelin, D. R. (2003). Magnetic quantum dots: Synthesis, spectroscopy, and magnetism of Co2+- and Ni2+-doped ZnO nanocrystals. *Journal of the American Chemical Society*, *125*(43), 13205–13218. doi:10.1021/ja036811v PMID:14570496

Shan, Y., Cao, L., Muhammad, B., Xu, B., Zhao, P., Cao, C., & Huang, Q. (2020). Iron-based porous metal–organic frameworks with crop nutritional function as carriers for controlled fungicide release. *Journal of Colloid and Interface Science*, *566*, 383–393. doi:10.1016/j.jcis.2020.01.112 PMID:32018178

Shephard, G. S., Stockenström, S., de Villiers, D., Engelbrecht, W. J., & Wessels, G. F. (2002). Degradation of microcystin toxins in a falling film photocatalytic reactor with immobilized titanium dioxide catalyst. *Water Research*, *36*(1), 140–146. doi:10.1016/S0043-1354(01)00213-5 PMID:11767740

Sheykhbaglou, R., Sedghi, M., Shishevan, M. T., & Sharifi, R. S. (2010). Effects of nano-iron oxide particles on agronomic traits of soybean. *Notulae Scientia Biologicae*, *2*(2), 112–113. doi:10.15835/nsb224667

Song, U., Shin, M., Lee, G., Roh, J., Kim, Y., & Lee, E. J. (2013). Functional analysis of TiO 2 nanoparticle toxicity in three plant species. *Biological Trace Element Research*, *155*(1), 93–103. doi:10.1007/s12011-013-9765-x PMID:23900645

Spadaro, J. A., Berger, T. J., Barranco, S. D., Chapin, S. E., & Becker, R. O. (1974). Antibacterial effects of silver electrodes with weak direct current. *Antimicrobial Agents and Chemotherapy*, *6*(5), 637–642. doi:10.1128/AAC.6.5.637 PMID:15825319

Staroń, A., Długosz, O., Pulit-Prociak, J., & Banach, M. (2020). Analysis of the Exposure of Organisms to the Action of Nanomaterials. *Materials (Basel)*, *13*(2), 349. doi:10.3390/ma13020349 PMID:31940903

Sun, X., Cai, W., & Chen, X. (2015). Positron emission tomography imaging using radiolabeled inorganic nanomaterials. *Accounts of Chemical Research*, *48*(2), 286–294. doi:10.1021/ar500362y PMID:25635467

Tahir, M., Zink, N., Eberhardt, M., Therese, H. A., Faiss, S., Janshoff, A., Kolb, U., Theato, P., & Tremel, W. (2007). Hierarchical assembly of TiO2 nanoparticles on WS2 nanotubes achieved through multifunctional polymeric ligands. *Small*, *3*(5), 829–834. doi:10.1002/smll.200600663 PMID:17407103

Tahir, M. N., Eberhardt, M., Theato, P., Faiß, S., Janshoff, A., Gorelik, T., Kolb, U., & Tremel, W. (2006b). Reactive polymers: A versatile toolbox for the immobilization of functional molecules on TiO2 nanoparticles. *Angewandte Chemie International Edition*, *45*(6), 908–912. doi:10.1002/anie.200502517 PMID:16381051

Tahir, M. N., Eberhardt, M., Therese, H. A., Kolb, U., Theato, P., Müller, W. E., Schröder, H.-C., & Tremel, W. (2006a). From Single Molecules to Nanoscopically Structured Functional Materials: Au Nanocrystal Growth on TiO2 Nanowires Controlled by Surface-Bound Silicatein. *Angewandte Chemie International Edition*, *45*(29), 4803–4809. doi:10.1002/anie.200503770 PMID:16795093

Teja, A. S., & Koh, P. Y. (2009). Synthesis, properties, and applications of magnetic iron oxide nanoparticles. *Progress in Crystal Growth and Characterization of Materials*, *55*(1-2), 22–45. doi:10.1016/j.pcrysgrow.2008.08.003

Tripathi, S., Mahra, S., Tiwari, K., Rana, S., Tripathi, D. K., Sharma, S., & Sahi, S. (2023). Recent Advances and Perspectives of Nanomaterials in Agricultural Management and Associated Environmental Risk: A Review. *Nanomaterials (Basel, Switzerland)*, *13*(10), 1604. doi:10.3390/nano13101604 PMID:37242021

Tseng, W. J., Chuang, Y. C., & Chen, Y. A. (2018). Mesoporous Fe3O4@ Ag@ TiO2 nanocomposite particles for magnetically recyclable photocatalysis and bactericide. *Advanced Powder Technology*, *29*(3), 664–671. doi:10.1016/j.apt.2017.12.008

Vaseashta, A., Vaclavikova, M., Vaseashta, S., Gallios, G., Roy, P., & Pummakarnchana, O. (2007). Nanostructures in environmental pollution detection, monitoring, and remediation. *Science and Technology of Advanced Materials*, *8*(1-2), 47–59. doi:10.1016/j.stam.2006.11.003

Vassileva, E. D., & Koseva, N. S. (2010). Sonochemically born proteinaceous micro-and nanocapsules. *Advances in Protein Chemistry and Structural Biology*, *80*, 205–252. doi:10.1016/B978-0-12-381264-3.00006-0 PMID:21109221

Vecitis, C. D., Zodrow, K. R., Kang, S., & Elimelech, M. (2010). Electronic-structure-dependent bacterial cytotoxicity of single-walled carbon nanotubes. *ACS Nano*, *4*(9), 5471–5479. doi:10.1021/nn101558x PMID:20812689

Walkey, C., Das, S., Seal, S., Erlichman, J., Heckman, K., Ghibelli, L., Traversa, E., McGinnis, J. F., & Self, W. T. (2015). Catalytic properties and biomedical applications of cerium oxide nanoparticles. *Environmental Science. Nano*, *2*(1), 33–53. doi:10.1039/C4EN00138A PMID:26207185

Wang, J., & Lin, Y. (2008). Functionalized carbon nanotubes and nanofibers for biosensing applications. *Trends in Analytical Chemistry*, *27*(7), 619–626. doi:10.1016/j.trac.2008.05.009 PMID:19122842

Wang, X., Zheng, K., Cheng, W., Li, J., Liang, X., Shen, J., Dou, D., Yin, M., & Yan, S. (2021). Field application of star polymer-delivered chitosan to amplify plant defense against potato late blight. *Chemical Engineering Journal*, *417*, 129327. doi:10.1016/j.cej.2021.129327

Wang, Y., & Herron, N. (1991). Nanometer-sized semiconductor clusters: Materials synthesis, quantum size effects, and photophysical properties. *Journal of Physical Chemistry*, *95*(2), 525–532. doi:10.1021/j100155a009

Weber, W. J. Jr. (2002). Distributed optimal technology networks: A concept and strategy for potable water sustainability. *Water Science and Technology*, *46*(6-7), 241–246. doi:10.2166/wst.2002.0685 PMID:12380997

Wei, C., Lin, W. Y., Zainal, Z., Williams, N. E., Zhu, K., Kruzic, A. P., Smith, R. L., & Rajeshwar, K. (1994). Bactericidal activity of TiO2 photocatalyst in aqueous media: Toward a solar-assisted water disinfection system. *Environmental Science & Technology*, *28*(5), 934–938. doi:10.1021/es00054a027 PMID:22191837

Wibowo, Y. G., Ramadan, B. S., Taher, T., & Khairurrijal, K. (2023). Advancements of Nanotechnology and Nanomaterials in Environmental and Human Protection for Combatting the COVID-19 During and Post-pandemic Era: A Comprehensive Scientific Review. *Biomedical Materials & Devices*, 1–24. doi:10.1007/s44174-023-00086-9 PMID:37363141

Wick, P., Manser, P., Limbach, L. K., Dettlaff-Weglikowska, U., Krumeich, F., Roth, S., Stark, W., & Bruinink, A. (2007). The degree and kind of agglomeration affect carbon nanotube cytotoxicity. *Toxicology Letters*, *168*(2), 121–131. doi:10.1016/j.toxlet.2006.08.019 PMID:17169512

William, W. Y., Chang, E., Drezek, R., & Colvin, V. L. (2006). Water-soluble quantum dots for biomedical applications. *Biochemical and Biophysical Research Communications*, *348*(3), 781–786. doi:10.1016/j.bbrc.2006.07.160 PMID:16904647

Wu, L. J., You, Q. S., Duan, J. L., Luo, Y. X., Liu, L. J., Li, X., Gao, Q., Zhu, H. P., He, Y., Xu, L., Jonas, J. B., Wang, W., & Guo, X. H. (2015). Prevalence and associated factors of myopia in high-school students in Beijing. *PLoS One*, *10*(3), e0120764. doi:10.1371/journal.pone.0120764 PMID:25803875

Wu, W., He, Q., & Jiang, C. (2008). Magnetic iron oxide nanoparticles: Synthesis and surface functionalization strategies. *Nanoscale Research Letters*, *3*(11), 397–415. doi:10.1007/s11671-008-9174-9 PMID:21749733

Xia, Y. (2014). Are we entering the nano era? *Angewandte Chemie International Edition*, *53*(46), 12268–12271. doi:10.1002/anie.201409223 PMID:25256914

Xiang, Y., Zhang, G., Chen, C., Liu, B., Cai, D., & Wu, Z. (2018). Fabrication of a pH-responsively controlled-release pesticide using an attapulgite-based hydrogel. *ACS Sustainable Chemistry & Engineering*, *6*(1), 1192–1201. doi:10.1021/acssuschemeng.7b03469

Xiu, Z. M., Ma, J., & Alvarez, P. J. (2011). Differential effect of common ligands and molecular oxygen on antimicrobial activity of silver nanoparticles versus silver ions. *Environmental Science & Technology*, *45*(20), 9003–9008. doi:10.1021/es201918f PMID:21950450

Xiu, Z. M., Zhang, Q. B., Puppala, H. L., Colvin, V. L., & Alvarez, P. J. (2012). Negligible particle-specific antibacterial activity of silver nanoparticles. *Nano Letters*, *12*(8), 4271–4275. doi:10.1021/nl301934w PMID:22765771

Xu, Y., Xu, C., Huang, Q., Cao, L., Teng, F., Zhao, P., & Jia, M. (2021). Size effect of mesoporous silica nanoparticles on pesticide loading, release, and delivery in cucumber plants. *Applied Sciences (Basel, Switzerland)*, *11*(2), 575. doi:10.3390/app11020575

Yamanaka, M., Hara, K., & Kudo, J. (2005). Bactericidal actions of a silver ion solution on Escherichia coli, studied by energy-filtering transmission electron microscopy and proteomic analysis. *Applied and Environmental Microbiology*, *71*(11), 7589–7593. doi:10.1128/AEM.71.11.7589-7593.2005 PMID:16269810

Yan, S., Hu, Q., Li, J., Chao, Z., Cai, C., Yin, M., Du, X., & Shen, J. (2019). A star polycation acts as a drug nanocarrier to improve the toxicity and persistence of botanical pesticides. *ACS Sustainable Chemistry & Engineering*, *7*(20), 17406–17413. doi:10.1021/acssuschemeng.9b04567

Yan, S., Hu, Q., Wei, Y., Jiang, Q., Yin, M., Dong, M., Shen, J., & Du, X. (2022). Calcium nutrition nanoagent rescues tomatoes from mosaic virus disease by accelerating calcium transport and activating antiviral immunity. *Frontiers in Plant Science*, *13*, 1092774. doi:10.3389/fpls.2022.1092774 PMID:36561462

Yan, S., Qian, J., Cai, C., Ma, Z., Li, J., Yin, M., Ren, B., & Shen, J. (2020a). Spray method application of transdermal dsRNA delivery system for efficient gene silencing and pest control on soybean aphid Aphis glycines. *Journal of Pest Science*, *93*(1), 449–459. doi:10.1007/s10340-019-01157-x

Yan, S., Ren, B., Zeng, B., & Shen, J. (2020b). Improving RNAi efficiency for pest control in crop species. *BioTechniques*, *68*(5), 283–290. doi:10.2144/btn-2019-0171 PMID:32202134

Yan, S., Ren, B. Y., & Shen, J. (2021). Nanoparticle-mediated double-stranded RNA delivery system: A promising approach for sustainable pest management. *Insect Science*, *28*(1), 21–34. doi:10.1111/1744-7917.12822 PMID:32478473

Yang, D., Cui, B., Wang, C., Zhao, X., Zeng, Z., Wang, Y., Sun, C., Liu, G., & Cui, H. (2017). Preparation and characterization of emamectin benzoate solid nanodispersion. *Journal of Nanomaterials*, *2017*, 2017. doi:10.1155/2017/6560780

Yasmeen, F., Raja, N. I., Razzaq, A., & Komatsu, S. (2017). Proteomic and physiological analyses of wheat seeds exposed to copper and iron nanoparticles. *Biochimica et Biophysica Acta (BBA)-. Biochimica et Biophysica Acta. Proteins and Proteomics*, *1865*(1), 28–42. doi:10.1016/j.bbapap.2016.10.001 PMID:27717896

Yguerabide, J., & Yguerabide, E. E. (1998). Light-scattering submicroscopic particles as highly fluorescent analogs and their use as tracer labels in clinical and biological applications: II. Experimental characterization. *Analytical Biochemistry*, *262*(2), 157–176. doi:10.1006/abio.1998.2760 PMID:9750129

Yin, J. M., Quan, M. P., Wang, Z., Wang, J., Yang, Z. K., Duan, L. S., Li, Z., Li, Q. X., Wang, H., & Tan, W. M. (2022). Cationic modified lignin: Regulation of synthetic microspheres for achieving anti-photolysis and sustained release of the abscisic acid. *Industrial Crops and Products*, *177*, 114573. doi:10.1016/j.indcrop.2022.114573

Yin, J. M., Wang, H. L., Yang, Z. K., Wang, J., Wang, Z., Duan, L. S., Li, Z., & Tan, W. M. (2020). Engineering lignin nanomicroparticles for the antiphotolysis and controlled release of the plant growth regulator abscisic acid. *Journal of Agricultural and Food Chemistry*, *68*(28), 7360–7368. doi:10.1021/acs.jafc.0c02835 PMID:32627551

Yu, J. C., Ho, W., Yu, J., Yip, H., Wong, P. K., & Zhao, J. (2005). Efficient visible-light-induced photocatalytic disinfection on sulfur-doped nanocrystalline titania. *Environmental Science & Technology*, *39*(4), 1175–1179. doi:10.1021/es035374h PMID:15773492

Yuan, Q., & Williams, R. A. (2007). Large scale manufacture of magnetic polymer particles using membranes and microfluidic devices. *China Particuology*, *5*(1-2), 26–42. doi:10.1016/j.cpart.2007.02.001

Zan, L., Fa, W., Peng, T., & Gong, Z. K. (2007). Photocatalysis effect of nanometer TiO2 and TiO2-coated ceramic plate on Hepatitis B virus. *Journal of Photochemistry and Photobiology. B, Biology*, *86*(2), 165–169. doi:10.1016/j.jphotobiol.2006.09.002 PMID:17055286

Zhang, F., Shi, Y., Sun, X., Zhao, D., & Stucky, G. D. (2009). Formation of hollow upconversion rare-earth fluoride nanospheres: Nanoscale kirkendall effect during ion exchange. *Chemistry of Materials*, *21*(21), 5237–5243. doi:10.1021/cm902231s

Zhang, S., Gao, H., & Bao, G. (2015). Physical principles of nanoparticle cellular endocytosis. *ACS Nano*, *9*(9), 8655–8671. doi:10.1021/acsnano.5b03184 PMID:26256227

Zhang, Y., Wu, B., Xu, H., Liu, H., Wang, M., He, Y., & Pan, B. (2016). Nanomaterials-enabled water and wastewater treatment. *NanoImpact*, *3*, 22–39. doi:10.1016/j.impact.2016.09.004

Zhao, L., Sun, Y., Hernandez-Viezcas, J. A., Servin, A. D., Hong, J., Niu, G., Peralta-Videa, J. R., Duarte-Gardea, M., & Gardea-Torresdey, J. L. (2013). Influence of CeO2 and ZnO nanoparticles on cucumber physiological markers and bioaccumulation of Ce and Zn: A life cycle study. *Journal of Agricultural and Food Chemistry*, *61*(49), 11945–11951. doi:10.1021/jf404328e PMID:24245665

Chapter 2
Characterization Methods for Assessing Hybrid Nanomaterial Performance

Prem Sagar
The Technological Institute of Textile Sciences, Bhiwani, India

Sushma Sangwan
Sumit Dhariwal Memorial Group of institutions, Hisar, India

ABSTRACT

The rapid development of nanotechnology has led to the emergence of hybrid nanomaterials with diverse applications across various fields. These materials combine different types of nanoparticles, organic molecules, or polymers to achieve enhanced properties and multifunctionality. To fully exploit the potential of hybrid nanomaterials, accurate and comprehensive characterization methods are crucial for assessing their performance and understanding their structure-function relationships. Characterization techniques such as electron microscopy, spectroscopy, and atomic force microscopy play a pivotal role in unravelling hybrid nanomaterials' intricate structures and properties, facilitating their integration into cutting-edge technologies and innovative solutions. This chapter aims to provide an in-depth exploration of the characterization techniques employed to evaluate the performance of hybrid nanomaterials, showcasing their significance in guiding material design and application development.

1. INTRODUCTION

Greenhouse gases (GHGs) are essential for maintaining the planet's climate at a level that is suitable for human survival, as they not only cause infectious diseases but

DOI: 10.4018/979-8-3693-1261-2.ch002

also increase and aggravate cardiovascular and respiratory problems in humankind. In the first quarter of 2022, Europe's (EU) economy's GHG emissions totalled 1029 million tons of Carbon dioxide (CO_2) equivalents (a 7% and 6% increase compared with the same quarter of 2020 and 2021) (Andrew, 2021), of which 27% of total EU-27 GHG emissions came from the transport sector. Research shows that one of the major causes of GHG emissions is the transport industry (Fontaras et al., 2017). Transportation-related emissions should be lowered by 94 per cent from 2005 levels by 2050 to achieve the National climate objectives of the EU- 2030 and Paris climate goals (Fontaras et al., 2017; Meireles et al., 2021). Study proves that in addition to improving aerodynamics and efficiency, reducing vehicle weight is indeed a pivotal approach to reducing carbon dioxide discharge (Kawajiri et al., 2020). When a vehicle's weight is lowered by 10%, CO_2 emissions can be reduced by 8%, according to a linear regression analysis of curb weight and CO_2 emissions (Lutsey, 2010). In order to create lightweight hybrid nanomaterials for cars with lower CO_2 emissions, research activities linked to their creation have significantly increased in recent years. The group of materials with structures and characteristics at the nanoscale, usually between 1 and 100 nanometers, are known as nanomaterials. Compared to bulk materials, this dimension enables special and improved physical, chemical, and biological capabilities. A noteworthy benefit of nanomaterials is their elevated surface area-to-volume ratio, potentially resulting in enhanced catalytic activity and reactivity. At the nanoscale, quantum effects also become significant and affect optical, electrical, and magnetic properties.

However, hybrid materials take advantage of the synergies and get around the constraints of each particular material by combining two or more different types of materials at the nanoscale. These hybrid materials frequently have improved multifunctionality, flexibility, and mechanical strength. When compared to conventional materials, hybrid constructions that incorporate nanoparticles can perform better. Applications for hybrid and nanomaterials are found in many different domains. Nanomaterials in electronics allow for the creation of more compact and effective devices. They can be used in medicine for imaging, diagnostics, and drug delivery. Nanomaterials' robustness and light weight are advantageous to the automotive sector, improving fuel economy. Nanomaterials are also used in energy conversion and storage technologies to increase performance and efficiency. Furthermore, as evidence of their adaptability, nanomaterials are used in water purification, environmental remediation, and catalysis. It is important to remember that the usage of nanomaterials raises questions regarding possible effects on the environment and human health. As a result, research is being done to better understand and reduce these dangers. Despite obstacles, research in a wide range of scientific and industrial fields is nevertheless spurred by the exceptional qualities and versatile applications of nanomaterials and hybrid materials.

1.1 Introduction to Hybrid Nanomaterials

The realm of nanotechnology has ushered in a new era of innovation in materials science, and at the forefront of this revolution lies the fascinating domain of hybrid nanomaterials. Hybrid nanomaterials are captivating not only for their diminutive size but also for their incredible potential (Rahman et al., 2022). At their core, they represent the art of combining distinct components, be it nanoparticles, organic molecules, or polymers, into a single entity. This amalgamation is not just a simple blending of materials; it's an alchemical process that gives birth to entirely new materials with properties that often transcend the sum of their individual parts. Hybrid nanomaterials' unmatched adaptability is one of its amazing benefits. These materials can be precisely tailored by researchers, who change their composition and structure to produce specific capabilities. It's similar to having a toolset of building bricks that can be put together in various ways to suit different demands and each offers a special capacity. Every element in the universe of materials has advantages and disadvantages. Hybrid nanomaterials offer a means to capitalize on the advantages while minimizing the disadvantages (Muhamad Azim et al., 2022). For instance, flexible polymers and nanoparticles recognized for their outstanding conductivity can be coupled to create materials that are both strong and electrically conductive. By merging disparate components, hybrid nanomaterials transcend the limitations that individual materials might pose.

The unique characteristics of hybrid nanomaterials differentiate them from traditional materials and offer a multitude of benefits in a wide variety of applications. The following are some essential characteristics of hybrid nanomaterials:

a) Multifunctionality: The capacity of hybrid nanomaterials to blend the special qualities of many materials is one of their main characteristics. The variety of possible applications is increased by this multifunctionality, which enables the integration of characteristics like mechanical strength, conductivity, magnetism, and optical activity into a single material.
b) Enhanced Performance: When comparing hybrid nanomaterials to their individual components, they frequently perform better. For example, blending nanoparticles with a polymer matrix can produce materials that are more flexible, durable, and mechanically strong—perfect for use in wearable technology or structural materials.
c) Synergistic Effects: Synergistic effects, in which the qualities of the hybrid material are much greater than the sum of its separate components, can result from the combining of diverse materials at the nanoscale. By utilizing this synergy, materials with enhanced thermal, electrical, or catalytic capabilities

can be produced, leading to improvements in catalysis, electronics, and energy storage.

d) Tailored Properties: The benefit of hybrid nanomaterials is their ability to modify their characteristics to satisfy particular needs. Researchers can accurately regulate the properties of a material by varying its composition, structure, and ratio of hybrid components. This allows for the production of materials with specified functionality for certain applications.

e) Improved Stability and Compatibility: Hybrid nanomaterials, which combine various materials, frequently have better compatibility and stability. This can be especially important in applications where materials are exposed to harsh environments, mechanical stress, or extreme temperatures.

f) Controlled Release and Drug Delivery: Hybrid nanomaterials are used in biomedicine to release drugs under controlled conditions. The precise control of release kinetics made possible by the combination of biocompatible materials and therapeutic agents enhances the effectiveness and minimizes side effects of drug delivery systems.

g) Versatility in Synthesis: Many techniques, such as chemical vapor deposition, sol-gel processes, and self-assembly methods, can be used to create hybrid nanomaterials. Due to the synthesis methods' versatility, hybrid materials with a variety of structures and compositions can be produced to meet the needs of particular applications.

h) Environmental Applications: Because of their special qualities, hybrid nanomaterials are used in environmental remediation projects. For instance, in water treatment processes, the efficiency of pollutant removal can be increased by combining nanoparticles with adsorbent materials.

All things considered, hybrid nanomaterials' unique qualities make them a promising class of materials for solving problems and fostering innovation in industries like electronics, health, energy, and environmental science.

As we delve deeper into this section, we'll explore the cutting-edge applications across various fields that are made possible by hybrid nanomaterials. These applications span from electronics and energy storage to healthcare and environmental remediation. The versatility and ingenuity of these materials have opened doors to solutions that were once considered out of reach. So, fasten your seatbelts as we embark on this enlightening journey through the realm of hybrid nanomaterials, where the marriage of components gives rise to materials with limitless possibilities, and where science fiction often becomes science fact.

2. TYPES OF HYBRID NANOMATERIALS

Various types of hybrid nanomaterials can be formed, including organic-inorganic hybrids, inorganic-inorganic hybrids, metal matrix nanomaterials and organic-organic hybrids. The diversity of these materials and their applications across fields such as electronics, medicine, energy, and catalysis will be explored. Hybrid nanomaterials now take us on a detailed expedition through the various types that exist, each possessing its unique characteristics and potential applications. Let's delve into the diverse categories and unearth the vast spectrum of possibilities they offer across fields as varied as electronics, medicine, energy, and catalysis. First on our itinerary are the organic-inorganic hybrids, a class of materials that bridge the realms of organic molecules and inorganic nanoparticles. These hybrids combine the structural stability of inorganics with the adjustable characteristics of organics to create a material that offers the best of both worlds. They have become major actors in industries like nanoelectronics, where they make it possible to make flexible, effective electronic devices. Furthermore, they are intriguing prospects for applications in medication delivery and bioimaging due to their compatibility with biological systems.

Following that, we explore the world of inorganic-inorganic hybrids. Here, different kinds of inorganic nanoparticles or structures combine to form new materials with unusual features. These hybrids are most useful in catalysis, where the special arrangement of catalysts can increase selectivity and reaction speeds. They show potential for increasing the effectiveness of batteries and solar cells in the energy sector. Their exquisite control at the nanoscale creates new possibilities for the creation and improvement of materials.

The universe of inorganic-inorganic hybrids is then investigated. Here, many inorganic nanoparticle or structure types interact to create novel materials with distinctive properties. These hybrids are especially helpful in catalysis, where the unique catalyst configuration can quicken reactions and boost selectivity. They have the potential to improve the energy sector's use of batteries and solar cells. The production and advancement of materials are given new possibilities by their exact control at the nanoscale.

We will uncover the practical uses of these hybrid nanomaterials as we proceed on our adventure. They are the secret to developing gadgets that are quicker and more energy-efficient. They offer advances in imaging and tailored drug administration in the medical field. They play a crucial role in the creation of effective solar panels and energy storage systems in the search for sustainable energy solutions. Furthermore, by removing pollutants and producing cleaner energy, their catalytic capability can aid in resolving environmental problems.

As we explore deeper into the complexities of these materials and their ever growing boundaries across the scientific landscape, our journey into the world of hybrid nanomaterials is sure to be an exciting one. So come along with us as we explore this fascinating intersection of creativity, technology, and science. A cutting-edge class of materials that combines the benefits of metals and nanotechnology is known as hybrid metal matrix nanomaterials. In these materials, nanoscale reinforcements like nanoparticles, nanowires, or nanotubes are usually infused into a metal matrix. These elements work together to produce special qualities and improved performance traits. Hybrid metal matrix nanomaterials' enhanced mechanical properties, such as greater strength, hardness, and wear resistance, are a significant benefit. The addition of nanoscale reinforcements makes them extraordinarily robust by assisting in the prevention of dislocation movement inside the metal matrix. Furthermore, because of the high aspect ratio of the nanoscale components, these materials frequently have higher thermal and electrical conductivity. In the aerospace, automotive, and electronics industries, where strong yet lightweight materials are crucial, hybrid metal matrix nanoparticles find applications. They are also being explored for their potential in biomedical implants and energy storage systems, thanks to their tailored properties and biocompatibility.

However, challenges like manufacturing scalability and cost remain, hindering their widespread adoption. Despite these challenges, the manufacturing of innovative materials with a wide range of commercial and technological uses is possible with the development of hybrid metal matrix nanomaterials.

Examining the performance properties of hybrid nanomaterials is essential to unlock their full potential, optimize their use across various industries, ensure safety, and drive technological advancements. It enables us to harness the unique properties of these materials and find innovative solutions to contemporary challenges.

3. PERFORMANCE PARAMETERS OF HYBRID NANOMATERIALS

Before delving into characterization methods, this section will define key performance parameters relevant to hybrid nanomaterials. Parameters such as mechanical strength, conductivity, optical properties, chemical reactivity, and stability will be highlighted.

It is important to note that the specific techniques, equipment, and parameters used for microstructural characterization and mechanical testing may vary depending on the nanomaterials being used, the equipment available, and the specific requirements of the study or application. The detailed varied approaches utilized to examine the fundamental mechanical and metallurgical features are listed below and can be seen in Figure 1.

Figure 1. Different types of nanomaterial characterization techniques

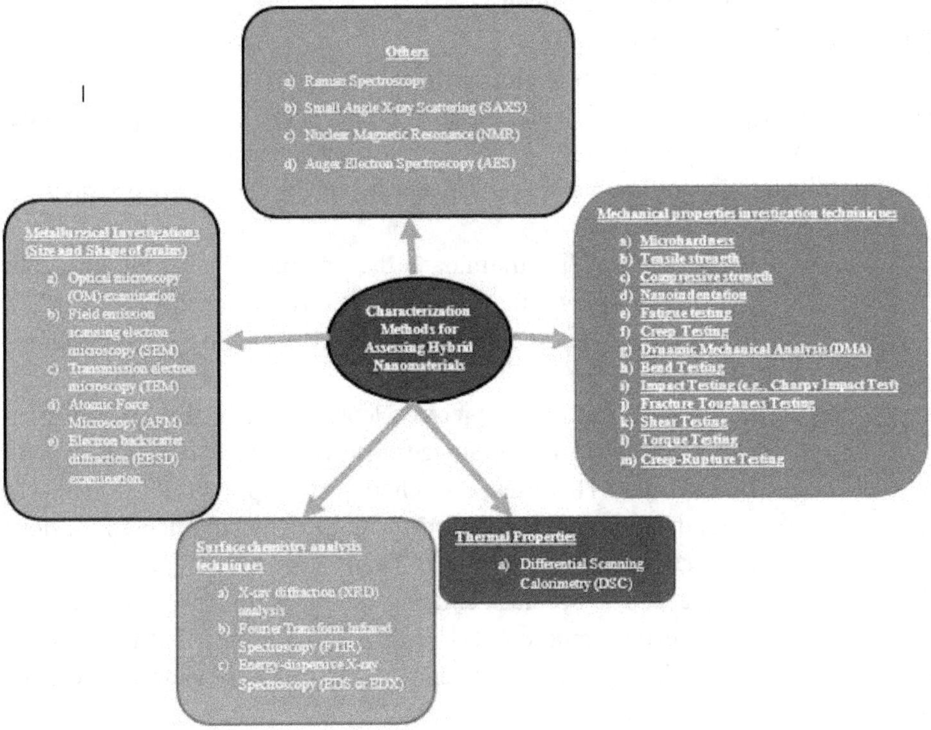

3.1 Microstructural Characterization

Standard metallurgical approaches are typically used to analyze the microstructure characterisation of hybrid nanomaterials. The fundamental five approaches to analyzing the microstructural characteristics of the hybrid nanomaterial are frequently used. These methods include:

a) Optical microscopy (OM) examination
b) X-ray diffraction (XRD) analysis
c) Scanning electron microscopy (SEM)
d) Transmission electron microscopy (TEM)
e) Electron backscatter diffraction (EBSD) examination.
f) Atomic Force Microscopy (AFM)
g) Fourier Transform Infrared Spectroscopy (FTIR)
h) Energy-dispersive X-ray Spectroscopy (EDS or EDX)
i) Raman Spectroscopy
j) Small Angle X-ray Scattering (SAXS)

k) Nuclear Magnetic Resonance (NMR)

l) Differential Scanning Calorimetry (DSC)

m) Auger Electron Spectroscopy (AES)

a) **Optical Microscopy examination:** Numerous disciplines, including biology, materials science, forensics, and many more, employ optical microscopy extensively. It includes viewing and examining microscopic material using visible light. An overview of optical microscope analysis is provided here:

- Light Source: To illuminate the sample, optical microscopes normally employ a light source, such as a halogen lamp or LED. The light is focused onto the sample after passing via a condenser lens.
- Objective Lens: The objective lens, which is the main lens in charge of enlarging the image, receives light that passes through the sample. The user can choose the appropriate amount of magnification and resolution thanks to objective lenses' various numerical apertures and magnification powers.
- Eyepiece: The eyepiece enlarges the image that the objective lens has already amplified. The eyepiece has a lens that projects the picture in front of the viewer's eyes so they may view and evaluate the sample.
- Setting Up the Stage and the Sample: The sample is put on a stage, a platform that can be moved both horizontally and vertically. Depending on the type of sample and the planned analysis, the sample may go through various preparation processes before being examined, such as staining, sectioning, or mounting on a glass slide.
- Focus and Magnification: To bring the desired area of the sample into sharp focus, the microscope user can adjust the focus by rotating the stage or by utilizing tiny focus knobs. The user can acquire various magnifications to view various aspects of the sample by altering the objective lens.
- Brightfield and Phase Contrast Microscopy: The most popular approach uses brightfield microscopy, which makes the sample look dark against a bright background. A specific condenser and objective lens are employed in phase contrast microscopy to improve the contrast between various sections of the sample, especially clear or translucent specimens.
- Polarized Light: Polarized light microscopy examines hybrid nanocomposite samples that respond to polarized light using

polarizers and specific filters. It divulges details on the sample's structural and optical properties.

- Fluorescence Microscopy: In this sort of microscopy, fluorescent dyes or markers are utilized, and when activated by a specific wavelength of light, they release light of various wavelengths. This method enables the visualization of certain molecules or sample structures, such as fluorescently marked DNA or proteins.
- Image Capture and Analysis: Digital cameras are frequently employed in optical microscopy to take pictures of the sample. Image processing technologies can be used to enhance, evaluate, and appraise these photographs. On the obtained photos, measurements, annotations, and quantitative analysis can all be done.
- Limitations: The resolution of optical microscopy is constrained by light diffraction, which reduces the level of detail that can be seen. Furthermore, a short depth of field may make it difficult to focus on several planes at once. These limitations can be overcome by using advanced techniques such as confocal microscopy or electron microscopy.
- Optical microscopy provides valuable insights into the structure, composition, and behavior of microscopic samples. It continues to be a fundamental tool in scientific research, medical diagnostics, quality control, and numerous other applications. The hybrid nanomaterial contains fine equiaxed grains, the investigation of the hybrid nanomaterial is done under an optical microscope. Before examining the samples, all the samples went through standard metallographic procedures as described by ASTM standards. Basically, this procedure follows the polishing and grinding of the samples followed by the cleaning of samples with a suitable etchant.

The specimen prepared can be examined under the magnification range of the optical microscope from 10x to 100x. One example of magnesium metal matrix hybrid nanomaterial grain evolution after performing friction stir processing and of the base, magnesium substrate is provided in Figure 2.

By the use of OM technique, this examination has been done. The main feature of hybrid nanomaterial is the evolution of fine grains (Moustafa, 2021). With the use of the OM technique, one can examine the size and morphology of grains along with the grain boundary and orientation of these grains.

Figure 2. Optical microscope images a) base magnesium alloy, b) hybrid magnesium nanomaterial reinforced with titanium carbide and fly ash particulates via friction stir processing

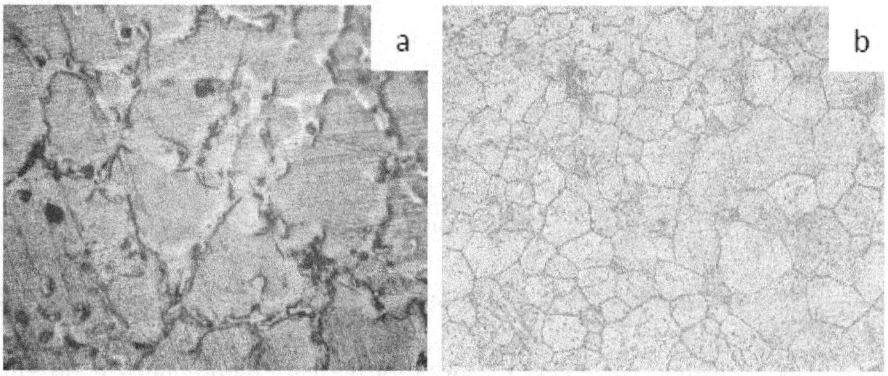

b) **X-ray diffraction (XRD) analysis:** The atomic and molecular structure of hybrid nanomaterials can be ascertained using the effective method known as XRD analysis. It is based on the idea that constructive and destructive interference occurs when X-rays interact with a sample of a nanomaterial, producing a distinctive diffraction pattern.

- Equipment: An X-ray diffractometer is a specialist equipment needed for XRD analysis. An X-ray tube that produces X-rays and a detector that gauges the intensity of the diffracted X-rays make up the diffractometer. The sample can be rotated at different angles because it is positioned on a goniometer.
- Sample Preparation: Powdered, small-particle samples of crystalline materials are frequently prepared. For a representative analysis, the material needs to be coarsely powdered and homogenized. Single crystals or thin film samples may be employed for certain applications.
- Bragg's Law: The basis for XRD analysis is Bragg's Law, which establishes a relationship between the angle of incidence, X-ray wavelength, and the distance between crystal lattice planes. Bragg's Law states that constructive interference happens when 2d sin = n, where n is an integer, d is the wavelength of X-rays, is the angle of incidence, and d is the distance between lattice planes.
- Diffraction Pattern: The diffracted X-rays are detected once the X-ray beam is steered onto the specimen. Different crystallographic planes scatter X-rays at various angles as the sample is rotated, producing a

diffraction pattern. A series of peaks that correspond to the angles at which constructive interference occurs make up the pattern.

- Data analysis: To learn more about the crystal structure, the diffraction pattern is captured and examined. The location, magnitude, and shape of the diffraction peaks reveal important information. The peaks' locations match the lattice spacing, and their intensities show how the atoms are arranged in the crystal. The size, flaws, and strain of the crystal can be inferred from the shape and width of the peaks.
- Indexing and Crystal Structure Determination: The diffraction pattern is compared to recognized crystal structures in a database to ascertain the crystal structure. Indexing is the process of comparing the measured diffraction angles to the theoretical angles for various crystal systems. Once indexing is accomplished, it is possible to discover the crystal structure by resolving mathematical equations and adjusting the parameters with the aid of specialist software.
- • Applications: Materials science, geology, chemistry, and many other disciplines use XRD analysis extensively. It is employed to explore phase transitions, assess crystal defects, identify unidentified compounds, ascertain phase composition, and research the crystalline structure of materials.
- XRD analysis is a non-destructive methodology that gives explained data for the atomic and molecular arrangement within crystalline materials. It has played a crucial role in advancing our understanding of materials and their properties, enabling the development of new materials and technologies.

In order to examine a material's crystallographic structure, chemical composition, and physical properties a nondestructive method known as XRD could provide accurate data. It is based on the constructive interference of crystalline samples and monochromatic X-rays. For the hybrid nanomaterial various crystallographic structures, chemical compositions, and physical properties of a material are essential to know. XRD analysis supports determining the basic chemical composition of the developed material.

Figure 3 illustrates the X-ray diffraction (XRD) patterns of the TiC/AZ31B magnesium matrix nanocomposite that was developed through friction stir processing (FSP). XRD is a powerful technique for analyzing the crystallographic structure of materials by measuring the diffraction of X-rays. In this context, the XRD patterns provide valuable insights into the crystalline phases present in the nanocomposite. Peaks and their positions in the XRD patterns correspond to specific crystallographic planes of the materials, allowing researchers to identify the composition and crystalline structure of the developed nanocomposite. By examining Figure 3, researchers can gain information about the formation of TiC (titanium carbide) within the AZ31B magnesium matrix through the FSP technique. The distinct peaks in the XRD patterns indicate the presence of specific crystalline phases, and their intensities can provide details about the relative amounts of each phase. This analysis is crucial for understanding the structural changes, phase transformations, and overall composition of the TiC/AZ31B magnesium matrix nanocomposite. The XRD patterns presented in Figure 3 serve as a valuable tool for researchers and readers to assess the success of the friction stir processing in achieving the desired nanocomposite structure and to further comprehend the material's potential properties and performance.

c) **Scanning electron microscopy (SEM):** It is a powerful analytical investigation approach to investigate a wide range of materials, at suitable magnification, and to produce high-resolution images. Especially in the case of composites processed via the FSP technique, it is essential to find the reinforcement distribution. SEM analysis helps in investigating the uniform dispersal of the reinforcement particles throughout the substrate. Moreover, Energy-dispersive X-ray spectroscopy (EDS) analysis usually taken in conjunction with SEM helps in determining various peaks of the presentation materials at a particular

Figure 3. XRD patterns of the TiC/AZ31B magnesium matrix nanocomposite developed through friction stir processing

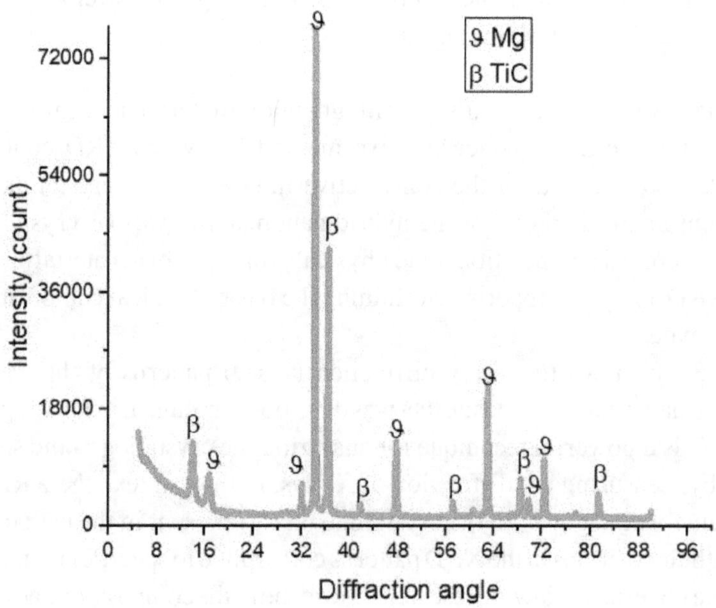

location. FESEM analysis results to examine the uniform distribution of titanium carbide, silicon carbide and fly ash particles in base metal steel SEM and EDS has been done. Figure 4 shows the result for magnesium base hybrid reinforced nanomaterial when developed via friction stir processing.

d) **Transmission electron microscopy (TEM):** Once after the preparation of the specimen via FSW or FSP the basic texture of the material undergoes various changes. All these changes need to be analyzed via suitable methods to examine the real evolution. TEMs provide topographical, morphological, compositional, and crystalline information. TEM microscopes have the potential to capture the images of specimens at an impressive higher resolution than light optical microscopy, attributed to the relatively smaller de Broglie wavelength of electrons. Since a single column of atoms is thousands of times smaller than a resolvable object observed in a light microscope, this allows the equipment to capture extremely fine detail. The images allow viewing samples on a molecular level, making it possible to analyze structure and texture. TEM instruments can operate in a variety of modes, including standard imaging, scanning TEM imaging (STEM), diffraction, spectroscopy, and combinations of these. Even within conventional imaging, there are numerous fundamentally distinct methods for producing contrast, known as "image contrast mechanisms". Contrast can arise from position-to-position differences in the thickness, atomic number, crystal structure, the slight quantum-mechanical phase shifts that individual atoms produce in electrons that pass through them ("phase contrast"), the energy lost by electrons on passing through the sample ("spectrum imaging") and more. Depending on the contrast mechanism as well as how the microscope is used—the settings of the lenses, apertures, and detectors—each mechanism provides the operator with a different kind of information. This means that a TEM is capable of delivering a remarkable diversity of information at the nanoscale and atomic levels, such as size, shape, and atom bonding. For this reason, TEM is regarded as a crucial tool for the materials sciences.

Figure 4. FESEM and EDS results of magnesium base hybrid nanomaterial

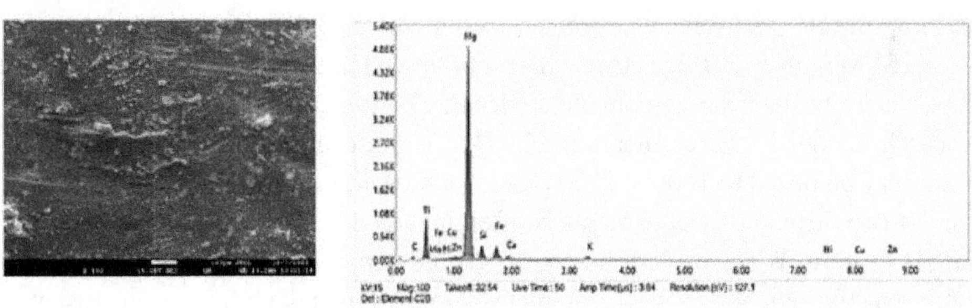

e) **Electron backscatter diffraction (EBSD) examination:** With its versatility in mapping orientation, crystal type, and perfection over a wide range of step sizes, electron backscatter diffraction (EBSD), a powerful microstructural analyzer tool, allows crystallographic information to be obtained from small volumes of material in a scanning electron microscope (SEM). These experiments have been carried out using X-ray, neutron, and/or electron diffraction in a transmission electron microscope, as well as spatially resolved acoustic spectroscopy (SRAS), which examines elastic waves rather than a diffraction event. The geographical resolution, the area or volume being studied, and whether the data are static or dynamic all influence the technique that is ultimately chosen. A technique called electron backscatter diffraction (EBSD) is used to examine the tiny crystallographic characteristics of materials. It is frequently used in the study of crystalline materials, including as metals, ceramics, and semiconductors, and is especially helpful in this regard. A focussed beam of extremely high-energy electrons is directed onto a sample's surface during an EBSD analysis. The material's atoms disperse as a result of the electrons' interactions with them. Backscattered electrons are some of the scattered electrons that are reflected back towards the detector. Backscattered electrons' pattern reveals details about the crystal structure and orientation of the substance.

Using a specialized detector, usually a phosphor screen or a CCD camera, the backscattered electrons are found. The intensity and position of the electrons are recorded by the detector, and this data is processed to produce the crystallographic data. To identify the crystal orientation, grain boundaries, and other microstructural properties, the electron patterns acquired from various areas of the sample are examined. The EBSD method can offer useful details regarding the material's grain size and shape, crystallographic texture, and phase identification. It is frequently used to explore phase transitions, characterize the deformation behavior of materials, and assess the quality of materials used in manufacturing processes. It also studies the microstructure of metals and alloys. In general, the EBSD analysis provides a potent tool for comprehending the crystallographic characteristics of materials, allowing researchers to learn more about the links between structure and property and improve the performance of materials in diverse applications.

An SEM with an EBSD detector that contains at least a phosphor screen, a small lens, and a low-light charge-coupled device (CCD) camera is used to conduct EBSD research. For quick measurements, the CCD chip's base resolution is 640 x 480 pixels, but it may be raised to 1600 × 1200 pixels for slower, more sensitive observations. One of two different CCD cameras is often included with commercially marketed EBSD systems. High-resolution detectors' fundamental advantage is that they are more sensitive, enabling a more complete examination of the information present in each diffraction pattern. To reduce their size and hasten processing, the diffraction patterns are binned for texture and orientation measurements. Modern CCD-based EBSD systems are capable of indexing patterns at speeds of up to 1800 patterns per second. This makes it possible to produce rich and quick microstructural maps.

f) **Atomic Force Microscopy (AFM):** AFM is a high-resolution imaging method that scans a sample's surface with a sharp tip. It can be utilized to investigate features at the nanoscale and offers details on the surface topography.
g) **Fourier Transform Infrared Spectroscopy (FTIR):** FTIR is a method used to quantify how much infrared light a material absorbs. It can be applied to hybrid nanomaterials to detect chemical bonds and functional groups.
h) **Energy-Dispersive X-ray Spectroscopy (EDS or EDX):** By identifying distinctive X-rays released by the sample, this method is frequently used in conjunction with SEM and TEM to examine the elemental composition of materials.
i) **Raman Spectroscopy:** Monochromatic light is scattered by a sample in Raman spectroscopy. It is useful for studying the structure and composition of nanomaterials and offers information on molecular vibrations.
j) **Small Angle X-ray Scattering (SAXS):** One method for determining the size and shape of nanoparticles is SAXS. It works especially well for describing nanomaterials in a solution.
k) **Nuclear Magnetic Resonance (NMR):** NMR is an effective method for researching the dynamics and structure of materials at the atomic level. It is particularly helpful for comprehending the chemical surroundings of nanomaterials' nuclei.
l) **Differential Scanning Calorimetry (DSC):** DSC calculates a sample's heat absorption or release based on temperature. Phase transitions and the thermal characteristics of hybrid nanomaterials can be studied with it.
m) **Auger Electron Spectroscopy (AES):** AES is a surface-sensitive technique that analyzes the energy of Auger electrons emitted from a sample. It provides information about the elemental composition of the surface.

When combined, these methods provide a thorough grasp of the composition, microstructure, and characteristics of hybrid nanomaterials. The particulars of the material and the information that researchers are looking for determine which technique is best.

3.2 Mechanical Testing

The basic mechanical properties that can be examined to investigate the real gain in the basic mechanical characteristic of hybrid nanomaterials are:

a) Microhardness
b) Tensile strength
c) Compressive strength
d) Nanoindentation
e) Fatigue testing
f) Creep Testing
g) Dynamic Mechanical Analysis (DMA)
h) Bend Testing
i) Impact Testing (e.g., Charpy Impact Test)
j) Fracture Toughness Testing
k) Shear Testing
l) Torque Testing
m) Creep-Rupture Testing

a) **Microhardness:** Indenter probes are displaced into the surfaces of the hybrid nanomaterials during hardness test procedures. The dwell time of the indentation is usually predetermined. To determine hardness in conventional mechanical testing, the size or depth of the indentation is measured. For smaller samples and plated surfaces, microhardness testing with applied stresses ≤ 10 N is commonly performed. In hybrid nanomaterials, microhardness is often examined along a horizontal line beneath, at a certain distance from the top surface, and at a certain number of evenly spaced spots between neighbouring sites. For a distance to either side of the central line that runs across the stir zone, any number of values may be chosen. More values may be taken and the mean value was taken into consideration at each point of indentation. Numerous studies have also shown the importance of measuring microhardness values for all hybrid nanomaterials (Acuña et al., 2019; Fotoohi Nezhad Khales et al., 2020).

b) **Tensile strength:** Tensile testing is a destructive engineering and materials science procedure in which a sample is subjected to controlled tension until it fully fails. Tensile testing is one of the most popular procedures for mechanical testing. It is used to assess a material's strength and the maximum stretch it can withstand before failing. This test method has been used to evaluate the yield strength, ultimate tensile strength, ductility, strain hardening characteristics, Young's modulus, and Poisson's ratio, according to numerous research studies (Sagar et al., 2021; Sagar and Handa, 2020a, 2020b; Singla et al., 2023a; Singla et al., 2023b; Sagar et al., 2023). In addition, in order to determine ductility and fracture mechanism, the tensile testing methods are very much important for materials.

c) **Compressive strength:** Techniques for composite compression testing provide a mechanism to compress a material without running the risk of buckling. Compression tests are carried out on composite materials that are in the shape of a relatively thin, flat, rectangular test specimen, such as laminate panels. For specimens, it became very much essential to investigate their compressive strength as suggested by (Hegde et al., 2011). As most of the specimens prepared have their potential application in the field where these materials are subjected to repeated loading.

d) **Nano-indention:** Nanoindentation is a technique used to measure the mechanical properties of materials at the nanoscale, including hardness, elastic modulus, and more. When it comes to hybrid nanomaterials, which are materials composed of multiple components at the nanoscale, measuring their mechanical properties can be particularly valuable for various applications. Here's a standard procedure for performing nanoindentation on hybrid nanomaterials:

(i) Sample Preparation: For testing, create a clean, flat surface from the hybrid nanomaterial. In order to do this, the hybrid material may need to be synthesized or deposited onto an appropriate substrate.

(ii) Instrument Setup: Utilize an atomic force microscope (AFM) with an integrated nanoindentation module or a separate nanoindenter as your instrument of choice for nanoindentation. For precise measurements, make sure the device is calibrated correctly.

(iii) Choosing the Indenter: Select a suitable indenter tip. Berkovich, conical, or spherical indenters are typical tips. The decision is based on the particular characteristics of the hybrid nanomaterial and the data you are seeking.

(iv) Parameters for Load and Unload: Set the parameters for the load and unload. The maximum load, loading speed, and holding period at maximum load are all included. Depending on the anticipated mechanical behavior of the hybrid material, the parameters may change.

(v) Carrying Out Indentation: Approach the indenter to the sample surface while keeping an eye on the force being applied and the depth of the indentation. To ensure representative data, make many indentations on the sample surface at various points.

(vi) Data Analysis: Examine the force-displacement curves to identify mechanical traits like elastic modulus and hardness. You might need to take into account how various components contribute to the overall mechanical properties of hybrid nanomaterials.

(vii) Statistical Analysis: If you've completed many indentations, perform statistical analysis on the data to determine mean values and standard deviations.

(viii) Reporting Results (viii) Clearly and succinctly summarize the measured mechanical properties and any pertinent analysis, including details on the structure and composition of the hybrid nanomaterial.

(ix) Interpretation: Consider the conclusions in light of the planned use or study objectives of the hybrid nanomaterial. It's crucial to remember that the individual steps may change based on the type of nanoindentation tool used and the properties of the hybrid nanomaterial. Additionally, care should be taken to account for any potential issues such as surface roughness, adhesion effects, and the influence of substrate properties on the measurements. Always refer to the manufacturer's guidelines for your specific nanoindentation instrument and consider consulting with experts in the field of nanomechanics to ensure accurate and meaningful results for your hybrid nanomaterials. (Bor et al., 2019; Mallikarjuna et al., 2017).

(e) **Fatigue testing**:

The design of hybrid nanomaterials may include many instances of fatigue testing. Nonetheless, focusing on methodologies for small-specimen level fatigue testing points to the necessity of standardisation of ongoing test method. A test specimen or a structure is subjected to cyclic loading during a fatigue test. The applied load is cycled between the defined maximum and minimum levels until a fatigue failure occurs, unlike monotonic tests where loading increases till failure. The output of the fatigue test is considered a run-out if the specified number of loading cycles exceeds and the failure does not occur. An S-N diagram, which plots the amplitude of the alternating stress against the number of cycles to failure, N, often represents the findings of several fatigue tests that were carried out using a variety of cyclic stress levels. The same minimum stress to maximum stress ratio—also known as the cyclic stress ratio, or R—is used for all of the fatigue tests represented in the plot. Furthermore, stiffness and strength decreases in composite materials and structures brought on by earlier cyclic loading are investigated using fatigue testing. A composite structure may undergo fatigue testing at various stages of its design, using test pieces of various sizes and complexity. Small-specimen and simple element-level tests are used to examine the fatigue behaviour of composite materials. In the case of materials developed via the examination of fatigue is very important, as suggested by (Fahad et al., 2021; Singh & Malik, 2023). The hybrid nanomaterials find their application in the field automotive and aerospace industry where materials undergo periodic fatigue, and the investigation of fatigue properties has become more vital.

f) **Creep Testing:** In creep testing, a material is subjected to a continuous load or stress for a predetermined amount of time, and the resulting deformation is measured. This is especially crucial to comprehending how materials react to extended exposure to stress as suggested by other studies (Yang et al., 2019).
g) **Dynamic Mechanical Analysis (DMA):** DMA measures a material's mechanical properties as a function of temperature, time, or frequency by putting it under oscillatory stress or strain. It is helpful in comprehending how materials behave in a viscoelastic manner (Bashir, 2021; Haris et al., 2022; Santos et al., 2020).
    i) **Bend Testing**: In order to ascertain a sample's flexural strength and modulus of elasticity, bend testing, also known as flexural testing, entails bending the sample. This is especially important for materials that are utilized in applications where bending is a frequent loading circumstance (Beigpour et al., 2021; Jeon et al., 2022; Nordin & Täljsten, 2004).
    j) **Impact Testing (e.g., Charpy Impact Test):** Impact testing assesses how resilient a material is to abrupt loads, like shock or impact. For instance, the energy absorbed by a material during fracture is measured by the Charpy impact test (Abidin et al., 2019; Hufenbach et al., 2008; Tanks et al., 2016).
    k) **Fracture Toughness Testing:** An indicator of a material's resistance to crack propagation is its fracture toughness. Fracture toughness is measured using tests like the single-edge notched bend (SENB) test (Abou El-Mal et al., 2015; Launey & Ritchie, 2009; Truong et al., 2020).
    l) **Shear Testing:** To determine a material's shear strength, shear testing entails applying a force parallel to the material's surface. This is especially important for applications where materials are subjected to shearing or sliding forces (ÖzkanÖzbek, 2023).
    m) **Torque Testing:** The resistance of a material to rotational forces is evaluated through torque testing. In applications where materials must endure torsional or twisting loads, it is crucial (Coşkun et al., 2021; Jiang et al., 2016).
    n) **Creep-Rupture Testing:** In order to determine a material's long-term creep behaviour and time to failure, creep-rupture testing entails applying a continuous load to it at high temperatures for an extended amount of time (Ennis, 2016; Reyngoud et al., 2015).

These methods offer a thorough comprehension of the mechanical characteristics of hybrid nanomaterials, assisting engineers and researchers in making defensible choices regarding their use in a variety of fields. Which testing technique is used for a particular application is determined by the particular mechanical properties that are required (Albalawi et al., 2021; Khan et al., 2022; Sagar & Handa, 2020c; Sagar & Handa, 2020d; Singla et al., 2023b).

The tables below showcase a comprehensive comparison of various characterization techniques in Table 1, highlighting their principles, advantages, and limitations for assessing hybrid nanomaterials. Additionally, Table 2 outlines the advantages and limitations of different methods employed to assess the mechanical properties of these hybrid nanomaterials.

5. CHALLENGES AND FUTURE DIRECTIONS

In conclusion, the characterization of hybrid nanomaterials has made significant strides in recent years, but it is not without its challenges and exciting future prospects.

Table 1. Comparison of characterization techniques

Technique	Principle	Advantages	Limitations
FTIR Spectroscopy	Infrared absorption	Non-destructive, qualitative analysis	Limited for certain materials
TEM Microscopy	Electron beam imaging	High-resolution, nanoscale visualization	Requires vacuum conditions
TGA	Thermogravimetric Analysis	Measures weight changes with temperature	Limited for volatile materials
XRD	X-ray Diffraction	Provides crystal structure information	Requires crystalline samples

Table 2 Advantages and limitations of mechanical properties assessment methods

Characterization Method	Advantages	Limitations
Tensile Testing	Direct measurement of material strength	Limited to macro-scale samples; may not capture nanoscale behavior
Nanoindentation	High spatial resolution; suitable for small samples	Limited to surface properties; may be influenced by substrate effects
Atomic Force Microscopy (AFM)	Nanoscale surface imaging; simultaneous mechanical property mapping	Limited to surface characterization; may not capture bulk properties
Scanning Electron Microscopy (SEM) with In-situ Mechanical Testing	Real-time imaging during mechanical testing	Sample preparation challenges; may require complex instrumentation

One of the key challenges lies in achieving a comprehensive understanding of the intricate interplay between different components within these materials. The synergistic effects and emergent properties that arise from the combination of diverse nanomaterials demand sophisticated characterization techniques that can provide precise insights into their structural, chemical, and physical properties at various length scales. Moreover, the advancement of characterization tools and methodologies will continue to be a driving force in this field. Researchers will be able to delve deeper into the characterization of hybrid nanomaterials thanks to the development of novel analytical techniques such as enhanced microscopy, spectroscopy, and computational modelling. These developments will make it possible to precisely manipulate and create hybrid nanomaterials with specialized features for a range of applications, from medicinal devices and environmental cleanup to renewable energy and catalysis.

Collaboration between researchers from several fields, such as materials science, chemistry, physics, and engineering, will be essential going ahead in order to address the complex problems related to hybrid nanomaterials. By using an interdisciplinary approach, we may better understand these materials and realize their full potential for a variety of technological breakthroughs. In conclusion, the characterisation of hybrid nanomaterials is a dynamic and ever-evolving field at the forefront of materials science, offering both problems and extraordinary opportunities.

6. TARGET AUDIENCE

Researchers, students, and professionals involved in nanotechnology, materials science, chemistry, physics, and engineering will find the proposed chapter useful. It will operate as a thorough reference for comprehending the characterisation techniques necessary for assessing the efficacy of hybrid nanomaterials in a range of applications.

REFERENCES

Abidin, N. M. Z., Sultan, M. T. H., Shah, A. U. M., & Safri, S. N. A. (2019). Charpy and Izod impact properties of natural fibre composites. *IOP Conference Series. Materials Science and Engineering*, *670*(1), 012031. Advance online publication. doi:10.1088/1757-899X/670/1/012031

Abou El-Mal, H. S. S., Sherbini, A. S., & Sallam, H. E. M. (2015). Mode II Fracture Toughness of Hybrid FRCs. *International Journal of Concrete Structures and Materials*, *9*(4), 475–486. doi:10.1007/s40069-015-0117-4

Acuña, R., Cristóbal, M. J., Abreu, C. M., & Cabeza, M. (2019). Microstructure and Wear Properties of Surface Composite Layer Produced by Friction Stir Processing (FSP) in AA2024-T351 Aluminum Alloy. *Metallurgical and Materials Transactions. A, Physical Metallurgy and Materials Science*, *50*(6), 2860–2874. Advance online publication. doi:10.1007/s11661-019-05172-6

Albalawi, F., Hussein, M. Z., Fakurazi, S., & Masarudin, M. J. (2021). Engineered Nanomaterials: The Challenges and Opportunities for Nanomedicines. *International Journal of Nanomedicine*, *16*, 161–184. doi:10.2147/IJN.S288236 PMID:33447033

Andrew, R. M. (2021). Towards near real-time, monthly fossil CO2 emissions estimates for the European Union with current-year projections. *Atmospheric Pollution Research*, *12*(12), 101229. doi:10.1016/j.apr.2021.101229 PMID:34938143

Bashir, M. A. (2021). Use of Dynamic Mechanical Analysis (DMA) for Characterizing Interfacial Interactions in Filled Polymers. In Solids (Vol. 2, Issue 1, pp. 108–120). doi:10.3390/solids2010006

Beigpour, R., Shokrollahi, H., & Khalili, S. M. R. (2021). Experimental and numerical analysis of a biodegradable hybrid composite under tensile and three-point bending tests. *Composite Structures*, *273*, 114255. doi:10.1016/j.compstruct.2021.114255

Bor, B., Giuntini, D., Domènech, B., Swain, M. V., & Schneider, G. A. (2019). Nanoindentation-based study of the mechanical behavior of bulk supercrystalline ceramic-organic nanocomposites. *Journal of the European Ceramic Society*, *39*(10), 3247–3256. doi:10.1016/j.jeurceramsoc.2019.03.053

Coşkun, Z., Çelik, T., & Kisioglu, Y. (2021). Design and Manufacture of the Torque Test Setup For Small and Shapeless Materials. *Turkish Journal of Engineering*, *6*(1), 81–86. Advance online publication. doi:10.31127/tuje.843320

Ennis, P. (2016). Creep and Stress Rupture Testing and Evaluation of Data. In Reference Module in Materials Science and Materials Engineering. doi:10.1016/B978-0-12-803581-8.03415-9

Fahad, N., Kadhim, A., Al-Khayat, R., & Al-Waily, M. (2021). Effect of SiO2 and Al2O3 Hybrid Nano Materials on Fatigue Behavior for Laminated Composite Materials Used to Manufacture Artificial Socket Prostheses. *Materials Science Forum*, *1039*, 493–509. . doi:10.4028/www.scientific.net/MSF.1039.493

Fontaras, G., Zacharof, N. G., & Ciuffo, B. (2017). Fuel consumption and CO2 emissions from passenger cars in Europe – Laboratory versus real-world emissions. In *Progress in Energy and Combustion Science* (Vol. 60, pp. 97–131). doi:10.1016/j.pecs.2016.12.004

Fotoohi Nezhad Khales, M., Sajjadi, S. A., & Kamyabi-Gol, A. (2020). Multipass Friction Stir Processing of Steel/SiC Nanocomposite: Assessment of Microstructure and Tribological Properties. *Journal of Materials Engineering and Performance*, *29*(7), 4241–4250. doi:10.1007/s11665-020-04947-y

Haris, N. I. N., Hassan, M. Z., Ilyas, R. A., Suhot, M. A., Sapuan, S. M., Dolah, R., Mohammad, R., & Asyraf, M. R. M. (2022). Dynamic mechanical properties of natural fiber reinforced hybrid polymer composites: A review. *Journal of Materials Research and Technology*, *19*, 167–182. doi:10.1016/j.jmrt.2022.04.155

Hegde, M. N., Hegde, P., Bhandary, S., & Deepika, K. (2011). An evalution of compressive strength of newer nanocomposite: An in vitro study. *Journal of Conservative Dentistry*, *14*(1), 36–39. doi:10.4103/0972-0707.80734 PMID:21691503

Hufenbach, W., Ibraim, F., Langkamp, A., Böhm, R., & Hornig, A. (2008). Charpy impact tests on composite structures – An experimental and numerical investigation. *Composites Science and Technology*, *68*(12), 2391–2400. doi:10.1016/j.compscitech.2007.10.008

Jeon, G., Ha, D., Park, Y., & Jeong, C. (2022). Three-Point Bending Properties of Hybrid Multi-Materials Using Adhesive Bonding Dependent on Strength Difference between Steel and Aluminum. In Materials (Vol. 15, Issue 9). doi:10.3390/ma15093328

Jiang, C., Hu, D., & Lu, Y. (2016). Digital Micromirror Device (DMD)-Based High-Cycle Torsional Fatigue Testing Micromachine for 1D Nanomaterials. *Micromachines*, *7*(3), 49. Advance online publication. doi:10.3390/mi7030049 PMID:30407421

Kawajiri, K., Kobayashi, M., & Sakamoto, K. (2020). Lightweight materials equal lightweight greenhouse gas emissions?: A historical analysis of greenhouse gases of vehicle material substitution. *Journal of Cleaner Production*, *253*, 119805. doi:10.1016/j.jclepro.2019.119805

Khan, Y., Sadia, H., Ali Shah, S. Z., Khan, M. N., Shah, A. A., Ullah, N., Ullah, M. F., Bibi, H., Bafakeeh, O. T., Khedher, N. B., Eldin, S. M., Fadhl, B. M., & Khan, M. I. (2022). Classification, Synthetic, and Characterization Approaches to Nanoparticles, and Their Applications in Various Fields of Nanotechnology: A Review. In Catalysts (Vol. 12, Issue 11). doi:10.3390/catal12111386

Launey, M. E., & Ritchie, R. O. (2009). On the fracture toughness of advanced materials. *Advanced Materials*, *21*(20), 2103–2110. doi:10.1002/adma.200803322

Lutsey, N. (2010). *Review of technical literature and trends related to automobile mass-reduction technology*. Institute of Transportation Studies, UC Davis, Institute of Transportation Studies, Working Paper Series.

Mallikarjuna, H. M., Ramesh, C. S., Koppad, P. G., Keshavamurthy, R., & Sethuram, D. (2017). Nanoindentation and wear behaviour of copper based hybrid composites reinforced with SiC and MWCNTs synthesized by spark plasma sintering. *Vacuum*, *145*, 320–333. doi:10.1016/j.vacuum.2017.09.016

Meireles, M., Robaina, M., & Magueta, D. (2021). The Effectiveness of Environmental Taxes in Reducing CO2 Emissions in Passenger Vehicles: The Case of Mediterranean Countries. In International Journal of Environmental Research and Public Health (Vol. 18, Issue 10). doi:10.3390/ijerph18105442

Moustafa, E. B. (2021). Hybridization effect of BN and Al2O3 nanoparticles on the physical, wear, and electrical properties of aluminum AA1060 nanocomposites. *Applied Physics. A, Materials Science & Processing*, *127*(9), 724. doi:10.1007/s00339-021-04871-5

Muhamad Azim, M. K., Arifutzzaman, A., Saidur, R., Khandaker, M. U., & Bradley, D. A. (2022). Recent progress in emerging hybrid nanomaterials towards the energy storage and heat transfer applications: A review. *Journal of Molecular Liquids*, *360*, 119443. doi:10.1016/j.molliq.2022.119443

Nordin, H., & Täljsten, B. (2004). Testing of hybrid FRP composite beams in bending. *Composites. Part B, Engineering*, *35*(1), 27–33. doi:10.1016/j.compositesb.2003.08.010

ÖzkanÖzbek. (2023). Shear and Fracture Characteristics of Nano-silica and GNP Hybrid Nanoparticle Reinforced Single Lap Joints. *Journal*, *13*(3), 1970–1982.

Rahman, A., Chowdhury, M. A., & Hossain, N. (2022). Green synthesis of hybrid nanoparticles for biomedical applications: A review. *Applied Surface Science Advances*, *11*, 100296. https://doi.org/https://doi.org/10.1016/j.apsadv.2022.100296

Reyngoud, B., Bishop, C., & Kral, M. (2015). Hybrid materials design to control creep in metallic pipes. *Materials & Design*, *84*, 25–35. doi:10.1016/j.matdes.2015.06.089

Sagar, P., & Handa, A. (2020a). Role of tool rotational speed on the tribological characteristics of magnesium-based az61a/tic composite developed via friction stir processing route. *Journal of Achievements in Materials and Manufacturing Engineering, 101*(2), 60–75. doi:10.5604/01.3001.0014.4921

Sagar, P., & Handa, A. (2020b). Prediction of wear resistance model for magnesium metal composite by response surface methodology using central composite design. *World Journal of Engineering*, ahead-of-print. doi:10.1108/WJE-08-2020-0379

Sagar, P., & Handa, A. (2020c). A comprehensive review of recent progress in fabrication of magnesium base composites by friction stir processing technique—A review. *AIMS Materials Science, 7*(5), 684–704. doi:10.3934/matersci.2020.5.684

Sagar, P., & Handa, A. (2020d). Selection of tool transverse speed considering trial run experimentations for AZ61/Tic composite developed via friction stir processing using triangular tool. *Materials Today: Proceedings, 38*, 198–203. doi:10.1016/j.matpr.2020.06.537

Sagar, P., & Handa, A. (2023). Exploring the mechanical, metallurgical, and fracture characteristics of hybrid-reinforced magnesium metal matrix composite synthesized via friction stir processing route. *Proceedings of the Institution of Mechanical Engineers, Part L: Journal of Materials: Design and Applications.* 10.1177/14644207231200640

Sagar, P., Handa, A., & Kumar, G. (2021). Metallurgical, mechanical and tribological behaviour of Reinforced magnesium-based composite developed Via Friction stir processing. *Proceedings of the Institution of Mechanical Engineers. Part E, Journal of Process Mechanical Engineering, 236*(4), 1440–1451. doi:10.1177/09544089211063099

Sagar, P., Kumar, G., & Handa, A. (2023). Progressive use of nanocomposite hydrogels materials for regeneration of damaged cartilage and their tribological mechanical properties. *Proceedings of the Institution of Mechanical Engineers, Part N: Journal of Nanomaterials, Nanoengineering and Nanosystems.* 10.1177/23977914231151487

Santos, P., Valvez, S., Monjon, A., & Reis, P. N. B. (2020). The hybridisation effect on the viscoelastic properties of polymeric composites. *Procedia Structural Integrity, 28*, 1816–1826. doi:10.1016/j.prostr.2020.11.004

Singh, V. V., & Malik, A. K. (2023). 3 - Approaches to produce hybrid nanomaterials for engineering applications. In *Micro and Nano Technologies* (pp. 55–69). Elsevier. https://doi.org/https://doi.org/10.1016/B978-0-323-98371-6.00013-6

Singla, S., Sagar, P., & Handa, A. (2023a, March 28). Magnesium-based nanocomposites synthesized using friction stir processing: An experimental study. *Materials and Manufacturing Processes*, 1–18. doi:10.1080/10426914.2023.2195909

Singla, S., Sagar, P., Handa, A., & Kang, A. S. (2023b). Recent Advances in Magnesium-Based Metal Matrix Surface Composites Developed via Friction Stir Processing Route—An Overview. *Metallography, Microstructure, and Analysis, 0123456789*(3), 385–400. Advance online publication. doi:10.1007/s13632-023-00969-9

Tanks, J., Sharp, S., & Harris, D. (2016). Charpy Impact Testing to Assess the Quality and Durability of Unidirectional CFRP Rods. *Polymer Testing*, *51*, 63–68. Advance online publication. doi:10.1016/j.polymertesting.2016.02.009

Truong, H. T. X., Martinez, M. J., Ochoa, O. O., & Lagoudas, D. C. (2020). Mode I fracture toughness of hybrid co-cured Al-CFRP and NiTi-CFRP interfaces: An experimental and computational study. *Composites. Part A, Applied Science and Manufacturing*, *135*, 105925. doi:10.1016/j.compositesa.2020.105925

Yang, H., Huang, Y., Song, B., Kainer, K. U., & Dieringa, H. (2019). Enhancing the creep resistance of AlN/Al nanoparticles reinforced Mg-2.85Nd-0.92Gd-0.41Zr-0.29Zn alloy by a high shear dispersion technique. *Materials Science and Engineering A*, *755*, 18–27. doi:10.1016/j.msea.2019.03.131

Chapter 3
Hybrid Nanomaterials:
A Sustainable Tool to Detect Environmental Problems

Archna Talwar
Isabella Thoburn College, Lucknow, India

Shruti Anand
https://orcid.org/0009-0005-8930-0652
Isabella Thoburn College, Lucknow, India

Avni Nayyar
Isabella Thoburn College, Lucknow, India

Faria Fatima
Integral University, India

Manaal Zahera
Era University, Lucknow, India

ABSTRACT

Hybrid nanomaterials, a novel class formed by integrating organic and inorganic components at the nanoscale, have gained attention in nanotechnology and materials science. This synergy creates structures with enhanced functionalities. In environmental detection, they play a crucial role due to their sustainable and cutting-edge attributes. This chapter examines their pivotal role in addressing environmental issues, leveraging diverse nanomaterial properties. With tunable properties, high surface area, and sensitivity, hybrid nanomaterials are ideal for sensing environmental parameters. Their sustainability aligns with the global push for eco-friendly solutions, extending to applications in renewable energy and waste reduction. The chapter explores synthesis methods, sensor design intricacies, and applications, highlighting their potential to revolutionize environmental monitoring and contribute to a sustainable future through advancements in detection technologies.
DOI: 10.4018/979-8-3693-1261-2.ch003

1. INTRODUCTION

In our rapidly evolving world, the critical need to address pressing environmental issues has never been more evident. From pollution and climate change to resource depletion, our planet faces unprecedented challenges that demand innovative solutions. Industrial operations are consistently increasing the amount of harmful substances in various parts of the environment. These pollutants can find their way into the food we eat and cause negative health impacts and environmental issues (Sharma et al., 2023). One of the most promising avenues for tackling these issues lies in the realm of nanotechnology, specifically, hybrid nanomaterials. These microscopic wonders have emerged as transformative tools with the potential to revolutionize environmental monitoring and management (Ibrahim, Hayyan, AlSaadi, Hayyan, & Ibrahim, 2016). Nanotechnology is a cutting-edge technology that operates at a scale of nanometers ranging from 1-100nm and creates devices, materials, and systems with unique properties and functions by precisely controlling the size and shape of the substances (Ibrahim, Hayyan, AlSaadi, Hayyan, & Ibrahim, 2016; Kumar, Sharma, Ahluwalia, Kaur, Pathak, Singh, & Kumar, 2017; Hussain et al., 2018). Hybrid nanomaterials consist of a combination of two or more organic and inorganic nanomaterials, such as nanoparticles, nanotubes, or nanowires, with unique properties and functionalities. These qualities make hybrid nanomaterials invaluable in addressing the shortcomings of conventional environmental monitoring techniques (Ananikov, 2019). Current challenges in environmental monitoring include the limited scope of measurements, high operational costs, and the inability to detect emerging contaminants swiftly. Moreover, conventional methods often require substantial energy inputs and generate significant waste, exacerbating the environmental problems they aim to mitigate (Salamanca et al., 2023). In contrast, hybrid nanomaterials offer a sustainable solution by enabling efficient and precise monitoring while minimizing resource consumption and waste generation.

Nanotechnology involves harnessing the distinctive characteristics of materials on a nanoscale level. Nanomaterials exhibit increased reactivity and, consequently, improved efficiency in comparison to larger counterparts because of their greater surface-to-volume ratio. Nanoparticles exhibit a notable capacity for adsorbing both inorganic and organic substances due to their distinctive attributes, which include a substantial surface area, a higher count of surface activation sites, and a strong affinity for other species (Baysal et al., 2018). Nanomaterials can utilize distinctive surface chemistry in contrast to conventional methods. This allows them to be modified or combined with functional groups that can precisely target particular molecules of interest such as pollutants for effective remediation. The materials utilized in pollution remediation mustn't become pollutants themselves after use. Consequently, the use of biodegradable materials is highly compelling in this application domain. Employing biodegradable materials not only enhances consumer trust and the adoption of a specific technology, as there is no production of residual material waste requiring disposal post-treatment, but it also offers a more environmentally friendly and safer alternative for remediating pollutants. In the realm of the environment, nanotechnology has demonstrated its versatility by not only detecting and sensing pollution but also contributing to the prevention and remediation of contamination (Kumar, Sharma, Ahluwalia, Kaur, Pathak, Singh, & Kumar, 2017). As a sustainable solution to the global imperative of safeguarding water, soil, and ensuring cleaner air, hybrid nanomaterials play a crucial role (Das et al., 2015). The chapter serves as a comprehensive guide, navigating the potential applications of hybrid nanomaterials in mitigating environmental challenges and advancing our collective understanding of the intersection between nanotechnology and environmental stewardship.

2. VARIOUS TYPES OF HYBRID NANOMATERIALS USED IN ENVIRONMENTAL DETECTION

Diverse nanomaterials, such as inorganic nanomaterials, carbon-based nanomaterials, and polymer-based nanomaterials find application in environmental remediation. They serve various functions including adsorption, catalysis, photocatalysis, membrane filtration, and sensing (Guerra, Attia, Whitehead, & Alexis, 2018; Ibrahim, Hayyan, AlSaadi, Hayyan, & Ibrahim, 2016; Zhang et al., 2016).

2.1. Carbon-Based Nanomaterials

Due to their exceptional physical and chemical characteristics, such as impressive acid stability, large surface area, and heat stability, carbon-based nanomaterials have served significant interest in pollution treatment applications (Liang, 2017). Additionally, these nanomaterials are used in photocatalytic methods to remove pollutants because of their distinctive electrical and porous structure, unique chemical stability, powerful absorption capacity, and nanoscale hollow tube features (Dutta et al., 2022). Nanostructured carbon materials are grouped into various categories based on their size, shape, and dimensions. These categories include:

2.1.1. Zero-dimensional Carbon: These materials have no specific shape and are essentially point-like structures at the nanoscale. Examples of zero-dimensional carbon materials (Figure 1) include fullerenes (Chun et al., 2015), carbon dots (CDs) (Han et al., 2018), nondiamond (ND), and graphene quantum dots (GQDs).

Figure 1. Zero-dimensional carbon

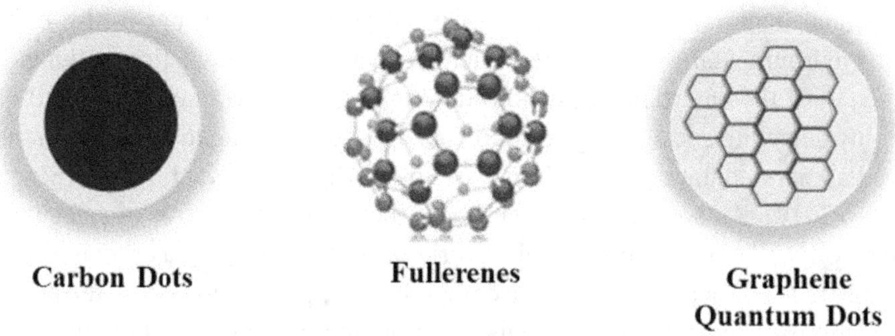

2.1.2. One-dimensional Carbon: These materials have a linear, tube-like structure at the nanoscale. The one-dimensional carbon materials are represented by carbon nanotubes (Liang et al., 2017) and carbon nanofibers (Kumar et al., 2019) (Figure 2).

2.1.3. Two-dimensional Carbon: Two-dimensional carbon materials encompass (Figure 3) graphene (Piumie et al., 2018), graphene derivatives, graphdiynes, and graphitic carbon nitride (g-C3N4). They have a flat, planar structure with carbon atoms arranged in a hexagonal lattice.

Carbon-based hybrid nanomaterials of various dimensions have shown significant attention due to their unique structural, electronic, and optical properties, as well as their ease of functionalization. These materials create efficient heterojunctions when different carbon-based matrices come into contact, leading to high-quality interfaces and efficient charge transfer. Heterostructures with various morphologies, such as point-to-point (0D-0D, 0D-1D), point-to-face (0D-2D), line-to-face (1D-1D, 1D-2D),

Figure 2. One-dimensional carbon

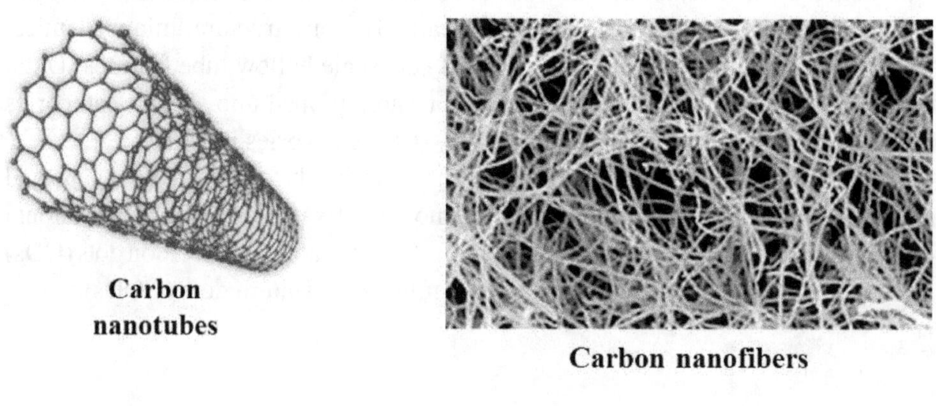

Figure 3. Two-dimensional carbon

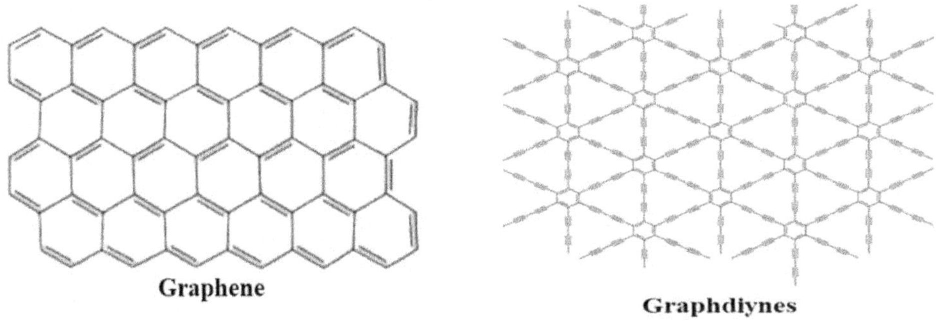

and face-to-face (2D-2D) contacts, serve as excellent support matrices. Efficient heterojunctions, especially point-to-face (0D-2D), line-to-face (1D-2D), and face-to-face (2D-2D), find extensive use in catalytic applications, promoting rapid charge transfer and improved catalytic dispersion, enhancing photocatalytic activity. Figure 4 illustrates the utilization of a photocatalytic method in the treatment of environmental pollutants. The morphology-based heterojunctions are particularly beneficial for photocatalytic water treatment (Mohapatra et al., 2023). Zero-dimensional carbon-based materials, when combined with two-dimensional carbon-based materials, exhibit remarkable potential due to their outstanding electrical conductivity and stable physicochemical properties, which favor charge-transfer kinetics. For instance, the C60/g-C3N4 composite and the 0D/2D-based CD/g-CN heterojunction exhibit enhanced photoactivity in wastewater treatment (Xu et al., 2013). Carbon nanotubes (CNTs) have unique properties that make them valuable in various composites for photocatalytic applications, particularly in combination with two-dimensional semiconductors, such as C3N4. The CNT/C3N4 hybrid catalyst shows improved performance in dye solution degradation, attributed to its light absorption capacity and suppression of charge carrier recombination (Sun et al., 2017). Functionalization is often necessary to create heterojunction catalysts, with graphene oxide (GO) being a suitable candidate due to its oxygen-containing functional groups. Combining CN with graphene derivatives results in a 2D/2D hybrid carbon-based catalyst, highly effective in enhancing photocatalytic performance (Zheng et al., 2020). 2D-2D CN/Graphene heterojunctions are particularly advantageous, offering a high catalytic surface area, abundant reaction sites, reduced charge transfer resistance, shorter diffusion paths, and improved light absorption capacity thanks to ultrathin 2D materials. In summary, these nanomaterials benefit from strong interfacial contacts, increased active sites, low charge transfer resistances, shorter diffusion paths, and enhanced light absorption capabilities, all contributing to their superior performance in various applications (Mohapatra et al., 2023).

Figure 4. Mechanism showing the photocatalytic degradation of pollutants

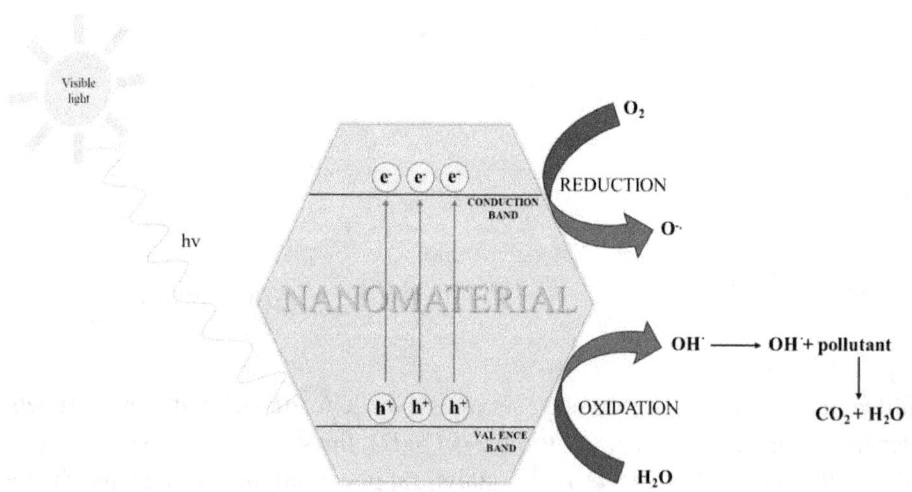

2.2. Inorganic Nanomaterials

Pollution resulting from heavy metals has emerged as a persistent and pervasive environmental challenge, posing threats to ecosystems and human health worldwide. In response to this issue, scientists have undertaken comprehensive research endeavors to gain a deeper understanding of heavy metals and their impact on the environment and human well-being. Metals and metal oxide nanomaterials have garnered significant attention for their effectiveness in the removal of heavy metals from various environmental matrices, including water and soil. These nanomaterials offer unique properties and surface chemistry that make them promising candidates for heavy metal remediation (Santhosh et al., 2016; Saravanan et al., 2011). Nanoparticles which are derived from metals such as silver and gold, as well as metal oxides like iron oxides, TiO_2, MgO, Fe_2O_3, and Al_2O_3, along with zero-valent metals including Fe^0, Zn^0, Sn^0, and Al^0, have received considerable attention in the realm of environmental remediation. This increased interest arises from their versatility in various applications, including water disinfection, the treatment of drinking water, groundwater remediation, wastewater purification, and air quality improvement. They exhibit a wide array of functions, including adsorption, antibacterial and antimicrobial properties, photocatalysis, reductive dehalogenation, desulfurization, and catalytic reduction activities (Mu et al., 2019; Gamage et al., 2021; Guerra, Attia, Whitehead, & Alexis, 2018; Khin et al., 2012). Table 1 provides an overview of various metals and metal oxide nanomaterials that researchers have explored for diverse environmental remediation purposes.

Table 1. An overview of metals and metal oxide nanomaterials for environmental remediation

S.No.	Nanomaterials	Size(nm)	Applications	References
1.	Au NPs	26.1 ± 4.7 nm 10-20nm	Nitro derivatives i.e., Methyl orange, 2-nitrophenol, 2,4-dinitrophenol, Eri chrome Black-T, Congo red Removal of Hg^{2+} ions from water	(Lisha et al., 2009; Qin et al., 2019)
2.	11-mercaptoundecanoic acid and 3-mercaptopropionic acid capped Au NPs	~13nm	Heavy metal-containing organic compounds	(Huang & Chang, 2007; Kim et al., 2001)
3.	TiO_2	50-60nm	Photodegradation of wastewater	(Abel et al., 2021)
4.	α-Fe_2O_3	02-55nm	1. reduction of Cadmium (II) 2. removal of toxic heavy metal ion, Cr^{6+} from water 3. photocatalytic activities for dye removal 4. methylene blue (MB) and Crystal violet (CV) removal degradation of dyes (reactive blue)	(Abdelrahman et al., 2019; Bibi et al., 2019; Gurlhosur & Sreekanth, 2018; Jerin & Remya, 2019; Wang et al., 2017)
5.	Bimetallic nanomaterials	~16nm	Removal of chlorinated and brominated contaminants from water and soil	(Abdelsattar et al., 2022; Bosetti et al., 2002; Chou et al., 2005; Cushing et al., 2004; Gupta & Silver, 1998; Kharisov et al., 2012; Kim et al., 2007; Nakamura et al., 2020; Rahn & Landry, 1973; Willis et al., 2005)

Various methods, such as thermal decomposition, hydrothermal synthesis, and co-precipitation, are used to synthesize metal and metal oxide nanomaterials (Cushing et al., 2004; Willis et al., 2005). Silver nanoparticles are renowned for their potent antibacterial properties against viruses, fungi, and bacteria, making them common in water disinfection (Bosetti et al., 2002; Chou et al., 2005; Gupta & Silver, 1998). Under UV irradiation, silver ions exhibit photoactivity, enhancing the UV-based inactivation of bacteria and viruses (Kim et al., 2007; Nakamura et al., 2020; Rahn & Landry, 1973). Ag nanoparticles effectively suppress bacterial biofilms and selectively detect Cd^{2+} ions in water samples (Abdelsattar et al., 2022). Silver nanoparticles are often combined with polymers and metal oxides to enhance overall effectiveness. Zero-valent iron nanomaterials (nZVI) are prominent in environmental remediation, effectively reducing and removing contaminants like chlorinated organic compounds, heavy metals, and nitrate through chemical reactions (Kharisov et al., 2012), as shown in Figure 5. Iron nanoparticles typically have a core-shell structure, featuring a zero-valent iron core and a shell with mixed-valent

Figure 5. Degradation of pollutants through various chemical reactions

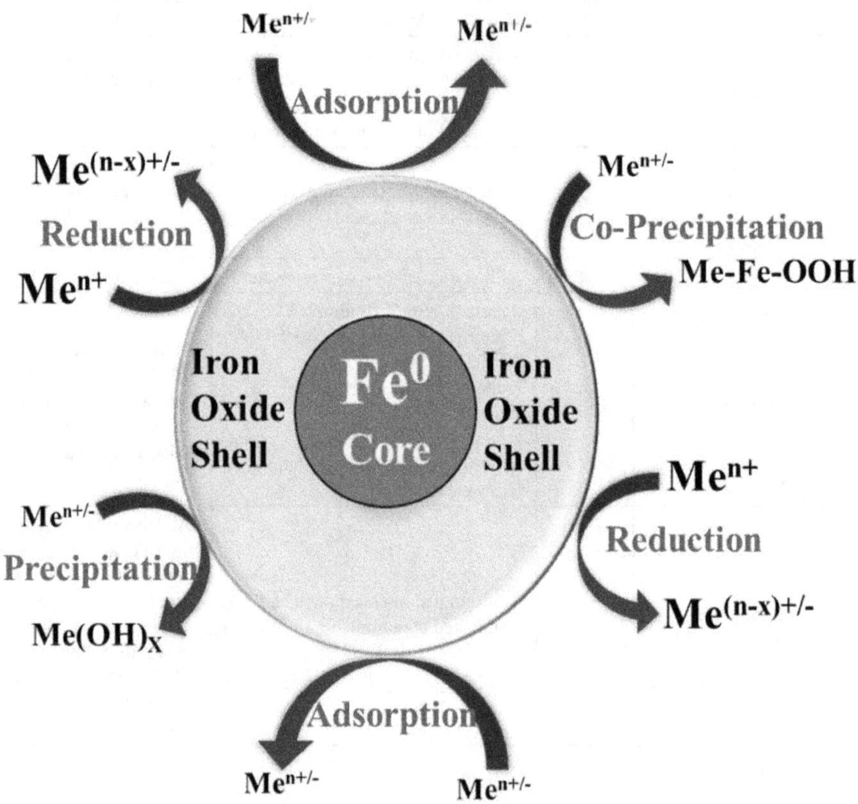

oxides containing Fe (II) and Fe (III) (Kharisov et al., 2012). The shell aids in the removal of contaminants, particularly heavy metals with a higher standard reduction potential than the Fe^{2+}/Fe couple (Li & Zhang, 2006).

In a recent study, researchers synthesized zinc oxide nanoparticles using the leaf extract of *Thymus vulgaris* for photocatalytic degradation applications. The study revealed that in the presence of 1.0 mg/mL ZnO NPs, both 50 mg L^{-1} Methylene blue (MB) and 50 mg L^{-1} Sudan red (SR) dyes were completely photodegraded under UV irradiation within 20 and 45 minutes, respectively, Additionally, when exposed to visible light, the degradation efficiency reached 91% within 60 minutes and 75% within 120 minutes for these days. Furthermore, the photocatalytic measurements of Rhodamine B (RTW) showed a degradation efficiency of 95% within 60 minutes under UV illumination and 79% within 90 minutes under visible illumination (45). These remarkable results highlight the potential of the synthesized zinc oxide nanoparticles for various applications. In a parallel study, Chakravarty et al. (Chakravarty et al., 2023) investigated the photocatalytic performance of TiO_2 nanoparticles and by observing the degradation of a Methylene blue (MB) solution under simulated solar irradiation, they found that the prepared TiO_2 photocatalyst exhibited superior photocatalytic efficacy compared to commercially available powdered TiO_2. Furthermore, the TiO_2 nanoparticles demonstrated significant antimicrobial effectiveness when tested against gram-positive bacteria. Collectively, these findings suggest that both the synthesized ZnO NPs and TiO_2 NPs hold significant promise for applications in technologies related to wastewater treatment and biomedical management.

2.3. Polymer-Based Nanomaterial

Polymer-based nanomaterials incorporate polymers as a fundamental component at the nanoscale, with engineered or modified polymer chains or networks to achieve specific properties and functions. Polymeric nanomembranes, like those made from chitosan using techniques like solvent casting, are employed for removing nanoparticles from aqueous solutions. These membranes guide particles through pores, facilitating chemical interactions that immobilize pollutants. Chitosan-based nanofiltration membranes are commonly used for purifying textile wastewater (Long et al., 2020), showing lower rejection rates for electroneutral and negatively charged dyes compared to positively charged ones. The efficiency of nanofiltration is also influenced by the physical size of the dyes (Weng et al., 2017).

Khare et al. (Khare et al., 2016) created composite films by incorporating iron oxide nanoparticles into chitosan-based carbon nanofibers (CNFs) in combination with polyvinyl alcohol. This novel composite material demonstrated excellent adsorption capabilities for removing Cr^{6+} from water, exhibiting a high metal uptake capacity of approximately 80 mg/g for the chitosan/iron-CNF composite. In a related study, Idris et al. (Sargin et al., 2020) explored the catalytic potential of a system consisting of chitosan-carbon nanotube-supported palladium nanoparticles. This catalyst proved highly effective in catalyzing the reduction of various nitroarenes, including 4-nitrophenol, 2-nitroaniline, 4-nitro-o-phenylenediamine, and 2,4-dinitrophenol. Moreover, this catalyst system exhibited exceptional performance in the degradation of diverse dyes, such as Congo red, Methyl orange, Methylene blue, and Methyl red. Additionally, this heterogeneous catalyst system showcased its capability in the eco-friendly and solvent-free synthesis of 18 different biaryl compounds through Suzuki-Miyaura coupling reactions, utilizing microwave irradiation for a brief duration of just six minutes.

In the realm of polymeric nanocomposites, two pivotal factors driving their utilization are their inherent biocompatibility and biodegradability. For instance, a sustainable hybrid nanomaterial adsorbent, composed of a hydrogel made from a mixture of wheat xylan and poly (acrylic acid) nanoparticles, incorporating enclosed Fe_3SO_4 nanoparticles with the unique characteristics of the magnetic hydrogel, achieves the removal of approximately 90% of methylene blue dye (Sun et al., 2015). Likewise, a nanogel formulated from sodium alginate loaded with tetrasodium thiacalixarene tetrasulfonate and Fe_3SO_4 demonstrated significant adsorption capacity for various ions, including Ni^{2+}, Co^{2+}, Pb^{2+}, Cd^{2+}, Cu^{2+}, and Cr^{2+}, from aqueous solutions (Lakouraj et al., 2014). It is evident that both polymer nanoparticles and polymer nanocomposites are pivotal in environmental remediation, primarily due to their distinct physiochemical properties that enable the efficient removal of various contaminants through diverse mechanisms.

3. APPLICATIONS OF HYBRID NANOMATERIALS IN POLLUTANTS OR CONTAMINANTS DETECTION IN ENVIRONMENTAL COMPONENTS

Environmental conservation and the mitigation of pollution are critical global concerns that demand immediate attention. The sheer volume of oil fields, landfills, private lands, military bases, manufacturing facilities, and industrial sites contaminated with pollutants is astounding on a worldwide scale. Consequently, nanomaterials offer various applications, such as water quality monitoring, air pollution detection, soil contamination analysis, and more, as elaborated.

3.1. Water Quality Monitoring

Hybrid nanomaterials, propelled by advanced nanotechnology, are leading the way in enhancing precision and sensitivity in water quality monitoring. Crafted by combining various nanoparticles and nanoscale structures, these hybrid nanomaterials create highly efficient and versatile sensors. Their capabilities span a wide spectrum, detecting water pollutants from heavy metals and organic compounds to microorganisms. These nanomaterial-based sensors play a crucial role in environmental monitoring, ensuring safe drinking water, and preserving aquatic ecosystems. Additionally, in water treatment, nanotechnology plays a pivotal role in developing nanoscale ceramic and polymer membranes. These membranes incorporate innovations like catalytic and zeolite nanoparticles on ceramic membranes, organic-inorganic hybrid nanocomposite membranes, and biomimetic membranes, contributing to advancements in water treatment techniques (Esmaeili & Beni, 2014). The utilization of nanofilters in water and wastewater treatment is another significant application of nanotechnology in environmental detection. The membrane employed in the nanofiltration process typically exhibits natural resistance to large molecules, and in comparison to alternative methods, it efficiently purifies well water or surface water with reduced energy consumption. This process is capable of eliminating various contaminants, including viruses, bacteria, organic pesticides, and calcium and magnesium salts from water (Mathew et al., 2019). Conventional water treatment methods often fall short of effectively eliminating contaminants, particularly metals and microorganisms. To address this challenge, researchers have explored the potential of nanomaterials in the removal of metals, microbes, and oil from polluted water sources. Nanoparticles, in particular, offer a high degree of versatility when it comes to pollutant treatment. Nanostructured particles, find application across various environmental facets, including soil and sediment remediation, solid waste management, water treatment, and liquid waste purification.

Nanostructured bimetallic particles, like iron-palladium, iron-silver, and zinc-palladium, have demonstrated high effectiveness in treating and purifying environmental pollutants such as organic chlorinated pesticides and halogenated organic solvents. Furthermore, employing nanofilters stands as a significant measure for environmental protection and energy conservation, categorized into microfilters, ultrafilters, and nanofilters based on pore size. Nanofiltration, in particular, operates at lower pressures compared to reverse osmosis, making nanofilters a cost-effective choice. Moreover, nanofilters possess the capability to eliminate viruses and bacteria, making them suitable for removing contaminants from both human drinking water and agricultural water sources (Mansas et al., 2020; Vendrame et al., 2019). This dual benefit safeguards the environment and promotes energy efficiency in water treatment processes. In contrast, zeolite and catalytic membranes, while offering some improvements over traditional membranes, have limited applications in water treatment. In the study by Jiang et al. (Jiang et al., 2018), nanotechnology-based treatment methods specifically adsorption and membrane filtration, were investigated. Adsorption was favored due to its simplicity and effectiveness in treating various contaminants, including organic materials, salts, bases, acids, and toxins. The efficiency of adsorption depends on factors like the adsorbent's pore structure and its interaction with contaminants (Yin et al., 2020). The study also investigated carbon-based nanoparticles, such as graphene-based nanoparticles and carbon nanotubes, which showed high performance in removing heavy metals and organic pollutants. However, carbon-based nanoparticles exhibited increased hydrophobicity and reduced surface area due to loose aggregates, which could be mitigated by adding functional groups or metal oxide particles. Notably, the study found that carbon nanoparticles could be reused by adjusting pH without losing their adsorption capacity. The study also delved into membrane filtration methods, including microfiltration, ultrafiltration, and nanofiltration, which differ based on membrane pore size. Membrane fouling emerged as a primary drawback, but this issue could be addressed through modifications involving silica, aluminum, zeolite, and titanium oxide. Additionally, silver could be incorporated to prevent biofilm growth and eliminate bacteria and viruses (Hsueh et al., 2015). One significant challenge is the detection of contaminants present in very low concentrations. Nanomaterials offer an advantage by concentrating pollutants to detectable levels, as demonstrated by using Au-TiO_2 nanoparticles to concentrate and remove low levels of insecticide in wastewater (Hadei et al., 2021). However, further advancements in effective sensors for detecting nanosized contaminants are crucial, as the problem of false detections remains prevalent in many treatment applications.

In the context of wastewater from oil refineries, which often contain a mixture of ions and metals like calcium (Ca^{2+}) and copper (Cu^{2+}), innovative approaches have been explored. He et al. (He et al., 2021) developed reusable nano adsorbents by employing Fe_3O_4/GO-COOH through magnetization and carboxylation of graphene oxide (GO). These nano adsorbents exhibited impressive removal capabilities, with a 78.4% removal rate for Ca^{2+} and a 51% removal rate for Cu^{2+} within just 60 minutes. Notably, even after undergoing five cycles of adsorption and desorption, these nano adsorbents maintained remarkable performance, retaining high recovery rates of 82.1% for Ca^{2+} and 91.8% for Cu^{2+}, along with substantial removal percentages of 72.3% for Ca^{2+} and 49.33% for Cu^{2+}. This research highlights the potential for nanotechnology to address complex wastewater treatment challenges effectively. Furthermore, photocatalysis has garnered extensive research attention in the scientific community within the context of wastewater treatment. This is primarily attributed to its capability to disintegrate a broad spectrum of substances, encompassing organic materials, estrogens, dyes, organic acids, pesticides, crude oil, and microorganisms, including resilient viruses and chlorine-resistant organisms. Additionally, it can target specific inorganic molecules like nitrous oxides. When coupled with filtration or precipitation techniques, photocatalysis proves effective in removing metals such as mercury from wastewater (Gómez-Pastora et al., 2017; McLain, 2019).

Nanomaterials, distinguished by their exceptional surface characteristics, mechanical robustness, chemical reactivity, electrical conductivity, magnetic attributes, and optical properties, exhibit unique responses compared to bulk materials. Moreover, nanomaterials introduce distinct quantum effects. Consequently, the utilization of nanomaterials as photocatalysts has piqued considerable interest among researchers due to their potential to enhance wastewater treatment processes significantly (Herrmann, 1999; Hot et al., 2017; Xu et al., 2012). The current demand for innovative wastewater treatment methods to effectively remove contaminants necessitates a closer look at nanotechnology, specifically its potential for nanomaterial-based wastewater treatment. While scientists have made significant progress in this field, there is still much uncharted territory to explore within nanotechnology. The promising prospects of nanotechnology in wastewater management are evident, but realizing its full potential will require dedicated efforts from the scientific community and government entities.

3.2. Air Pollution Detection

Air pollution involves the release of particles, biological elements, or harmful substances into the atmosphere in various forms—solid, liquid, or gaseous—posing serious environmental threats to ecosystems and living organisms. Sources can be natural or human-induced, categorized into primary (directly emitted) and secondary (formed through chemical reactions) pollutants. Primary pollutants result from natural events or human activities, such as fossil fuel combustion, including carbon monoxide, sulfur, nitrogen oxides, and volatile organic compounds. Secondary pollutants, like peroxyacetyl nitrate, form through chemical reactions involving primary pollutants. Continuous air quality monitoring is crucial in addressing environmental challenges. Nanotechnology offers innovative solutions, developing nanomaterials with unique properties for sustainable pollution control. Researchers have explored nanomaterials like TiO2/Ag tissue nanocomposites effectively removing toluene and airborne pathogens, and Cu-Cd/TiO2-Pt nanostructured hollow spheres breaking down isopropanol in the gas phase under solar irradiation (Mohamed & Awad, 2020; Mohamed & Do, 2020; Sha & Bhattacharyya, 2020).

The emission of CO_2 plays a significant role in global warming and environmental climate change. Consequently, extensive worldwide research efforts have been made to develop practical materials for capturing CO_2. It has been reported that the global concentration of environmental CO_2 has increased from approximately 280 ppm in the preindustrial era to 390 ppm at present. The primary contributors to this observed rise in CO_2 levels include emissions from vehicles, fossil fuel-fired power plants, deforestation, and chemical processes (Samanta et al., 2012). Carbonaceous adsorbents, with attributes like hydrophobic nature, high thermal stability, chemical resistance, adjustable pore structures, and cost-effectiveness, are widely considered the most promising adsorbents for CO_2 capture (Wickramaratne & Jaroniec, 2013). Given that approximately 60% of global climate change impact is attributed to CO_2 emissions, particularly from industrial exhaust gases rich in nitrogen (N_2) and CO_2, capturing and separating CO_2, especially from the nitrogen-rich stream, is crucial. In industrial environments, toxic pollutants like benzene, dioxin, toluene, ethyl benzene, and p-xylene are present in the atmosphere (Su et al., 2010). Carbon nanotubes, with a porous wall of 2.9 nm in diameter, show a significant correlation with benzene dioxin rings, forming a strong connection and increasing adsorption potential within the pore. Additionally, carbon nanotubes exhibit high oxidation resistance, making them valuable for regenerating adsorbents at elevated temperatures (Mubeen et al., 2021). These findings emphasize the importance of advanced materials and technologies in addressing environmental challenges related to pollution and greenhouse gas emissions. Scientists have long known about the adverse health effects of exposure to particulate matter and heavy metals, contributing to conditions like heart disease and lung cancer, particularly in urban areas where airborne particles fall within the 100-300 nm size range, and heavy metals persist without biodegradation (Numan et al., 2021). The urgency to address these issues highlights the need for sensors capable of detecting these ions before reaching hazardous levels (Kharwar & Singh, 2021). Quantum dot nanomaterials emerge as a promising solution, providing precise and advanced sensors for heavy metal detection due to their unique physicochemical attributes, elevated specific surface areas, and heightened reactivity (Pooja & Chowdhury, 2021). By leveraging quantum dot nanomaterials and integrating them with optical or chemical sensor converters, robust detection systems can be created to identify multiple metals in complex scenarios (Wang et al., 2021). Certain biomass components are integrated into quantum dot nanomaterial nanotechnology. These nanobiosensors are employed as attractive fluorescent nanosensors due to their high environmental compatibility and biocompatibility. For instance, carbon quantum dots extracted from green algae waste are utilized for the detection of Fe (III) in effluents (Liu et al., 2020). This innovative approach enhances the potential of quantum dot nanomaterials in monitoring and managing heavy metal contaminants in diverse environmental settings.

According to the World Health Organization (WHO) (Abu-Hani, 2017), key air pollutants include particulate matter, nitrogen dioxide, surface ozone, and carbon monoxide. Green nanotechnology emerges as a promising solution to combat air pollution, offering innovative nanomaterial-based sensors with advantages such as high sensitivity, low power consumption, and the ability to detect pollutants at minimal concentrations. These sensors operate efficiently in close proximity to target objects and can assess the environmental toxicity of nanoparticles, enhancing air quality monitoring. Traditional gas sensors, relying on metal oxide semiconductors (MOS) and polymeric materials, often face drawbacks like high costs and limited sensitivity. In contrast, MOS-based gas sensors show outstanding performance with rapid response times, high sensitivity, and stability (Jimenez-Cadena et al., 2007; Liu, 2012). Carbon-based nanomaterials like single-walled carbon nanotubes (SWCNT) and graphene exhibit unique properties, making them promising for sensing air contaminants. SWCNT-based sensors demonstrate rapid response and recovery, particularly effective for substances like dimethyl methyl phosphonate (DMMP) (Wang et al., 2009). Graphene and its composites, known for excellent sensing capabilities through gas molecule adsorption, have been explored for air pollution detection. Researchers have even developed reduced graphene oxide-based sensors, incorporating various metal oxides to create metal oxide-graphene nanohybrids (Cui, 2013; Gutés, 2012; Hu et al., 2012; Zhang et al., 2014).

Recent studies have showcased remarkable advancements in adsorption materials for air pollution control. For instance, Chao Yang et al. (Yang et al., 2020) developed $ZnFe_2O_4$/activated carbon adsorbents with exceptional H_2S removal capacity, presenting a significant advancement in efficient H_2S removal. Similarly, Gao et al. (Gao, 2019) introduced a novel adsorption material for efficient SO_2 removal. Song et al. (Song, Li, Li et al, 2019) improved phenol adsorption characteristics significantly by modifying silica nanoparticles with imidazolium-based ionic liquid moieties. Lastly, Faghihi-Zarandi et al. (Faghihi-Zarandi et al., 2019) demonstrated the effectiveness of various ILs for toluene elimination from the air, with certain ILs exhibiting up to 99.5% toluene removal capacity. Collectively, these studies underscore the immense potential of green-synthesized nanoparticles and modified materials in advancing air pollution control methods. They offer substantial improvements in adsorption capacities for various pollutants, including H_2S, SO_2, phenol, and toluene. This research heralds a promising era for more efficient and sustainable solutions to combat air pollution, making it a critical area of exploration in environmental science and technology.

Biologically derived metal oxide nanoparticles, like titanium, zinc, and silver, are gaining attention for their photocatalytic ability to eliminate environmental pollutants. Table 2 highlights some of these biologically inspired nanoparticles used in environmental remediation, known for their stability, cost-effectiveness, and rapid oxidation processes. Green-synthesized silver nanoparticles effectively decontaminate various organic dyes, while gold nanoparticles synthesized with intracellular protein extract from *Pycnoporus sanguineus* catalytically degrade 4-nitroaniline (Shi et al., 2015). Titanium dioxide nanoparticles from plant and bacterial sources serve as nanocatalysts for degrading both inorganic and organic pollutants. For air purification, effective immobilization of photocatalytic nanomaterials is crucial, requiring attributes like mass transfer, strong bonding, stability under light exposure, high surface area, good adsorption, and cost-effectiveness. A thin film combining TiO_2 photocatalyst, sulfur, and trifluoroacetic acid applied onto ceramic tiles serves as effective support (O'Keeffe et al., 2013). Research by Liang et al. demonstrated the substantial removal (85%) of Zn^{2+} contamination from aqueous solutions using NZVI particles and dissolved oxygen. NZVI particles, coated with an iron hydroxide shell through a corrosion method involving dissolved oxygen, proved effective in eliminating various organic chemical pollutants, emphasizing the potential of biologically synthesized nanoparticles in effectively addressing environmental pollution challenges (Liang et al., 2014; Raychoudhury & Scheytt, 2013).

Table 2. Green synthesis of inorganic and metal oxide nanoparticles, their characteristics with environmental applications

S.No.	Biological Source	Nanoparticles	Size (nm)	Morphology	Applications	References
1.	*Aspergillus terreus*	Zinc oxide	8	Spherical	1. Biosensors 2. Catalysis 3. Molecular diagnostics 4. Solar cells 5. Optoelectronics 6. Cell labelling 7. Cell imaging	(Raliya & Tarafdar, 2014)
2.	*Aloe barbadensis*	Indium oxide	5-50	Spherical	1. Gas sensors 2. Solar cells	(Maensiri et al., 2008)
3.	*Morus*	Silver	15–20	Spherical	Antimicrobial activity against *E. coli, B. subtilis*	(Singh et al., 2017)
4.	*Aspergillus favus*	Titanium dioxide	12-15	Spherical	Plant nutrient fertilizer	(Raliya et al., 2015)
5.	*Sedum alfredi*	Zinc Oxide	53.7	Hexgonal-wurtzite and pseudo-spherical	Nanoelectronics	(Qu et al., 2011)
6.	*Rhizopus nigricans*	Silver	35-40	Round	1. Catalysis 2. Bactericidal	(Ravindra & Rajasab, 2014)

3.3. Soil Contamination Analysis

Urbanization's expansion has intensified human activities, leading to a higher prevalence of soil contaminants. The global concern for soil remediation has grown, with nanotechnology being considered a crucial solution. For instance, the substantial annual use of 187 million tonnes of fertilizer and 4 million tonnes of pesticides in agriculture has significantly contributed to soil contamination (Usman, Farooq, Wakeel, Nawaz, Cheema, Rehman, Ashraf, & Sanaullah, 2020). Heavy metals, a primary source of soil contaminants, have garnered attention due to their diverse origins, resistance to biodegradation, and tendency to accumulate over time (Fajardo et al., 2020). In soil remediation, common in situ techniques include adsorption, immobilization, Fenton and Fenton-like oxidation, reduction reactions, and innovative combinations involving nanotechnology and bioremediation. The fusion of nanotechnology and bioremediation has become a subject of significant interest (Cecchin et al., 2017). Nanoparticles are typically used for adsorption of inorganic contaminants like heavy metals, while reduction reactions and degradation catalyzed by substances are employed for organic contaminants. Nanomaterials, including carbon nanomaterials, nanoscale zero-valent iron (nZVI), titanium oxide

(TiO_2), iron (III) oxide (Fe_3O_4), zinc oxide (ZnO), and various nanocomposites, enhance the effectiveness of adsorption and oxidation processes, facilitating the removal of persistent micro-pollutants from the soil. Notably, nZVI stands out as the preferred nanoparticle for eradicating heavy metal pollutants, converting toxic metals and chlorinated organic compounds into less harmful forms (Fajardo et al., 2012).

3.3.1. Immobilization

In situ immobilization of contaminants, a cost-effective and environmentally friendly approach to remediating contaminated soil (Usman, Farooq, Wakeel, Nawaz, Cheema, Rehman, Ashraf, & Sanaullah, 2020), employs a variety of nanosized additives like carbon nanomaterials (CNTs, fullerenes, C70, and graphene) and metal oxide nanomaterials (Fe_3O_4 and TiO_2), along with nanocomposites and engineered nanomaterials (ENMs). These nanomaterials use diverse adsorption mechanisms. Carbon nanomaterials, especially carbon nanotubes, leverage their hydrophobic surface properties for potent adsorption, effectively eliminating organic pollutants (Guerra, Attia, Whitehead, & Alexis, 2018). Carbon nanotubes, in particular, excel in adsorbing organic compounds, surpassing the organic matter in soils and exhibiting a specific affinity for ionizable organic compounds like pesticides (Zhang et al., 2019). Metal oxide nanomaterials, such as Titanium dioxide and iron (III) oxide, primarily use surface complexation as their adsorption mechanism. Iron (III) oxide effectively removes heavy metals like Arsenic and Cadmium, achieving a maximum adsorption capacity of 37.03 mg/g for Cadmium remediation under pH 6.0 conditions (Qian et al., 2020). Titanium dioxide shows a remarkable 94% sorption rate in removing Th (IV) under pH 4.0 conditions, with acidic soil conditions promoting surface complexation as the dominant adsorption mechanism (Tan et al., 2007).

3.3.2. Reduction Reaction

The use of nZVI nanoparticles in reduction reactions shows promise for removing heavy metals and organic compounds from contaminated soil and addressing water and groundwater contamination (Chrysochoou et al., 2012; Stefaniuk et al., 2016; Zhao et al., 2016). With their small size and substantial surface area, nZVI particles enhance remediation efficiency by directly interacting with contaminants. Typically introduced into contaminated soil through wall injection (Hou et al., 2019; Song, Fang, Zhu et al, 2019), nZVI particles exhibit strong reduction capabilities, converting harmful contaminants, such as Chromium (VI), into less toxic compounds like Chromium (III) and ferrous chromite (Hou et al., 2019; Qian, Shang, Zhang et al, 2019). When combined with biochar or compost, nZVI's reactivity is boosted, preventing particle agglomeration and improving dispersion, leading to a remarkable reduction of soil-bound Cr (VI) (Galdames et al., 2017; Qian, Liu, Zhang et al, 2019). For instance, green tea-modified nZVI injection reduced soil-bound Cr (VI) by 66% from an initial concentration of 100 mg kg-1 soil (Chrysochoou et al., 2012). Moreover, 1.00 g of nZVI effectively reduced approximately 280 mg of Cr (VI) in 1 kg of contaminated soil, with 98% removal achieved within 24 hours under specific treatment conditions (Franco et al., 2008). nZVI also efficiently removes organic contaminants like trichloroethylene, TBBPA, and pesticides (DDT and 2,4-D) from soils. CMC-stabilized nZVI injected into potting soil with 9.2% organic matter resulted in the dichlorination of 44% of TCE within 30 hours. Injecting 20% aqueous nZVI into soil with an initial DDT concentration of 24 mg per kg removed 25% of DDT within 72 hours. Larger quantities of nZVI are required for remediating soils contaminated over longer periods (El-Temsah & Joner, 2013; El-Temsah et al., 2016; Zhang et al., 2011).

3.3.3. Fenton-Like Reaction

Fenton reagents, composed of H_2O_2 and Fe (II), are commonly used for treating organic wastewater. This method relies on OH radicals generated from H_2O_2 to oxidize organic substances in the presence of ferrous ions. Iron minerals and engineered nanomaterials like Fe_3O_4 have gained prominence as catalysts for underground Fenton-like oxidation processes due to their cost-effectiveness, environmental abundance, reusability, and ease of synthesis (Garrido-Ramírez et al., 2010; Hou et al., 2016). Fe_3O_4-catalyzed Fenton-like reactions have shown promise in remediating PAH-contaminated soils without generating harmful by-products or requiring pH adjustments (Usman et al., 2012). The remediation process's efficiency can be enhanced by incorporating pre-treatment steps with availability-enhancement agents like ethanol, cyclodextrin, or nitrilotriacetic acid (Su, 2017; Sun et al., 2014; Usman et al., 2012). Additionally, including organic chelating agents such as catechol, gallic acid, and EDTA in the Fenton system extends the optimal pH range to 6.0-9.0, increasing its applicability in real-world field implementations (Lu et al., 2010).

4. SUSTAINABLE ADVANTAGES OF HYBRID NANOMATERIALS OVER TRADITIONAL DETECTION METHODS

The use of hybrid nanomaterials is a significant advancement towards sustainability goals, offering the potential to address various environmental and resource efficiency concerns. These nanomaterials provide distinct advantages over traditional detection methods, leading to reduced resource consumption, improved energy efficiency, and minimized waste generation.

4.1. Reduced Resource Consumption

4.1.1. Material Efficiency: Hybrid nanomaterials are engineered at the nanoscale, allowing precise control over their composition and properties. This precision reduces the need for bulk materials, as smaller quantities of nanomaterials can achieve the same or even superior results compared to larger quantities of conventional materials.

4.1.2. Raw Material Utilization: The synthesis of hybrid nanomaterials often involves using readily available, low-impact raw materials. This contrasts with resource-intensive processes required to extract and produce many traditional materials, such as metals and chemicals.

4.2. Energy Efficiency

4.2.1. Low Energy Synthesis: The fabrication of hybrid nanomaterials typically involves energy-efficient techniques like sol-gel synthesis, chemical vapor deposition, or electrodeposition. These processes often consume less energy compared to energy-intensive methods in traditional material manufacturing.

4.2.2. Enhanced Properties: hybrid nanomaterials can exhibit superior properties, such as high surface area, improved conductivity, and enhanced catalytic activity. These attributes can lead to more energy-efficient processes and devices, including sensors and catalysts.

4.3. Minimal Waste Generation

4.3.1. Precise Manufacturing: The precise nature of nanomaterial synthesis minimizes waste by reducing the likelihood of defects or undesirable byproducts. This precision results in more efficient use of raw materials and fewer waste products.

4.3.2. Recyclability: Many hybrid nanomaterials can be designed to be recyclable, promoting a circular economy. Their recoverability and reusability can further reduce waste generation.

4.4. Improved Sensitivity and Selectivity

4.4.1. Reduced Sample Size: Hybrid nanomaterial-based sensors often require smaller sample sizes, reducing the amount of material or substances needed for analysis. This can be especially advantageous in applications like environmental monitoring, where sample availability is limited.

4.4.2. Enhanced Detection: The unique properties of hybrid nanomaterials, such as their high surface-to-volume ratio and tailored surface chemistry, can enhance the sensitivity and selectivity of detection methods. This reduces false positives and negatives, leading to more accurate and efficient measurements.

4.5. Longevity and Durability

4.5.1. Extended Lifespan: Hybrid nanomaterial-based devices, such as sensors and catalysts, tend to have longer lifespans and require less frequent replacement or maintenance. This reduces the overall environmental impact associated with the production and disposal of these devices.

Hybrid nanomaterials contribute to sustainability goals by enhancing resource efficiency, conserving energy, reducing waste, and improving environmental performance across various applications. They are a promising avenue for advancing technologies while minimizing the environmental impact of scientific and industrial processes, playing a crucial role in building a more sustainable future.

5. CONCLUSION

Hybrid nanomaterials are a promising tool for detecting environmental issues, offering enhanced sensitivity, selectivity, and efficiency in monitoring pollutants, contaminants, and pathogens. Their adaptability to various detection techniques, such as sensors and biosensors, makes them suitable for diverse environmental settings, providing real-time data for monitoring and early warning. Their specificity reduces false positives, improving accuracy. These nanomaterials offer sustainability benefits, being designed to be environmentally friendly, non-toxic, and cost-effective. As the world faces environmental challenges, the development of sustainable technologies like hybrid nanomaterials is crucial. The future of hybrid nanomaterials in environmental detection holds exciting possibilities. Researchers can design materials with multifunctional properties, enabling simultaneous sensing of multiple parameters for more comprehensive monitoring. Exploring new nanomaterial combinations, including 2D materials like graphene and carbon nanotubes, can enhance sensing performance. Integrating hybrid nanomaterial-based sensors with IoT and big data analytics enables real-time data collection and proactive responses to environmental issues. Miniaturization and portability democratize environmental monitoring, involving individuals and communities in citizen science efforts. Research into biodegradable nanomaterials addresses environmental impact concerns. Hybrid nanomaterials may not only detect but also remediate environmental issues, with research focusing on capturing and neutralizing pollutants directly. Addressing regulatory and ethical considerations, including privacy and data security, is essential as the field advances. Overall, hybrid nanomaterials represent a promising future in addressing and mitigating environmental challenges while promoting sustainability and responsible technological development.

REFERENCES

Abdelrahman, E. A., Hegazey, R. M., Kotp, Y. H., & Alharbi, A. (2019). Facile synthesis of Fe2O3 nanoparticles from Egyptian insecticide cans for efficient photocatalytic degradation of methylene blue and crystal violet dyes. *Spectrochimica Acta. Part A: Molecular and Biomolecular Spectroscopy, 222*, 117195. doi:10.1016/j.saa.2019.117195 PMID:31176151

Abdelsattar, A. S., Gouda, S. M., Hassana, Y. Y., Farouka, W. M., Makky, S., Nasr, A., Hakim, T. A., & El-Shibiny, A. (2022). In vitro bacteriophage-mediated synthesis of silver nanoparticles for antibacterial applications and heavy metal detection. *Materials Letters, 318*, 132184. doi:10.1016/j.matlet.2022.132184

Abel, S., Jule, L. T., & Belay, F. (2021). Application of titanium dioxide nanoparticles synthesized by sol-gel methods in wastewater treatment. *Journal of Nanomaterials, 2021*, 1–6.

Abu-Hani, A. F. S. (2017). Design, fabrication, and characterization of low-power gas sensors based on organic-inorganic nano-composite. *Organic Electronics C, 42*, 284–292. https://www. infona.pl//resource/bwmeta1.element.elsevier-eb8437f5-ad53-3bb2-83ec-852938746948

Ananikov, V. P. (2019). Organic–inorganic hybrid nanomaterials. *Nanomaterials (Basel, Switzerland), 9*(9), 1197. doi:10.3390/nano9091197 PMID:31454924

Baysal, A., Kuznek, C., & Ozcan, M. (2018). Starch coated titanium dioxide nanoparticles as a challenging sorbent to separate and preconcentrate some heavy metals using graphite furnace atomic absorption spectrometry. *International Journal of Environmental Analytical Chemistry, 98*(1), 45–55. doi:10.1080/03067319.2018.1427741

Bibi, I., Nazar, N., Ata, S., Sultan, M., Ali, A., Abbas, A., Jilani, K., Kamal, S., Sarim, F. M., Khan, M. I., Jalal, F., & Iqbal, M. (2019). Green synthesis of iron oxide nanoparticles using pomegranate seeds extract and photocatalytic activity evaluation for the degradation of textile dye. *Journal of Materials Research and Technology, 8*(6), 6115–6124. doi:10.1016/j.jmrt.2019.10.006

Bosetti, M., Masse, A., Tobin, E., & Cannas, M. (2002). Silver coated materials for external fixation devices: In vitro biocompatibility and genotoxicity. *Biomaterials, 23*(3), 887–892. doi:10.1016/S0142-9612(01)00198-3 PMID:11771707

Cecchin, I., Reddy, K. R., Thome, A., Tessaro, E. F., & Schnaid, F. (2017). Nanobioremediation: Integration of nanoparticles and bioremediation for sustainable remediation of chlorinated organic contaminants in soil. *International Biodeterioration & Biodegradation, 119*, 419–428. doi:10.1016/j.ibiod.2016.09.027

Chakravarty, P., Deka, H., & Chowdhury, D. (2023). Anthracene removal potential of green synthesized titanium dioxide nanoparticles (TiO2-NPs) and Alcaligenes faecalis HP8 from contaminated soil. *Chemosphere, 321*, 138102. doi:10.1016/j.chemosphere.2023.138102 PMID:36764617

Chou, K.-S., Lu, Y.-C., & Lee, H.-H. (2005). Effect of alkaline ion on the mechanism and kinetics of chemical reduction of silver. *Materials Chemistry and Physics, 94*(2-3), 429–433. doi:10.1016/j.matchemphys.2005.05.029

Chrysochoou, M., Johnston, C. P., & Dahal, G. (2012). A comparative evaluation of hexavalent chromium treatment in contaminated soil by calcium polysulfide and green-tea nanoscale zero-valent iron. *Journal of Hazardous Materials, 201*, 33–42. doi:10.1016/j.jhazmat.2011.11.003 PMID:22169240

Chun, C. K., Sofer, Z., Simek, P., Jankovsky, O., Klimova, K., Bakardjieva, S., Kuckova, S. H., & Pumera, M. (2015). Synthesis of Strongly Fluorescent Graphene Quantum Dots by Cage-Opening Buckminsterfullerene. *ACS Nano, 9*(3), 2548–2555. doi:10.1021/nn505639q PMID:25761306

Cui, S. (2013). Indium-doped SNO2 nanoparticle-graphene nanohybrids: simple one-pot synthesis and their selective detection of NO2. *J Mater Chem A, 1*(14), 4462–4467. www.rsc. org/MaterialsA

Cushing, B. L., Kolesnichenko, V. L., & O'Connor, C. J. (2004). Recent advances in the liquid-phase syntheses of inorganic nanoparticles. *Chemical Reviews, 104*(9), 3893–3946. doi:10.1021/cr030027b PMID:15352782

Das, S., Sen, B., & Debnath, N. (2015). Recent trends in nanomaterials applications in environmental monitoring and remediation. *Environmental Science and Pollution Research International, 22*(23), 18333–18344. doi:10.1007/s11356-015-5491-6 PMID:26490920

Durgalakshmi, D., Rajendran, S., & Naushad, M. (2019). Current role of nanomaterials in environmental remediation. In N. Mu, S. Rajendran, & F. Gracia (Eds.), *Adv. Nanostructured Mater. Environ. Remediat* (pp. 1–20). Springer International Publishing. doi:10.1007/978-3-030-04477-0_1

Dutta, V., Verma, R., Gopalkrishnan, C., Yuan, M.-H., Batoo, K. M., Jayavel, R., Chauhan, A., Lin, K.-Y. A., Balasubramani, R., & Ghotekar, S. (2022). Bio-Inspired Synthesis of Carbon-Based Nanomaterials and Their Potential Environmental Applications: A State-of-the-Art Review. *Inorganics*, *10*(10), 169. doi:10.3390/inorganics10100169

El-Temsah, Y. S., & Joner, E. J. (2013). Effects of nano-sized zero-valent iron (nZVI) on DDT degradation in soil and its toxicity to collembola and ostracods. *Chemosphere*, *92*(1), 131–137. doi:10.1016/j.chemosphere.2013.02.039 PMID:23522781

El-Temsah, Y. S., Sevcu, A., Bobcikova, K., Cernik, M., & Joner, E. J. (2016). DDT degradation efficiency and exotoxicological effects of two types of nano-sized zero-valent iron (nZVI) in water and soil. *Chemosphere*, *144*, 2221–2228. doi:10.1016/j.chemosphere.2015.10.122 PMID:26598990

Esmaeili, A., & Beni, A. A. (2014). A novel fixed-bed reactor design incorporating an electrospun PVA/chitosan nanofiber membrane. *Journal of Hazardous Materials*, *280*, 788–796. doi:10.1016/j.jhazmat.2014.08.048 PMID:25240648

Faghihi-Zarandi, A., Shirkhanloo, H., & Jamshidzadeh, C. (2019). A new method for removal of hazardous toluene vapor from air based on ionic liquid-phase adsorbent. *International Journal of Environmental Science and Technology*, *16*(6), 2797–2808. doi:10.1007/s13762-018-1975-5

Fajardo, C., Ortiz, L. T., Rodríguez-Membibre, M. L., Nande, M., Lobo, M. C., & Martin, M. (2012). Assessing the impact of zero-valent iron (ZVI) nanotechnology on soil microbial structure and functionality: A molecular approach. *Chemosphere*, *86*(8), 802–808. doi:10.1016/j.chemosphere.2011.11.041 PMID:22169206

Fajardo, C., Sánchez-Fortún, S., Costa, G., Nande, M., Botias, P., García-Cantalejo, J., Mengs, G., & Martin, M. (2020). Evaluation of nanoremediation strategy in a Pb, Zn and Cd contaminated soil. *The Science of the Total Environment*, *706*, 136041. doi:10.1016/j.scitotenv.2019.136041 PMID:31855644

Franco, D. V., Da Silva, L. M., & Jardim, W. F. (2008). Reduction of hexavalent chromium in soil and groundwater using zero-valent iron under batch and semi-batch conditions. *Water, Air, and Soil Pollution*, *197*(1-4), 49–60. doi:10.1007/s11270-008-9790-0

Galdames, A., Mendoza, A., Orueta, M., de Soto García, I. S., Sánchez, M., Virto, I., & Vilas, J. L. (2017). Development of new remediation technologies for contaminated soils based on the application of zero-valent iron nanoparticles and bioremediation with compost. *Resource-Efficient Technologies, 3*(2), 166–176. doi:10.1016/j.reffit.2017.03.008

Gamage, A., Punniamoorthy, T., & Madhujith, T. (2021). *Starch-based hybrid nanomaterials for environmental remediation*. Starch-Evolution and Recent Advances.

Gao, S. (2019). Ionic liquid functionalized 3D mesoporous FDU-12 for effective SO2 capture. *ACS Sustain Chem Eng, 8*(1), 586–593. doi:10.1021/acssuschemeng.9b06129

Garrido-Ramírez, E. G., Theng, B. K. G., & Mora, M. L. (2010). Clays and oxide minerals as catalysts and nanocatalysts in Fenton-like reactions—A review. *Applied Clay Science, 47*(3-4), 182–192. doi:10.1016/j.clay.2009.11.044

Gómez-Pastora, J., Dominguez, S., Bringas, E., Rivero, M. J., Ortiz, I., & Dionysiou, D. D. (2017). Review and perspectives on the use of magnetic nanophotocatalysts (MNPCs) in water treatment. *Chemical Engineering Journal, 310*, 407–427. doi:10.1016/j.cej.2016.04.140

Guerra, F. D., Attia, M. F., Whitehead, D. C., & Alexis, F. (2018). Nanotechnology for environmental remediation: Materials and applications. *Molecules (Basel, Switzerland), 23*(7), 1760. doi:10.3390/molecules23071760 PMID:30021974

Gupta, A., & Silver, S. (1998). Molecular genetics: Silver as a biocide: Will resistance become a problem. *Nature Biotechnology, 16*(10), 888. doi:10.1038/nbt1098-888 PMID:9788326

Gurlhosur, S. H., & Sreekanth, B. (2018). Synthesis, characterization of iron oxide (α-Fe2O3) nanoparticles and its application in photocatalytic reduction of chromium (VI). *Rasayan Journal of Chemistry, 11*, 1678–1685. doi:10.31788/RJC.2018.1145051

Gutés, A. (2012) Graphene decoration with metal nanoparticles: towards easy integration for sensing applications. *Nanoscale, 4*(2), 438–440. https://pubs.rsc.org/en/content/articlehtml/2012/ nr/c1nr11537e

Hadei, M., Mesdaghinia, A., Nabizadeh, R., Mahvi, A. H., Rabbani, S., & Naddafi, K. (2021). A comprehensive systematic review of photocatalytic degradation of pesticides using nano TiO2. *Environmental Science and Pollution Research International, 28*(11), 13055–13071. doi:10.1007/s11356-021-12576-8 PMID:33483929

Han, M., Zhu, S. J., Lu, S. Y., Song, Y. B., Feng, T. L., Tao, S. Y., Liu, J. J., & Yang, B. (2018). Recent progress on the photocatalysis of carbon dots: Classification, mechanism and applications. *Nano Today, 19*, 201–218. doi:10.1016/j.nantod.2018.02.008

He, L., Wang, L., Zhu, H., Wang, Z., Zhang, L., Yang, L., Dai, Y., Mo, H., Zhang, J., & Shen, J. (2021). A reusable Fe_3O_4/GO-COOH nanoadsorbent for Ca^{2+} and Cu^{2+} removal from oilfield wastewater. *Chemical Engineering Research & Design, 166*, 248–258. doi:10.1016/j.cherd.2020.12.019

Herrmann, J.-M. (1999). Heterogeneous photocatalysis: Fundamentals and applications to the removal of various types of aqueous pollutants. *Catalysis Today, 53*(1), 115–129. doi:10.1016/S0920-5861(99)00107-8

Hot, J., Topalov, J., Ringot, E., & Bertron, A. (2017). Investigation on Parameters Affecting the Effectiveness of Photocatalytic Functional Coatings to Degrade NO: TiO_2 Amount on Surface, Illumination, and Substrate Roughness. *International Journal of Photoenergy, 2017*, 1–14. doi:10.1155/2017/6241615

Hou, L., Wang, L., Royer, S., & Zhang, H. (2016). Ultrasound-assisted heterogeneous Fenton-like degradation of tetracycline over a magnetite catalyst. *Journal of Hazardous Materials, 302*, 458–467. doi:10.1016/j.jhazmat.2015.09.033 PMID:26521091

Hou, S., Wu, B., Peng, D., Wang, Z., Wang, Y., & Xu, H. (2019). Remediation performance and mechanism of hexavalent chromium in alkaline soil using multi-layer loaded nano-zero-valent iron. *Environmental Pollution, 252*, 553–561. doi:10.1016/j.envpol.2019.05.083 PMID:31181500

Hsueh, Y. H., Lin, K. S., Ke, W. J., Hsieh, C. T., Chiang, C. L., Tzou, D. Y., & Liu, S. T. (2015). The antimicrobial properties of silver nanoparticles in bacillus subtilis are mediated by released Ag+ ions. *PLoS One, 10*(12), e0144306. doi:10.1371/journal.pone.0144306 PMID:26669836

Hu, N., Wang, Y., Chai, J., Gao, R., Yang, Z., Kong, E. S.-W., & Zhang, Y. (2012). Gas sensor based on P-Phenylenediamine reduced graphene oxide. *Sensors and Actuators. B, Chemical, 163*(1), 107–114. doi:10.1016/j.snb.2012.01.016

Huang, C.-C., & Chang, H.-T. (2007). Parameters for selective colorimetric sensing of mercury(II) in aqueous solutions using mercaptopropionic acid-modified gold nanoparticles. *Chemical Communications (Cambridge)*, (12), 1215–1217. doi:10.1039/B615383F PMID:17356760

Ibrahim, R. K., Hayyan, M., AlSaadi, M. A., Hayyan, A., & Ibrahim, S. (2016). Environmental application of nanotechnology: Air, soil, and water. *Environmental Science and Pollution Research International*, *23*(14), 13754–13788. doi:10.1007/s11356-016-6457-z PMID:27074929

Jerin, V. M., & Remya, R. (2019). Investigation on the removal of toxic chromium ion from waste water using Fe2O3 nanoparticles. *Materials Today: Proceedings*, *9*, 27–31. doi:10.1016/j.matpr.2019.02.032

Jiang, M., Qi, Y., Liu, H., & Chen, Y. (2018). The Role of Nanomaterials and Nanotechnologies in Wastewater Treatment: A Bibliometric Analysis. *Nanoscale Research Letters*, *13*(1), 233. doi:10.1186/s11671-018-2649-4 PMID:30097816

Jimenez-Cadena, G., Riu, J., & Xavier Rius, F. (2007). Gas sensors based on nanostructured materials. *Analyst*, *132*(11), 1083–1099. doi:10.1039/b704562j PMID:17955141

Kaur, J., Pathak, T., Singh, A., & Kumar, K. (2017). Application of nanotechnology in the environment biotechnology. In R. Kumar, A. K. Sharma, & S. S. Ahluwalia (Eds.), *Adv. Environ. Biotechnol* (pp. 155–165). Springer. doi:10.1007/978-981-10-4041-2_9

Khare, P., Yadav, A., Ramkumar, J., & Verma, N. (2016). Microchannel-embedded metal–carbon–polymer nanocomposite as a novel support for chitosan for efficient removal of hexavalent chromium from water under dynamic conditions. *Chemical Engineering Journal*, *293*, 44–54. doi:10.1016/j.cej.2016.02.049

Kharisov, B. I., Rasika Dias, H. V., Kharissova, O. V., Manuel Jiménez-Pérez, V., Olvera Pérez, B., & Muñoz Flores, B. (2012). Iron-containing nanomaterials: Synthesis, properties, and environmental applications. *RSC Advances*, *2*(25), 9325–9358. doi:10.1039/c2ra20812a

Kharwar, S., & Singh, S. (2021). First-principles investigation of zigzag graphene nanoribbons based nanosensor for heavy metal detector. *Materials Today: Proceedings*, *47*, 2227–2231. Advance online publication. doi:10.1016/j.matpr.2021.04.183

Khin, M. M., Nair, A. S., Babu, V. J., Murugan, R., & Ramakrishna, S. (2012). A review on nanomaterials for environmental remediation. *Energy & Environmental Science*, *5*(8), 8075–8109. doi:10.1039/c2ee21818f

Kim, J. S., Kuk, E., Yu, K. N., Kim, J.-H., Park, S. J., Lee, H. J., Kim, S. H., Park, Y. K., Park, Y. H., Hwang, C.-Y., Kim, Y.-K., Lee, Y.-S., Jeong, D. H., & Cho, M.-H. (2007). Antimicrobial effects of silver nanoparticles. *Nanomedicine; Nanotechnology, Biology, and Medicine*, *3*(1), 95–101. doi:10.1016/j.nano.2006.12.001 PMID:17379174

Kim, Y., Johnson, R. C., & Hupp, J. T. (2001). Gold nanoparticle-based sensing of "spectroscopically silent" heavy metal ions. *Nano Letters*, *1*(4), 165–167. doi:10.1021/nl0100116

Kumar, Hietala, & Oksman. (2019). Lignin-Based Electrospun Carbon Nanofibers. *Front. Mater.*, *6*, 62.

Lakouraj, M. M., Mojerlou, F., & Zare, E. N. (2014). Nanogel and superparamagnetic nanocomposite based on sodium alginate for sorption of heavy metal ions. *Carbohydrate Polymers*, *106*, 34–41. doi:10.1016/j.carbpol.2014.01.092 PMID:24721048

Li, X. Q., & Zhang, W. X. (2006). Iron nanoparticles: The core–shell structure and unique properties for Ni(II) sequestration. *Langmuir*, *22*(10), 4638–4642. doi:10.1021/la060057k PMID:16649775

Liang, P. C. (2017). Adsorption of Radionuclide Uranium onto Carbon-Based Nanomaterials from Aqueous Systems. *Huaxue Jinzhan*, *29*(9), 1062–1071.

Liang, W., Dai, C., Zhou, X., & Zhang, Y. (2014). Application of zero-valent iron nanoparticles for the removal of aqueous zinc ions under various experimental conditions. *PLoS One*, *9*(1). https:// pubmed.ncbi.nlm.nih.gov/24416439/

Liang, Y., Gu, P. C., Yao, W., Yu, S. J., Wang, J., & Xiang, X. K. (2017). Adsorption of Radionuclide Uranium onto Carbon-Based Nanomaterials from Aqueous Systems. *Huaxue Jinzhan*, *29*(9), 1062–1071.

Lisha, K. P., Anshup, & Pradeep, T. (2009). Towards a practical solution for removing inorganic mercury from drinking water using gold nanoparticles. *Gold Bulletin*, *42*(2), 144–152. doi:10.1007/BF03214924

Liu, F., Zhu, S., Li, D., Chen, G., & Ho, S. H. (2020). Detecting ferric iron by microalgal residuederived fluorescent nanosensor with an advanced kinetic model. *iScience*, *23*(6), 101174. doi:10.1016/j.isci.2020.101174 PMID:32498017

Liu, X. (2012). A survey on gas sensing technology. *Sensors*, *12*(7), 9635–9665. http://www. mdpi.com/1424-8220/12/7/9635

Long, Q., Zhang, Z., Qi, G., Wang, Z., Chen, Y., & Liu, Z.-Q. (2020). Fabrication of Chitosan Nanofiltration Membranes by the Film Casting Strategy for Effective Removal of Dyes/Salts in Textile Wastewater. *ACS Sustainable Chemistry & Engineering*, *8*(6), 2512–2522. doi:10.1021/acssuschemeng.9b07026

Lu, M., Zhang, Z., Qiao, W., Guan, Y., Xiao, M., & Peng, C. (2010). Removal of residual contaminants in petroleum-contaminated soil by Fenton-like oxidation. *Journal of Hazardous Materials*, *179*(1-3), 604–611. doi:10.1016/j.jhazmat.2010.03.046 PMID:20378246

Maensiri, S., Laokul, P., & Klinkaewnarong, J. (2008). Indium oxide (in 2O3) nanoparticles using aloe vera plant extract: Synthesis and optical properties. *Journal of Optoelectronics and Advanced Materials*, *10*, 161–165.

Mansas, C., Atfane-Karfane, L., Petit, E., Mendret, J., Brosillon, S., & Ayral, A. (2020). Functionalized ceramic nanofilter for wastewater treatment by coupling membrane separation and catalytic ozonation. *Journal of Environmental Chemical Engineering*, *8*(4), 104043. doi:10.1016/j.jece.2020.104043

Mathew, J., Joy, J., & George, S. C. (2019). Potential applications of nanotechnology in transportation: A review. *Journal of King Saud University. Science*, *31*(4), 586–594. doi:10.1016/j.jksus.2018.03.015

McLain, A. A. (2019). *Photocatalytic Properties of Zinc Oxide and Graphene Nanocomposites.* Public Knowledge Project. doi:10.17307/wsc.v1i1.253

Mohamed, E. F., & Awad, G. (2020). Photodegradation of gaseous toluene and disinfection of airborne microorganisms from polluted air using immobilized TiO_2 nanoparticle photocatalyst–based filter. *Environmental Science and Pollution Research International*, *27*(19), 24507–24517. doi:10.1007/s11356-020-08779-0 PMID:32307677

Mohamed, E.F., & Do, T.-O. (2020). Synthesis of new hollow nanocomposite photocatalysts: sunlight applications for removal of gaseous organic pollutants. *J. Taiwan Ins. Chem. E. 6*(10).

Mohapatra, L., Cheon, D., & Yoo, S. H. (2023). Carbon-Based Nanomaterials for Catalytic Wastewater Treatment: A Review. *Molecules (Basel, Switzerland)*, *28*(4), 1805. doi:10.3390/molecules28041805 PMID:36838793

Mubeen, I., Tulaphol, S., Shengyong, L., Pan, D., Zhang, P., Sajid, M., Yan, M., & Stevens, W. R. (2021). Online measurement of 1, 2, 4-trichlorobenzene as dioxin indicator on multi-walled carbon nanotubes. *Environmental Pollution*, *268*, 115329. doi:10.1016/j.envpol.2020.115329 PMID:33137681

Nakamura, S., Ando, N., Sato, M., & Ishihara, M. (2020, April 30). Ultraviolet Irradiation Enhances the Microbicidal Activity of Silver Nanoparticles by Hydroxyl Radicals. *International Journal of Molecular Sciences*, *21*(9), 3204. doi:10.3390/ijms21093204 PMID:32366059

Numan, A., Gill, A. A. S., Rafique, S., Guduri, M., Zhan, Y., Maddiboyina, B., Li, L., Singh, S., & Nguyen Dang, N. (2021). Rationally engineered nanosensors: A novel strategy for the detection of heavy metal ions in the environment. *Journal of Hazardous Materials*, *409*, 124493. doi:10.1016/j.jhazmat.2020.124493 PMID:33229259

O'Keeffe, C., Gannon, P., Gilson, P., Kafizas, A., Parkin, I. P., & Binions, R. (2013). Air purification by heterogeneous photocatalytic oxidation with multi-doped thin film titanium dioxide. *Thin Solid Films*, *537*, 131–136. doi:10.1016/j.tsf.2013.04.016

Piumie, R. P., Power, A., Chandra, S., & Chapman, J. (2018). Graphene, electrospun membranes and granular activated carbon for eliminating heavy metals, pesticides and bacteria in water and wastewater treatment processes. *Analyst*, *143*(23), 5629–5645. doi:10.1039/C8AN00922H PMID:30357153

Pooja, P., & Chowdhury, P. (2021). Functionalized CdTe fluorescence nanosensor for the sensitive detection of water borne environmentally hazardous metal ions. *Optical Materials*, *111*, 110584. doi:10.1016/j.optmat.2020.110584

Qian, L., Liu, S., Zhang, W., Chen, Y., Da Ouyang, L. H., Yan, J., & Chen, M. (2019). Enhanced reduction and adsorption of hexavalent chromium by palladium and silicon-rich biochar supported nanoscale zero-valent iron. *Journal of Colloid and Interface Science*, *533*, 428–436. doi:10.1016/j.jcis.2018.08.075 PMID:30172153

Qian, L., Shang, X., Zhang, B., Zhang, W., Su, A., Chen, Y., Da Ouyang, L. H., Yan, J., & Chen, M. (2019). Enhanced removal of Cr (VI) by silicon rich biochar-supported nanoscale zero-valent iron. *Chemosphere*, *215*, 739–745. doi:10.1016/j.chemosphere.2018.10.030 PMID:30347367

Qian, Y., Qin, C., Chen, M., & Lin, S. (2020). Nanotechnology in soil remediation - applications vs. implications. *Ecotoxicology and Environmental Safety*, *15*(201), 110815. doi:10.1016/j.ecoenv.2020.110815 PMID:32559688

Qin, L., Huang, D., Xu, P., Zeng, G., Lai, C., Fu, Y., Yi, H., Li, B., Zhang, C., Cheng, M., Zhou, C., & Wen, X. (2019). In-situ deposition of gold nanoparticles onto polydopamine-decorated g-C3N4 for highly efficient reduction of nitroaromatics in environmental water purification. *Journal of Colloid and Interface Science*, *534*, 357–369. doi:10.1016/j.jcis.2018.09.051 PMID:30243177

Qu, J., Luo, C., & Hou, J. (2011). Synthesis of ZnO nanoparticles from Zn-hyperaccumulator (Sedum alfredii Hance) plants. *Micro & Nano Letters*, *6*(3), 174–176. doi:10.1049/mnl.2011.0004

Rahn, R. O., & Landry, L. C. (1973). Ultraviolet irradiation of nucleic acids complexed with heavy atoms-II. Phosphorescence and photodimerization of DNA complexed with Ag. *Photochemistry and Photobiology*, *18*(1), 29–38. doi:10.1111/j.1751-1097.1973.tb06389.x PMID:4744137

Raliya, R., Biswas, P., & Tarafdar, J. C. (2015). TiO2 nanoparticle biosynthesis and its physiological effect on mung bean (Vigna radiata L.). *Biotechnology Reports (Amsterdam, Netherlands)*, *5*, 22–26. doi:10.1016/j.btre.2014.10.009 PMID:28626678

Raliya, R., & Tarafdar, J. C. (2014). Biosynthesis and characterization of zinc, magnesium and titanium nanoparticles: An eco-friendly approach. *International Nano Letters*, *4*(1), 93. doi:10.1007/s40089-014-0093-8

Ravindra, B. K., & Rajasab, A. H. (2014). A comparative study on biosynthesis of silver nanoparticles using four different fungal species. *International Journal of Pharmacy and Pharmaceutical Sciences*, *6*(1), 372–376.

Raychoudhury, T., & Scheytt, T. (2013). Potential of zerovalent iron nanoparticles for remediation of environmental organic contaminants in water: A review. *Water Sci Technol, 68*(7), 1425–1439. https://pubmed.ncbi.nlm.nih.gov/24135090/

Salamanca, M., Mar Peña, A. H., Prádanos, P., & Palacio, L. (2023). Forward Osmosis Application for the Removal of Emerging Contaminants from Municipal Wastewater: A Review. *Membranes*, *13*(7), 655. doi:10.3390/membranes13070655 PMID:37505021

Samanta, A., Zhao, A., Shimizu, G. K. H., Sarkar, P., & Gupta, R. (2012). Post-Combustion CO 2 Capture Using Solid Sorbents. *RE:view*.

Santhosh, C., Velmurugan, V., Jacob, G., Jeong, S. K., Grace, A. N., & Bhatnagar, A. (2016). Role of nanomaterials in water treatment applications: A review. *Chemical Engineering Journal*, *306*, 1116–1137. doi:10.1016/j.cej.2016.08.053

Saravanan, R., Shankar, H., Prakash, T., Narayanan, V., & Stephen, A. (2011). ZnO/CdO composite nanorods for photocatalytic degradation of methylene blue under visible light. *Materials Chemistry and Physics*, *125*(1-2), 277–280. doi:10.1016/j.matchemphys.2010.09.030

Sargin, I., Baran, T., & Arslan, G. (2020). Environmental remediation by chitosan-carbon nanotube supported palladium nanoparticles: Conversion of toxic nitroarenes into aromatic amines, degradation of dye pollutants and green synthesis of biaryls. *Separation and Purification Technology*, *247*, 116987. doi:10.1016/j.seppur.2020.116987

Sha, R., & Bhattacharyya, T. K. (2020). MoS2-based nanosensors in biomedical and environmental monitoring applications. *Electrochimica Acta*, *349*, 136370. doi:10.1016/j.electacta.2020.136370

Sharma, A., Chauhan, A. K., Kataria, N., & Garg, V. K. (2023). Hybrid nanomaterials for the removal of organic pollutants from wastewater. In *Hybrid Nanomaterials for Sustainable Applications* (pp. 71–102). Elsevier. doi:10.1016/B978-0-323-98371-6.00022-7

Shi, C., Zhu, N., Cao, Y., & Wu, P. (2015). Biosynthesis of gold nanoparticles assisted by the intracellular protein extract of Pycnoporus sanguineus and its catalysis in degradation of 4-nitroaniline. *Nanoscale Research Letters*, *10*(1), 1–8. doi:10.1186/s11671-015-0856-9 PMID:25852436

Singh, J., Singh, N., & Rathi, A. (2017). Facile approach to synthesize and characterization of silver nanoparticles by using mulberry leaves extract in aqueous medium and its application in antimicrobial activity. *J Nanostructures*, *7*, 134–40. . doi:10.22052/jns.2017.02.007

Singh, P. P. (2018). Environmental remediation by nanoadsorbents-based polymer nanocomposite. In C. M. Hussain & A. K. Mishra (Eds.), *New Polym. Nanocomposites Environ. Remediat* (pp. 223–241). Elsevier. doi:10.1016/B978-0-12-811033-1.00010-X

Song, X., Li, D., Li, R., Chen, Y., Huang, L., Wei, M., Pan, X., Zhou, H., & Tang, H. (2019). Ionic liquid modified inorganic nanoparticles for gaseous phenol adsorption. *Journal of Wuhan University of Technology. Materials Science Edition*, *34*(4), 787–790. doi:10.1007/s11595-019-2118-7

Song, Y., Fang, G., Zhu, C., Zhu, F., Wu, S., Chen, N., Wu, T., Wang, Y., Gao, J., & Zhou, D. (2019). Zero-valent iron activated persulfate remediation of polycyclic aromatic hydrocarbon-contaminated soils: An in situ pilot-scale study. *Chemical Engineering Journal*, *355*, 65–75. doi:10.1016/j.cej.2018.08.126

Stefaniuk, M., Oleszczuk, P., & Ok, Y. S. (2016). Review on nano zerovalent iron (nZVI): From synthesis to environmental applications. *Chemical Engineering Journal*, *287*, 618–632. doi:10.1016/j.cej.2015.11.046

Su, C. (2017). Environmental implications and applications of engineered nanoscale magnetite and its hybrid nanocomposites: A review of recent literature. *Journal of Hazardous Materials, 322*, 48–84. doi:10.1016/j.jhazmat.2016.06.060 PMID:27477792

Su, Lu, & Hu. (2010). *Colloids and surfaces A: physicochemical and engineering aspects adsorption of benzene, toluene, ethylbenzene and p-xylene by NaOCl oxidized carbon nanotubes.* https://doi.org/.colsurfa.2009.10.025 doi:10.1016/j

Sun, L., Du, T., Hu, C., Chen, J., Lu, J., Lu, Z., & Han, H. (2017). Antibacterial activity of graphene oxide/g-C3N4 composite through photocatalytic disinfection under visible light. *ACS Sustainable Chemistry & Engineering, 5*(10), 8693–8701. doi:10.1021/acssuschemeng.7b01431

Sun, S.-P., Zeng, X., Li, C., & Lemley, A. T. (2014). Enhanced heterogeneous and homogeneous Fenton-like degradation of carbamazepine by nano-Fe3O4/H2O2 with nitrilotriacetic acid. *Chemical Engineering Journal, 244*, 44–49. doi:10.1016/j.cej.2014.01.039

Sun, X.-F., Liu, B., Jing, Z., & Wang, H. (2015). Preparation and adsorption property of xylan/poly (acrylic acid) magnetic nanocomposite hydrogel adsorbent. *Carbohydrate Polymers, 118*, 16–23. doi:10.1016/j.carbpol.2014.11.013 PMID:25542101

Tan, X., Wang, X., Chen, C., & Sun, A. (2007). Effect of soil humic and fulvic acids, pH and ionic strength on Th(IV) sorption to TiO2 nanoparticles. *Applied Radiation and Isotopes, 65*(4), 375–381. doi:10.1016/j.apradiso.2006.10.014 PMID:17157512

Taymaz, H., Bircan, M. D., Kamış, H., Orhan, H., Aydoğan, Z., & Akıllı, A. (2022). Facile and green synthesis of ZnO nanoparticles for effective photocatalytic degradation of organic dyes and real textile wastewater. *International Journal of Phytoremediation*, 1–12. PMID:36437748

Usman, M., Farooq, M., Wakeel, A., Nawaz, A., Cheema, S. A., Rehman, H., Ashraf, I., & Sanaullah, M. (2020). Nanotechnology in agriculture: Current status, challenges and future opportunities. *The Science of the Total Environment, 721*, 137778. doi:10.1016/j.scitotenv.2020.137778 PMID:32179352

Usman, M., Faure, P., Ruby, C., & Hanna, K. (2012). Remediation of PAH-contaminated soils by magnetite catalyzed Fenton-like oxidation. *Applied Catalysis B: Environmental, 117*, 10–17. doi:10.1016/j.apcatb.2012.01.007

Vendrame, L. F. O., Zuchetto, T., Fagan, S. B., & Zanella, I. (2019). Nanofilter based on functionalized carbon nanostructures for the adsorption of pentachlorophenol molecules. *Computational & Theoretical Chemistry*, *1165*, 112561. doi:10.1016/j.comptc.2019.112561

Wang, J., Shao, X., Zhang, Q., Tian, G., Ji, X., & Bao, W. (2017). Preparation of mesoporous magnetic Fe2O3 nanoparticle and its application for organic dyes removal. *Journal of Molecular Liquids*, *248*, 13–18. doi:10.1016/j.molliq.2017.10.026

Wang, X., Kong, L., Zhou, S., Ma, C., Lin, W., Sun, X., Kirsanov, D., Legin, A., Wan, H., & Wang, P. (2021). Development of QDs-based nanosensors for heavy metal detection: A review on transducer principles and in-situ detection. *Talanta*, *122903*. Advance online publication. doi:10.1016/j.talanta.2021.122903 PMID:34857381

Wang, Y., Zhou, Z., Yang, Z., Chen, X., Xu, D., & Zhang, Y. (2009). Gas sensors based on deposited single-walled carbon nanotube networks for DMMP detection. *Nanotechnology*, *20*(34), 345502. doi:10.1088/0957-4484/20/34/345502 PMID:19652278

Weng, R., Chen, L., Lin, S., Zhang, H., Wu, H., Liu, K., Cao, S., & Huang, L. (2017). Preparation and Characterization of Antibacterial Cellulose/chitosan Nanofiltration Membranes. *Polymers*, *9*(12), 1–13. doi:10.3390/polym9040116 PMID:30970796

Wickramaratne & Jaroniec. (2013). *Activated Carbon Spheres for CO 2 Adsorption*. Academic Press.

Willis, A. L., Turro, N. J., & O'Brien, S. (2005). Spectroscopic characterization of the surface of iron oxide nanocrystals. *Chemistry of Materials*, *17*(24), 5970–5975. doi:10.1021/cm051370v

Xu, P., Zeng, G. M., Huang, D. L., Feng, C. L., Hu, S., Zhao, M. H., Lai, C., Wei, Z., Huang, C., Xie, G. X., & Liu, Z. F. (2012). Use of iron oxide nanomaterials in wastewater treatment: A review. *The Science of the Total Environment*, *424*, 1–10. doi:10.1016/j.scitotenv.2012.02.023 PMID:22391097

Xu, Y., Xu, H., Wang, L., Yan, J., Li, H., Song, Y., Huang, L., & Cai, G. (2013). The CNT modified white C 3 N 4 composite photocatalyst with enhanced visible-light response photoactivity. *Dalton Transactions (Cambridge, England)*, *42*(21), 7604–7613. doi:10.1039/c3dt32871f PMID:23535886

Yang, C., Florent, M., de Falco, G., Fan, H., & Bandosz, T. J. (2020). ZnFe2O4/activated carbon as a regenerable adsorbent for catalytic removal of H2S from air at room temperature. *Chemical Engineering Journal*, *394*, 124906. doi:10.1016/j.cej.2020.124906

Yin, Z., Cui, C., Chen, H., Duoni, Yu, X., & Qian, W. (2020). The Application of Carbon Nanotube/Graphene-Based Nanomaterials in Wastewater Treatment. *Small, 16*(15), 1902301. doi:10.1002/smll.201902301

Zhang, H., Feng, J., Fei, T., Liu, S., & Zhang, T. (2014). SnO2 nanoparticles-reduced graphene oxide nanocomposites for NO2 sensing at low operating temperature. *Sensors and Actuators. B, Chemical, 190*, 472–478. doi:10.1016/j.snb.2013.08.067

Zhang, M., He, F., Zhao, D., & Hao, X. (2011). Degradation of soil-sorbed trichloroethylene by stabilized zero valent iron nanoparticles: Effects of sorption, surfactants, and natural organic matter. *Water Research, 45*(7), 2401–2414. doi:10.1016/j.watres.2011.01.028 PMID:21376362

Zhang, T., Lowry, G., Capiro, N. L., Chen, W., Chen, Y., Dionysiou, D. D., Elliott, D., Choshal, S., Hofmann, T., Hsu-Kim, H., Hughes, J., Jiang, C., Jiang, G., Jing, C., Kavanaugh, M., Li, Q., Liu, S., Ma, J., Pan, B., ... Alvarez, P. J. J. (2019). In situ remediation of subsurface contamination: Opportunities and challenges for nanotechnology and advances materials. *Environmental Science. Nano, 6*(5), 1283–1302. doi:10.1039/C9EN00143C

Zhang, Y., Wu, B., Xu, H., Liu, H., Wang, M., He, Y., & Pan, B. (2016). Nanomaterials-enabled water and wastewater treatment. *NanoImpact, 3-4*, 22–39. doi:10.1016/j.impact.2016.09.004

Zhao, X., Liu, W., Cai, Z., Han, B., Qian, T., & Zhao, D. (2016). An overview of preparation and applications of stabilized zero-valent iron nanoparticles for soil and groundwater remediation. *Water Research, 100*, 245–266. doi:10.1016/j.watres.2016.05.019 PMID:27206054

Zheng, Y., Liu, Y., Guo, X., Chen, Z., Zhang, W., Wang, Y., Tang, X., Zhang, Y., & Zhao, Y. (2020). Sulfur-doped g-C3N4/rGO porous nanosheets for highly efficient photocatalytic degradation of refractory contaminants. *Journal of Materials Science and Technology, 41*, 117–126. doi:10.1016/j.jmst.2019.09.018

Chapter 4
Integration of Hybrid Nanomaterials and Artificial Intelligence for Sustainable Agriculture

Madhu Bala
https://orcid.org/0000-0003-2478-3532
Maharaja Agrasen University, India

Ritika Sharma
Maharaja Agrasen University, India

Shilpa Gupta
https://orcid.org/0000-0002-0848-3954
Maharaja Agrasen University, India

ABSTRACT

The agricultural industry is a primary source of livelihood for the world's population. The importance of agriculture in ensuring a sustainable future is becoming even more evident as we deal with the problems generated by a growing global population. To address the challenges of food security, environmental preservation, and resource efficiency, the integration of hybrid nanomaterials with artificial intelligence (AI) represents a transformative approach to attain sustainable agriculture. This chapter will concentrate on the applications of hybrid nanomaterials and AI in agriculture, emphasising their functions in sustainable and smart farming. Further, researchers working in this domain will get an insight that how these two cutting edge techniques can be used for characterizing and quantifying the interactions between nanomaterials, plants, soil, and pests thereby paving the way for a more sustainable and productive future for agriculture.

DOI: 10.4018/979-8-3693-1261-2.ch004

1. INTRODUCTION

Agriculture is the field that provides raw materials not only to industrial sector but it also fulfils the ongoing food demands of increased population. It is observed that the rapid increase in population has raised so many problems in agriculture sector like soil degradation, climate change, environmental pollution, scarcity of food etc. (Pal & Kumar, 2023; Sankaran et al., 2010). The two cutting edge technologies, artificial intelligence and nanotechnology are the new saviours that provide solution to these problems. To understand the integration of these technologies firstly it is needed to understand them separately. Conventional agricultural practices such as excessive use of pesticides and fertilizers are causing environmental hazards. The soil is losing its fertility, problem of water logging, human health issues etc. are the side effects of such practices (Lutz & Coradi, 2022). Nanotechnology has become a crucial enabling technology for agriculture production due to its better efficiency and precise target delivery. It refers to the use of nanoscale materials and tools to overcome the numerous agricultural issues and enhance farming methods. From the past few years nanotechnology has given promising solutions to various agricultural problems such as soil improvement, water management, crop protection and many more (Zhao et al., 2022). It has opened new vistas for the development of hybrid nanomaterials with specialised features which are appropriate for a range of agricultural applications. Organic and inorganic elements combine to form hybrid nanomaterials which have special qualities like increased surface area, greater reactivity and increased mechanical strength. These resources provide ground-breaking approaches to post-harvest technologies, crop protection, soil enhancement and water management (Fu et al., 2022). Additionally, they show potential for improving the management of the environment and livestock, supporting pervasive sustainability in agriculture. Nano-packaging materials also increase the shelf life of agricultural goods, lowering food waste. These advantages work together to create a more sustainable and resilient agricultural system. Scientists and researchers are creating novel approaches to increase crop yields, maximise resource utilisation, and improve food safety by working at the nanoscale. Hence it is attracting the attention towards itself of not only the researchers but of the farmers too.

On the other hand Artificial Intelligence (AI), helps the farmers in gathering and analysing data from many sources, including sensors, satellites, and drones, to help them decide on planting, irrigation, fertilisation, and pest control in an informed manner. It makes it possible to manage resources precisely, cutting down on waste and increasing crop yields (Orillo et al., 2014). AI enabled systems have the power to analyse images, find the diseases if any and guides the farmers to take action in time to optimize the usage of resources and to maximize crop yields. AI deployment can ensure that the transition to more contemporary agricultural practises is as profitable and effective as possible. Agriculture could be made more sustainable in the future with the wise use of AI to the sector. Farmers are given real-time insights using AI-powered sensor networks and decision support systems which optimise resource allocation and increases productivity of crops. The efficiency and effectiveness of farming practises driven by AI are further improved by integration with nanomaterials.

The fusion of AI with nanotechnology has the power to accelerate innovation and influence the direction of many different industries. Researchers can create unique solutions to some of the most important problems confronting humanity today by combining the skills of both professions (Danai-Varsou et al., 2023). Nanobots controlled by AI algorithms can deliver nutrients, fertilizers and pesticides to plants with more precision. Early disease detection is also feasible with nanoscale biosensors and AI driven recommendations. Despite these immense positive points, the integration of nanomaterials and AI has challenges too. To protect the safety of the environment and people, regulatory concerns surrounding the usage of nanomaterials must be carefully examined. Thorough risk assessments are required due to ethical concerns over the potential long-term effects of nanomaterials on the environment and food chain. Sustainable agriculture encourages social fairness and economic success while helping to protect the environment. In addition to lowering the environmental effect of the conventional agricultural system, it also increases yields and provides consumers with healthier goods.

1.1 Why Integrate Nanotechnology and Artificial Intelligence?

- Agriculture's environmental effect can be diminished by using nanotechnology to administer nutrients and agrochemicals accurately. By ensuring that these resources are used only when and where they are required, AI can further optimise their utilisation while reducing runoff and pollution.
- Early disease and pest identification by nanoscale sensors and AI can assist farmers in taking preventative actions to stop epidemics and lessen the need for chemical interventions.
- A reduction in the requirement for manual labour in agriculture is possible because AI-driven machinery with nano sensors can perform different task with high precision and better efficiency.

In the realm of nanotechnology, artificial intelligence is essential, especially when creating novel materials and putting together nanostructures. Researchers can analyse large information on material properties and precisely forecast how they will behave under various settings thanks to machine learning techniques.

1.2 Importance of Sustainability in Agriculture

The resilience and durability of our food system are fundamentally supported by agriculture's sustainability.

- It exemplifies the value of caring for our land, water, and ecosystems, guaranteeing that our farming methods can feed the world's expanding population while preserving environment for coming generations.
- With the goal of minimising the negative effects of climate change and cutting greenhouse gas emissions, sustainable agriculture places a high priority on resource efficiency, soil health, and biodiversity preservation are given top priority in sustainable agriculture, which lessens the harmful effects of climate change and cuts greenhouse gas emissions.
- In a time of increased environmental consciousness, it supports farmers' ability to make a living off their land, corresponds with customer desire for food produced responsibly, and encourages regulatory compliance.

In short, sustainability in agriculture is essential for ensuring food security, protecting our natural resources, and promoting a healthy environment for all.

2. NANOTECHNOLOGY IN AGRICULTURE

Nanotechnology has a large and developing role in agriculture, with a wide range of possible uses (as shown in Figure 1) and benefits in the sector. Nanomaterials, with their unique properties at the nanoscale, have opened up exciting opportunities for innovation and advancement in various fields, including agriculture (Bala et al., 2023). Real-time monitoring of soil conditions, moisture content, nutrient levels, and even plant health is possible with nanoscale sensors. Informed decisions about irrigation, fertilisation, and pest management may be made by farmers with the use of this data, which improves resource efficiency and increases crop yields. Herbicides and insecticides can have their efficiency increased while using less of them because of nano formulations. Targeted distribution is made possible by nanoencapsulation and controlled release technologies, which also minimises harm to unintended creatures and the environment. Nanoscale fertiliser delivery devices can increase plant nutrient uptake, decrease nutrient loss through leaching, and increase nutrient-use effectiveness (Institute of Electrical and Electronics Engineers. Philippines Section et al., 2014). This can boost agricultural output while reducing the negative effects of excess fertiliser runoff on the environment.

Figure 1. Nanotechnology in agriculture

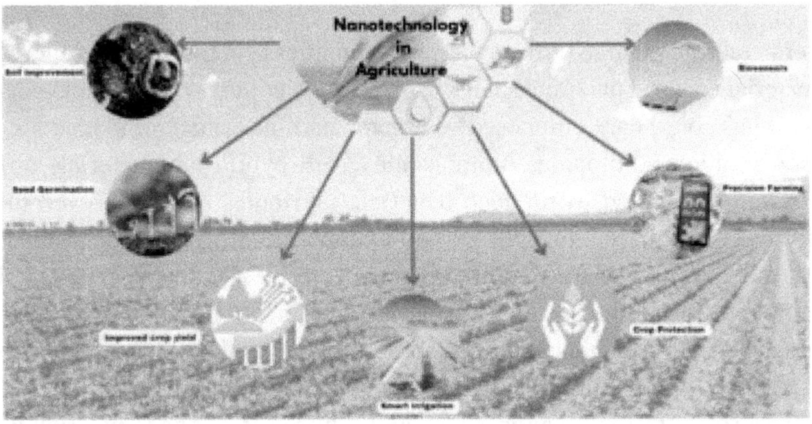

2.1 Benefits and Challenges of Nanomaterials in Agriculture

Nanomaterials have the potential to improve agricultural productivity, regulate nutrients, and lessen environmental impact in agriculture. To ensure the appropriate and sustainable use of nanomaterials in agriculture, it is essential to address the associated issues, including regulatory barriers, health and environmental hazards, and ethical issues. The benefits of nano materials in the field of agriculture includes improved nutrient delivery, enhanced crop production, disease management, improve soil quality (A. Ahmed et al., 2023) by enhancing its water-holding capacity, nutrient retention, and overall fertility. Nano sensors and nanomaterials can enable real-time monitoring of soil and crop conditions, allowing for more precise and data-driven farming practices. However, there are certain challenges too that needs to be focused.

Challenges:

- If not adequately controlled and regulated, some nanomaterials, particularly those with unique features, may pose health concerns to consumers, farmworkers, and the environment.
- The cost of creating and utilising nanoparticles in agriculture can prevent some small-scale farmers from utilising them.
- It's possible that farmers lack the skills and knowledge needed to use nanomaterials safely and successfully.
- Agriculture-related nanomaterial regulations are presently being developed. It might be difficult to ensure their safe usage while resolving issues with labelling, testing, and approval.

2.2 Types of Hybrid Nanomaterials Used in Agriculture

In order to take advantage of the synergistic qualities of multiple different types of nanomaterials, hybrid nanomaterials are used in agriculture. Hybrid nano materials (depicted in Table 1) are composed of organic and inorganic components such as metal ions, polymers, sulphides, biomolecules etc. In hybrid nanomaterials, different materials are combined to produce beneficial attributes such improved thermal stability, mechanical strength, light emission, gas permeability, electron conductivity, and controlled wetting characteristics (Gupta et al., 2022; Kumar et al., 2023) These elements come together through some particular interactions, which synergistically improve their useful qualities. Innumerable sorts of materials with distinctive features can be produced by combining various components and their interactions. Significant role is played by metal oxide nanoparticles and polymer-coated nanomaterials in plant development and defence against pests and diseases(Nishu & Kumar, 2023). Hybrid nanomaterials like magnetite and maghemite can be used as effective adsorbents and as special nano-sized materials in the biotech and biomedical areas.

As researchers look for novel solutions to meet agricultural difficulties while minimising environmental impacts, the development and application of such materials continue to advance.

Table 1. Role of hybrid nano materials along with their composition and use in Agriculture

Hybrid Nanomaterial	Composition	Use	References
Magnetite	Mixture of two iron oxides, FeO and Fe2O3	Eliminate toxic heavy metals from aqueous solutions, also inhibits fungus infection of plants	(Tohamy et al., 2022)
Silica-Coated Nanoparticles	Core nanoparticles made of materials like iron oxide (Fe3O4) or zinc oxide (ZnO), coated with a layer of silica (SiO2).	Used for controlled release of nutrients and pesticides in agriculture	(Rastogi et al., 2019)
Carbon Nanotube-Graphene Hybrid	Combination of carbon nanotubes (CNTs) and graphene sheets.	Used to enhance soil quality and water retention.	(Yin et al., 2020)
Nanoclay-Hydrogel Composites	Nanoclays incorporated into hydrogel matrices.	Improve soil structure, increase water retention. used for soil amendment and moisture retention.	(Merino et al., 2020)
Silver Nanoparticle-Cellulose Nanofiber Hybrid	Silver nanoparticles embedded in cellulose nanofibers	Inhibit the growth of pathogenic microorganisms in soil and on plants, reducing the need for chemical pesticides.	(Salama et al., 2021)
Magnetic Nanoparticle-Activated Carbon Hybrid	Magnetic nanoparticles combined with activated carbon	Used for soil remediation by adsorbing and removing heavy metals and other contaminants from the soil	(Livani et al., 2018)

3. ARTIFICIAL INTELLIGENCE IN AGRICULTURE

In the current scenario, agricultural industry is turning towards artificial intelligence in order to increase and improve the crop yield. Real time monitoring of crops, harvesting and production of healthy crops is possible due to AI. Machine learning and deep learning are poised to play an increasingly important role in determining the future of farming and food production as agriculture faces mounting challenges in providing the world's food demand while conserving resources (Kiranmai et al., 2021). In agriculture, these technologies have become potent tools that can alter the sector's productivity, sustainability, and efficiency. In order to give farmers useful insights, machine learning algorithms can evaluate enormous amounts of data from sensors, satellites, and drones (Fasiolo et al., 2023). For example, they may anticipate crop yields, optimise irrigation plans, and spot crop diseases or pest infestations. Deep learning, a kind of machine learning, excels at tasks like image identification, allowing the creation of sophisticated systems that can use computer vision to identify and classify problems like weeds and illnesses in crops (K. Ahmed et al., 2019). Through the use of AI-driven robotics and autonomous machinery, these technologies not only facilitate better decision-making but also enable the automation of processes like planting and harvesting. Figure 2 is representing the various fields of agriculture in which AI is playing an important role.

Figure 2. Artificial intelligence in agriculture

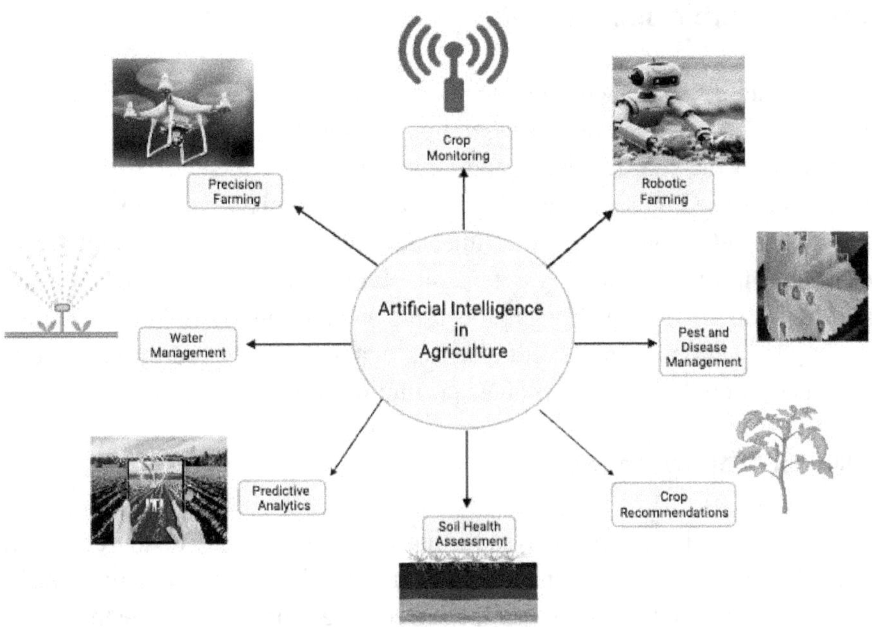

3.1 Precision Farming

Precision agriculture is not only the adoption of new technology rather, it is a revolution in information made possible by new technologies that produces a more sophisticated and accurate farm management system. The aim of precision farming is to increase the production, reduce labour time and input cost, effective management of irrigation system and fertilizers, and to improve the environmental sustainability (Divya et al., 2023). To achieve this aim precision farming works on 5R principles: Right Input should be applied at the Right Amount, to the Right Place, at the Right Time, and in the Right Manner. AI plays a crucial role in precision agriculture to achieve its aims specified above. It can analyse the data collected from drones and sensors to monitor soil conditions, crop health and weather patterns. Based on soil moisture levels, AI can calculate the optimal irrigation schedules thereby helping in reducing the water wastage. Further, machine learning algorithms are capable of identifying the target pests or weeds and hence helps in reducing the need for chemical inputs. Deep learning algorithms are gaining attention in finding the plant leaf diseases and its severity among the plants. Many applications have been developed using these algorithms because of which farmers are able to know what kind of disease a plant is having and what are the necessary remediation for it (Shewale & Daruwala, 2023). In general it can be said that AI is causing a paradigm shift in agriculture that will increase its productivity, sustainability, and ability to feed a growing global population while reducing its environmental impact.

3.2 Soil Health Assessment

Large amounts of soil-related data, such as soil samples, sensor data, satellite photos, and historical records, can be processed by AI algorithms. This enables a thorough examination of the characteristics and circumstances of the soil. To forecast soil quality indicators (Figure 3) including nutrient levels, pH, organic matter content, and moisture retention ability, machine learning models can be trained. These forecasts aid farmers in making defensible choices about fertilisation, irrigation, and crop selection. Through image processing and data analysis, AI algorithms can spot early indications of pests and diseases that are transmitted through the soil. This reduces crop damage and enables prompt response.

3.3 Water Management

Data on water quality can be analysed using AI using a variety of sources, such as sensors, drones, and remote monitoring systems. Real-time detection of impurities, diseases, and pollutants enables prompt corrective action to guarantee the safety of

Figure 3. Soil quality indicators measured by machine learning

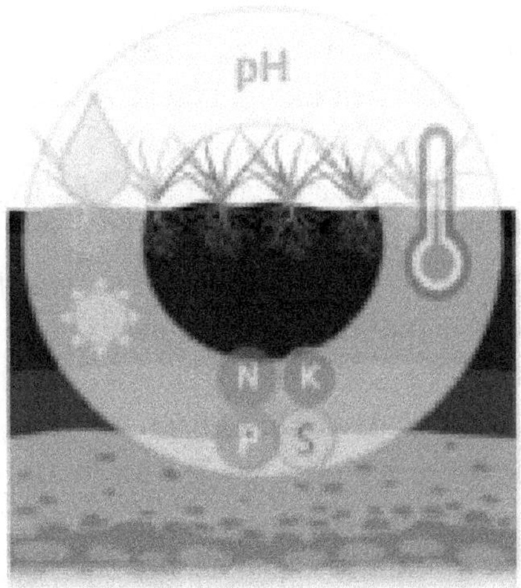

irrigation water. AI-driven irrigation systems have the ability to accurately regulate the quantity, timing, and duration of water supplied to crops. This reduces water waste, stops over-irrigation, and aids in water resource conservation. AI can help with the design of systems for the treatment and recycling of agricultural process wastewater. It can improve the treatment procedure and guarantee that the water is of a calibre suitable for reusing in irrigation.

3.4 Crop Monitoring and Recommendations

In order to track crop health, growth, and stress levels, AI systems examine satellite and drone footage. This makes it possible to identify problems like pest infestations, disease outbreaks, or nutrient deficits early on. Based on past data, current circumstances, and other environmental factors, AI models can forecast agricultural yields. Farmers can use this information to plan their marketing and harvesting plans. Through picture recognition and sensor data analysis, AI can recognise and evaluate crop illnesses and pests. Early detection minimises crop loss and enables prompt response (Bala & Mehan, 2021, 2022).AI evaluates how well-suited certain crop types are to the soil, climate, and pest problems of the region. In order to maximise productivity and quality, it suggests the crop varieties that perform the best in a given environment. When recommending crops, AI takes market trends, pricing history, and demand projections into account. It aids farmers in selecting the crops that will yield the greatest economic benefit.

3.5 Robotic Farming

Precision farming, commonly referred to as agri-robotics or robotic farming, is a developing agricultural practise that uses robotics, automation, and technology to streamline various farming processes. Its objectives include raising agricultural output, decreasing resource waste, and enhancing the overall effectiveness and sustainability of farming activities.

Robotic tractors, harvesters, and other pieces of equipment are capable of working independently while carrying out activities like planting, weeding, harvesting, and even crop spraying. These robots can precisely move around fields and complete jobs thanks to GPS, sensors, and AI algorithms (Basheer, 2023). Robots on the ground and drones gather information about the health of the crops, the land, and the surroundings. These data are processed by AI to produce insights for smarter decision-making. Robotic farming lessens the amount of time farmworkers spend in hazardous situations and in abrasive environments. It enhances agricultural workplace safety and conditions.

4. INTEGRATING AI AND NANOMATERIALS FOR PRECISION AGRICULTURE

Using AI and nanomaterials in agriculture is a promising new direction for conventional farming techniques. For precision farming, AI's data analytics and decision-making abilities can leverage the power of nanoscale materials. By enabling the development of sophisticated sensors, delivery mechanisms, and imaging agents, nanomaterials have the potential to fundamentally alter how we monitor soil health, control pests, utilise resources efficiently, and improve crop health. The combination of AI and nanomaterials has the potential to address the problems of resource conservation and food security while influencing the direction of agriculture.

4.1 Nanomaterials and AI for Soil Improvement

In addition to being a source of water and nutrients, soil gives agricultural plants the structural support they need to grow. Fertile soil produces nourishing and bountiful food supply. Copper, zinc, iron, magnesium and sulphur are the fundamental essential nutrients for healthy soil (Kumar et al., 2023). However, it is observed that quality of soil is degrading from the past few years and growing population is one of the main reason for it. Further, biotic agents (cold, heat, drought etc.) and abiotic agents (bacteria, fungi, viruses, insects) are also responsible for environmental threats and produces adverse effects on health of the soil. According to United Nations 670 million people will be suffering from hunger by 2030 ("The State of Food Security and Nutrition in the World 2022," 2022). At the same time, a roadmap created by academics combines AI and machine learning with smart and nano-enabled agriculture to potentially lower this figure. Since they have the potential to increase soil fertility and increase agricultural yield, nanocomposites have drawn interest from the fields of agriculture and soil science. The materials known as nanocomposites are made of nanoparticles scattered in a matrix and can be created to release nutrients, enhance soil structure, and encourage plant development (Ji et al., 2022). Table 2 is depicting some nanomaterials used for enhancing the fertility of soil.

Artificial intelligence (AI) and nanocomposites are two different but related strategies that can be utilised to increase soil fertility. In terms of improving soil fertility, these two technologies are contrasted as presented in Table 3.

4.2 Nanomaterial-Based Sensors

A biosensor is a tool that detects the presence of the analyte through contact with bioreceptors (DNA, enzymes etc.). Transducers convert the altered bioreceptor's physicochemical characteristics into a detectable signal as a result of this interaction between the analyte and the bioreceptors. Figure 4 is representing different types of biosensors based on transducers and based on the biological recognition elements. Nanosensors are capable of detecting various environmental parameters such as soil moisture, nutrient levels, pH, plant disease detection and pest infestations at the nanoscale (Ramachandran et al., 2022). They often generate large datasets and AI can be employed to retrieve data from multiple sensors and extract meaningful information from them (J. Zhang et al., 2022; Sharma & Dhadly, 2023). For example to check environmental conditions, data from nanosensors measuring temperature,

Table 2. Nanocomposites for soil fertility enhancement

Nanocomposites	Effect on Soil
Nano clays	Enhance the cation exchange capacity of soils, allowing them to retain more nutrients and water(Merino et al., 2020).
Magnesium oxide nanoparticles	Such liming materials containing nanoscale particles can be applied to acidic soils to raise their pH and make them more suitable for plant growth(Kanjana, 2020).
Biopolymer-based nanocomposites	It can be used to deliver pesticides or fungicides to protect crops from pests and diseases (Machado et al., 2022).
Titanium dioxide, zinc oxide, silicon oxide, magnesium oxide, gold	Nanocomposite-based soil stabilizers can be used to prevent soil erosion and maintain soil fertility (Duhan et al., 2017).
Nano biochar	It is a carbon-rich material that can improve soil structure and nutrient retention(Nepal et al., 2023).

Table 3. Nanocomposites vs AI and integration of both for soil enhancement

Factors	Nanocomposites	AI	Nanocomposites + AI
Mechanism	Nanocomposites are physical materials that consist of nanoparticles dispersed within a matrix. They are designed to release nutrients, improve soil structure, and promote plant growth through physical and chemical processes.	AI involves the use of algorithms and data analysis to optimize agricultural practices, make data-driven decisions, and improve soil management.	AI has the potential to guide nanomaterials that how much amount of nutrients a particular type of soil needed.
Application	Nanocomposites are applied directly to the soil, where they release nutrients slowly, improve soil structure, or protect organic matter, depending on their design and purpose.	By considering real-time soil data, crop growth stages, and weather conditions, AI helps farmers apply the right amount of nutrients at the right time and in the right place.	AI along with nano composites can minimize over-fertilization, reduces nutrient wastage, and optimizes crop yields.
Data Analysis	Nano sensors can detect and quantify the levels of essential nutrients in the soil, including nitrogen, phosphorus, and potassium	It identifies patterns, trends, and correlations related to soil nutrient levels and their impact on crop performance.	Large amounts of data are produced by nano sensors. AI can interpret data from soil nano sensors in the context of soil improvement to deliver useful insights into nutrient levels, pH, moisture content, and other soil factors.
Energy Efficiency	Batteries in nano sensors consume more power specially when large amount of data needs to be transmitted and processed.	AI can efficiently receive data from nano sensors using wireless protocols like Zigbee or Wi-Fi.	AI algorithms can optimize the power consumption of nano sensors. This is especially important in remote or autonomous monitoring systems where energy efficiency is critical for extended sensor lifetimes.

humidity and pressure can be combined to assess whether the current environmental conditions are suitable for particular crop or not. Also, AI algorithms are utilized to process data from these sensors in real-time, allowing for precise and timely decisions regarding irrigation, fertilization, and pest control.

4.3 Nanomaterials and AI for Enhanced Nutrient Delivery

Nanomaterials can be used to develop nano encapsulation systems that are capable of releasing nutrients slowly and improving nutrient uptake by plants. Such systems encapsulate vitamins, minerals and other bioactive compounds at the desired site. Furthermore, according to (P. Zhang et al., 2021) nanomaterials act as protective shields, safeguarding nutrients from these degrading factors until they are ready for absorption. This is particularly important in the case of sensitive nutrients.AI can

Figure 4. Types of nanosensors

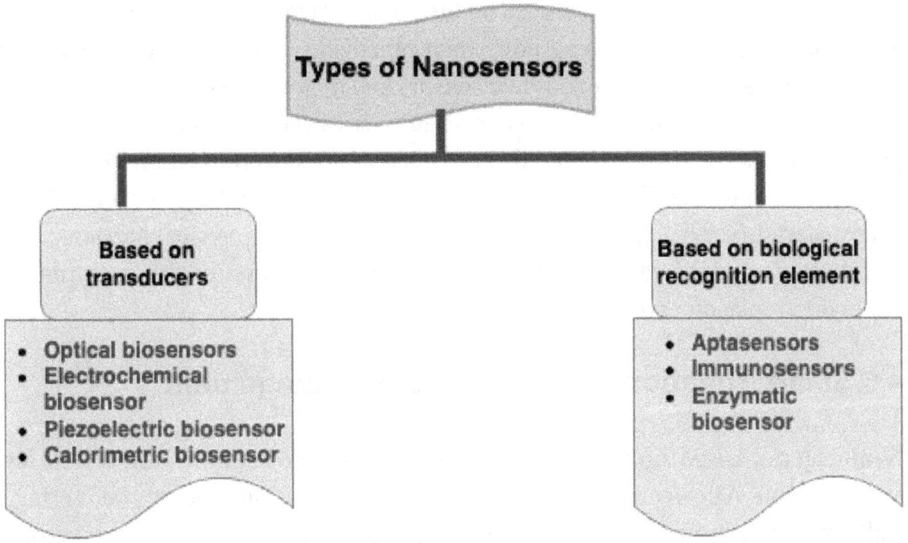

optimize nutrient release patterns based on plant growth stages, weather conditions, and soil characteristics to maximize plant health and minimize waste.

4.4 Smart Nanofertilizers

Nutrients can be delivered to plant roots more effectively with nanofertilizers. Their small size enables better penetration of the cell walls of the soil and the roots, enhancing nutrient uptake and decreasing waste. Such nutrients are more stable and less prone to degradation due to exposure to environmental factors. With the advancements in technology, smart nanofertilizers are designed that can be controlled remotely or autonomously through AI systems. They are designed with more advanced features and capabilities (Solanki et al., 2015). Even sensors or indicators that keep track of plant health or nutrient levels can be included in some intelligent nanofertilizers. Farmers can receive this information, which enables accurate nutrient application modifications and real-time feedback.AI can analyse data from soil sensors, weather forecasts, and crop models to determine the optimal timing and dosage of nano fertilizer application.

4.5 Nanomaterials and AI for Pest and Disease Management

Develop nanomaterial-based delivery systems for pesticides and fungicides to reduce environmental contamination and health risks. Nanocarriers can shield the active components and release them gradually, requiring fewer chemicals and having a smaller negative influence on the environment. This targeted delivery can also reduce the risk of harming non-target organisms. To find specific biomarkers linked to diseases or pests, nanoscale sensors and materials can be used in the field. Real-time information on the existence and seriousness of the problems can be provided by these sensors(Sen, 2021). AI can be used to predict and identify pest and disease outbreaks by analysing the data received from these sensors. Creating nanomaterial-based coatings or films for crops that act as physical barriers to pests and diseases and AI can assist in monitoring the condition of these coatings and deploying replacements when necessary.

4.6 Smart Irrigation and Waste Water Management

Nanomaterial-based moisture sensors can be employed to monitor soil moisture levels at a fine scale.AI can analyse this data to control irrigation systems, delivering the right amount of water to specific areas of the field, thus conserving water resources (Bhagat et al., 2020). On the other hand, in case of, waste water management, 2D graphene, which includes GO, reduced GO, and pure graphene, is one of the essential nanomaterials for water filtration (Tewari et al., 2023; Khalid et al., 2020). These nanoparticles are well known for their superior abilities as photocatalysts, antifouling agents, and adsorbents. Comparing conventional polyamide membranes to graphene membranes, the latter have atomically thin membranes that provide greater filtration. Figure 5 is showing different nanomaterials for waste water management along with their benefits.

4.7 Crop Monitoring and Yield Prediction

In many cases, before any outward signs of stress become apparent, nanosensors can identify stress causes such nutritional shortages, pathogen infections, and drought stress. To find trends and early signs of crop stress, AI can analyse environmental and nanosensor data together (Lutz & Coradi, 2022). The ability of machine learning algorithms to learn from prior data allows them to identify small indicators of stress that a human eye might miss. Employ AI-powered drones equipped with hyperspectral or multispectral cameras to monitor crop health and development. Nanomaterial-based imaging agents can enhance the accuracy of these assessments by providing detailed information about plant physiology. Using both historical and current data, AI can create predictive models to anticipate crop yields. These models are capable of taking into account a wide range of factors, such as soil quality, pest and disease pressures, and weather conditions.

Figure 5. Nanomaterials used in waste water treatment

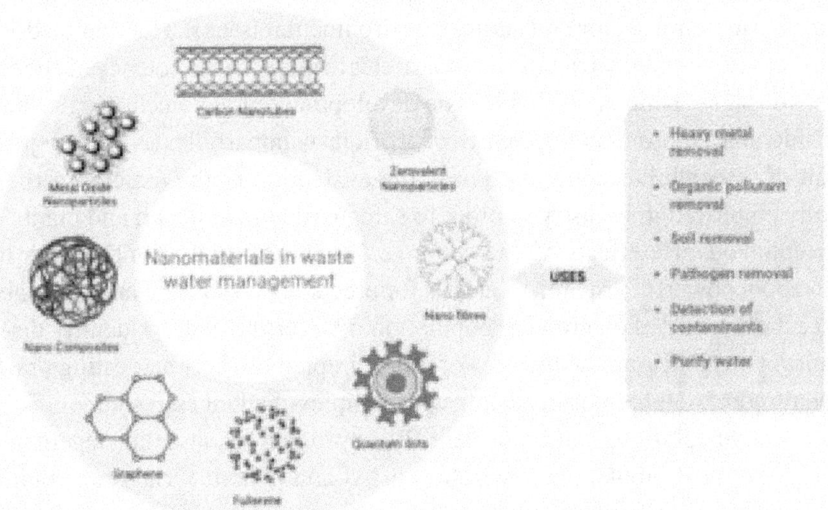

4.8 Data Integration and Decision Support

Making decisions is made easier with the help of nanosensors, which can provide thorough information about the soil, plants and ambient conditions. AI can interpret the complex data produced by nanosensors and other sources to give farmers useful information, give recommendations and optimizing farm management strategies. In predictive modelling both technologies can play a crucial role. Nanosensors can provide information on the state of a system right now, which can be utilised as input for forecasting models. Using real-time and historical data, AI may create predictive models that foretell future outcomes or events. These models might be useful in situations when it is necessary to forecast market trends, disease outbreaks, or equipment failures.

5. ENVIRONMENTAL SUSTAINABILITY

In agriculture, environmental sustainability refers to produce crops, food and fibre in a way that minimizes the negative impact on environment while maintaining the viability of land and resources. The sustainable practices in agriculture includes avoiding the overuse of pesticides, no-till farming and cover cropping, drip irrigation, reducing soil erosion, improve soil fertility etc. These practices can be well implemented using nanotechnology and artificial intelligence. Through the development of nanopesticides and the control of plant diseases, nanotechnology can help the agricultural sector and address environmental issues. This can be done by utilising environmentally friendly nanoparticles to increase the potency of chemicals at lower dosages (Khin et al., 2012). The development of nanotechnologies has led to considerable environmental releases of artificial nanoparticles (NPs). An growing amount of research has concentrated on the evaluation of the toxicity of the NPs typically employed in industry in order to safeguard human health and plants from the potential adverse effects of a wide range of nanomaterials. On the other hand, AI can apply powerful predictive models for precision agriculture and is capable to manage the demand of on growing population. Farmers are able to identify diseases of plants by just clicking the pictures of crops, implement best harvesting practices and many more. AI is helping in dealing with complex challenges, resource efficiency and innovative solutions can be driven. By improving resource management and reducing chemical inputs, the integration of AI and nanomaterials can contribute to more sustainable farming practices, minimizing the environmental impact of agriculture (Saha & Bharadwaj, 2023).

*Figure 6. Analysis of research publications in AI and Nanotechnology in agriculture from year 2014 to current (*publication data taken from https://app.dimensions.ai)*

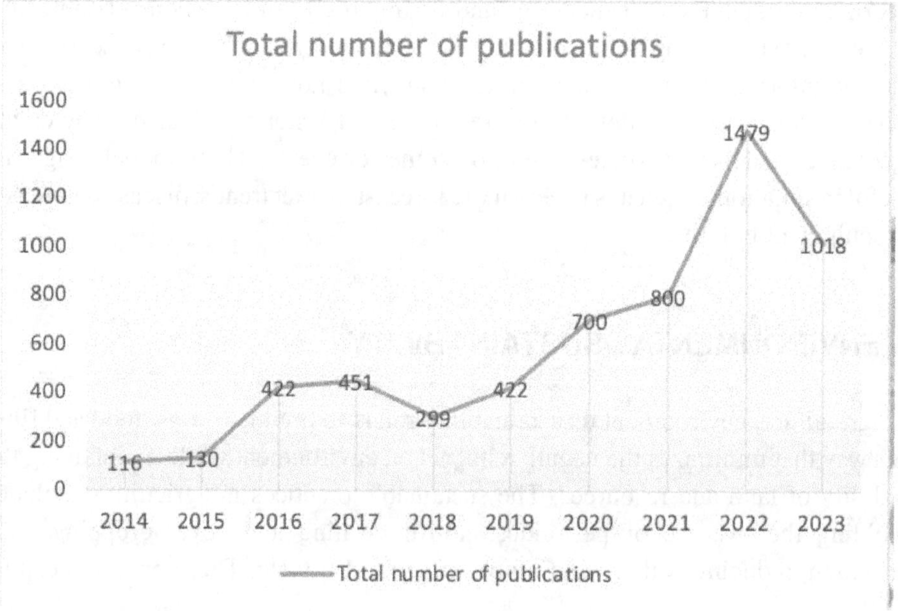

6. CONCLUSION AND FUTURE PERSPECTIVES

The fusion of AI and nanotechnology has enormous potential to make drastic changes in the field of agriculture. It is observed that from the past few years integration of these two technologies has become an area of interest for the research community and the progress can be seen in Figure 6. Researchers can create innovative and efficient solutions to most important problems like disease identification in crops, maintaining soil fertility, waste water management etc. using these technologies. The nano sensors and AI predictive analytics could analyse the real time conditions of soil such as soil moisture level detection and automated irrigation system to optimize the water usage through AI. Furthermore, the fusion of AI's technologies with nanotechnology's microscopic precision offers potential solutions in various sectors of agriculture like waste reduction, pollination enhancement, AI based system for harvesting etc. to increase efficiency, sustainability, cost- efficiency and customizability. However, while the advantages of increases yield, reduction in consumption of resources and environmental footprints of farming practices, the hybridization of AI and nanotechnology in agriculture could be limited. Because, the complexity introduced by such advanced technologies could divide the farmers on the large scale and small scale. Also, genetic modifications can't be ignored, and the potential environmental implications of extensive nanotechnology usage remain a subject for investigation.

As we stride towards this high-tech agricultural future, it's imperative to balance innovation with caution, ensuring that the pursuit of efficiency doesn't overshadow long-term sustainability and ethical considerations. It should be emphasised that the machine learning and deep learning models depend on the availability of sufficiently large volumes of data that will be used for training and validation purposes. These data are currently limited and dispersed across numerous databases and formats. Further, certain AI systems might not be well matched to the unique requirements of certain farms. Customization can be difficult and may need for a high level of knowledge.

REFERENCES

Ahmed, A., He, P., He, P., Wu, Y., He, Y., & Munir, S. (2023). Environmental effect of agriculture-related manufactured nano-objects on soil microbial communities. In *Environment International* (Vol. 173). Elsevier Ltd. doi:10.1016/j.envint.2023.107819

Ahmed, K., Shahidi, T. R., & Irfanul Alam, S. Md., & Momen, S. (2019). Rice Leaf Disease Detection Using Machine Learning Techniques. *2019 International Conference on Sustainable Technologies for Industry 4.0 (STI)*, 1–5. 10.1109/STI47673.2019.9068096

Bala, M., Kumar Bansal, S., & Fatima, F. (2023). Nanotechnology: A boon for agriculture. *Materials Today: Proceedings*, *73*(October), 267–270. doi:10.1016/j.matpr.2022.09.498

Bala, M., & Mehan, V. (2021). Identification of Rice Plant Diseases Using Image Processing, Machine Learning & Deep Learning: A Review. *CEUR Workshop Proceedings, 3058*, 0–3.

Bala, M., & Mehan, V. (2022). Metaheuristic Techniques for Classification Used in Identification of Plant Diseases. *ECS Transactions*, *107*(1), 13473–13480. doi:10.1149/10701.13473ecst

Basheer, A. A. (2023). Graphene materials for fabrication of robots. In *Materials Chemistry and Physics* (Vol. 302). Elsevier Ltd. doi:10.1016/j.matchemphys.2023.127781

Bhagat, S. K., Tung, T. M., & Yaseen, Z. M. (2020). Development of artificial intelligence for modeling wastewater heavy metal removal: State of the art, application assessment and possible future research. In *Journal of Cleaner Production* (Vol. 250). Elsevier Ltd. doi:10.1016/j.jclepro.2019.119473

Danai-Varsou, D., Zhang, P., Afantitis, A., Guo, Z., Lynch, I., & Melagraki, G. (2023). Nanoinformatics and artificial intelligence for nano-enabled sustainable agriculture. In *Nano-Enabled Sustainable and Precision Agriculture* (pp. 503–531). Elsevier. doi:10.1016/B978-0-323-91233-4.00015-6

Divya, S., Panda, S., Hajra, S., Jeyaraj, R., Paul, A., Park, S. H., Kim, H. J., & Oh, T. H. (2023). Smart data processing for energy harvesting systems using artificial intelligence. In *Nano Energy* (Vol. 106). Elsevier Ltd. doi:10.1016/j.nanoen.2022.108084

Duhan, J. S., Kumar, R., Kumar, N., Kaur, P., Nehra, K., & Duhan, S. (2017). Nanotechnology: The new perspective in precision agriculture. In Biotechnology Reports (Vol. 15, pp. 11–23). Elsevier B.V. doi:10.1016/j.btre.2017.03.002

Fasiolo, D. T., Scalera, L., Maset, E., & Gasparetto, A. (2023). Towards autonomous mapping in agriculture: A review of supportive technologies for ground robotics. *Robotics and Autonomous Systems*, *169*, 104514. Advance online publication. doi:10.1016/j.robot.2023.104514

Fu, X., Zheng, Z., Sha, Z., Cao, H., Yuan, Q., Yu, H., & Li, Q. (2022). Biorefining waste into nanobiotechnologies can revolutionize sustainable agriculture. In Trends in Biotechnology (Vol. 40, Issue 12, pp. 1503–1518). Elsevier Ltd. doi:10.1016/j.tibtech.2022.09.013

Gupta, R., Chauhan, H., Garg, V. K., & Kataria, N. (2022). Chemical and physical properties of nanoparticles and hybrid materials. In *Sustainable Nanotechnology for Environmental Remediation* (pp. 199–220). Elsevier. doi:10.1016/B978-0-12-824547-7.00024-2

Ji, M., AL-Huqail, A. A., Majdi, A., Liu, H., Ali, I., & Marzouki, R. (2022). Remediation of mine polluted soil with nano-enhanced materials: Development of extreme learning machine approaches. *Chemosphere*, *307*, 135772. Advance online publication. doi:10.1016/j.chemosphere.2022.135772 PMID:35931267

Kanjana, D. (2020). Foliar application of magnesium oxide nanoparticles on nutrient element concentrations, growth, physiological, and yield parameters of cotton. *Journal of Plant Nutrition*, *43*(20), 3035–3049. doi:10.1080/01904167.2020.1799001

Khalid, Z. B., Nasrullah, M., Nayeem, A., Wahid, Z. A., Singh, L., & Krishnan, S. (2020). Application of 2D Graphene-Based Nanomaterials for Pollutant Removal from Advanced Water and Wastewater Treatment Processes. In *ACS Symposium Series* (*Vol. 1353*, pp. 191–217). American Chemical Society. 10.1021/bk-2020-1353.ch009

Khin, M. M., Nair, A. S., Babu, V. J., Murugan, R., & Ramakrishna, S. (2012). A review on nanomaterials for environmental remediation. In Energy and Environmental Science (Vol. 5, Issue 8, pp. 8075–8109). Royal Society of Chemistry. doi:10.1039/c2ee21818f

Kiranmai, B., Venu Vasantha, S., & Rama Krishna, S. (2021). Techniques for Rice Leaf Disease Detection using Machine Learning Algorithms. *International Journal of Engineering and Technical Research, 9*(8), 162–166. www.ijert.org

Kumar, A., Gupta, G., Bapna, K., & Shivagan, D. D. (2023). Semiconductor-metal-oxide-based nano-composites for humidity sensing applications. In *Materials Research Bulletin* (Vol. 158). Elsevier Ltd. doi:10.1016/j.materresbull.2022.112053

Livani, M. J., Ghorbani, M., & Mehdipour, H. (2018). Preparation of an activated carbon from hazelnut shells and its composites with magnetic NiFe2O4 nanoparticles. *Xinxing Tan Cailiao. New Carbon Materials*, *33*(6), 578–586. doi:10.1016/S1872-5805(18)60358-0

Lutz, É., & Coradi, P. C. (2022). Applications of new technologies for monitoring and predicting grains quality stored: Sensors, Internet of Things, and Artificial Intelligence. In *Measurement: Journal of the International Measurement Confederation* (Vol. 188). Elsevier B.V. doi:10.1016/j.measurement.2021.110609

Machado, T. O., Grabow, J., Sayer, C., de Araújo, P. H. H., Ehrenhard, M. L., & Wurm, F. R. (2022). Biopolymer-based nanocarriers for sustained release of agrochemicals: A review on materials and social science perspectives for a sustainable future of agri- and horticulture. In *Advances in Colloid and Interface Science* (Vol. 303). Elsevier B.V. doi:10.1016/j.cis.2022.102645

Merino, D., Tomadoni, B., Salcedo, M. F., Mansilla, A. Y., Casalongué, C. A., & Alvarez, V. A. (2020). Nanoclay as Carriers of Bioactive Molecules Applied to Agriculture. In *Handbook of Nanomaterials and Nanocomposites for Energy and Environmental Applications* (pp. 1–22). Springer International Publishing. doi:10.1007/978-3-030-11155-7_62-1

Nepal, J., Ahmad, W., Munsif, F., Khan, A., & Zou, Z. (2023). Advances and prospects of biochar in improving soil fertility, biochemical quality, and environmental applications. In *Frontiers in Environmental Science* (Vol. 11). Frontiers Media S.A. doi:10.3389/fenvs.2023.1114752

Nishu, & Kumar, S. (2023). Smart and innovative nanotechnology applications for water purification. *Hybrid Advances, 3*, 100044. doi:10.1016/j.hybadv.2023.100044

Orillo, J. W., Dela Cruz, J., Agapito, L., Satimbre, P. J., & Valenzuela, I. (2014). Identification of diseases in rice plant (oryza sativa) using back propagation Artificial Neural Network. *2014 International Conference on Humanoid, Nanotechnology, Information Technology, Communication and Control, Environment and Management (HNICEM)*, 1–6. 10.1109/HNICEM.2014.7016248

Pal, A., & Kumar, V. (2023). AgriDet: Plant Leaf Disease severity classification using agriculture detection framework. *Engineering Applications of Artificial Intelligence, 119*, 105754. doi:10.1016/j.engappai.2022.105754

Ramachandran, R. P., Erkinbaev, C., & Vellaichamy, C. (2022). Smart nano-biosensors in sustainable agriculture and environmental applications. In Food, Medical, and Environmental Applications of Nanomaterials (pp. 527–542). Elsevier. doi:10.1016/B978-0-12-822858-6.00019-4

Rastogi, A., Tripathi, D. K., Yadav, S., Chauhan, D. K., Živčák, M., Ghorbanpour, M., El-Sheery, N. I., & Brestic, M. (2019). Application of silicon nanoparticles in agriculture. In 3 Biotech (Vol. 9, Issue 3). Springer Verlag. doi:10.1007/s13205-019-1626-7

Saha, S., & Bharadwaj, A. (2023). A step towards smart agriculture using metallic nanostructures. In *Plant Stress* (Vol. 10). Elsevier B.V. doi:10.1016/j.stress.2023.100216

Salama, A., Abouzeid, R. E., Owda, M. E., Cruz-Maya, I., & Guarino, V. (2021). Cellulose–silver composites materials: Preparation and applications. In Biomolecules (Vol. 11, Issue 11). MDPI. doi:10.3390/biom11111684

Sankaran, S., Mishra, A., Ehsani, R., & Davis, C. (2010). A review of advanced techniques for detecting plant diseases. *Computers and Electronics in Agriculture*, *72*(1), 1–13. doi:10.1016/j.compag.2010.02.007

Sen, K. (2021). Application of nanotechnology in pesticides adsorption with statistical optimization and modeling. In *Cognitive Data Models for Sustainable Environment* (pp. 65–99). Elsevier. doi:10.1016/B978-0-12-824038-0.00005-5

Sharma, S., & Dhadly, D. K. (2023). Nano-inspired biosensors and plant diseases: recent advances and challenges. In *Nanoparticles and Plant-Microbe Interactions: An Environmental Perspective* (pp. 135–162). Elsevier. doi:10.1016/B978-0-323-90619-7.00002-3

Shewale, M. V., & Daruwala, R. D. (2023). High performance deep learning architecture for early detection and classification of plant leaf disease. *Journal of Agriculture and Food Research*, *14*, 100675. Advance online publication. doi:10.1016/j.jafr.2023.100675

Solanki, P., Bhargava, A., Chhipa, H., Jain, N., & Panwar, J. (2015). Nano-fertilizers and their smart delivery system. In *Nanotechnologies in Food and Agriculture* (pp. 81–101). Springer International Publishing. doi:10.1007/978-3-319-14024-7_4

Tewari, C., Tatrari, G., Kumar, S., Pathak, M., Rawat, K. S., Kim, Y. N., Saha, B., Jung, Y. C., Mukhopadhyay, P., & Sahoo, N. G. (2023). Can graphene-based composites and membranes solve current water purification challenges - a comprehensive review. In *Desalination* (Vol. 567). Elsevier B.V. doi:10.1016/j.desal.2023.116952

The State of Food Security and Nutrition in the World. (2022). 2022. In *The State of Food Security and Nutrition in the World 2022*. FAO. doi:10.4060/cc0639en

Tohamy, H. A. S., El-Sakhawy, M., & Kamel, S. (2022). Development of Magnetite/Graphene Oxide Hydrogels from Agricultural Wastes for Water Treatment. *Journal of Renewable Materials*, *10*(7), 1889–1909. doi:10.32604/jrm.2022.019211

Yin, Z., Cui, C., Chen, H., Duoni, Y. X., & Qian, W. (2020). The Application of Carbon Nanotube/Graphene-Based Nanomaterials in Wastewater Treatment. In Small (Vol. 16, Issue 15). Wiley-VCH Verlag. doi:10.1002/smll.201902301

Zhang, J., Huang, H., Song, G., Huang, K., Luo, Y., Liu, Q., He, X., & Cheng, N. (2022). Intelligent biosensing strategies for rapid detection in food safety: A review. *Biosensors & Bioelectronics*, *202*, 114003. Advance online publication. doi:10.1016/j.bios.2022.114003 PMID:35065479

Zhang, P., Guo, Z., Ullah, S., Melagraki, G., Afantitis, A., & Lynch, I. (2021). Nanotechnology and artificial intelligence to enable sustainable and precision agriculture. In Nature Plants (Vol. 7, Issue 7, pp. 864–876). Nature Research. doi:10.1038/s41477-021-00946-6

Zhao, W., Liu, Y., Zhang, P., Zhou, P., Wu, Z., Lou, B., Jiang, Y., Shakoor, N., Li, M., Li, Y., Lynch, I., Rui, Y., & Tan, Z. (2022). Engineered Zn-based nano-pesticides as an opportunity for treatment of phytopathogens in agriculture. *NanoImpact*, *28*, 100420. Advance online publication. doi:10.1016/j.impact.2022.100420 PMID:36038133

Chapter 5
Energy Harvesting and Storage Applications for Hybrid Nanomaterials

Jagadeesh Ramadoss
Bharathidasan University, India

Arumugam Sonachalam
Bharathidasan University, India

ABSTRACT

Hybrid nanomaterials (HNs) have developed as promising candidates for advancing energy harvesting and storage technologies due to their unique combination of properties. This chapter focuses on developments in the design, synthesis, and applications of HNs for energy storage/harvesting applications. These materials, formed by integrating distinct nanoscale components such as nanoparticles, nanowires, and nanosheets, exhibit synergistic effects that enhance energy conversion, storage capacity, and overall device performance. These strategies encompass bottom-up and top-down approaches, self-assembly methods, and surface functionalization techniques. It aims to understand how these combinations affect charge transport, catalytic activity, and electrochemical behavior. These applications span a wide range, including solar cells, thermoelectric generators, electrochemical energy storage, and conversion.

1. INTRODUCTION

The development of electrochemistry where very slow as nearly 200 years were taken to explore the chemical and electrical relation. The relationship between the interfacial

DOI: 10.4018/979-8-3693-1261-2.ch005

characteristics of electrochemical processes and their reaction rates was discovered in the 20th century. After conducting further investigation into electrochemistry, it became evident that the choice of electrode material plays a crucial role in facilitating efficient electrochemical responses by enabling charge transfer reactions (Muhamad Azim et al., 2022). The field of electrochemical energy has undergone extensive research in the last few decades to address the global energy crisis. Energy storage, production, harvesting, conversion, and other processes are all included in electrochemical energy. An essential material of the energy processes in the entire landscape is energy storage, which operates in conjunction with energy generation and harvesting. One of the primary goals of sustainable growth is to transform people's lifestyles by ensuring access to clean and affordable energy. Renewable energy must replace non-renewable energy in the energy sector (Ibrahim et al., 2020). Energy storage devices based on nanoengineering and material processes are significantly advancing the development of energy technology and impacting human lifestyles. This is due to the increasing demand for clean and renewable energy. Supercapacitors, batteries, solar cells, fuel cells, and other energy storage and generation devices fall under the category of energy devices. Supercapacitors, batteries, fuel cells, and other energy storage and generation technologies are extensively researched and developed by scientists to meet the growing demand for sustainable energy systems.

The urgent need to find sustainable and eco-friendly alternatives to fossil fuels arises from the rapid growth of the world's population and the resulting environmental issues. Batteries, fuel cells, water splitting, supercapacitors, and other electrochemical energy conversion and storage technologies are promising other sources for achieving a sustainable environment. The electrodes required for these cutting-edge electrochemical energy systems must possess low potential, a high active surface area, affordability, manageability, and good conductivity. Modern electrode materials that are effective and efficient include hybrid nanostructured nanomaterials and their composites (Chen et al., 2017; Mez-Romero & Cuentas-Gallegos, 2005). Their scalability is, however, constrained by their exorbitant price and scarcity. Large efforts have been made to develop non-metallic catalysts, including transition metal phosphides, metal sulfides, and their alloys, with enhanced activity for energy storage and conversion, to tackle these challenges in the Figure 1.

Nanostructured materials are an efficient substitute for the materials currently used in energy storage and can be employed as alternatives. Nanostructured composite materials can take various forms, such as carbon nanomaterials, coordination polymers, metal-based nanomaterials, and nanostructured inorganic materials (Dragoman, 2023; Shahinul Islam, 2023). This chapter focuses on the utilization of nanostructured materials and their composites, highlighting the advantages and challenges associated with their application in energy storage and conversion. Because of this, after years of study and development, the majority of

Figure 1. Society applications of energy storage and conversion
Source: Yu et al. (2023)

electrodes in today's energy systems are made of HNs. These electrodes are made by developing nanostructures and nanoscale electrode materials, including metals, metal oxide nanomaterials, and carbon materials, among others (Pasquini, 2022). These HNs still exhibit unsatisfactory durability, especially in complex environments, making them unsuitable for direct use in industrial applications. Despite being a trendy research topic in the scientific and industrial communities, the limitations of these technologies prevent their practical implementation. To tackle the global energy problem, researchers are increasingly concentrating on developing hybrid nanostructures and nanomaterials that have improved properties for use in energy generation and storage applications. According to this viewpoint, nature can serve as a guiding light to help address issues related to nanoengineering and nano energy materials. In this review, the emphasis has been on nanostructured materials that were created using natural organisms as inspiration (Yun et al., 2019). These materials were then used in energy devices to enable sustainable accommodations and applied functional improvements in various applications, as shown in Figure 2. In this study, we investigate numerous research studies that have produced hybrid nanostructures, which imitate natural objects found in nature, for different applications.

Figure 2. Applications of hybrid nanomaterials

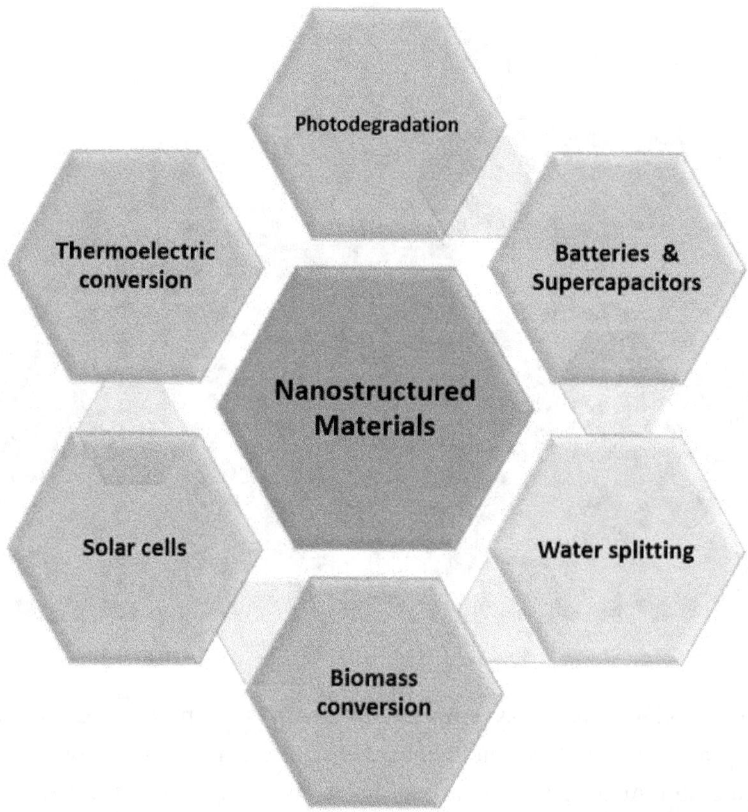

2. HYBRID NANOMATERIALS FOR ENERGY STORAGE AND CONVERSION

A class of nanocomposites, known as hybrid nanostructured materials, is composed of two or more molecularly distinct components, each of which has at least one nanoscale dimension. They are combined in a way that gives them qualities that no single ingredient could possess on its own. In HNs, metal ions, metal clusters, particles, oxides, sulfides, non-metallic elements, and their derivatives are often combined through specific interactions that enhance their surface functional properties synergistically (Shahinul Islam, 2023). From the creation of molecules to nanoscale binding, self-assembly, and microstructure, various intermolecular interactions such as electrostatic interactions, dispersion interactions, and H-bonding play crucial roles in the assembly of hybrid materials. The Figure 3 illustrates the different applications of HNs. Additionally, Class I hybrid materials are those that are created through weak chemical interactions, including dipole-dipole forces, H_2 bonds, and Coulomb

Figure 3. General classification of hybrid nanomaterials

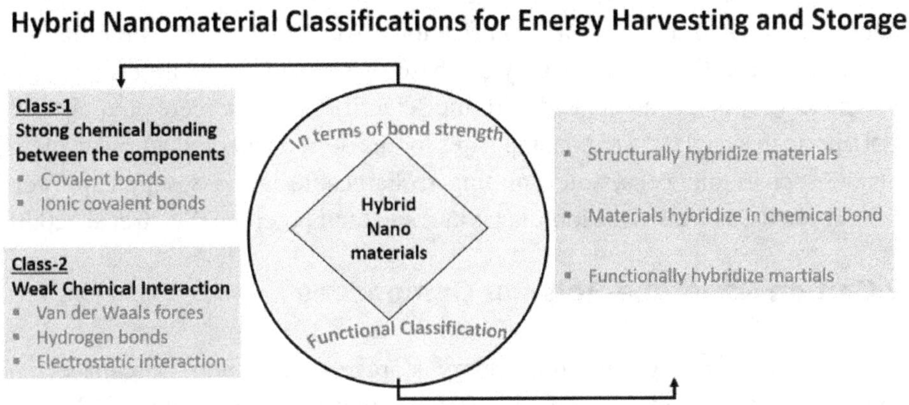

forces. Class II HNs, on the other hand, are created by fusing the constituent parts through strong chemical bonds, such as covalent or ionic-covalent connections.

Types of Hybrid Nanomaterials

1. Polymer-Nanoparticle Hybrids
2. Carbon Nanotube-Polymer Composites
3. Nanoparticle-Nanowire Hybrids
4. Nanoparticle-Quantum Dot Hybrids
5. Nanoparticle-Nanocarbon Hybrids
6. Nanoparticle-Metal-Organic Framework (MOF) Hybrids
7. Nanoparticle-Biomolecule Hybrids

2.1 Polymer-Nanoparticle Hybrids

Due to their distinct benefits over other materials, polymers have drawn a lot of interest. These advantages include universal and controllable electrical conductivity, a straightforward invention process, higher mechanical strength, low weight, low cost, and ease of processing. Polymer nanocomposites have been eagerly investigated as exceptional energy generators for low-power electronic strategies and conformed increasing surface area, electronic conductivity, and spectacular electrochemical behavior by extending the potential of utilization. Conducting polymers were combined with inorganic, organic, and HNs to create multifunctional hybrid nanocomposites that have improved performance in a range of devices, including energy storage, energy conversion, and defensive ones (Sonika et al., 2022). The ability and support of contemporary developments for the creation of multifunctional nanomaterials and nanocomposites, as well as the methods for creating nanostructured polymers, have been discussed. Investigated are the mechanics underlying their electrical conductivity as well as potential modifications. Also highlighted is the ongoing study towards creating superior polymer-based NPs. This evaluation focuses on the significant projects associated with the scientific and commercial application of polymeric materials and nanocomposites for the design and production of energy strategies such as solar cells, water splitting, rechargeable batteries, supercapacitors, and energy cells, as well as the looming issues and their potential (Saleh et al., 2020).

2.2 Carbon Nanotube-Polymer Composites

The fields of energy generation, energy storage, conversion, aerospace, and automotive industries utilize carbon-based polymer nanocomposites (CPNCs) in various applications. These key traits of nanostructured materials include their ease of processing, adaptability in configuration, lightweight, and flexibility in requirements. The future of renewable energy plans is expected to be revolutionized with the use of supercapacitors and fuel cells for energy storage. Novel nanocomposite materials have been discovered for automotive and electric energy storage applications. The demand for affordable devices with high energy and power densities is met by these materials. Insulating polymers with high aspect ratios that are loaded with conductive nanofillers, such as carbon nanotubes (CNT) and graphene nanoplatelets (GNP), have been shown to have potential as dielectric components. Space charge polarization is significantly improved due to the performance of dispersed nanocomposites acting as nanocapacitor electrodes in an insulating arrangement (Siwal et al., 2020). To enhance the interfacial polarization at the polymer interface and increase the density of interfacial polarization, it is assumed that the nanofiller will exhibit high conductivity. The polymer matrix, nanofiller, and polymer interface can all be polarized in a polymer nanocomposite.

2.3 Nanoparticle-Nanowire Hybrids

The research investigations on energy conversion and storage have focused on composition metal oxides into or onto a template with high electrical conductivity. This is done to enhance the capacitance and catalytic performance of a particular metal oxide. The incorporation of active materials into a porous and conductive template, which can provide superior electronic and ionic conductivity, is an additional strategy for enhancing electrochemical capacitive performance (Vernickaite et al., 2016). This approach complements the composite method used to create the well-known core-shell nanowire structure. Carbon supports were more commonly used for energy conversion and storage applications compared to porous noble metal templates. To create a metal oxide with a one-dimensional (1D) structure, such as nanorods and nanowires, in the template and then investigate its electrochemical performance, however, has received little attention. In contrast to nanowires, nanoparticles have a larger surface area at the interface between the active materials and the electrolyte (Arava, Gowda, Shaijumon, & Ajayan, 2012; Pescaglini & Iacopino, 2015). Due to their superior electrical conductivity compared to nanoparticles, nanowires have instead been largely regarded as a suitable form for capacitive applications. The capacitive capabilities of a specific metal oxide can be enhanced by combining the advantages of nanowires and nanoparticles, which have a high specific surface area. Adding a metal oxide with a hybrid nanowire and nanoparticle structure to a porous, conductive substrate is expected to enhance the metal oxide's capacitive capabilities.

2.4 Nanoparticle-Quantum Dot Hybrids

In contrast to their electrochemical inertness, nanoparticle-quantum dot hybrids exhibit configurable band gaps and redox potentials. They can be used in a variety of energy storage and harvesting applications by modifying the synthetic conditions (Liu et al., 2017). It is relatively easy to create hybrid nanoparticle-quantum dots. Methods include bottom-up carbonization of molecular precursors, as well as top-down methods such as oxidative cutting of larger graphitic carbons. The exciting physicochemical characteristics of nanoparticle-quantum dot hybrids in optoelectronics, catalysis, and sensors, among others, include redox behavior, as well as optical and electrical properties. Nanoparticle-quantum dot hybrids are superior to other materials of this kind when considering solubility, strong chemical inertness, ease of modification, and high resistance to photobleaching. Nanoparticle-quantum dot hybrids are effective materials, such as photocatalysts, for selective oxidation, light-driven acid catalysis, and H_2 bond catalysis, due to their rich photoluminescence (PL) and photochemical capabilities. To enhance the performance of energy storage devices and energy conversion, they have also been used as highly effective additives (Xu et al., 2022). Nanoparticle-quantum dot hybrids are ideal light absorbers for mesoscopic solar cells because of their size-dependent optical absorption and the various exciton production effects they exhibit.

2.5 Nanoparticle-Nanocarbon Hybrids

Nanoparticle-nanocarbon hybrids have been meticulously designed and synthesized as exceptional electrodes and catalysts for electrochemical energy storage/conversion. These HNs, which are created through nucleation and development of NPs on oxidized nanocarbon under carefully regulated synthetic circumstances without free particle growth in solution, contain inorganic nanoparticles that are firmly anchored on oxidized graphene sheets or carbon nanotubes. This technology was originally fabrication for graphene and has recently been applied to carbon nanotubes produced using the same oxidation process as GO (Mahmudul Hasan et al., 2021; Subrata Ghosh, 2022). The ability to oxidize the outer walls of multi-walled nanotubes to produce an abundance of functional groups and defects necessary for the nucleation and growth of nanoparticles while maintaining the high electrical conductivity of nanotubes provided by the relatively undamaged inner tubes makes nanoparticle-nanocarbon hybrids more advantageous than their graphene counterparts in some cases. Strong coupling between electrochemically active materials and nano-carbons is facilitated by the selective and uniform distribution of anchored inorganic materials on the surface of nano-carbons. This approach differs fundamentally from the conventional method of fabricating battery materials and electrocatalysts, where active inorganic materials are usually physically mixed with conducting carbon to achieve electrical conductivity (Miyako et al., 2016; Nandi et al., 2020). The size and surface structures of the hybrid electrode particles should be controlled to optimize catalytic and capacitance effects. Oxidized graphene and CNTs can serve as clean growth substrates for controlling the morphology and size of hybrid electrode particles. They also act as electrically conducting substrates, enhancing charge transport in electrocatalysis. This contrasts with surfactants or other capping agents, which may obstruct the active sites on the catalyst surface.

2.6 Nanoparticle-Metal-Organic Framework (MOF) Hybrids

Metal nanoparticles (NPs) and MOFs have a well-developed and rapidly expanding role in energy conversion and storage. This gives us even more justification to research metal NPs and MOFs, and it is anticipated that the number of metal NP@ MOF hybrids will continue to rise. Metal NPs and MOFs were combined to create a new composite material, capitalizing on their complementary properties (Wang & Dai, 2013). These composites benefit from the framework created by the metal NPs and MOFs, and the intercalation of MOFs between metal NPs prevents sheet restacking, thereby enhancing the stability and performance of the composites in various applications. The advantages and disadvantages shown in Figure 4. They have shown great application trends in energy storage and catalytic applications. Metal NPs@MOFs composite hybrid materials. We aim to summarize and analyze the benefits of hybrid materials consisting of metal NPs incorporated into MOFs for applications in energy storage and electrocatalysis (Xu et al., 2023). Additionally, we will thoroughly examine the current understanding of the relationship between structure and performance, as well as rational design strategies, to promote further advancements in this emerging research field.

Figure 4. Advantages and disadvantages of MOF/NPs material

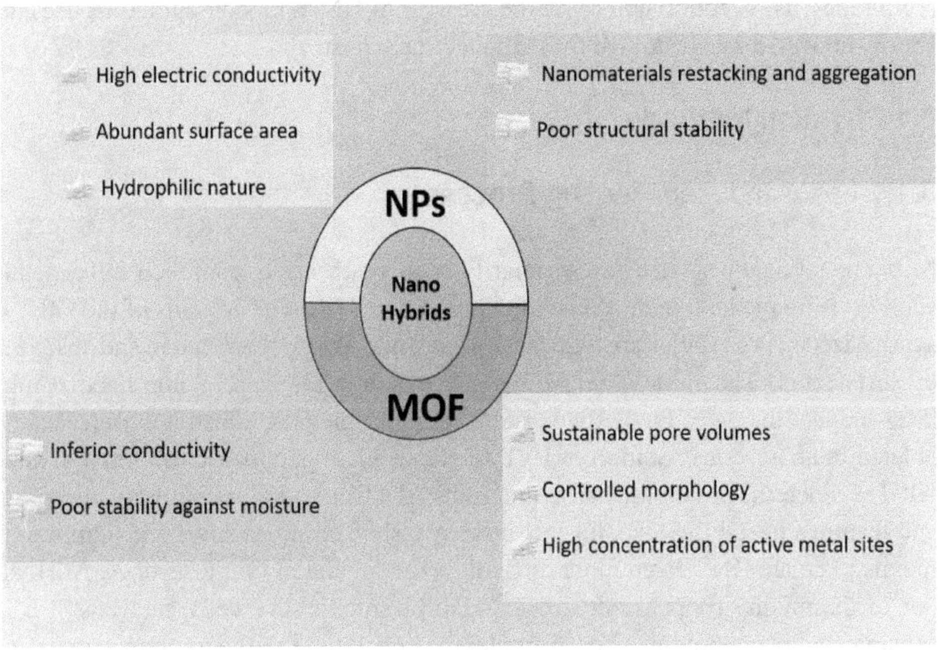

2.7 Nanoparticle-Biomolecule Hybrids

The fabrication of 2D and 3D NPs architectures on surfaces is made possible by the chemical functionalities connected to nanoparticles. Because supramolecular chemistry has been so successful, complex hybrid systems made of NPs functionalized with different molecular and biomolecular components have been assembled (Ramadoss et al., 2023). On surfaces, composite layered or aggregated structures of (macro)molecule-cross-linked NPs have been created, resulting in increased electrochemistry, and enhanced electrochemical energy conversion/storage properties. Nanoscale devices, such as single-electron transistors, molecular switches based on NPs, and metal-insulator-nanoparticle-insulator-metal capacitors, have all been made possible through the development of NP architectures on surfaces.

3. SYNTHESIS OF HYBRID NANOMATERIALS

Physical and chemical methods can both be used to synthesize HNs. In general, physical synthesis approaches provide a simpler method for hybridizing nanomaterials without the need to strictly adhere to complex procedures. Contrarily, chemical synthesis pathways need to undergo chemical processing and are subject to strict regulations. This section will cover the creation of HNs and their numerous uses in energy conversion and storage through electrochemistry.

3.1 Physical Methods

3.1.1 Laser-Induced Heating Process

Noble metal nanoparticles have primarily been synthesized from their bulk metal sources using the laser-induced heating process (LIHP). The colloids of the LIHP-synthesized nanoparticles are highly stable and pure due to their charge and absence of surfactants. The rattle-type Ag@Al_2O_3 nanohybrids were synthesized using laser-induced heating. First, silver and aluminum metallic substrates were laser-ablated in an aqueous solution of PVP to create silver nanoparticles (AgNPs) and AlNPs. Since the structural phase diagrams of silver and aluminum are complex and feature a miscibility gap, it is not an easy task to combine silver and aluminum metals to create HNs. Even more difficult is the regulated synthesis of AgAl HNs due to aluminum's propensity to oxidize. For plasmonic solar cells, catalysis SERS sensing, and nanophotonics, for example, the cohabitation of aluminum and Ag in the same nanostructure is interesting (Katz & Willner, 2004; Tri et al., 2019). Here, passing thermodynamic restrictions using the LIHP technique is used to describe the preparation of Al-Ag HNs. The synthetic technique is broadly applicable and practically enables the synthesis of a wide range of multielement HNs in a single step. One can explore non-equilibrium synthetic methods to incorporate immiscible metals within the same nanostructure. This approach is effective in situations where the growth kinetics of NPs are rapid, often leading to the formation of metastable or intermediate products that are not thermodynamically favoured. To create plasmonic Ag-Al HNs in a single step, we utilized laser irradiation to achieve selective heating.

3.1.2 Mechanical Mixing

The dispersion of two distinct nanomaterials can be facilitated by mechanical mixing, which involves the use of mechanical tools such as an ultrasonic dispersion system and magnetic stirrers. By utilizing these mechanical tools, the produced nanomaterials can prevent stacking or agglomeration.

3.1.3 Self-Assembly Method

The self-assembly reaction is a chemical pathway that has attracted significant attention for the preparation of organic-inorganic HNs. This technique allows for tremendous flexibility in the size, shape, and crystallinity of the final product. It achieves this by controlling the nucleation and growth of HNs by encapsulating the inorganic precursors in a uniform organic gel (Arava, Gowda, Shaijumon, & Ajayan, 2012). However, careful adjustment of variables such as the choice of solvents, temperature, precursor concentration, pH, and agitation are necessary to control the kinetics and growth in this type of synthesis.

3.2 Chemical Methods

3.2.1 Sonochemical Synthesis Method

Acoustic cavitation is the mechanism used in sonochemical synthesis. Free radical polymerization was started using ultrasound; these radicals are created during the development, growth, and collapse of acoustic cavitation. Since the radicals produced during sonication assist in activating the polymerization reaction, an external initiator is not necessary during the ultrasound-assisted polymerization process. The NP's monomer globule was exposed to free radicals, which accelerated the polymerization process. Acid-functionalized nanoparticles were added during the partial polymerization of the NPs (Singh, 2018). The PMMA had functionalized nanoparticles added to it. Due to the physical and chemical effects of ultrasound, such as intense turbulence and the generation of high-velocity micro-jet free radicals, the size of the sonochemical synthesis technique was significantly reduced.

3.2.2 Hydrothermal Method

The solubility of aqueous solution at high temperatures and hot water makes the hydrothermal method one of the most popular ways to synthesize HNs. It requires an autoclave, which is a robust container filled with a solution that can withstand high temperatures and pressures, to operate effectively. Typically, the precursor is placed into the autoclave at a specific temperature for a specific period, and any required post-treatment is then applied. These techniques have the benefit of simplifying the manipulation of the size and shape of nanoparticles (Catauro & Ciprioti, 2021). While thermal decomposition processes may involve multiple phases, hydrothermal processes often require elevated reaction temperatures and pressures. The presence or absence of ligands, stabilizers, or surfactants can affect the hydrothermal synthesis of HNs.

3.2.3 Sol-gel Method

In comparison to the original components, hybrids are new materials that have improved qualities. The particle sizes of each phase in hybrid materials range from 1 to 100 nm, and these materials exhibit well-defined phase distributions. To combine the exceptional and adaptable characteristics of various polymer classes with those of inorganic components for diverse applications, polymer scientists have recently developed innovative and demanding methods for creating hybrid materials. The simplest way to create a hybrid material is through solution blending (Potdara, 2020). This process involves dissolving a polymer in a suitable solvent and then adding the appropriate number of inorganic nanoparticles to create a dispersion. This dispersion can be achieved through mechanical stirring or ultrasonic techniques. This method is relatively straightforward and inexpensive, but it has drawbacks because it encourages the tendency of nanoparticles to aggregate. Due to non-homogeneity and the resulting low physical, chemical, and mechanical properties, nanoparticle agglomeration negatively affects the quality of materials.

4. HYBRID NANOMATERIALS FOR ENERGY STORAGE AND CONVERSION

The chemistry of the materials used as electrodes and electrocatalysts plays a critical role in energy storage/conversion systems, including supercapacitors, batteries, and HER. The ability of these HNs to store electrical energy as chemical energy depends on their chemical bonds. The electrochemical activity, conductivity, and structural stability of materials are some of the variables that influence the storage and conversion efficiency of charges. The system is more effective at storing and converting energy when the electrochemical characteristics of the HNs are improved (Cao et al., 2024; Yu et al., 2023). Therefore, a crucial component of scientific research on HNs is the design and development of customized hybrid nanoparticles to meet specific energy storage needs. HNs have been thoroughly studied for energy storage and conversion applications over the past few decades, as shown in Figure 5. First, the development of nanostructured hybrid materials for use in these hot industries has exploded since the discovery of HNs. HNs are considered excellent materials for electrochemical energy storage/conversion due to their appealing crystal structural characteristics, significant thermal conductivity, high electrical conductivity, superior intrinsic carrier mobility, and impressive mechanical strength (Bahadur et al., 2024; Kothandam et al., 2023). Advanced techniques, such as heteroatom doping, can be used to further enhance the characteristics of HNs. The advantages and disadvantages are listed

Figure 5. Hybrid nanomaterials for energy storage and conversion
Source: Yu et al. (2023)

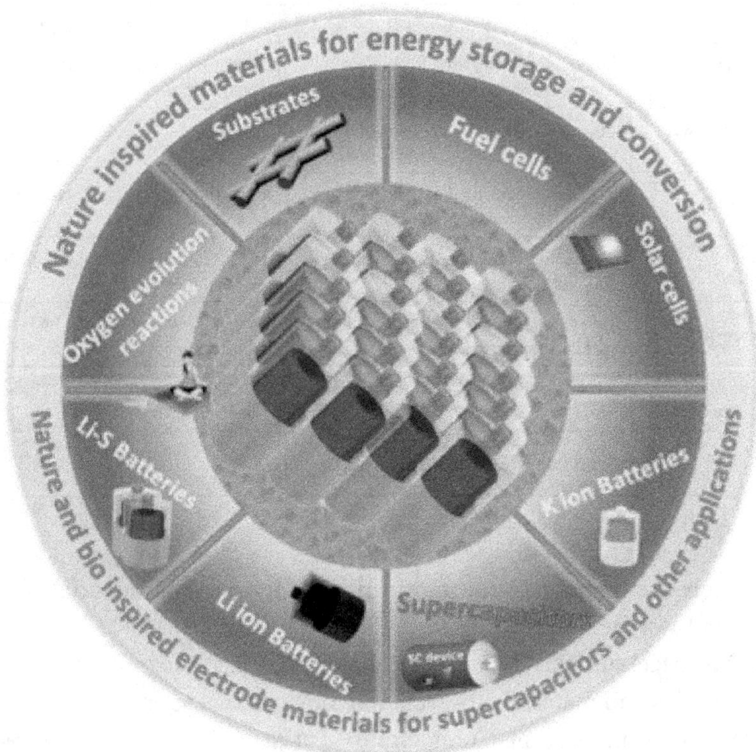

in Table 1. These techniques are particularly effective in increasing active sites by introducing structural flaws.

The energy storage/conversion system requires a significant surface area for diffusion-controlled reactions, and manipulating the surface can enhance these characteristics.

5. REAL-TIME APPLICATIONS OF ENERGY STORAGE AND CONVERSION

Energy harvesting and storage applications for hybrid nanomaterials have the potential to revolutionize various industries and technologies. These materials combine the advantages of nanotechnology with energy storage and harvesting mechanisms to create more efficient and versatile systems (Fayaz et al., 2024; Kulurumotlakatla et

Table 1. HNs advantages and disadvantages for energy storage/conversion applications

HNs Properties	Advantages	Disadvantages
Enhanced Energy Storage Capacity	Increased surface area for higher energy density.	Synthesis and fabrication can be complex and costly.
Improved Charge and Discharge Rates	Enhanced ion and electron transport properties.	Stability issues, especially with cycling.
Longer Cycle Life	Reduced electrode degradation and wear.	Potential for particle agglomeration and loss of properties.
High Thermal Conductivity	Improved heat dissipation for safety and performance.	May require additional engineering to manage thermal issues.
Tunable Properties	Tailored nanomaterial design for specific applications.	Design optimization can be time-consuming.
Compatibility with Various Energy Sources	Suitable for diverse energy conversion processes.	Compatibility challenges with different materials.
Potential Environmental Impact	Reduced material usage and waste.	Environmental concerns regarding nanomaterial toxicity.

al., 2024). Here are some real-time applications of hybrid nanomaterials for energy harvesting and storage:

Wearable Electronics: Hybrid nanomaterials can be used to create energy-harvesting fabrics and materials for wearable devices. These materials can harness energy from body movements, temperature variations, or even solar radiation, offering a consistent power supply for wearables such as smartwatches, fitness trackers, and medical sensors.

Self-Powered Sensors: Energy harvesters based on nanomaterials can provide power to sensors used in remote or inaccessible locations. These sensors can monitor environmental conditions, structural integrity, and more. The harvested energy ensures their continuous operation without the need for battery replacement.

Internet of Things (IoT) devices: Many IoT devices are designed for long-term, remote deployment. Nanomaterial-based energy harvesters can power these devices, reducing the maintenance and cost associated with battery replacement in large-scale IoT networks.

Remote Sensing and Monitoring: In industries such as agriculture, environmental monitoring, and infrastructure, hybrid nanomaterials can be used to power remote sensors and monitoring equipment. They can harness energy from ambient sources such as vibrations, light, or temperature gradients.

Medical Implants: Nanomaterials can be incorporated into medical implants to harness energy from the body's physiological processes, such as blood flow or temperature gradients. This can extend the lifespan of implants such as pacemakers and neurostimulators, reducing the need for invasive surgeries to replace batteries.

Smart Buildings: Nanomaterial-based energy harvesters can be integrated into building structures to capture energy from vibrations, temperature fluctuations, and solar radiation. This energy can be used to power sensors, lighting, or other building systems.

Energy-Efficient Electronics: Hybrid nanomaterials can enhance the energy efficiency of electronic devices, thereby reducing power consumption and prolonging battery life. This is particularly useful in applications such as smartphones, laptops, and electric vehicles.

Wireless Charging: Nanomaterial-based energy storage systems can enable efficient wireless charging for various devices. They can store harvested energy and release it wirelessly to charge electronic devices, making the charging process more convenient.

Environmental Monitoring: Hybrid nanomaterials can be utilized in autonomous environmental monitoring devices to monitor pollution levels, air quality, and other related factors. The stored energy can power data transmission and remote communication.

Aerospace Applications: In the aerospace industry, nanomaterials can be utilized to harness energy from vibrations and temperature gradients in aircraft, thereby reducing the dependence on traditional power sources and enhancing the energy efficiency of aerospace systems.

Transportation: Energy harvesting and storage materials can be applied in electric vehicles to capture energy during braking or from temperature fluctuations. This stored energy can extend the vehicle's range and improve its overall efficiency.

6. CONCLUSION AND FUTURE DIRECTION

In this chapter, we conclude that the composition of HNs in energy harvesting and storage applications has significant potential for addressing the growing demand for sustainable energy solutions. These innovative materials offer unique advantages, such as enhanced energy conversion and storage capabilities, improved efficiency, and scalability. By harnessing the synergistic properties of various nanomaterials, we can pave the way for more efficient and environmentally friendly energy technologies, ultimately contributing to a greener and more sustainable future. As research in this field continues to advance, we can expect to see increasingly practical and impactful applications emerge, revolutionizing the way we generate, store, and utilize energy for the upcoming generation in society.

In this research, we focus on the future of advanced energy harvesting and storage devices for HNs. This field holds great promise in addressing the growing energy and environmental challenges facing the world. Continued research and innovation in this field are expected to lead to more efficient and sustainable energy solutions, which will power a cleaner and greener future for the world.

REFERENCES

Arava, L. M. R., Gowda, S. R., Shaijumon, M. M., & Ajayan, P. M. (2012). Hybrid Nanostructures for Energy Storage Applications. *Advanced Materials, 24*(37), 5045–5064. Advance online publication. doi:10.1002/adma.201104502 PMID:22740354

Bahadur, Singh, Li, Singh, Srivastava, Sakamoto, Chang, Murugavel, & Vinu. (2024). Hybrid nanoarchitectonics of ordered mesoporous C_{60}–BCN with high surface area for supercapacitors and lithium-ion batteries. *Carbon, 216*(5), 118568. doi:10.1016/j.carbon.2023.118568

Cao, J., Yan, C., Chai, Z., Wang, Z., Du, M., Li, G., Wang, H., & Deng, H. (2024). Laser-induced transient conversion of rhodochrosite/polyimide into multifunctional MnO_2/graphene electrodes for energy storage applications. *Journal of Colloid and Interface Science*, *653*, 606–616. doi:10.1016/j.jcis.2023.09.083 PMID:37738933

Catauro, M., & Ciprioti, S. V. (2021). Characterization of Hybrid Materials Prepared by Sol-Gel Method for Biomedical Implementations. A Critical Review. *Materials (Basel), 14*(7), 1788. doi:10.3390/ma14071788 PMID:33916333

Chen, D., Wang, D., Yang, Y., Huang, Q., Zhu, S., & Zheng, Z. (2017). Self-Healing Materials for Next-Generation Energy Harvesting and Storage Devices. *Advanced Energy Materials*, *7*(23), 1700890. doi:10.1002/aenm.201700890

Dragoman, D. (2023). Nanomaterials for Energy Harvesting. *Nanomaterials (Basel, Switzerland)*, *13*(7), 1154. doi:10.3390/nano13071154 PMID:37049248

Fayaz, M., Lai, W., Li, J., Chen, W., Luo, X., Wang, Z., Chen, Y., De Li, S. M. A., & Chen, Y. (2024). Prussian blue analogues and their derived materials for electrochemical energy storage: Promises and Challenges. *Materials Research Bulletin, 112593*. Advance online publication. doi:10.1016/j.materresbull.2023.112593

Ibrahim, I. D., Sadiku, E. R., Jamiru, T., Hamam, Y., Alayli, Y., & Eze, A. A. (2020). Prospects of nanostructured composite materials for energy harvesting and storage. *Journal of King Saud University. Science*, *32*(1), 758–764. doi:10.1016/j.jksus.2019.01.006

Katz, E., & Willner, I. (2004). Integrated Nanoparticle–Biomolecule Hybrid Systems: Synthesis, Properties, and Applications. *Angewandte Chemie International Edition*, *43*(45), 6042–6108. doi:10.1002/anie.200400651 PMID:15538757

Kothandam, G., Singh, G., Guan, X., Lee, J. M., Ramadass, K., Joseph, S., Benzigar, M., Karakoti, A., Yi, J., Kumar, P., & Vinu, A. (2023). Recent Advances in Carbon-Based Electrodes for Energy Storage and Conversion. *Advancement of Science*, *10*(18), 2301045. doi:10.1002/advs.202301045 PMID:37096838

Kulurumotlakatla, D. K., Ramachandran, T., Kumar, Y. A., Mohammed, A. A. A., & Kang, M. C. (2024). Hierarchically fabricated nano flakes-rod-like CoMoO–S supported Ni-foam for high-performance supercapacitor electrode material. *Journal of Physics and Chemistry of Solids*, *185*, 111735. doi:10.1016/j.jpcs.2023.111735

Liu, X., Sheng, G., Zhong, M., & Zhou, X. (2017). Hybrid nanowires and nanoparticles of WO_3 in carbon aerogel for supercapacitor applications. *Nanoscale*, *00*, 1–9. doi:10.1039/C7NR07191D PMID:29350741

Mahmudul Hasan, Hasan, Reza, Islam, & Susan. (2021). Carbon dots as nano-modules for energy conversion and storage. *Materials Today. Communications*, *29*.

Mez-Romero, P. G. O., & Cuentas-Gallegos, K. (2005). Hybrid nanocomposite materials for energy storage and conversion applications. *Journal of Materials Science*, *4*(0), 1423–1428. doi:10.1007/s10853-005-0578-y

Miyako, Pichon, Menard-Moyon, Vacchi, Lefevre, Begin-Colin, & Bianco. (2016). Design, synthesis, characterization and properties of magnetic nanoparticle nanocarbon hybrids. *Carbon, 96*, 49-56. doi:10.1016/j.carbon.2015.09.045

Muhamad Azim, M. K., Arifutzzaman, A., Saidur, R., Khandaker, M. U., & Bradley, D. A. (2022). Recent progress in emerging hybrid nanomaterials towards the energy storage and heat transfer applications: A review. *Journal of Molecular Liquids*, *360*, 119443. doi:10.1016/j.molliq.2022.119443

Nandi, D., Mohan, V. B., Bhowmick, A. K., & Bhattacharyya, D. (2020). Metal/metal oxide decorated graphene synthesis and application as supercapacitor: A review. *Journal of Materials Science*, *55*(15), 6375–6400. doi:10.1007/s10853-020-04475-z

Pasquini, L. (2022). Nanostructured Materials for Energy Storage and Conversion. *Nanomaterials (Basel, Switzerland)*, *12*(9), 1583. doi:10.3390/nano12091583 PMID:35564292

Pescaglini, A., & Iacopino, D. (2015). Metal nanoparticle–semiconductor nanowire hybrid nanostructures for plasmon-enhanced optoelectronics and sensing. *Journal of Materials Chemistry. C, Materials for Optical and Electronic Devices, 3*(45), 11785–11800. doi:10.1039/C5TC01820J

Potdara, S. B. (2020). Sonochemical approach for synthesis of zinc oxide-poly methyl methacrylate hybrid nanoparticles and its application in corrosion inhibition. *Ultrasonics Sonochemistry, 68,* 105200. doi:10.1016/j.ultsonch.2020.105200 PMID:32512431

Ramadoss, J., Mariadass, M. C., Karuppasamy, J., Sonachalam, A., & Govindasamy, M. (2023). Enhancing Electrochemical Influence of Nanocomposite $Ti_3C_2T_x/MoS_2$ Layered Electrode for Hybrid Supercapacitor by Synergetic Effect, ACS. *The Journal of Physical Chemistry. C, Nanomaterials and Interfaces, 127*(34), 16828–16840. doi:10.1021/acs.jpcc.3c04271

Saleh, T. A., Shetti, N. P., Shanbhag, M. M., Raghava Reddy, K., & Aminabhavi, T. M. (2020). Recent trends in functionalized nanoparticles loaded polymeric composites: An energy application. *Materials Science for Energy Technologies, 3,* 515–525. doi:10.1016/j.mset.2020.05.005

Shahinul Islam, M. (2023). Hybrid Nanostructured Materials as Electrodes in Energy Storage Devices. *Inorganics, 11*(5), 183. doi:10.3390/inorganics11050183

Singh, G. P. (2018). Hydrothermal Synthesis of Hybrid Nanoparticles for Future Directions of Renewal Energy Applications, Exploring the Realms of Nature for Nanosynthesis. doi:10.1007/978-3-319-99570-0_14

Siwal, S. S., Zhang, Q., Devi, N., & Thakur, V. K. (2020). Carbon-Based Polymer Nanocomposite for High-Performance Energy Storage Applications. *Polymers, 12*(3), 505. doi:10.3390/polym12030505 PMID:32110927

Sonika, S. K. V., Samanta, S., Srivastava, A. K., Biswas, S., Alsharabi, R. M., & Rajput, S. (2022). Conducting Polymer Nanocomposite for Energy Storage and Energy Harvesting Systems. *Advances in Materials Science and Engineering, 23,* 1–23. Advance online publication. doi:10.1155/2022/2266899

Subrata Ghosh, S. R. (2022). Nanoparticle-enhanced multifunctional nanocarbons-recent advances on electrochemical energy storage applications. *Journal of Physics. D, Applied Physics, 55*(41), 413001. doi:10.1088/1361-6463/ac7bb5

Tri, Ouellet-Plamondon, Rtimi, Assadi, & Nguyen. (2019). *Methods for Synthesis of Hybrid Nanoparticles, Micro and Nano Technologies.* doi:10.1016/B978-0-12-814134-2.00003-6

Vernickaite, E., Bubniene, U., Cesiulis, H., Ramanavicius, A., & Podlaha, E. J. (2016). A Hybrid Approach to Fabricated Nanowire-Nanoparticle Composites of a Co-W Alloy and Au Nanoparticles. *Journal of the Electrochemical Society*, *163*(7), 344–348. doi:10.1149/2.1401607jes

Wang, H., & Dai, H. (2013). strongly coupled inorganic–nano-carbon hybrid materials for energy storage. *Chemical Society Reviews*, *42*(7), 3088–3113. doi:10.1039/c2cs35307e PMID:23361617

Xu, Niu, Li, Yang, Gao, Ding, Ni, Zhu, Liu, Tang, Lv, Peng, Hu, Zhou, & Xu. (2022). Recent progress of quantum dots for energy storage applications. *Carbon Neutrality*, 1-13. doi:10.1007/s43979-022-00002-y

Xu, T., Wang, Y., Xue, Y., Li, J., & Wan, Y. (2023). MXenes@metal-organic framework hybrids for energy storage and electrocatalytic application: Insights into recent advances. *Chemical Engineering Journal*, *470*, 144247. doi:10.1016/j.cej.2023.144247

Yu, H., Zhang, J., Shaheen, I., Ahmad, M., Chen, X., Akkinepally, B., & Hussain, I. (2023). Nature-resembled nanostructures for energy storage/conversion applications. *Journal of Industrial and Engineering Chemistry*. Advance online publication. doi:10.1016/j.jiec.2023.08.041

Yun, S., Zhang, Y., Xu, Q., Liu, J., & Qin, Y. (2019). Recent advance in new-generation integrated devices for energy harvesting and storage. *Nano Energy*, *60*, 600–619. doi:10.1016/j.nanoen.2019.03.074

Chapter 6
Unleashing the Potential:
Exploring Hybrid Nanomaterials for Advanced Energy Harvesting and Storage Systems

Pooja Kapoor
School of Basic and Applied Sciences, Maharaja Agrasen University, Baddi, India

Yogyata Pathania
DAV Post Graduate College, Chandigarh, India

ABSTRACT

The fast-rising market for portable electronic devices and the development of hybrid electric vehicles require the use of clean energy sources at a much higher level than presently in use. Nanomaterials have the potential to revolutionize energy harvesting and storage, providing a sustainable solution for our ever-growing energy needs. These materials combine the unique properties of different nanomaterials, leading to the development of new hybrid nanomaterials, to create synergistic effects, enhancing their overall performance and efficiency. Owing to their tunability and versatility, scientists can tailor the composition, morphology, and structure of these materials to optimize their performance. This technology has led to the development of advanced supercapacitors, fuel cells, solar cells, supercapacitors, rechargeable batteries, etc. with high energy density, fast charging capabilities, and long cycle life. This chapter is focused on specially designed hybrid nanomaterials for energy harvesting and storage applications.

DOI: 10.4018/979-8-3693-1261-2.ch006

INTRODUCTION

Burning fossil fuels is widely recognized as the leading contributor to climate change, resulting in significant impacts on terrestrial and aquatic ecosystems and posing health risks for both humans and the environment (Crutzen et. al., 2008b). A promising method to resolve this challenge is to replace fossil fuel energy resources with alternative options that are environmental friendly, readily available, cost-effective, and environmentally sustainable. Renewable energy sources, such as solar energy harvesting and storage, are becoming more crucial in order to meet energy challenges (Lee et. al., 2018b). Scientists are developing materials that can effectively harness solar energy. Solar cells should be made with materials that have the proper band gap to get the most out of the sun's energy, have good charge transport, are highly stable, have low manufacturing costs, and of course, be eco-friendly.

In the current times, the use of bulk materials is not sufficient to meet the growing energy demand. As a result, various nano-structured materials have been explored. However, their inherent properties (less conductance, poor mechanical strength, and side reactions due to unprotected surfaces) make them unsuitable for meeting the requirements (Liu et. al., 2010; Kang & Ceder, 2009; Fu et. al., 2006). To overcome the shortcomings of single-phase nano-materials, various hybrid nano-structures have been proposed (Wang et. al., 2008; Ran & Lee, 2008; Zhang et. al., 2008; Zhang et. al., 2009; Ran et. al., 2010). Hybrid nanomaterials (HNMs) are promising candidates for energy harvesting and storage applications due to their unique functionalities at the nanoscale. These materials integrate different components, such as nanomaterials, polymers, and organic molecules, to achieve enhanced performance and address the limitations of conventional energy storage systems. They offer several benefits which include high surface area, tunable properties, and synergistic effects that enhance overall performance. The high surface area provides more sites for electrochemical reactions to occur which results in better energy storage efficiency. Tailoring the properties of HNMs by tuning the composition, structure, and morphology of its constituents allows the designing of materials with desirable properties, such as conductivity, stability, and ion transport, required for different energy storage applications. The combination of a conductive material with a high-capacity material can improve both conductivity and energy storage capacity.

Over the past several decades, various energy harvesting technologies have undergone a great deal of research for their practical applications. Piezo (Xu et. al., 2010; Wu et. al., 2014) and triboelectricity (Zhu et. al., 2014; Jing et. al., 2015) are used to capture mechanical energy, pyroelectric (Morozovska et. al., 2010; Yang et. al., 2012), and thermoelectric (Poudel et. al., 2008), (Kim et. al., 2015) effects are used to capture heat, and photovoltaics (You et. al., 2013) are used to capture solar energy. All of these technologies are designed to turn energy from the environment into electricity.

Energy storage devices are important for powering electronic devices/systems. They enable us to use energy to drive them, and so the integration of energy harvesting and storage devices is the ultimate goal in terms of technology development. This is a relatively new area of energy research that focuses on the use of nanomaterials and nano-technologies to capture energy to drive micro-/nanoelectronic systems. It can be used as a substitute for batteries or to extend their life.

This chapter is structured into nine sections. The Introduction section provides a concise overview of energy harvesting and storage, as well as the role of nanomaterials in energy applications and the emergence of HNMs. In next section, the different types of HNMs and their fabrication methods are introduced. The third section focuses on the characterization techniques. The utilization of HNMs for energy harvesting and energy storage is discussed in sections four and five respectively. The sixth section delves into the performance and efficiency of hybrid materials. The emerging trends in HNMs are presented in Section seven. The eighth section addresses the environmental implications associated with the use of hybrid HNMs. Finally, section nine concludes the chapter and provides insights into future prospects.

TYPES AND SYNTHESIS OF HYBRID NANOMATERIALS

The three main types of hybrid materials are classified as illustrated in Figure 1. Based on their utility, there are three kinds of hybrid materials: Structurally hybridized materials, Functionally hybridized materials, and Chemically hybridized materials (Guan & Han, 2019; Srivatsan, 2009).

On the basis of the type of binding that governs the interaction between the constituents, hybrid materials can also be divided into two types: class 1 and class 2 materials (Sánchez et. al., 1999; Adnan et. al., 2018). Class 1 hybrid materials are weakly bonded via Vander Waals interactions, electrostatic interactions, or hydrogen interactions between the components. Class 2 hybrid materials are strongly bonded through covalent and ionic interactions. Stable and strong binding class 2 hybrid materials have been developed recently (Nicole et. al., 2010; Mir et. al., 2018). Class 2 hybrid materials with covalent bonding can be used to create novel materials from functionalized alkoxides, lessen phase gaps, and randomly define interfaces between organic-inorganic components (Nicole et. al., 2010).

Hybrid materials can be further classified into two types based on the composition (addition of organic and inorganic components). Organic-inorganic materials are made up of organic moieties that are used to change inorganic materials, and inorganic-organic materials are made up of inorganic materials and matrices (Nicole et. al.,

Figure 1. Types of hybrid nanomaterials

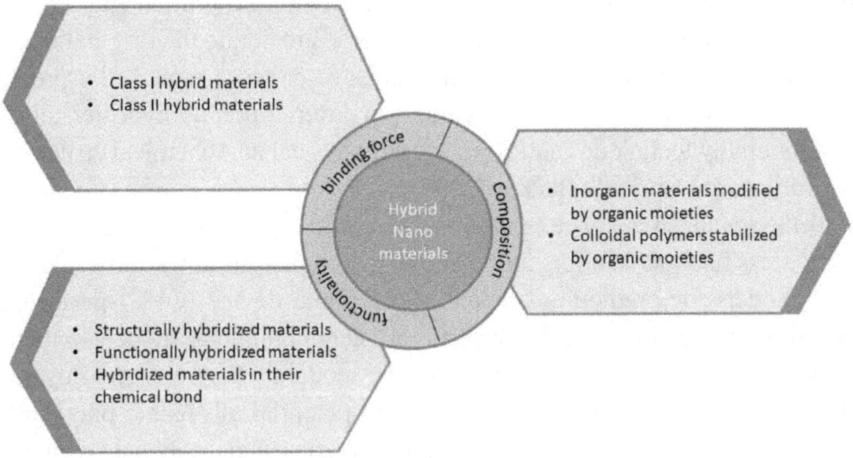

2010). Colloidal polymers are made up of moieties that are stabilized by organic molecules. Inorganic materials are changed by surface charges or by ligatures.

In the past few decades, there has been massive progress in the development of various synthesis methods to fabricate HNMs with different architecture and dimensionalities. Such hybrid materials may range from 0D nanoparticles to 2D nanosheets, including core-shell hybrid nano-structures (for instance, a NiSe@C nanocomposite derived from metal-organic frameworks (MOFs) (Shi et. al., 2022), nanoparticles-based hybrid nano-materials such as a Graphene-silver hybrid nanoparticle (Kalidasan et. al., 2022), well-defined heterostructures (CoZn-Se@ NMX heterostructured hybrid; Ye et. al., 2021), hierarchical heterostructures (such as hierarchical MXene/TMC [SnS/NiS/MoS2]; Jin et. al., 2021) and more. Key challenges for researchers include the development of these high-quality materials, as well as nanostructures, nano- /micro-combination, and hybridization techniques, as well as to control pore shapes, surface modification, and composition optimization.

The *Sol-gel process* is a highly versatile technique used to synthesize homogeneous molecules of organic and inorganic compounds by utilizing particular bridged, polymorphic precursors in combination with a hydrothermal synthesis process. This process is applicable to the synthesis of crystalline micro-porous hybrid solids, including zeolites, MOFs, and other materials (Chae et. al., 2004). Hybrid nanostructures are also obtained by growing one layer of material on another layer with similar crystal structure. This technique is known as *Epitaxial growth.* Core-shell nanoparticles, nanowires, and nanotubes are grown using this technique. By *controlling various synthesis parameters* such as chemical composition, treatment temperature, seeding material, etc., the properties of the hybrid nanomaterial can be tailored to meet specific requirements. *Self-assembling techniques* enable the control and adjustment of various nanomaterials within a unified framework for synthesizing new materials. The resulting materials maintain the original integrity of their nano building blocks. By assembling two or more organically produced nanoparticles, metal oxides, chalcogenides, or layered double hydrides together, the self-assembly technique reinforces overall physical and chemical structures to create truly unique options (Backov, 2006).

With the advancement of computational power, computational science is making great progress in understanding and manipulating materials at the molecular level. Current models for predicting organic photovoltaic device (OPV) performance are based on two main approaches: self-consistent continuum models, which use electronic structure calculations, and atomistic modeling like molecular dynamics or Monte Carlo. The key to unlocking the full potential of organic photovoltaics lies in developing further advances in data informatics, which can help connect the continuum and atomistic models of how materials interact, as well as expand the range of organic materials that can be used to build OPV devices.

CHARACTERIZATION TECHNIQUES FOR HYBRID NANOMATERIALS

To obtain information about material properties, various characterization techniques have been used for structural analysis, surface and interface analysis, and chemical composition analysis as illustrated in Figure 2.

Structural Analysis Techniques

The characterization of nanoparticles involves the utilization of *X-ray diffraction* (XRD), *transmission electron microscopy* (TEM), and *scanning electron microscopy* (SEM) to examine their size, morphology, and surface charge (Hasanin et. al., 2023; Roy, A., 2017). EDS, or *energy-dispersive X-ray spectroscopy*, is commonly used in conjunction with TEM and SEM to analyze the composition and spatial distribution of elements in HNMs (Quadri et. al., 2022). Fourier transform infrared (FTIR) spectroscopy facilitates the identification of functional groups in the sample (Hasanin et. al., 2023). The utilization of the Raman spectroscopy (RS) technique is essential in the examination of the vibrational characteristics of HNMs (Ferri et. al., 2014). TGA (*Thermogravimetric analysis*) (Alzate et. al., 2022) is used for thermal stability analysis and degradation potential analysis of HNMs. NMR (*Nuclear magnetic resonance*) spectroscopy (Agarwal et. al. 2017) is used to look at the atomic structure and chemical environment of HNMs.

Figure 2. Different characterization techniques of hybrid nanomaterials

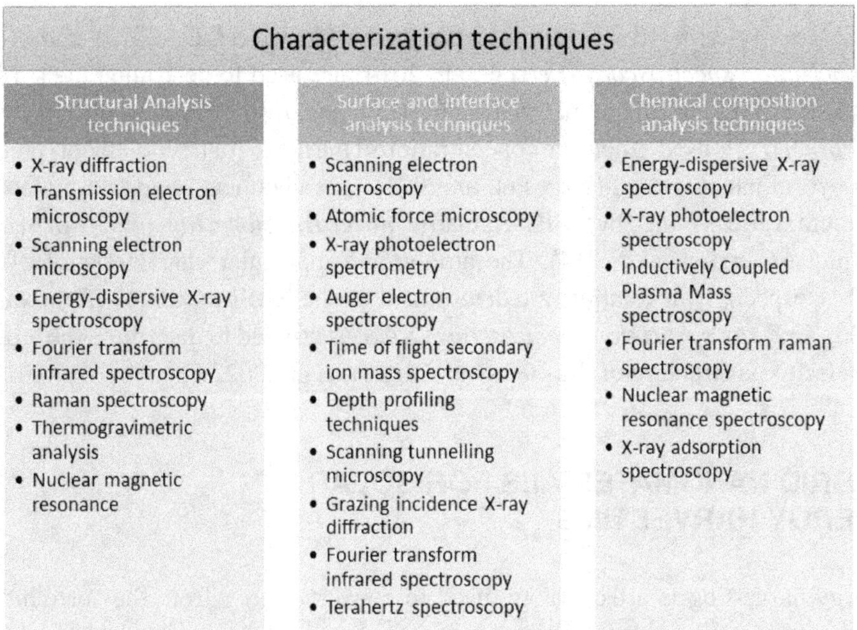

Surface And Interface Analysis Techniques

SEM (Quadri et. al., 2022) and *Atomic Force Microscopy* (AFM) (Wright et. al., 2013) are high-resolution imaging techniques that convey information about the surface topography of HNMs at the nanoscale. XPS, or *X-ray photoelectron spectrometry* and *AES (Auger electron spectroscopy)*, provide elemental composition and chemical state information (Chang et. al., 2022). TOF-SIMS or, *Time-of-Flight Secondary Ion Mass Spectrometry* provides information on the distribution of surface species and their chemical bonding (Priebe & Michler, 2023). The characterization of layer-by-layer composition and chemical profile of HNMs is done through *Depth Profiling Techniques* (Chang et. al., 2022). *Scanning Tunneling Microscopy* (STM) and *Grazing Incidence X-ray Diffraction* (GIXRD) provide insight into the surface structure (Altamura et. al., 2012). FTIR or *Fourier Transform Infrared Spectroscopy* reveals information on surface modifications and interatomic interactions (Hasanin et. al., 2023). *Terahertz spectroscopy* is an important technique that eliminates the need for sample contact and is capable of elucidating the dynamics of charge carriers that are photo-injected on a femto-to-nanosecond time scale (Wang et. al., 2022).

Chemical Composition Analysis Techniques

EDS (*Energy-dispersive X-ray spectroscopy*) (Quadri et. al., 2022) and *X-ray photoelectron spectroscopy* (Ferri et. al., 2014) are used to determine the surface chemistry of a hybrid nanomaterial. ICP -MS (*Inductively Coupled Plasma Mass Spectrometry*) is used to note the concentration of the metal (or trace element) present in a hybrid nanomaterial (Flores et. al, 2021). The chemical bond and molecular structure are determined by the FT-Raman (*Fourier Transform Raman spectroscopy)* technique (Ferri et. al., 2014). The atomic and molecular characteristics of the HNMs, especially in solution, are determined by the NMR technique (Agarwal et. al. 2017). *X-ray absorption spectroscopy* (XAS) is utilized to determine the atomic and electronic structure of the elements (Quadri et. al., 2022).

HYBRID NANOMATERIALS FOR SOLAR ENERGY HARVESTING

Energy harvesting is a technology used to convert energy from the surrounding environment, such as solar radiations, geothermal gradients, or mechanical vibrations, into usable forms like electricity. HNMs can improve energy harvesting systems by enhancing light absorption, charge separation, and transport in photovoltaic devices.

Integrating nanoscale semiconductor materials with plasmonic nanoparticles can improve light absorption and generate localized electric fields. By combining these plasmonic nanoparticles with semiconductor materials, the HNMs can exhibit improved light harvesting efficiency and increased photocurrent generation. Xu et. al. (2020) have provided an in-depth analysis of perovskite hybrid organic-inorganic perovskite solar cells, including its synthesizing technique, materials, morphologies of photoanode, and prospects and complications for these solar cells. Singh et. al. (2019) reported Ag-TiO_2 hybrid nanomaterial where the plasmonic Ag nanoparticles improved the photo-catalytic property of TiO_2 by varying its concentration.

Graphene-based hybrids are formed by integrating graphene, a two-dimensional carbon material with exceptional electrical conductivity, with other materials to improve their energy harvesting capabilities. These hybrid materials can be utilized for applications ranging from solar cells and thermoelectric generators to self-powered sensors and wearable devices. A Fe_2O_3/g-C_3N_4 hybrid nanocomposite for photocatalysis application has shown the enhanced absorption of visible light and more charge carriers (Christoforidis et. al., 2016).

Another strategy for solar energy harvesting involves the incorporation of organic or polymeric materials with nanoparticles. Organic molecules or polymers can provide the advantages of low-cost processing, lightweight, and flexibility, while nanoparticles can enhance charge transport and collection efficiency. HNMs in combination with organic/polymeric components have shown promise for application in organic solar cells and dye-sensitized solar cells. Recently, Shiyani et. al. (2020) have outlined the most recent advances in hybrid solar cell technology, which combine natural photosynthesis proteins with semiconducting thin films, nano-wires, nano-dots, and quantum dots, to create hybrid photovoltaic systems and photovoltaic cells. Hybrid solar cells use a wide range of inorganic materials such as semiconductors and nano-materials as well as organic materials. Natural pigments are also used in these hybrid devices, which results in environmentally friendly and low-cost devices that capture solar energy and produce electricity using these materials. Biofuel solar cells are being created by the hybridization of semiconductors with photosynthetic proteins. The absorption of nearly 100% of light by these biofuel solar cells has been observed to generate electricity. To initiate chemical reactions, including water splitting or photodegradation of organic pollutants, solar energy can be harnessed in a number of ways. To catalyze chemical reactions, photochemical processes (such as artificial photosynthesis) can be utilized to transform sunlight into chemical energy. The formation of an Ag_3PO_4 triangular prism using a straightforward photocatalyst method (Dong et. al., 2013) is an example of a chemical reaction that can be catalyzed by solar energy.

Furthermore, energy can also be harvested from thermal gradients by the utilization of HNMs. Some researchers are interested in thermoelectric materials that have the ability to convert thermal energy into electrical energy to achieve their improved thermoelectric performance. The combination of HNMs, such as conducting polymers, and inorganic nanomaterials like bismuth telluride or lead chalcogenides (Liu et. al., 2021) leads to higher thermoelectric efficiency, which is due to the enhanced electrical conductivity and reduced thermal conductivity. Additionally, semiconductor nanomaterials combined with plasmonic nanoparticles enable efficient light harvesting and subsequent heat generation.

Owing to excellent thermal conductivity and efficient heat transfer, carbon-based nanomaterials can be incorporated into HNMs to enhance thermal absorption and heat dissipation for solar thermal energy conversion. Selvaraj and Haribabu (2020) introduced graphene (G) into aluminum oxide (Al_2O_3) to create G/Al_2O_3 hybrid nano-fluids. These nanofluids have a high surface area, which means they can carry more heat and show a better heat transfer enhancement. Basically, nanofluids are suspended particles in a heat transfer liquid. By dispersing nanoparticles of metals, metal oxides, and carbon nanostructures, into the nanofluid, the thermal properties of the fluid can be significantly enhanced. Tong et. al. (2020) incorporated MWCNT (Multi-walled Carbon NanoTube) into Iron Oxide (Fe_3O_4) nanofluid in order to improve the energy efficiency of the conversion of solar energy into heat energy. MWCNT increased thermal conductivity and also led to an increase in photo-thermal energy convergence efficiency, which was twice that of Single-Walled Carbon Nano Fluids due to its superior thermal and optical properties.

Phase Change Materials (PCMs) store and discharge large quantities of heat energy during phase transformations. By addition of nanoparticles into PCMs, their thermal conductivity can be improved, allowing for more rapid heat transfer and efficient solar thermal energy storage. This leads to improved heat transfer and, consequently, more efficient solar thermal energy conversion. Overall, HNMs can enhance light absorption, facilitate heat transfer, and improve thermal energy storage, ultimately leading to increased conversion efficiencies and lower costs of solar thermal technologies. Composite phase change materials such as fatty acid/ fatty acid-grafted CNT and fatty acid/CNT were experimentally studied for heat storage applications (Al-Ahmed et. al., 2020). Avid et. al. (2019) examined the impact of surface alterations by MWCNT on phase change properties of paraffin / MWCNT nanocomposite.

HNMs offer great potential for solar energy harvesting, but there are several challenges and limitations that need to be addressed before widespread deployment can take place. The synthesis and fabrication of HNMs is costly, which limits their scalability in large-scale industrial applications. Many HNMs are unstable under high-temperature and high-intensity light conditions, which can affect their performance in solar energy harvesting applications. Most HNMs are restricted to a limited bandwidth of absorption. They only absorb a specific range of wavelengths. Achieving broadband absorption with HNMs presents a formidable challenge. Certain HNMs incorporate rare or toxic elements, which, if mishandled during synthesis and manufacturing, could result in adverse environmental effects. Moreover, the susceptibility of HNMs to degradation over time, caused by environmental exposure or elevated temperatures, can significantly affect their functionality and necessitate frequent replacement.

HYBRID NANOMATERIALS FOR ENERGY STORAGE

In addition to energy harvesting, HNMs are crucial for energy storage systems, like batteries and supercapacitors. Carbon-based nanomaterials graphene or carbon nanotubes, integrated with metal oxides or conducting polymers improve the efficiency of electrode structures. Another approach to energy storage involves the combination of various types of nanomaterials, like metal nanoparticles, metal oxides, or conducting polymers, to achieve better efficiency. HNMs exhibit enhanced charge storage capacity and improved electrochemical performance (Islam et. al. 2023).

Energy storage systems are divided into four types: mechanical, thermal, electrochemical, and electromagnetic systems (Figure 3). Pumped hydro energy

Figure 3. Different types of energy storage systems

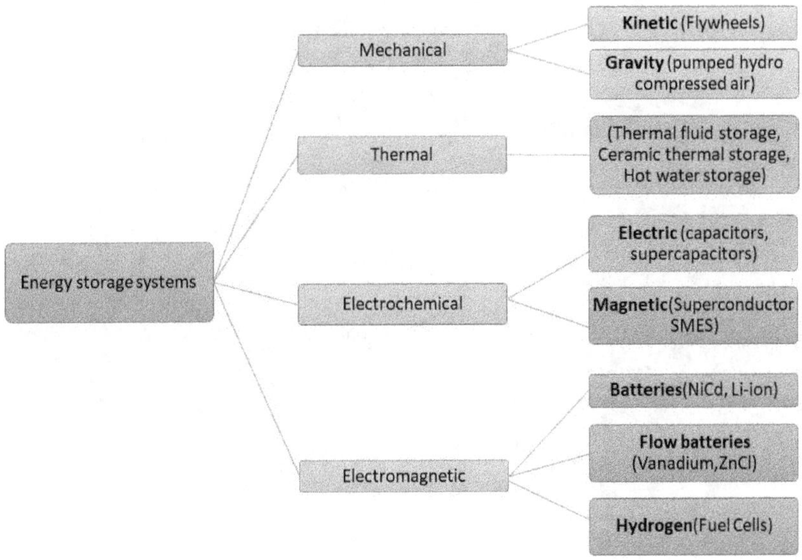

storage systems (Ahearne, 2004), compressed air energy storage systems (Sears, 2005), and flywheels (Chen et. al., 2009) which make use of freely available sources (water and air) come under the mechanical energy storage systems. Electromagnetic energy sources include superconducting magnetic energy storage systems (Xue et. al., 2006) and supercapacitors (Chen, 2006) while chemical energy sources include fuel cells, batteries (Kim et. al., 2019; Yamada, 2009), etc.

Hybrid Nanomaterials for Supercapacitors

Supercapacitors are high-capacity capacitors having high capacitance values. Different types of supercapacitors are categorized in Figure 4. Supercapacitors store energy electrostatically in an electric double layer. This layer is formed by the separation of charges at the interface (Tehrani et. al., 2017; Häggström & Delsing, 2018), which creates an electric field. When the capacitor is charged, the ions in the electrolyte are attracted respectively towards the negative electrode and the positive electrode. They can charge and discharge very quickly, have a longer lifespan, and are inexpensive.

HNMs can be used to develop supercapacitors with higher energy density and power density. HNMs based on carbon nanotubes and conducting polymers have demonstrated enhanced performance in supercapacitors. Sheberla et. al (2017) have developed a MOF-based device that can be used to create a new type of supercapacitor that can charge and discharge in seconds.

Figure 4. Different types of supercapacitors

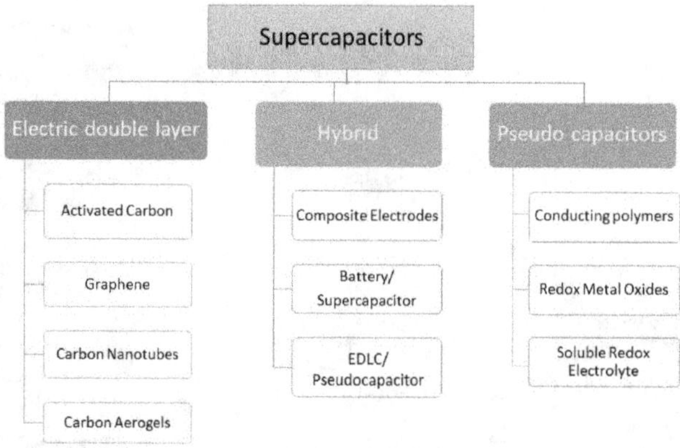

It is becoming very crucial to achieve higher power and current densities than available currently in the case of supercapacitors for their more vast applications. Carbon nanotubes (CNTs) (Portet et. al. 2006) have excellent electronic properties and higher surface area and have gained huge attention as an electrode material for supercapacitors as compared to other materials (Frąckowiak, & Béguin, 2002; An et. al., 2002; Frąckowiak et. al., 2000). However, a major issue for CNT-based supercapacitors is the higher value of resistance present among the electrode and the current collector (An et. al., 2002; Simon & Gogotsi, 2008). So exploration is going on to find better electrode material or to improve the performance of CNT-based electrodes.

Ultra-high power supercapacitor electrodes are fabricated by using multi-segmented CNT/AuNW hybrid structures (Li, Bando, & Golberg, 2004). Here the problem of contact resistance is reduced and hence leads to better performance of the supercapacitor. Here both the CNT acting as electrode and AuNW acting as current collector are present in a single nanowire. A high power density value of 48 kW/kg is observed here. Pure CNT electrodes have a lower performance while CNT/AuNW hybrids show outstanding capacitance properties (Li, Bando, & Golberg, 2004).

Hybrid Nanomaterials for Batteries

Figure 5 illustrates two types of batteries: Physical and Chemical. Chemical batteries are further divided into primary and secondary batteries.

Lithium-ion batteries (secondary chemical batteries) are a type of rechargeable batteries which have become increasingly popular. They can be used in a wide variety of devices, including laptops, smartphones, and electric vehicles (Kim et. al., 2019; Yamada, 2009). It is charged and discharged by the flow of lithium ions between the electrodes. The negative electrode consists of carbon or graphite while the other electrode is metallic and the electrolyte taken is of any lithium salt. These batteries possess a higher energy density and a longer lifetime.

HNMs can be used to develop lithium-ion battery electrodes with higher energy density and power density. HNMs based on graphene and metal oxides show encouraging results for lithium-ion batteries (Kaskhedikar, & Maier, 2009; Shaijumon et. al., 2008; Chan et. al., 2007; Whittingham, 2004).

Various oxides of transition metal are largely investigated as electrode material for lithium-ion batteries because of their cost effectiveness, eco-friendliness, and large capacity (Nam et. al., 2006; Dillon et. al., 2008). For Li-ion batteries, hetero transition metal oxide coaxial nanowires having cores and shells of different metal oxides (e.g. tin oxide core and indium oxide shell) appear as favorable electrode materials (Kim et. al., 2007). Hybrid nanowires /nanotubes formed by tailoring the desirable physio-chemical properties of combining substances possess better

Figure 5. Different types of batteries

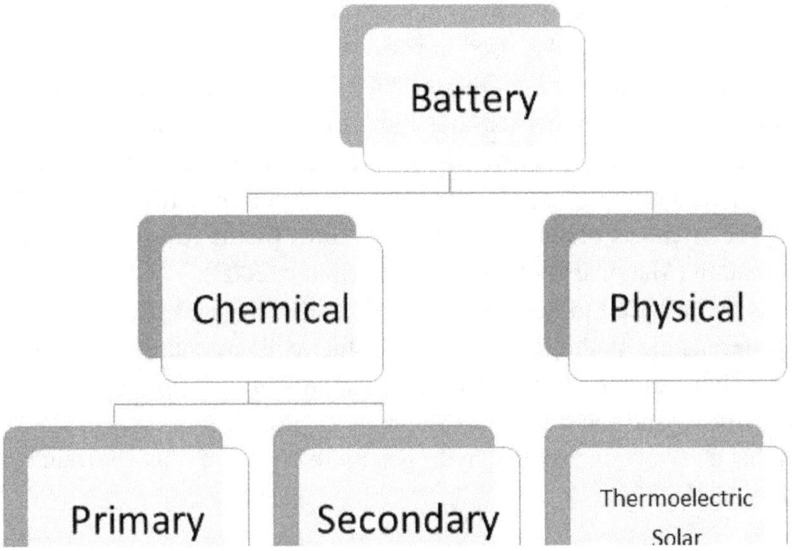

applicabilities. Hybrid nanotubes having desirable properties are outstanding candidates to achieve enhanced performance for Li-ion batteries.

Co-axial CNT /MnO_2 hybrid nanotubes show higher energy and power density than their individual components. Au/SnO_2/CNT nanowire electrodes show a high charge-discharge cycle and more capacity (Chen et. al., 2008). Au improves the speed of lithium ion insertion and extraction reactions and the SnO_2 layer shortens the length of the diffusion path for the flow of ions. Chen et. al. (2008) observed a large capacity value and outstanding cycle stability of SWNT with SnO_2 core-shell structures. Kim et. al., (2008) observed the performance of Au/V2O5/CNT core-shell nanorods as electrodes for Li batteries. Here, enhanced performance is achieved by the smooth transport of electrons, better Li-ion dispersion, and efficient electrolyte movement.

An efficient technique to improve the Li-ion cell efficiency is the fabrication of nanostructured battery components due to enhanced charge transfer and ion diffusion (Taberna et. al., 2006), (Ripenbein et.al., 2010). Nanostructured Li-ion batteries consist of hybrid electrodes and hybrid nanomaterial electrolytes (Rhodes et. al., 2011; Gowda et. al., 2012). There is a rising demand for nanoscale energy storage devices (Golodnitsky et. al., 2006; Nathan et. al., 2005; Kuwata et. al., 2004). Superfast charge/discharge rates and outstanding cyclic stabilities are shown by micro/nano-sized electrodes (Rho et. al., 2002; Xia et. al., 2006).

Hybrid Nanomaterials for Fuel Cell

A fuel cell is a device that converts the chemical energy of a fuel into electricity using reactions (Saikia, Kakati, Boro, & Verma, 2018, pp. 303–337; Khurmi, 2014). Fuel cells are used to supply power and as a backup for industries and residences in remote areas. They are also used in various vehicles used in air, water, and land transport (Singla, Nijhawan, & Oberoi, 2021). These cells contain electrodes and an electrolyte which allows ions, mostly protons, to move across the cell. At the anode, a catalyst is used to allow oxidation of fuel which generates electrons and ions (protons). Ion movement occurs across the electrolyte from the anode to the cathode. Electricity is produced at the same time, by the flow of electrons in the external circuit (Manoharan et. al., 2019; Larminie, 2003).

HNMs can be used to develop fuel cells that are more efficient and durable. These materials are finding applications in fuel cells as catalysts and membrane materials. They can improve the efficiency and durability of fuel cells by enhancing catalytic activity and ion transport (Singh et. al., 2020). Researchers have created a HNM that can store and release hydrogen more efficiently than current materials (Ezendam et. al. 2022).

MOF clusters have been studied the world around and are excellent materials for hydrogen adsorption and hence show better efficiency (Yang et. al., 2011; Yoo et. al., 2011). Poly-oxo-metalate based nanohybrids have excellent catalytic properties for photo-electrochemical reactions. The distinct properties of HNMs can increase the efficiency of fuel cells. Carbon nanostructures like nanofibres, aerogels, nanotubes, graphene, and fullerenes (Liu, & Li, 2020; Darkrim et. al., 2002) and their hybrids in combination with metals have been used in hydrogen fuel cells for increased efficiency (Yu et. al., 2015; Adams, & Chen, 2011). Conducting polymer-based nanohybrids and magnetic nanohybrids also shows improved catalytic properties and increased efficiency for fuel cells.

Technical Challenges in Energy Storage

There is a quest for improved material in order to enhance efficiency. New materials for electrodes and electrolytes are required in order to improve energy densities, charge-discharge rate and durability. These materials will not lead to energy losses and degradation due to side reactions at the electrodes. Fast reaction rates to produce ions and faster ion motion between the electrodes are required. So a quest is going on to find better materials for energy storage systems and is becoming an active research area.

The material used has to be environmentally friendly and easy to dispose of. A material that can be recycled easily without becoming a threat to the environment. Environmentally, Li-ion recycling is expected to reduce the emission of poisonous gasses to the environment. High energy density storage systems can pose safety risks, such as thermal runaway. Managing heat generation and dissipation is crucial to ensure safe and reliable operation. Na-ion batteries use sodium metal which is flammable and operates in high-temperature conditions, which poses an issue of safety/fire risk (Kinter-Meyer et. al., 2012; Yang et. al., 2011; Li, 2013). Proper recycling and disposal of materials used in energy storage systems becomes essential.

Efforts are made to reduce the cost of the storage systems. The material used has to be abundant and cheap. Li-ion recycling by hydro-metallurgical or direct recycling routes will reduce the total cost of producing a cell. The techniques involved in the production of energy storage systems and their recycling have to be cost-effective (Yao et. al., 2016; Shaqsi et. al., 2020). The use of precious metals like gold may improve efficiency but may also increase the overall cost.

PERFORMANCE AND EFFICIENCY OF HYBRID NANOMATERIALS

Due to porous structures, high surface area, high electrical conductivity, charge accommodation capacity, and tunable electronic structures, HNMs show the improved performance of energy storage systems.

Energy Density and Specific Energy: Energy density is the amount of energy stored per unit volume, while specific energy is the energy stored per unit mass. Power density is the amount of energy stored per unit volume. Higher energy density and specific energy allow for compact and lightweight storage systems, enabling the use of HNMs in various applications most importantly in electric transport and electronic devices (Sharma et. al., 2020). High surface area and porosity enable them to store more energy and thus increase energy density.

Cycle life and Durability: Cycle life refers to the number of charge-discharge cycles a storage system can undergo while maintaining its performance within acceptable limits. Durability relates to the overall lifespan and operational reliability of the system. High values are desired for cycle life and durability for better energy storage (Townsend & Gouws, 2022). By combining different materials with high energy densities and long cycle lives, HNMs can provide superior performance to traditional batteries.

Cost and Techno-economic viability: The cost-effectiveness of energy storage technologies is crucial for their widespread adoption and integration into the energy infrastructure (Yao et. al., 2016; Shaqsi et. al. 2020). Techno-economic viability involves optimizing the design, manufacturing, and operation of storage systems to achieve competitive costs and meet performance requirements. The cost-effectiveness and techno-economic viability both can be achieved by utilizing new processes for the production of HNMs.

Safety and Environmental Considerations: Ensuring the safety of energy storage systems during operation, maintenance, and end-of-life stages is critical for protecting people, property, and the environment (Kinter-Meyer et. al., 2012; Yang et. al., 2011b; Li, 2013). Additionally, minimizing the environmental impact of energy storage technologies is crucial for achieving a sustainable energy transition. The use of hybrid nanomaterial-based solar cells and batteries can reduce the need for fossil fuels, leading to a cleaner, more sustainable energy system.

Grid formation and control: The efficient and dependable use of energy stored in energy storage systems is essential and dependent on the successful operation of the grid infrastructure. Appropriate management and control of energy flows is essential for preserving grid stability and maximizing the advantages of energy storage. HNMs can be integrated with other technologies like sensors and actuators to create smart systems that can respond to changes in the environment. This could lead to new applications in areas like smart grids.

EMERGING TRENDS IN HYBRID NANOMATERIALS FOR ENERGY STORAGE

One significant trend in HNMs is the development of synthesis methods, with more novel and distinct materials being created each year. Templates and non-template methods are being utilized to synthesize HNMs (Li et. al., 2005; Luo, & Zhu, 2006; Maschmann et. al., 2006; Qiu et. al., 2008). Coating the carbon nanotubes with Au/Ag nanoparticles and their encapsulation with polymers have been thoroughly investigated (Qiu et. al., 2008; Liu et. al., 2005; Yu et. al., 2005).

The other trend is the combination of novel materials to produce HNMs that exhibit superior and enhanced performance. Metal oxides, polymers, and other materials are being combined with carbon nanomaterials to create HNMs that possess higher energy density, longer lifespans, and cost-effectiveness (Taberna et. al., 2006; Linko et. al., 2022). These materials could potentially facilitate faster electron transfer and aid in ion transport.

Self-assembled HNMs are formed spontaneously from the interaction of different types of materials at the nanoscale. This is an optimistic approach for the development of new materials with complex structures and functionalities (Wiggers et. al., 2020; Linko et. al., 2022). HNMs are investigated for various biomedical uses. HNMs are used to develop new drug delivery systems, tissue engineering, and imaging agents (Rahman et. al., 2022).

The integration of HNMs into devices is a significant area of focus for researchers, as it holds immense potential for numerous applications.

ENVIRONMENTAL IMPLICATIONS OF USING HYBRID NANOMATERIALS

Ongoing investigations are shedding light on the environmental implications associated with the utilization of HNMs in energy storage, revealing a range of potential concerns. Among these concerns is the possibility of HNMs being released into the environment during the manufacturing, usage, or disposal of energy storage devices. In the event of such release, these nanomaterials could negatively impact ecosystems and human health. Notably, certain HNMs have been found to be toxic to aquatic life. Griffitt et. al. (2008) have studied the effect of toxicity of metallic nanomaterials on aquatic life. They exposed different organisms like zebrafish, daphnids, and algae to metallic nanoparticles like Ag, Cu, Al, Co, etc., and showed that nano-Ag and Cu are toxic for all organisms. The toxicity of Ag for zebrafish and ZnO particles for freshwater algae has been reported by Asharani et.al. (2008) and Franklin et. al., (2008) respectively.

The other concern is the accumulation of HNMs in the body over time. Humans can ingest or inhale the HNMs from the environment, which can lead to adverse health effects, including respiratory problems, inflammation, and cancer. It can lead to gene alterations and cellular changes. Their deposition in the respiratory system causes nanostructured toxicity (Anderson et. al., 1990), (Kreyling et, al, 2006) and affects lung functionality, leading to breathing difficulties and throat irritation (White et. al., 2003). Karlsson et. al. (2008) have reported inflammation and apoptosis in human T cells due to SWCNTs and MWCNTs can cause the activation of genes involved in cell transport, metabolism, and stress response. Xia et. al. (2008) have investigated that the cause of DNA damage to the body cells is the highly toxic CuO nanoparticles.

Steps to Control Environmental Risks

Figure 6 shows the various preventive measures against potential environmental hazards of HNMs, which include the release of toxic material into the environment, which must be accounted for during their synthesis process. To ensure a sustainable production process, abundant and less hazardous materials should be used. Rahman et. al. (2022) have emphasized the potential of the green synthesis method. They have used water instead of toxic solvents for synthesizing carbon-based HNMs. Dutta & Das (2021) have also reviewed the synthesis and applications of green nanotechnology towards a cleaner environment. To ensure safety and minimize the release of HNMs into the environment, devices should be designed with sealed casings.

Understanding the environmental and health impacts of HNMs is essential for mitigating any adverse effects. Various researchers have investigated the toxicity of HNMs to aquatic life. Graffiti et. al. (2008) found that nanoparticles of silver, copper, and nickel cause toxicity in all aquatic organisms. The soluble forms of nanometals are found to be less toxic than nanoparticulate forms. Asharani et. al. (2008) studied that silver nanoparticles induce dose-dependent toxicity and hinder the development of zebrafish embryos. Franklin et. al. (2007) conducted a toxicity experiment using freshwater algae and found that nanoparticulate zinc ores and bulk zinc are equally toxic.

HNMs composed of fine and ultrafine particles adversely affect human health (Anderson et. al., 1990; White et. al., 2003; Kreyling et. al., 2006; Karlsson et. al., 2008; Xia et. al., 2008). The total amount of ultrafine particles in the respiratory tract of individuals with different lung diseases was measured by Anderson et. al.

Figure 6. Preventive measures to combat environmental risks

(1990). The individuals with obstructive lung disease were found to have more ultrafine particles compared with individuals with restrictive lung disease. Karlsson et. al. (2008) found CuO nanoparticles to be highly cytotoxic and DNA damaging.

The development of efficient and effective recycling and disposal methods for devices made up of nanohybrid materials should take place. Sun et. al. (2021) have proposed an environmentally friendly and industrial feasible recycling process for waste Lithium-ion batteries that can recover up to 95% of the materials, including HNMs.

Lastly, proper education about the potential risks and benefits of HNMs should be imparted to the general public to ensure the use of these materials in a responsible and sustainable way. It may be effective through educational programs and public outreach campaigns.

CONCLUSIONS AND FUTURE PROSPECTS

The research in the field of HNMs is continuously rising, due to their fascinating features and the freedom to tune their properties by controlling the mixture composition and morphology. The synergistic effect results in the enhanced physicochemical properties of these materials. The special structure and enhanced physico-chemical properties of HNMs have made them an integral material for energy harvesting, storage, and heat transfer applications. By using HNMs, improved capacitance, energy, and power density, charging and discharging rates, and the life cycle of energy storage systems can be obtained which results in their outstanding performance in energy storage. Considerable research efforts are needed for their commercialization or use in real-life applications. However, the present advancements made in the use of HNMs and their synthesis are minimal. Various new advanced production methods like green synthesis and the combinations of different nanomaterials to form new materials can be explored to achieve increased efficiency in energy conversion and longer battery life. The new HNMs can revolutionize the energy industry and help us to move towards a more sustainable future.

Conflicts of Interest

The authors declare no conflict of interest.

Author's Contribution

Both the authors contributed equally to this chapter and have read and approved the final version.

REFERENCES

Adams, B. D., & Chen, A. (2011). The role of palladium in a hydrogen economy. *Materials Today*, *14*(6), 282–289. doi:10.1016/S1369-7021(11)70143-2

Adnan, M. M., Dalod, A. R. M., Balcı, M. H., Glaum, J., & Einarsrud, M. (2018). In situ synthesis of hybrid Inorganic–polymer nanocomposites. *Polymers*, *10*(10), 1129. doi:10.3390/polym10101129 PMID:30961054

Agarwal, N., Nair, M. S., Mazumder, A., & Poluri, K. M. (2017). Characterization of Nanomaterials Using Nuclear Magnetic Resonance Spectroscopy. *Characterization of Nanomaterials*, 61-102. doi:10.1016/B978-0-08-101973-3.00003-1

Ahearne, J. (2004). Storage of electric energy. In *Report on research and development of energy technologies* (p. 22). International Union of Pure and Applied Physics. https://www.iupap.org/wg/energy/report-a.pdf

Al-Ahmed, A., Sarı, A., Mazumder, M. J., Salhi, B., Hekimoğlu, G., Al-Sulaiman, F. A., & Inamuddin. (2020). Thermal energy storage and thermal conductivity properties of fatty acid/fatty acid-grafted-CNTs and fatty acid/CNTs as novel composite phase change materials. *Scientific Reports*, *10*(1), 15388. Advance online publication. doi:10.1038/s41598-020-71891-1 PMID:32958838

Altamura, D., Sibillano, T., Siliqi, D., Caro, L. D., & Giannini, C. (2012). *Assembled Nanostructured Architectures Studied by Grazing Incidence X-Ray Scattering*. Nanomaterials and Nanotechnology. doi:10.5772/55777

Alzate, M., Gamba, O., Daza, C., Santamaria, A., & Gallego, J. (2022). Iron/multiwalled carbon nanotube (Fe/MWCNT) hybrid materials characterization: Thermogravimetric analysis as a powerful characterization technique. *Journal of Thermal Analysis and Calorimetry*, *147*(22), 12355–12363. doi:10.1007/s10973-022-11446-w

An, K. H., Jeon, K. K., Heo, J. K., Lim, S. C., Bae, D., & Lee, Y. H. (2002). High-Capacitance supercapacitor using a nanocomposite electrode of Single-Walled carbon nanotube and polypyrrole. *Journal of the Electrochemical Society*, *149*(8), A1058. doi:10.1149/1.1491235

Anderson, P., Wilson, J. D., & Hiller, F. C. (1990). Respiratory Tract Deposition of Ultrafine Particles in Subjects with Obstructive or Restrictive Lung Disease. *Chest*, *97*(5), 1115–1120. doi:10.1378/chest.97.5.1115 PMID:2331906

Asharani, P. V., Wu, Y., Gong, Z., & Valiyaveettil, S. (2008). Toxicity of silver nanoparticles in zebrafish models. *Nanotechnology*, *19*(25), 255102. doi:10.1088/0957-4484/19/25/255102 PMID:21828644

Avid, A., Jafari, S. H., Khonakdar, H. A., Ghaffari, M., Krause, B., & Pötschke, P. (2019). Surface modification of MWCNT and its influence on properties of paraffin/MWCNT nanocomposites as phase change material. *Journal of Applied Polymer Science*, *137*(9), 48428. Advance online publication. doi:10.1002/app.48428

Backov, R. (2006). Combining soft matter and soft chemistry: Integrative chemistry towards designing novel and complex multiscale architectures. *Soft Matter*, *2*(6), 452. doi:10.1039/b602579j PMID:32680243

Chae, H. K., Siberio-Pérez, D. Y., Kim, J., Go, Y., Eddaoudi, M., Matzger, A. J., O'Keeffe, M., & Yaghi, O. M. (2004). A route to high surface area, porosity, and inclusion of large molecules in crystals. *Nature*, *427*(6974), 523–527. doi:10.1038/nature02311 PMID:14765190

Chan, C. K., Peng, H., Liu, G., McIlwrath, K., Zhang, X. F., Huggins, R. A., & Cui, Y. (2007). High-performance lithium battery anodes using silicon nanowires. *Nature Nanotechnology*, *3*(1), 31–35. doi:10.1038/nnano.2007.411 PMID:18654447

Chang, H.-Y., Lin, W.-C., Chu, P.-C., Wang, Y.-K., Sogo, M., Iida, S., Peng, C.-J., & Miyayama, T. (2022). X-ray Photoelectron Spectroscopy Equipped with Gas Cluster Ion Beams for Evaluation of the Sputtering Behavior of Various Nanomaterials. *ACS Applied Nano Materials*, *5*(3), 4260–4268. doi:10.1021/acsanm.2c00202

Chen, G., Wang, Z., & Xia, D. (2008). One-Pot Synthesis of Carbon Nanotube@ SnO_2–Au Coaxial Nanocable for Lithium-Ion Batteries with High Rate Capability. *Chemistry of Materials*, *20*(22), 6951–6956. doi:10.1021/cm801853c

Chen, H. (2006). *WO2007096656A1 - A method of storing energy and a cryogenic energy storage system -Google Patents.* https://patents.google.com/patent/WO2007096656A1/en

Chen, H., Cong, T. N., Yang, W., Tan, C., Li, Y., & Ding, Y. (2009). Progress in electrical energy storage system: A critical review. *Progress in Natural Science*, *19*(3), 291-312. doi:10.1016/j.pnsc.2008.07.014

Chen, Y., Zhu, C., Xue, X., Shi, X., & Cao, M. (2008). High capacity and excellent cycling stability of single-walled carbon nanotube/SnO2 core-shell structures as Li-insertion materials. *Applied Physics Letters*, *92*(22), 223101. Advance online publication. doi:10.1063/1.2937839

Christoforidis, K. C., Montini, T., Bontempi, E., Zafeiratos, S., Jaén, J. J. D., & Fornasiero, P. (2016). Synthesis and photocatalytic application of visible-light active β -Fe2O3/g-C3N4 hybrid nanocomposites. *Applied Catalysis B: Environmental*, *187*, 171–180. doi:10.1016/j.apcatb.2016.01.013

Crutzen, P. J., Mosier, A. R., Smith, K. A., & Winiwarter, W. (2008b). N_2O release from agro-biofuel production negates global warming reduction by replacing fossil fuels. *Atmospheric Chemistry and Physics*, *8*(2), 389–395. doi:10.5194/acp-8-389-2008

Darkrim, F., Malbrunot, P., & Tartaglia, G. (2002). Review of hydrogen storage by adsorption in carbon nanotubes. *International Journal of Hydrogen Energy*, *27*(2), 193–202. doi:10.1016/S0360-3199(01)00103-3

Dillon, A. C., Mahan, A. H., Deshpande, R., Parilla, P. A., Jones, K. M., & Lee, S. (2008). Metal oxide nano-particles for improved electrochromic and lithium-ion battery technologies. *Thin Solid Films*, *516*(5), 794–797. doi:10.1016/j.tsf.2007.06.177

Dong, P., Wang, Y., Li, H., Li, H., Ma, X., & Han, L. (2013). Shape-controllable synthesis and morphology-dependent photocatalytic properties of Ag3PO4 crystals. *Journal of Materials Chemistry. A, Materials for Energy and Sustainability*, *1*(15), 4651. doi:10.1039/c3ta00130j

Dutta, D., & Das, B. M. (2021). Scope of green nanotechnology towards amalgamation of green chemistry for cleaner environment: A review on synthesis and applications of green nanoparticles. *Environmental Nanotechnology, Monitoring & Management*, *15*, 100418. doi:10.1016/j.enmm.2020.100418

Ezendam, S., Herran, M., Nan, L., Gruber, C., Kang, Y., Gröbmeyer, F., Lin, R., Gargiulo, J., Sousa-Castillo, A., & Cortés, E. (2022). Hybrid Plasmonic Nanomaterials for Hydrogen Generation and Carbon Dioxide Reduction. *ACS Energy Letters*, *7*(2), 778–815. doi:10.1021/acsenergylett.1c02241 PMID:35178471

Ferri, M., Dash, M., Cometa, S., De Giglio, E., Sabbatini, L., & Chiellini, F. (2014). Preparation and characterization of hybrid nanoparticles based on chitosan and po ly(methacryloylglycylglycine). *Journal of Nanoparticle Research*, *16*(5), 2422. doi:10.1007/s11051-014-2422-2

Flores, K., Turley, R. S., Valdes, C., Ye, Y., Cantu, J., Hernandez-Viezcas, J. A., Parsons, J. G., & Gardea-Torresdey, J. L. (2021). Environmental applications and recent innovations in single particle inductively coupled plasma mass spectrometry (SP-ICP-MS). *Applied Spectroscopy Reviews*, *56*(1), 1–26. doi:10.1080/05704928.2019.1694937

Frąckowiak, E., & Béguin, F. (2002). Electrochemical storage of energy in carbon nanotubes and nanostructured carbons. *Carbon*, *40*(10), 1775–1787. doi:10.1016/S0008-6223(02)00045-3

Frąckowiak, E., Méténier, K., Bertagna, V., & Béguin, F. (2000). Supercapacitor electrodes from multiwalled carbon nanotubes. *Applied Physics Letters*, *77*(15), 2421–2423. doi:10.1063/1.1290146

Franklin, N. M., Rogers, N. J., Apte, S. C., Batley, G. E., Gadd, G. M., & Casey, P. (2007). Comparative Toxicity of Nanoparticulate ZnO, Bulk ZnO, and $ZnCl_2$ to a Freshwater Microalga (Pseudokirchneriella subcapitata): The Importance of Particle Solubility. *Environmental Science & Technology*, *41*(24), 8484–8490. doi:10.1021/es071445r PMID:18200883

Fu, L., Liu, H., Li, C., Wu, Y., Rahm, E., Holze, R., & Wu, H. (2006). Surface modifications of electrode materials for lithium ion batteries. *Solid State Sciences*, *8*(2), 113–128. doi:10.1016/j.solidstatesciences.2005.10.019

Golodnitsky, D., Nathan, M., Yufit, V., Strauss, E., Freedman, K., Burstein, L., Gladkich, A., & Peled, E. (2006). Progress in three-dimensional (3D) Li-ion microbatteries. *Solid State Ionics, 177*(26–32), 2811–2819. doi:10.1016/j.ssi.2006.02.048

Gowda, S. R., Reddy, A. L. M., Zhan, X., Jafry, H. R., & Ajayan, P. M. (2012). 3D nanoporous nanowire current collectors for thin film microbatteries. *Nano Letters, 12*(3), 1198–1202. doi:10.1021/nl2034464 PMID:22313375

Griffitt, R. J., Luo, J., Gao, J., Bonzongo, J. J., & Barber, D. S. (2008). Effects of particle composition and species on toxicity of metallic nanomaterials in aquatic organisms. *Environmental Toxicology and Chemistry, 27*(9), 1972–1978. doi:10.1897/08-002.1 PMID:18690762

Guan, G., & Han, M. (2019). Functionalized hybridization of 2D nanomaterials. *Advancement of Science, 6*(23), 1901837. Advance online publication. doi:10.1002/advs.201901837 PMID:31832321

Häggström, F., & Delsing, J. (2018). IoT Energy Storage - A Forecast. *Energy Harvesting and Systems, 5*(3–4), 43–51. doi:10.1515/ehs-2018-0010

Hasanin, M. S., Hashem, A. H., Al-Askar, A. A., Haponiuk, J., & Saied, E. (2023). A novel nanocomposite based on mycosynthesized bimetallic zinc-copperoxide nanoparticles, nanocellulose and chitosan: Characterization, antimicrobial and photocatalytic activities. *Electronic Journal of Biotechnology, 65*, 45–55. doi:10.1016/j.ejbt.2023.05.001

Islam, M. S., Mubarak, M., & Lee, H. (2023). Hybrid Nanostructured Materials as Electrodes in Energy Storage Devices. *Inorganics, 11*(5), 183. doi:10.3390/inorganics11050183

Jin, J., Xiao, T., Zhang, Y., Han, Z., Wang, H., Wang, R., Gong, Y., He, B., Liu, X., & Zhou, K. (2021). Hierarchical MXene/transition metal chalcogenide heterostructures for electrochemical energy storage and conversion. *Nanoscale, 13*(47), 19740–19770. doi:10.1039/D1NR05799E PMID:34821248

Jing, Q., Xie, Y., Zhu, G., Han, R. P. S., & Wang, Z. L. (2015). Self-powered thin-film motion vector sensor. *Nature Communications, 6*(1), 8031. doi:10.1038/ncomms9031 PMID:26271603

Kalidasan, B., Pandey, A. K., Saidur, R., Yadav, A., Samykano, M., & Tyagi, V. (2022). Graphene–Silver Hybrid Nanoparticle-based Organic Phase Change Materials for Enhanced Thermal Energy Storage. *Sustainability (Basel), 14*(20), 13240. doi:10.3390/su142013240

Kang, B., & Ceder, G. (2009). Battery materials for ultrafast charging and discharging. *Nature*, *458*(7235), 190–193. doi:10.1038/nature07853 PMID:19279634

Karlsson, H., Cronholm, P., Gustafsson, J., & Möller, L. (2008). Copper Oxide Nanoparticles Are Highly Toxic: A Comparison between Metal Oxide Nanoparticles and Carbon Nanotubes. *Chemical Research in Toxicology*, *21*(9), 1726–1732. doi:10.1021/tx800064j PMID:18710264

Kaskhedikar, N., & Maier, J. (2009). Lithium storage in carbon nanostructures. *Advanced Materials*, *21*(25–26), 2664–2680. doi:10.1002/adma.200901079 PMID:36751065

Khurmi, R. S. (2014). *Material Science*. Chand Publishers.

Kim, D. W., Hwang, I. S., Kwon, S., Kang, H. Y., Park, K. S., Choi, Y. J., Choi, K. J., & Park, J. (2007). Highly Conductive Coaxial SnO_2–In_2O_3 Heterostructured Nanowires for Li Ion Battery Electrodes. *Nano Letters*, *7*(10), 3041–3045. doi:10.1021/nl0715037 PMID:17760477

Kim, S. I., Lee, K. H., Mun, H. A., Kim, H. S., Hwang, S. W., Roh, J. W., Yang, D. J., Shin, W. H., Li, X. S., Lee, Y. H., Snyder, G. J., & Kim, S. W. (2015). Dense dislocation arrays embedded in grain boundaries for high-performance bulk thermoelectrics. *Science*, *348*(6230), 109–114. doi:10.1126/science.aaa4166 PMID:25838382

Kim, T., Song, W., Son, D.-Y., Ono, L. K., & Qi, Y. (2019). Lithium-ion batteries: Outlook on present, future, and hybridized technologies. *Journal of Materials Chemistry. A, Materials for Energy and Sustainability*, *7*(7), 2942–2964. doi:10.1039/C8TA10513H

Kim, Y., Ahn, H., Nam, S. H., Lee, S. H., Shim, H., & Kim, W. (2008). Honeycomb pattern array of vertically standing core-shell nanorods: Its application to Li energy electrodes. *Applied Physics Letters*, *93*(10), 103104. Advance online publication. doi:10.1063/1.2977862

Kinter-Meyer, M., Colella, W., Elizondo, M., Nguyen, T., Viswanathan, V., & Zhang, Y. (2012). *Book National Assessment of Energy Storage for Grid Balancing and Arbitrage: Phase 1, WECC* (M. B. P. Kinter-Meyer, Ed.). Pacific Northwest National Laboratory. doi:10.2172/1131386

Kreyling, W. G., Semmler-Behnke, M., & Möller, W. (2006). Ultrafine Particle–Lung interactions: Does size matter? *Journal of Aerosol Medicine*, *19*(1), 74–83. doi:10.1089/jam.2006.19.74 PMID:16551218

Kuwata, N., Kawamura, J., Toribami, K., Hattori, T., & Sata, N. (2004). Thin-film lithium-ion battery with amorphous solid electrolyte fabricated by pulsed laser deposition. *Electrochemistry Communications*, *6*(4), 417–421. doi:10.1016/j.elecom.2004.02.010

Larminie, J. (2003). *Fuel Cell Systems Explained* (2nd ed.). SAE International. doi:10.1002/9781118878330

Lee, N. A., Gilligan, G. E., & Rochford, J. (2018). Solar energy conversion. In Elsevier eBooks (pp. 881–918). doi:10.1016/B978-0-12-809270-5.00030-3

Li, C., Guo, Y., Li, B., Wang, C., Li, W., & Chen, B. (2005). Template synthesis of SC@C82(I) nanowires and nanotubes at room temperature. *Advanced Materials*, *17*(1), 71–73. doi:10.1002/adma.200400155

Li, J. (2013). Addressing the grand challenges in energy storage. *Advanced Functional Materials*, *23*(8), 924–928. doi:10.1002/adfm.201203058

Li, N., Martin, C. R., & Scrosati, B. (1999). A High-Rate, High-Capacity, nanostructured tin oxide electrode. *Electrochemical and Solid-State Letters*, *3*(7), 316. doi:10.1149/1.1391134

Li, Y., Bando, Y., & Golberg, D. (2004). Indium-Assisted growth of aligned Ultra-Long Silica nanotubes. *Advanced Materials*, *16*(1), 37–40. doi:10.1002/adma.200305732

Linko, V., Zhang, H., Nonappa, N., Kostiainen, M. A., & Ikkala, O. (2022). From precision colloidal hybrid materials to advanced functional assemblies. *Accounts of Chemical Research*, *55*(13), 1785–1795. doi:10.1021/acs.accounts.2c00093 PMID:35647700

Liu, C., Li, F., Ma, L., & Cheng, H. (2010). Advanced materials for energy storage. *Advanced Materials*, *22*(8), E28–E62. doi:10.1002/adma.200903328 PMID:20217798

Liu, H., & Li, Y. (2020). Modified carbon nanotubes for hydrogen storage at moderate pressure and room temperature. *Fullerenes, Nanotubes, and Carbon Nanostructures*, *28*(8), 663–670. doi:10.1080/1536383X.2020.1738396

Liu, J., Wang, T., Uchida, T., & Kumar, S. (2005). Carbon nanotube core-polymer shell nanofibers. *Journal of Applied Polymer Science*, *96*(5), 1992–1995. doi:10.1002/app.21662

Liu, R., Duay, J., Lane, T., & Lee, S. B. (2010). Synthesis and characterization of RuO2/poly(3,4-ethylenedioxythiophene) composite nanotubes for supercapacitors. *Physical Chemistry Chemical Physics*, *12*(17), 4309. doi:10.1039/b918589p PMID:20407700

Liu, Y., Li, P., Jiang, Q., Jiang, F., Liu, J., Liu, G., Liu, C., Du, Y., & Xu, J. (2021). Organic/inorganic hybrid for flexible thermoelectric fibers. *Chemical Engineering Journal*, *405*, 126510. doi:10.1016/j.cej.2020.126510

Luo, J., & Zhu, J. (2006). Arrays of one-dimensional metal/silicon and metal/carbon nanotube heterojunctions. *Nanotechnology*, *17*(11), S262–S270. doi:10.1088/0957-4484/17/11/S06

Manoharan, Y., Hosseini, S. E., Butler, B., Alzhahrani, H., Senior, B. T. F., Ashuri, T., & Krohn, J. (2019). Hydrogen Fuel Cell Vehicles; Current Status and Future Prospect. *Applied Sciences (Basel, Switzerland)*, *9*(11), 2296. doi:10.3390/app9112296

Maschmann, M. R., Franklin, A. D., Amama, P. B., Zakharov, D. N., Stach, E. A., Sands, T. D., & Fisher, T. S. (2006). Vertical single- and double-walled carbon nanotubes grown from modified porous anodic alumina templates. *Nanotechnology*, *17*(15), 3925–3929. doi:10.1088/0957-4484/17/15/052

Mir, S. H., Nagahara, L. A., Thundat, T., Mokarian-Tabari, P., Furukawa, H., & Khosla, A. (2018). Review—Organic-Inorganic Hybrid Functional Materials: An Integrated Platform for Applied Technologies. *Journal of the Electrochemical Society*, *165*(8), B3137–B3156. doi:10.1149/2.0191808jes

Morozovska, A. N., Eliseev, E. A., Svechnikov, G. S., & Kalinin, S. V. (2010). Pyroelectric response of ferroelectric nanowires: Size effect and electric energy harvesting. *Journal of Applied Physics*, *108*(4), 042009. doi:10.1063/1.3474964

Nam, K. T., Kim, D. W., Yoo, P. J., Chiang, C. Y., Meethong, N., Hammond, P. T., Chiang, Y. M., & Belcher, A. M. (2006). Virus-Enabled synthesis and assembly of nanowires for lithium ion battery electrodes. *Science*, *312*(5775), 885–888. doi:10.1126/science.1122716 PMID:16601154

Nathan, M., Golodnitsky, D., Yufit, V., Strauss, E., Ripenbein, T., Shechtman, I., Menkin, S., & Peled, E. (2005). Three-dimensional thin-film Li-ion microbatteries for autonomous MEMS. *Journal of Microelectromechanical Systems*, *14*(5), 879–885. doi:10.1109/JMEMS.2005.851860

Nicole, L., Rozes, L., & Sánchez, C. (2010). Integrative approaches to hybrid multifunctional materials: From multidisciplinary research to applied technologies. *Advanced Materials*, *22*(29), 3208–3214. doi:10.1002/adma.201000231 PMID:20552603

Portet, C., Taberna, P., Simon, P., & Flahaut, E. (2006). Modification of al current Collector/Active Material Interface for power improvement of electrochemical capacitor electrodes. *Journal of the Electrochemical Society*, *153*(4), A649. doi:10.1149/1.2168298

Poudel, B., Hao, Q., Ma, Y., Lan, Y., Minnich, A. J., Yu, B., Yan, X., Wang, D., Muto, A., Vashaee, D., Chen, X., Liu, J., Dresselhaus, M. S., Chen, G., & Ren, Z. (2008). High-Thermoelectric performance of nanostructured bismuth antimony telluride bulk alloys. *Science*, *320*(5876), 634–638. doi:10.1126/science.1156446 PMID:18356488

Priebe, A., & Michler, J. (2023). Review of Recent Advances in Gas-Assisted Focused Ion Beam Time-of-Flight Secondary Ion Mass Spectrometry (FIB-TOF-SIMS). *Materials (Basel)*, *16*(5), 2090. doi:10.3390/ma16052090 PMID:36903205

Qiu, Y., Yu, J., Fang, G., Shi, H., Zhou, X., & Bai, X. (2008). Synthesis of Carbon/Carbon Core/Shell Nanotubes with a High Specific Surface Area. *The Journal of Physical Chemistry. C, Nanomaterials and Interfaces*, *113*(1), 61–68. doi:10.1021/jp806971e

Quadri, T. W., Fayemi, O. E., Olasunkanmi, L. O., & Ebenso, E. E. (2022). Survey of different electrochemical and analytical techniques for corrosion measurements. *Electrochemical and Analytical Techniques for Sustainable Corrosion Monitoring*, 293-323. doi:10.1016/B978-0-443-15783-7.00012-8

Rahman, A., Chowdhury, M. A., & Hossain, N. (2022). Green synthesis of hybrid nanoparticles for biomedical applications: A review. *Applied Surface Science Advances*, *11*, 100296. doi:10.1016/j.apsadv.2022.100296

Ran, L., Duay, J., & Lee, S. B. (2010). Redox Exchange Induced MnO_2 Nanoparticle Enrichment in Poly(3,4-ethylenedioxythiophene) Nanowires for Electrochemical Energy Storage. *ACS Nano*, *4*(7), 4299–4307. doi:10.1021/nn1010182 PMID:20590128

Ran, L., & Lee, S. B. (2008). MnO_2/Poly(3,4-ethylenedioxythiophene) Coaxial Nanowires by One-Step Co electrodeposition for Electrochemical Energy Storage. *Journal of the American Chemical Society*, *130*(10), 2942–2943. doi:10.1021/ja7112382 PMID:18275200

Rho, Y. H., Kanamura, K., Fujisaki, M., Hamagami, J., Suda, S., & Umegaki, T. (2002). Preparation of Li4Ti5O12 and LiCoO2 thin film electrodes from precursors obtained by sol–gel method. *Solid State Ionics*, *151*(1–4), 151–157. doi:10.1016/S0167-2738(02)00594-5

Rhodes, C. P., Long, J. W., Pettigrew, K. A., Stroud, R. M., & Rolison, D. R. (2011). Architectural integration of the components necessary for electrical energy storage on the nanoscale and in 3D. *Nanoscale*, *3*(4), 1731. doi:10.1039/c0nr00731e PMID:21327256

Ripenbein, T., Golodnitsky, D., Nathan, M., & Peled, E. (2010). Novel porous-silicon structures for 3D-interlaced microbatteries. *Electrochimica Acta*, *56*(1), 37–41. doi:10.1016/j.electacta.2009.01.057

Roy, A. (2017). Silver Nanoparticles Synthesis from a Pharmaceutically Important Medicinal Plant Plumbago Zeylanica. *MOJ Bioequivalence & Bioavailability*, *3*(5). Advance online publication. doi:10.15406/mojbb.2017.03.00046

Saikia, K., Kakati, B. K., Boro, B., & Verma, A. (2018). Current advances and applications of fuel cell technologies. In Springer eBooks (pp. 303–337). doi:10.1007/978-981-13-1307-3_13

Sánchez, C., Ribot, F., & Lebeau, B. (1999). Molecular design of hybrid organic-inorganic nanocomposites synthesized via sol-gel chemistry. *Journal of Materials Chemistry*, *9*(1), 35–44. doi:10.1039/a805538f

Sears, J. (2005). TEX: the next generation of energy storage technology. *Telecommunications Energy Conference, INTELEC 2004*, 218–222. 10.1109/INTLEC.2004.1401469

Selvaraj, V., & Haribabu, K. (2020). Synthesis of graphene encased alumina and its application as nanofluid for cooling of heat-generating electronic devices. *Powder Technology*, *363*, 665–675. doi:10.1016/j.powtec.2020.01.050

Shaijumon, M. M., Ou, F. S., Ci, L., & Ajayan, P. M. (2008). Synthesis of hybrid nanowire arrays and their application as high power supercapacitor electrodes. *Chemical Communications*, *20*(20), 2373. doi:10.1039/b800866c PMID:18473074

Shaqsi, A. Z. A., Sopian, K., & Al-Hinai, A. (2020). Review of energy storage services, applications, limitations, and benefits. *Energy Reports*, *6*, 288–306. doi:10.1016/j.egyr.2020.07.028

Sharma, S., Panwar, A. K., & Tripathi, M. (2020). Storage technologies for electric vehicles. [English Edition]. *Journal of Traffic and Transportation Engineering*, *7*(3), 340–361. doi:10.1016/j.jtte.2020.04.004

Sheberla, D., Bachman, J. C., Elias, J. S., Sun, C. J., Shao-Horn, Y., & Dincă, M. (2017). Conductive MOF electrodes for stable supercapacitors with high areal capacitance. *Nature Materials*, *16*(2), 220–224. doi:10.1038/nmat4766 PMID:27723738

Shi, Y., Fang, L., Han-Dong, P., Deng, X., & Sun, Z. (2022). Metal-Organic Framework-Derived NiSe Embedded into a Porous Multi-Heteroatom Self-Doped Carbon Matrix as a Promising Anode for Sodium-Ion Battery. *Nanomaterials (Basel, Switzerland)*, *12*(19), 3345. doi:10.3390/nano12193345 PMID:36234473

Shiyani, T., & Bagchi, T. (2020). Hybrid nanostructures for solar-energy-conversion applications. *Nanomaterials and Energy*, *9*(1), 39–46. doi:10.1680/jnaen.19.00029

Simon, P., & Gogotsi, Y. (2008). Materials for electrochemical capacitors. *Nature Materials*, *7*(11), 845–854. doi:10.1038/nmat2297 PMID:18956000

Singh, J., Tripathi, N., & Mohapatra, S. (2019). Synthesis of Ag–TiO_2 hybrid nanoparticles with enhanced photocatalytic activity by a facile wet chemical method. *Nano-Structures and Nano-Objects*, *18*, 100266. doi:10.1016/j.nanoso.2019.100266

Singh, R., Altaee, A., & Gautam, S. (2020). Nanomaterials in the advancement of hydrogen energy storage. *Heliyon*, *6*(7), e04487. doi:10.1016/j.heliyon.2020.e04487 PMID:32743097

Singla, M. K., Nijhawan, P., & Oberoi, A. S. (2021). Hydrogen fuel and fuel cell technology for cleaner future: A review. *Environmental Science and Pollution Research International*, *28*(13), 15607–15626. doi:10.1007/s11356-020-12231-8 PMID:33538968

Srivatsan, T. S. (2009). A Review of: Hybrid Materials: Synthesis, Characterization, and Application, G. Kickelbick (Ed.). Materials and Manufacturing Processes, 24(10–11), 1231. https://doi.org/ doi:10.1080/10426910902984058

Sun, S., Jin, C., He, W., Li, G., Zhu, H., & Huang, J. (2021). Management status of waste lithium-ion batteries in China and a complete closed-circuit recycling process. *The Science of the Total Environment*, *776*, 145913. doi:10.1016/j.scitotenv.2021.145913 PMID:33639457

Taberna, P., Mitra, S., Poizot, P., Simon, P., & Tarascon, J. (2006). High rate capabilities Fe3O4-based Cu nano-architectured electrodes for lithium-ion battery applications. *Nature Materials*, *5*(7), 567–573. doi:10.1038/nmat1672 PMID:16783360

Tehrani, Z., Thomas, D. J., Korochkina, T., Phillips, C., Lupo, D., Lehtimäki, S., O'Mahony, J., & Gethin, D. T. (2017). Large-area printed supercapacitor technology for low-cost domestic green energy storage. *Energy*, *118*, 1313–1321. doi:10.1016/j.energy.2016.11.019

Tong, Y., Boldoo, T., Mr, J. H., & Cho, H. (2020). Improvement of photo-thermal energy conversion performance of MWCNT/Fe3O4 hybrid nanofluid compared to Fe_3O_4 nanofluid. *Energy, 196*, 117086. doi:10.1016/j.energy.2020.117086

Townsend, A., & Gouws, R. (2022). A Comparative Review of Lead-Acid, Lithium-Ion and Ultra-Capacitor Technologies and Their Degradation Mechanisms. *Energies, 15*(13), 4930. doi:10.3390/en15134930

Wang, B., Qin, X., Meng, K., Zhu, L., & Li, Z. (2022). Classification of Amino Acids Using Hybrid Terahertz Spectrum and an Efficient Channel Attention Convolutional Neural Network. *Nanomaterials (Basel, Switzerland), 12*(12), 2114. doi:10.3390/nano12122114 PMID:35745458

Wang, Y., Wang, Y., Hosono, E., Wang, K., & Zhou, H. (2008). The Design of a $LiFePO_4$/Carbon Nanocomposite With a Core–Shell Structure and Its Synthesis by an In Situ Polymerization Restriction Method. *Angewandte Chemie International Edition, 47*(39), 7461–7465. doi:10.1002/anie.200802539 PMID:18720357

White, J. M. L., Powell, A., Brady, K., & Russell-Jones, R. (2003). Severe generalized argyria secondary to ingestion of colloidal silver protein. *Clinical and Experimental Dermatology, 28*(3), 254–256. doi:10.1046/j.1365-2230.2003.01214.x PMID:12780705

Whittingham, M. S. (2004). Lithium batteries and cathode materials. *Chemical Reviews, 104*(10), 4271–4302. doi:10.1021/cr020731c PMID:15669156

Wiggers, H., Sehlleier, Y. H., Kunze, F., Xiao, L., Schnurre, S. M., & Schulz, C. (2020). Self-assembled nano-silicon/graphite hybrid embedded in a conductive polyaniline matrix for the performance enhancement of industrial applicable lithium-ion battery anodes. *Solid State Ionics, 344*, 115117. doi:10.1016/j.ssi.2019.115117

Wright, C., Powell, L., Johnson, D., & Hilal, N. (2013). MICROSCOPY | Atomic Force Microscopy. Encyclopedia of Food Microbiology (Second Edition), 666-675. doi:10.1016/B978-0-12-384730-0.00217-2

Wu, W., Wang, L., Li, Y., Zhang, F., Lin, L., Niu, S., Chenet, D., Zhang, X., Hao, Y., Heinz, T. F., Hone, J., & Wang, Z. L. (2014). Piezoelectricity of single-atomic-layer MoS_2 for energy conversion and piezotronics. *Nature, 514*(7523), 470–474. doi:10.1038/nature13792 PMID:25317560

Xia, H., Li, L., & Ceder, G. (2006). Li diffusion in LiCoO2 thin films prepared by pulsed laser deposition. *Journal of Power Sources, 159*(2), 1422–1427. doi:10.1016/j.jpowsour.2005.12.012

Xia, T., Kovochich, M., Liong, M., Mädler, L., Gilbert, B., Shi, H., Yeh, J. I., Zink, J. I., & Nel, A. E. (2008). Comparison of the mechanism of toxicity of zinc oxide and cerium oxide nanoparticles based on dissolution and oxidative stress properties. *ACS Nano*, *2*(10), 2121–2134. doi:10.1021/nn800511k PMID:19206459

Xu, S., Qin, Y., Xu, C., Wei, Y., Yang, R., & Wang, Z. L. (2010). Self-powered nanowire devices. *Nature Nanotechnology*, *5*(5), 366–373. doi:10.1038/nnano.2010.46 PMID:20348913

Xu, W., Gao, Y., Ming, W., He, F., Li, J., Zhu, X., Kang, F., Li, J., & Wei, G. (2020). Suppressing Defects-Induced Nonradiative Recombination for Efficient Perovskite Solar Cells through Green Antisolvent Engineering. *Advanced Materials*, *32*(38), 2003965. Advance online publication. doi:10.1002/adma.202003965 PMID:32767422

Xue, X., Cheng, K. W. E., & Sutanto, D. (2006). A study of the status and future of superconducting magnetic energy storage in power systems. *Superconductor Science and Technology*, *19*(6), R31–R39. doi:10.1088/0953-2048/19/6/R01

Yamada, H. (2009). Secondary Batteries – Lithium Rechargeable Systems – Lithium-Ion | Lithium Vanadium Oxide/Niobium Oxide Batteries. In Elsevier eBooks (pp. 368–374). doi:10.1016/B978-044452745-5.00220-3

Yang, J., Grzech, A., Mulder, F. M., & Dingemans, T. J. (2011). Methyl modified MOF-5: A water stable hydrogen storage material. *Chemical Communications*, *47*(18), 5244. doi:10.1039/c1cc11054c PMID:21451855

Yang, Y., Guo, W., Pradel, K. C., Zhu, G., Zhou, Y., Zhang, Y., Hu, Y., Lin, L., & Wang, Z. L. (2012). Pyroelectric nanogenerators for harvesting thermoelectric energy. *Nano Letters*, *12*(6), 2833–2838. doi:10.1021/nl3003039 PMID:22545631

Yang, Z., Zhang, J., Kintner-Meyer, M. C., Lu, X., Choi, D., Lemmon, J. P., & Li, J. (2011b). Electrochemical energy storage for green grid. *Chemical Reviews*, *111*(5), 3577–3613. doi:10.1021/cr100290v PMID:21375330

Yao, L., Yang, B., Cui, H., Zhuang, J., Ye, J., & Xue, J. (2016). Challenges and progresses of energy storage technology and its application in power systems. *Journal of Modern Power Systems and Clean Energy*, *4*(4), 519–528. doi:10.1007/s40565-016-0248-x

Ye, Z., Jiang, Y., Li, L., Wu, F., & Chen, R. (2021). Self-Assembly of 0D–2D Heterostructure Electrocatalyst from MOF and MXene for Boosted Lithium Polysulfide Conversion Reaction. *Advanced Materials*, *33*(33), 2101204. Advance online publication. doi:10.1002/adma.202101204 PMID:34245063

Yoo, Y. K., Varela-Guerrero, V., & Jeong, H. (2011). Isoreticular Metal–Organic Frameworks and Their Membranes with Enhanced Crack Resistance and Moisture Stability by Surfactant-Assisted Drying. *Langmuir*, *27*(6), 2652–2657. doi:10.1021/la104775d PMID:21299194

You, J., Dou, L., Yoshimura, K., Kato, T., Ohya, K., Moriarty, T., Emery, K., Chen, C. C., Gao, J., Li, G., & Yang, Y. (2013). A polymer tandem solar cell with 10.6% power conversion efficiency. *Nature Communications*, *4*(1), 1446. Advance online publication. doi:10.1038/ncomms2411 PMID:23385590

Yu, W., Liu, C., Hou, P., Zhang, L., Shan, X., Li, F., & Cheng, H. (2015). Lithiation of silicon nanoparticles confined in carbon nanotubes. *ACS Nano*, *9*(5), 5063–5071. doi:10.1021/acsnano.5b00157 PMID:25869474

Yu, Y., Ouyang, C., Gao, Y., Si, Z., Chen, W., Wang, Z., & Xue, G. (2005). Synthesis and characterization of carbon nanotube/polypyrrole core–shell nanocomposites via *in situ* inverse microemulsion. *Journal of Polymer Science. Part A, Polymer Chemistry*, *43*(23), 6105–6115. doi:10.1002/pola.21114

Zhang, H., Cao, G., Wang, Z., Yang, Y., Shi, Z., & Gu, Z. (2008). Growth of manganese oxide nanoflowers on Vertically-Aligned carbon nanotube arrays for High-Rate Electrochemical capacitive Energy Storage. *Nano Letters*, *8*(9), 2664–2668. doi:10.1021/nl800925j PMID:18715042

Zhang, H., Chen, F., Zhai, Y., Jiang, K., Li, Q., & Fan, S. (2009). Cross-Stacked Carbon Nanotube Sheets Uniformly Loaded with SnO_2 Nanoparticles: A Novel Binder-Free and High-Capacity Anode Material for Lithium-Ion Batteries. *Advanced Materials*, *21*(22), 2299–2304. doi:10.1002/adma.200802290

Zhu, G., Chen, J., Tie-Jun, Z., Jing, Q., & Wang, Z. L. (2014). Radial-arrayed rotary electrification for high performance triboelectric generator. *Nature Communications*, *5*(1), 3426. doi:10.1038/ncomms4426 PMID:24594501

Chapter 7
Biomedical Applications of Hybrid Nanomaterials

Bancha Yingngam
https://orcid.org/0000-0001-7215-9123
Faculty of Pharmaceutical Sciences, Ubon Ratchathani University, Thailand

Benjabhorn Sethabouppha
Faculty of Pharmaceutical Sciences, Ubon Ratchathani University, Thailand

Krit Yingngam
Faculty of Pharmaceutical Sciences, Ubon Ratchathani University, Thailand

Rojjares Netthong
https://orcid.org/0000-0001-9751-7564
Faculty of Pharmaceutical Sciences, Ubon Ratchathani University, Thailand

Jeerisuda Khumsikiew
Faculty of Pharmaceutical Sciences, Ubon Ratchathani University, Thailand

Sisira Donsamak
Faculty of Pharmaceutical Sciences, Ubon Ratchathani University, Thailand

Abhiruj Navabhatra
https://orcid.org/0000-0003-4129-1302
College of Pharmacy, Rangsit University, Thailand

ABSTRACT

The ever-evolving field of nanotechnology has led to groundbreaking advancements in biomedical science, particularly through the development of hybrid nanomaterials. This chapter explores the synthesis, characterization, and diverse biomedical applications of these materials, which combine the advantageous properties of both organic and inorganic constituents for optimized functionality. The text delves into

DOI: 10.4018/979-8-3693-1261-2.ch007

the use of hybrid nanomaterials in areas such as drug delivery, diagnostics, tissue engineering, and biosensors. A key case study demonstrating the wound-healing applications of silver nanoparticle-loaded nanofibrous scaffolds is included. The chapter also addresses the challenges and ethical considerations associated with the clinical translation of these nanomaterials. Therefore, this chapter serves as a comprehensive guide for readers interested in understanding and harnessing the immense potential of hybrid nanomaterials in biomedical applications.

INTRODUCTION

Nanotechnology, defined as the manipulation of materials at the atomic or molecular scale, has become a cornerstone in multiple scientific disciplines (Deshmukh, Thorat, Shirsat, & Ramanavicius, 2023). The concept of nanotechnology was first introduced by physicist Richard Feynman in his famous 1959 talk, "There's Plenty of Room at the Bottom," although he did not use the term "nanotechnology" itself. This term was later coined by Norio Taniguchi in 1974 to describe precision engineering at the nanometer scale. The origin of the term "nanoparticles" emerged as scientists and engineers began to recognize and work with materials at the nanoscale, where unique properties distinct from bulk materials were observed. The term gained prominence in scientific literature during the late 20th century as research in the field of nanomaterials expanded, particularly with the advent of advanced microscopy techniques that allowed for the direct observation and manipulation of materials at the atomic and molecular scale. Key references for these developments include Feynman's original lecture and Taniguchi's 1974 paper on nanotechnology (Lal et al., 2017). In the biomedical sciences, it has emerged as a transformative platform offering unparalleled opportunities. The most striking feature of nanotechnology in this domain is its capability to manipulate biological interactions at both the cellular and molecular levels (Beemkumar et al., 2023). This precision enables advancements in diagnostics, therapeutic interventions, and even preventive medicine (Alghamdi, Fatima, Hussain, Nisar, & Alghamdi, 2023; Paul et al., 2023). Traditional therapies often encounter challenges such as systemic side effects, suboptimal targeting, and limited bioavailability. Nanotechnology addresses these challenges by facilitating targeted drug delivery systems, enhancing diagnostic sensitivity, and developing biocompatible materials for medical applications (Jeyasubramanian, Sakthivel, Thangagiri, & Raja, 2023).

Hybrid nanomaterials, combining the advantageous properties of both organic and inorganic components, have attracted significant attention in recent years. These nanomaterials exhibit a synergy of properties unattainable by either constituent alone (Karri et al., 2023). For instance, the organic component may enhance biocompatibility, while the inorganic component could contribute structural integrity or unique electronic properties (Sanchis-Gual, Coronado-Puchau, Mallah, & Coronado, 2023). This functional versatility enables a wide range of biomedical applications, not limited to targeted drug delivery, biosensing, tissue engineering, and diagnostic imaging (Sharmila, Gowri, Karthikeyan, & Faiyazuddin, 2023). Hybrid nanomaterials allow for the customization of properties to meet specific biomedical needs, thereby facilitating more effective and safer healthcare solutions (Beemkumar et al., 2023). Moreover, these materials have the potential to overcome limitations inherent to purely organic or inorganic nanomaterials, such as poor stability or restricted functionalization options (Nadaf et al., 2023).

The objectives of this chapter are multifaceted and aim to provide a comprehensive understanding of the role and potential of hybrid nanomaterials in biomedical applications.

(1) To describe the methodologies for synthesizing hybrid nanomaterials, focusing on optimizing the choice of organic and inorganic components for specific biomedical applications.
(2) To elucidate the tools and techniques used for characterizing these nanomaterials, emphasizing the importance of each technique in confirming their properties and functionalities.
(3) To explore the versatile roles that hybrid nanomaterials can play in fields such as drug delivery, diagnostics, tissue engineering, and biosensors.
(4) To examine key examples that demonstrate the efficacy and potential of hybrid nanomaterials in addressing real-world biomedical challenges.
(5) To discuss the technical challenges that must be navigated for the successful translation of these materials from the laboratory to the clinic.
(6) To outline emerging trends and future research avenues that could further expand the utility of hybrid nanomaterials in the biomedical sciences.

The organization of this chapter is as follows: Section 2 offers an in-depth discussion of the methodologies for synthesizing hybrid nanomaterials, focusing on the optimal selection of organic and inorganic components for various biomedical applications. Section 3 delves into the characterization techniques of hybrid nanomaterials, outlining critical methods for confirming their properties and functionalities. Section 4 explores the wide range of biomedical applications of these materials, from targeted drug delivery systems to diagnostic imaging. Section 5 presents a case study that showcases the potential and effectiveness of silver nanoparticle-loaded nanofibrous scaffolds in wound healing applications.

Section 6 highlights the challenges in clinical translation, discussing the technical hurdles that must be overcome to bring these materials from the bench to the bedside. Finally, Section 7 concludes the chapter by summarizing key takeaways and outlining future research opportunities.

OVERVIEW OF HYBRID NANOMATERIALS

The classification of hybrid nanomaterials is both intricate and expansive, reflecting the broad array of biomedical applications for which these materials are well suited. Characterized by the integration of two or more distinct types of materials at the nanoscale, these complex composites often exhibit properties superior to those of their individual components (D. Zhu et al., 2023). The development of various hybrid nanomaterials, including organic–inorganic hybrids, metallic hybrids, polymer hybrids, and carbon-based hybrids, is a collective achievement without a singular pioneer. Organic–inorganic hybrids saw significant advancements through the work of Geoffrey A. Ozin since the 1980s (Lofgreen and Ozin, 2014). The evolution of metallic hybrid nanomaterials was shaped by contributions from many, including key figures such as Robert Langer and Joseph M. Jacobson in the late 20th century. In the realm of polymer hybrid nanomaterials, advancements grew from the broader field of polymer science, with pioneers such as Stephanie Kwolek, who invented Kevlar in 1965, playing a crucial role (Mathur, 2013). The field of carbon-based hybrid nanomaterials gained momentum following the discovery of fullerenes by Richard Smalley, Harold Kroto, and Robert Curl in 1985 and the discovery of carbon nanotubes by Sumio Iijima in 1991 (Iijima, 1994; Curl Jr et al., 1996). These developments highlight the collaborative and evolutionary nature of scientific research in the field of nanomaterials.

This section provides a detailed classification of hybrid nanomaterials, taking into account various parameters, such as the types of constituent materials—whether they are organic, inorganic, or a combination of both. Additionally, this section offers a comprehensive discussion of the multifaceted methodologies used in the synthesis and fabrication of these unique structures. Furthermore, this section expands the dialog to include essential considerations related to the biocompatibility and toxicity of hybrid nanomaterials. Such scrutiny is critical, especially given the intended applications of these materials in biologically sensitive environments. Therefore, thorough *in vitro* and *in vivo* assessments are essential for establishing a robust safety profile. Table 1 presents a comprehensive analysis that categorizes these materials into four main types: organic–inorganic hybrids, metallic hybrid nanomaterials, polymer hybrid nanomaterials, and carbon-based hybrid nanomaterials. Each category is evaluated based on factors such as biocompatibility, cost, mechanical strength, and potential toxicity, among others. This comparative framework offers a well-rounded perspective, enabling researchers and practitioners to make informed choices about which types of hybrid nanomaterials are most suitable for their specific applications in the biomedical field.

Table 1. Comparative analysis of the advantages and disadvantages of hybrid nanomaterials for biomedical applications, categorized by organic–inorganic hybrids, metallic hybrid nanomaterials, polymer hybrid nanomaterials, and carbon-based hybrid nanomaterials

Type of Hybrid Nanomaterials	Advantages	Disadvantages	References
Organic–inorganic hybrid nanomaterials	1) High biocompatibility: Organic components often enhance the biocompatibility of these hybrids, making them suitable for *in vivo* applications. 2) Tunable functionality: The combination of organic and inorganic materials allows for customizable chemical and physical properties, such as drug delivery or imaging capabilities. 3) Enhanced mechanical strength: The inorganic components can contribute to greater mechanical stability compared to purely organic materials. 4) Multifunctionality: The ability to incorporate different types of molecules can result in multifunctional nanomaterials capable of simultaneous imaging, drug delivery, and other tasks. 5) Controlled release: Organic components can be engineered to provide controlled drug release over extended periods, thereby improving therapeutic efficacy.	1) Complex synthesis: Creating organic–inorganic hybrids often involves complex synthesis procedures that can be laborious and costly. 2) Potential instability: The interface between the organic and inorganic components can sometimes be unstable, potentially leading to the premature release of drugs or disintegration of the nanomaterial. 3) Limited scalability: Due to their complexity, scaling up the production of these nanomaterials for commercial applications can be challenging. 4) Potential toxicity: While the organic components are generally biocompatible, the inorganic components can sometimes introduce toxicity. 5) Batch-to-batch variability: The complex synthesis can result in variations from one batch to another, affecting reproducibility and reliability.	X. Qi et al. (2023)
Metallic hybrid nanomaterials	1) Excellent electrical conductivity: Metallic components often provide superior electrical conductivity, which can be crucial for applications like biosensors or electrical stimulation of tissues. 2) Photothermal properties: Some metallic nanomaterials are excellent at converting light to heat, making them ideal candidates for photothermal therapies. 3) High surface area: Metallic hybrids can have high surface areas that allow for the loading of a significant amount of drugs, antibodies, or other active molecules. 4) Magnetic properties: Certain metallic hybrids exhibit magnetic properties that can be useful for applications like magnetic resonance imaging or targeted drug delivery. 5) Chemical stability: Metals are generally less susceptible to chemical degradation, providing long-lasting functionality.	1) Potential toxicity: Metals like silver or cadmium can be toxic at certain concentrations, posing risks for *in vivo* applications. 2) Cost: High-quality metallic materials and their hybrids can be expensive to manufacture and purchase. 3) Limited biodegradability: Many metallic components are not biodegradable, which can pose long-term biocompatibility issues. 4) Heavy: Metallic components can add significant weight to the hybrid material, which may be a concern for some applications. 5) Oxidation risks: Some metals are prone to oxidation, which can alter their properties and potentially reduce their effectiveness in biomedical applications.	Sanchis-Gual et al. (2023)

continued on following page

Table 1. Continued

Type of Hybrid Nanomaterials	Advantages	Disadvantages	References
Polymer hybrid nanomaterials	1) Biodegradability: Many polymers used in these hybrids are biodegradable, making them suitable for *in vivo* applications where material breakdown and clearance are essential. 2) Flexibility: Polymers offer structural flexibility, allowing for various shapes and sizes, which can be advantageous for specific biomedical applications like targeted drug delivery. 3) Tunable drug release: The polymer matrix can be engineered to achieve controlled, sustained release of therapeutic agents, improving treatment efficacy. 4) Good biocompatibility: Polymers like poly(lactic-co-glycolic acid) are the Food and Drug Administration-approved for human use, offering high biocompatibility. 5) Ease of functionalization: Polymer surfaces are often easy to modify, allowing for the attachment of various functional groups or bioactive molecules.	1) Potential use of toxic solvents: Some polymers require toxic solvents for their processing, which could pose challenges for *in vivo* applications. 2) Mechanical weakness: Certain polymers may lack the mechanical strength needed for some biomedical applications, like tissue engineering scaffolds. 3) Temperature sensitivity: Polymers can be sensitive to temperature changes, which could affect their structural integrity and performance. 4) Batch-to-batch variability: Like other types of nanomaterials, polymers may suffer from batch-to-batch variability, affecting their reproducibility in biomedical applications. 5) Cost: While generally cheaper than metallic or inorganic materials, the costs can add up when specialized, functionalized, or high molecular weight polymers are needed.	Solangi, Karri, Mubarak, and Mazari (2023)
Carbon-based hybrid nanomaterials	1) High surface area: Carbon-based materials like graphene and carbon nanotubes offer extremely high surface areas, which can be beneficial for drug loading and sensing applications. 2) Mechanical strength: Carbon-based nanomaterials often possess remarkable mechanical strength, making them suitable for applications like tissue engineering scaffolds or bone implants. 3) Electrical conductivity: Some carbon-based materials exhibit excellent electrical conductivity, which can be beneficial for applications like neural interfaces or biosensors. 4) Chemical inertness: Carbon-based materials are generally chemically inert, reducing the likelihood of unwanted chemical reactions when deployed in biological environments. 5) Versatility: Carbon-based materials can be functionalized with a variety of chemical groups, making them highly versatile for different biomedical applications.	1) Potential toxicity: Some forms of carbon, especially in the nanoscale, have shown potential toxicity to cells and biological systems, raising safety concerns. 2) Challenging functionalization: While versatile, the functionalization of carbon-based materials can be challenging and may require harsh chemical treatments. 3) Environmental impact: The production of certain carbon-based materials can have a negative environmental impact due to the use of toxic chemicals or high-energy processes. 4) Cost: High-quality carbon-based nanomaterials can be expensive to produce, particularly when specific functionalities or high purity are needed. 5) Limited biodegradability: Unlike some organic materials, many carbon-based materials are not biodegradable and may persist in biological systems, raising long-term safety concerns.	Mallakpour, Azadi, and Hussain (2023)

Organic–Inorganic Hybrids

Organic–inorganic hybrids are nanocomposites in which organic polymers or molecules are combined with inorganic elements. These combinations often occur via covalent bonding, ion-dipole interactions, or van der Waals forces (Sanchis-Gual et al., 2023). This synergy gives rise to materials that possess both the robustness of inorganic compounds and the flexibility of organic materials, thereby outperforming traditional materials. The synthesis of organic–inorganic hybrid nanomaterials can be broadly classified into two approaches: "top-down" and "bottom-up" methods. The top-down approach involves breaking down larger structures into their constituent nanocomponents; the bottom-up method involves the assembly of nanoscale materials from atomic or molecular units (Ashfaq et al., 2022; El-Khawaga, Zidan, & El-Mageed, 2023). Common synthesis techniques include sol-gel processes, hydrothermal methods, and layer-by-layer assembly (Lima et al., 2022). The inherent advantages of organic–inorganic hybrids lie in their multifunctionality. For instance, the organic component can confer biocompatibility or flexibility, while the inorganic component may contribute to structural rigidity, thermal stability, or electrical conductivity (Jiang, Gao, Zhang, & Chen, 2023). This versatility makes them ideal candidates for diverse applications, such as drug delivery, bioimaging, and tissue engineering. One notable example of an organic–inorganic hybrid is the combination of poly(ε-caprolactone) and glycine betaine with calcium carbonate, which is widely used for drug delivery. The calcium carbonate core provides structural integrity, while the poly(ε-caprolactone) layer enhances solubility and biocompatibility (Butto-Miranda, Cabrera-Barjas, Ibáñez, & Neira-Carrillo, 2023). Similarly, graphene oxide combined with metallic nanoparticles, such as gold or silver, has shown promise in applications such as biosensing and photothermal therapy (Qi, Sun, Liu, & Zhang, 2023).

Metallic Hybrid Nanomaterials

Metallic hybrid nanomaterials represent another subcategory of hybrid nanomaterials that have garnered significant attention in biomedical research. These nanomaterials typically combine metal nanoparticles with organic polymers or inorganic substrates to achieve enhanced functionalities. Generally, metallic hybrid nanomaterials comprise a metallic core, often made from metals such as gold, silver, or iron, which is surrounded by organic or inorganic shells. These shells can be composed of polymers, lipids, or other metal oxides (Thakur et al., 2017). The interaction between the core and shell is commonly facilitated through covalent bonds, electrostatic interactions, or van der Waals forces (J. Li et al., 2023). A variety of synthetic methods are available for producing metallic hybrid nanomaterials, including chemical reduction, sol-gel processes, and coprecipitation (Fernandes, Nayak, Garg, & Nayak, 2023). More advanced methods, such as laser ablation and green synthesis using biological entities, are also gaining momentum (Rahman, Chowdhury, & Hossain, 2022; Tarasenka et al., 2023). Metallic hybrids leverage the electrical, optical, and catalytic properties of metals while incorporating the versatility of organic or inorganic shells. For example, a gold nanoparticle core may provide excellent conductivity and photothermal properties, while a surrounding polymer shell could offer biocompatibility and functional groups for further modification (Alvarez-Lorenzo, Vivero-Lopez, & Concheiro, 2023). A prominent example of metallic hybrid nanomaterials in biomedicine is the use of gold nanoparticle-antibody conjugates for targeted drug delivery or photothermal therapy in cancer treatment (Fithri et al., 2023). Iron oxide-gold hybrids have also been studied for their potential applications in magnetic resonance imaging and localized hyperthermia treatments (Kharey et al., 2023).

Polymer Hybrid Nanomaterials

Polymer hybrid nanomaterials represent a specific category of hybrid materials that combine the unique characteristics of polymeric substances with other types of nanomaterials, such as metals, ceramics, or carbon-based materials (Lavrova et al., 2023). These hybrids capitalize on the versatility, lightweight nature, and ease of functionalization of polymers while also benefiting from the distinct properties of the secondary materials. The structure of polymer hybrid nanomaterials typically involves a polymeric matrix in which nanoparticles of the secondary material are embedded or attached. The types of bonding that can occur range from covalent and ionic interactions to hydrogen bonding and van der Waals forces. This versatility affords tremendous scope for designing nanocomposites tailored for specific biomedical applications. Various techniques are available for synthesizing polymer hybrid nanomaterials, including *in situ* polymerization, melt blending, and solvent casting. Advanced methodologies, such as electrospinning and layer-by-layer assembly, are also being employed to produce nanocomposites with high precision and uniformity (Niu, Chen, Zhang, & Tan, 2022). Polymer hybrid nanomaterials combine the advantageous properties of their constituent materials. For example, a polymer matrix can offer biocompatibility and mechanical flexibility, while embedded nanoparticles may provide electrical conductivity, magnetic properties, or enhanced mechanical strength (Wang, Li, Li, & Zhang, 2022). The resulting hybrids can thus exhibit multifunctional capabilities that are advantageous in biomedical applications. Some of the most common applications for polymer hybrid nanomaterials in biomedicine include targeted drug delivery systems, scaffolds for tissue engineering, and sensors for diagnostics. For instance, polylactic-co-glycolic acid combined with magnetic nanoparticles can be used for both magnetic resonance imaging and targeted drug delivery (Panda et al., 2023). Similarly, polyethylene glycol combined with silica nanoparticles has shown promise in enhancing bioimaging applications (Lavrova et al., 2023).

Carbon-based Hybrid Nanomaterials

Carbon-based hybrid nanomaterials have garnered increasing attention due to their exceptional electrical, mechanical, and thermal properties. By integrating carbon structures such as graphene, carbon nanotubes, or fullerenes with other materials such as metals, polymers, or ceramics, these hybrids offer synergistic properties that are advantageous for biomedical applications (Y. Zhang et al., 2023). The typical structure of carbon-based hybrid nanomaterials consists of a carbon-based core or matrix into which the secondary material is embedded, layered, or bonded. The interactions between the constituents can be facilitated through covalent bonds, hydrogen bonds, or van der Waals interactions, depending on the specific materials involved and the targeted applications (Huo, Xiu, Meng, & Quan, 2023).

Various synthesis routes are available for crafting these advanced materials. Chemical vapor deposition, solvothermal methods, and functionalization techniques are among the commonly used methods (Palacio-Vergara, Álvarez-Gómez, Gallego, & López, 2023). Additionally, electrospinning and hydrothermal methods have been employed to create carbon-based hybrid nanofibers and nanostructures (Ilyas, Shamsuddin, Kai Xiang, Estellé, & Pendyala, 2023). Carbon-based hybrid nanomaterials often possess multifaceted properties suitable for biomedical applications. For example, the exceptional electrical conductivity of carbon nanotubes can be combined with the antimicrobial properties of silver nanoparticles, yielding hybrids useful for creating infection-resistant medical devices (Umair Wani et al., 2023). These carbon-based hybrids find applications in a myriad of biomedical areas, including drug delivery, sensing, imaging, and tissue engineering. For instance, graphene oxide combined with magnetic nanoparticles has been explored for use in magnetic resonance imaging-guided drug delivery (Mukhortova et al., 2022). Similarly, carbon nanotube-polymer hybrids have been investigated for their potential use in cardiac and neural tissue engineering (Pathak, Punetha, Bhatt, & Punetha, 2023).

CHARACTERIZATION TECHNIQUES

Microscopic Techniques

The utility and functionality of hybrid nanomaterials in biomedical applications are closely linked to their structural and compositional characteristics. Therefore, accurate characterization is essential. Table 2 summarizes the advantages and disadvantages of various commonly used characterization techniques for hybrid nanomaterials. A range of microscopic techniques provides valuable insights into the morphology, composition, and interfacial structures of these materials at the nanoscale. Transmission electron microscopy is particularly effective at delivering high-resolution images that reveal the internal structures of core-shell or composite nanomaterials. This technique can also be employed to study the encapsulation efficiency of drug molecules within nanocarriers (Suh et al., 2023). Scanning electron microscopy (SEM) offers a lower-resolution perspective focused on surface morphology. It is often employed to investigate the interface between nanomaterials and biological tissues—a critical consideration for applications such as tissue engineering scaffolds and implantable devices (Sun et al., 2023). Further techniques include atomic force microscopy, which excels at characterizing surface roughness, mechanical properties, and functional groups on the nanomaterial surface (Gulati & Adachi, 2023). The capabilities of atomic force microscopy are instrumental in understanding factors such as cellular adhesion, which are vital for designing medical implants and tissue engineering scaffolds (F. C. Oliveira et al., 2021). Confocal laser scanning microscopy is employed for three-dimensional imaging and is especially useful when nanomaterials are tagged with fluorescent markers. This aids in tracking drug distribution within biological tissues (Potara et al., 2022). Finally, scanning tunneling microscopy maps the atomic or molecular arrangements on the surface. This provides important data on the stability of molecules adsorbed on hybrid nanomaterial surfaces, which in turn influences release profiles in drug delivery systems (Rodríguez-Galván & Contreras-Torres, 2022).

Spectroscopic Techniques

Accurate characterization of hybrid nanomaterials is essential for understanding their functionalities, properties, and potential applications in the biomedical field. While microscopic techniques provide insights into structural attributes, spectroscopic methods offer complementary information on the composition, chemical bonds, electronic states, and other features of these nanomaterials. Ultraviolet–visible spectroscopy is commonly used for quality control and for analyzing unique optical properties, such as those found in quantum dots (Ganguly & Margel, 2023). Fourier transform infrared spectroscopy (FTIR) excels at identifying types of chemical bonds by measuring the absorption of infrared radiation. This technique is often applied to study biocompatibility and functionalization (Shevtsova et al., 2022). In the study conducted by the author's research group on nanocapsule-nanofibrous scaffolds designed for targeted drug delivery in cancer treatment using doxorubicin, Fourier transform infrared spectroscopy with attenuated total reflectance (FTIR-ATR) proved invaluable for molecular-level characterization of the nanocomposites (Figure 1). This nondestructive, highly sensitive technique is capable of providing critical molecular-level information about the structure, chemical composition, and

Table 2. Comparative analysis of the advantages and disadvantages of various characterization techniques for hybrid nanomaterials

Characterization Techniques	Advantages	Disadvantages	References
Microscopic techniques			
Scanning electron microscopy	High resolution, easy sample preparation	Limited to surface features, costly	Yingngam, Kacha, et al. (2019)
Transmission electron microscopy	Extremely high resolution, can study internal structures	Complex sample preparation, costly	Yingngam, Chiangsom, et al. (2019)
Atomic force microscopy	High resolution, can study mechanical	Limited scan size, complex data interpretation	Yingngam, Chiangsom, et al. (2019)
Spectroscopic techniques			
X-ray diffraction	Phase identification, crystalline structure	Limited to crystalline materials, sample preparation	Yingngam, Chiangsom, et al. (2019)
Fourier transform infrared spectroscopy	Functional group identification, quick, minimal sample preparation	Limited to organic materials, spectral overlap	Navabhatra, Maniratanachote, and Yingngam (2022)
Nuclear magnetic resonance	Detailed molecular information, isotopic differentiation	Expensive, sample quantity, complex data interpretation	Solangi et al. (2023)
Mechanical and thermal techniques			
Tensile testing	Simple, direct measure of strength and elasticity	Destructive, can be time-consuming	Charitos et al. (2021)
Dynamic mechanical analysis	Measures mechanical properties as a function of temperature	Complex setup, limited to specific geometries	Bashir (2021)
Differential scanning calorimetry and thermogravimetric analysis	Measures thermal properties, identifies phase changes	Limited sample size, require baseline correction	Yingngam, Kacha, et al. (2019)

interactions within hybrid nanomaterials. Specifically, in a project where nanocapsule-nanofibrous scaffolds were synthesized to encapsulate doxorubicin, FTIR-ATR spectra effectively identified the functional groups present in the formulation. Peaks corresponding to ester groups—found in both the lipid and the polymer—were clearly visible at approximately 1735 cm^{-1}, indicating their presence. Additionally, amide bonds, markers for potential protein interactions, were detectable at approximately 1650 cm^{-1}. Notably, new peaks emerged in the spectra after drug encapsulation, corresponding to the functional groups of the drug and thus confirming successful loading. The spectra showed no significant changes after several weeks of storage, indicating structural stability.

Nuclear magnetic resonance spectroscopy provides extensive information on the composition and chemical structure and is particularly useful for understanding the encapsulation efficacy of drugs in nanoparticle systems (Revathy, Sajini, Augustine, & Joseph, 2023). X-ray photoelectron spectroscopy focuses on surface chemical analysis and is instrumental in investigating surface coatings intended for improved biocompatibility or targeted drug delivery. Raman spectroscopy offers detailed information on vibrational, rotational, and other low-frequency modes. It is sensitive to the chemical composition and molecular structure of the sample and is employed to study the physicochemical stability and compositional changes of hybrid nanomaterials under physiological conditions. Mass spectrometry is primarily used to determine the elemental composition of nanomaterials, particularly nanocomposites and core-shell nanoparticles (Revathy et al., 2023). Collectively, these spectroscopic techniques provide a comprehensive approach to the characterization of hybrid nanomaterials. Choosing the appropriate spectroscopic method can illuminate various aspects, such as chemical composition, structural integrity, and interactions with biological systems. This multifaceted understanding is critical for guiding better design and optimizing applications in the biomedical sector.

Mechanical and Thermal Techniques

The unique mechanical and thermal properties of hybrid nanomaterials are pivotal for their effective deployment in biomedical applications. While optical and spectroscopic methods are indispensable for examining structural and chemical attributes, specialized techniques are essential for assessing mechanical and thermal

Figure 1. Fourier transform infrared spectroscopy with an attenuated total reflectance (FTIR-ATR) instrument at the Central Scientific Laboratory of Ubon Ratchathani University was used to study the FTIR-ATR spectra of nanocapsule-loaded nanofibrous scaffolds

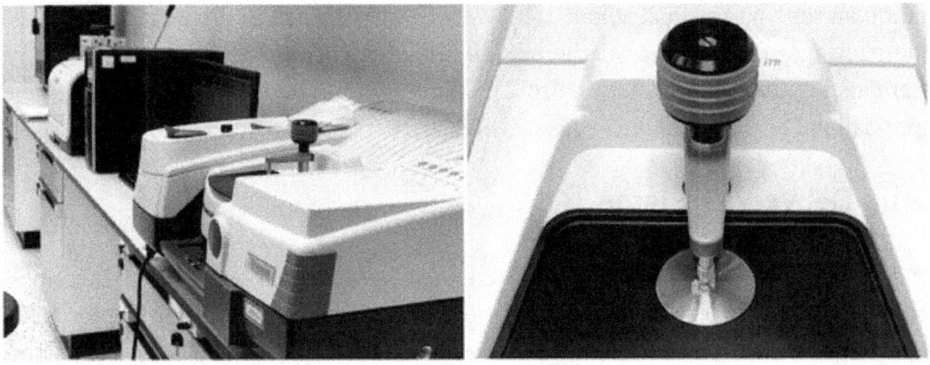

characteristics. Nanoindentation provides high-resolution data on hardness and elastic modulus at the nanoscale, making it crucial for evaluating the mechanical stability of materials intended for load-bearing applications such as orthopedic implants and dental fillings (Nezhad et al., 2021). Tensile testing, on the other hand, yields insights into a material's mechanical strength, elasticity, and yield strength. This method is particularly useful for evaluating materials designed for sutures, ligaments, or tissue scaffolds. Rheological measurements further complement these evaluations by examining the flow and deformation characteristics of materials, such as hydrogels, commonly used in drug delivery and tissue engineering (Las-Casas et al., 2023). Thermal properties are equally vital, given their impact on the stability and efficacy of biomedical materials. Differential scanning calorimetry is a standard technique for analyzing thermal properties such as melting points, glass transition temperatures, and heat capacity. This technique is particularly valuable for assessing the thermal stability of drug-loaded nanocarriers or polymers, ensuring that they maintain their integrity under physiological conditions. Thermogravimetric analysis measures mass changes relative to temperature and provides essential data on thermal stability and composition—critical parameters for materials intended for *in vivo* applications (Las-Casas et al., 2023). Thermal conductivity measurements are used to evaluate how well a material conducts heat, a critical factor for specialized applications such as thermal ablation therapy and photothermal therapy targeting cancer cells.

BIOMEDICAL APPLICATIONS

The application of nanotechnology in medicine, commonly known as nanomedicine, has opened new frontiers in the field of biomedicine. Within this domain, hybrid nanomaterials have emerged as versatile tools with unprecedented capabilities. Their unique physicochemical properties—such as high surface area, tunable morphology, and functional diversity—make them exceptionally well suited for a wide array of biomedical applications. This section explores the transformative impact of hybrid nanomaterials on medical science, focusing on their roles in drug delivery, tissue engineering, regenerative medicine, photothermal and photodynamic therapy, theranostics, and biosensors. Figure 2 displays a diagram detailing the biomedical applications of these nanomaterials, which are explained further below.

Drug Delivery Systems

In drug delivery, hybrid nanomaterials are engineered to achieve synergistic effects in target specificity, drug release, and biocompatibility. This section explores the use of hybrid nanomaterials in drug delivery systems, highlighting key mechanisms

Figure 2. Diagram illustrating the diverse biomedical applications of hybrid nanomaterials

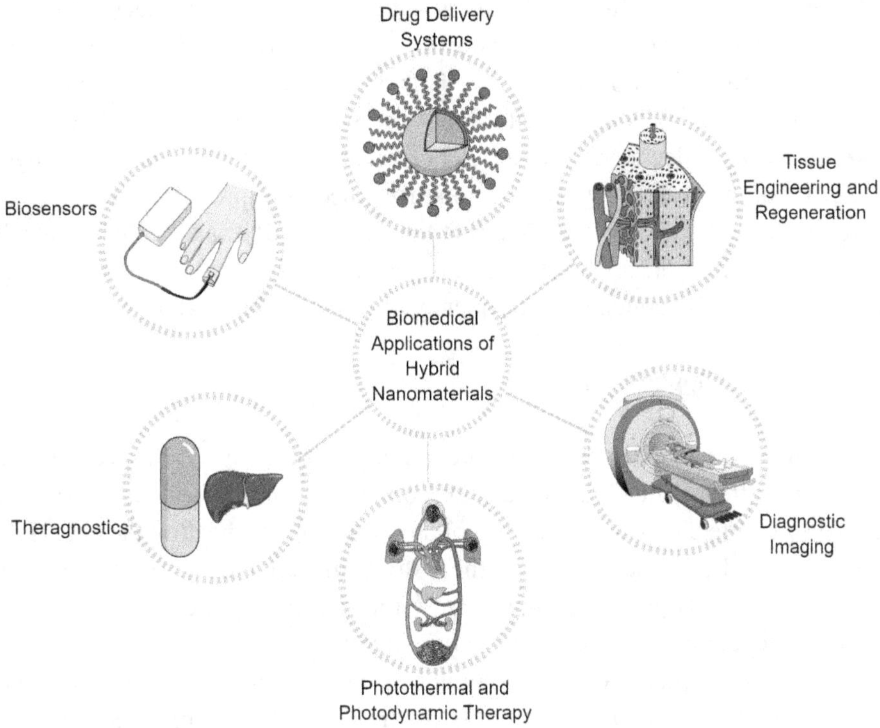

and examples. One primary advantage of employing hybrid nanomaterials for drug delivery is their capacity for targeted action. Unlike traditional drug administration methods, which distribute medication throughout the body and affect both healthy and diseased tissues, hybrid nanomaterials offer enhanced stability for therapeutic agents. Their robust structure and composition ensure stability in biological media, shielding encapsulated drugs from premature release or degradation. This feature is particularly beneficial for extending the shelf life of pharmaceuticals and maintaining the integrity of sensitive drugs, such as biological molecules or proteins, under challenging physiological conditions (Flimelová, Ryabchikov, Behrends, & Bulgakova, 2023). Beyond stability, hybrid nanomaterials enable targeted drug delivery through surface modifications, which increase the therapeutic index and minimize adverse effects on healthy cells. Their engineered design allows for controlled drug release, adding further to their efficacy (Kabay et al., 2023). Biocompatibility is another critical advantage. The inclusion of organic components such as polymers in the hybrid structure often enhances biocompatibility, reducing the risk of triggering an immune response. This feature makes them highly suitable for *in vivo* applications and long-term treatments, particularly in sensitive areas such as autoimmune diseases or organ transplants. Finally, hybrid nanomaterials offer the intriguing benefit of multifunctionality. They can be engineered to include imaging agents, thus creating a theragnostic platform that allows for both therapeutic drug delivery and diagnostic imaging. This dual functionality enables healthcare professionals to monitor treatment effectiveness in real time, adjust dosages, and make data-driven decisions promptly. The emergence of multifunctional hybrid nanomaterials heralds a new era in personalized medicine, providing the prospect of highly individualized, effective, and safe treatments. Additionally, their controlled release mechanisms ensure that therapeutics are dispensed in a slow and sustainable manner, potentially reducing the frequency of drug administration.

1) Anticancer applications

One of the most extensively researched applications of hybrid nanomaterials is in cancer treatment. Traditional chemotherapy drugs often lack specificity, leading to severe side effects due to collateral damage to healthy tissues. Hybrid nanomaterials, however, can deliver these drugs in a targeted manner, specifically to tumor cells. This precision significantly minimizes damage to healthy tissues, potentially reducing side effects and improving the overall quality of life for cancer patients. For example, magnetic nanohybrids can be guided to tumor sites using magnetic fields (Beemkumar et al., 2023). In 2023, Beemkumar and colleagues conducted a study on hybrid nanomaterials made from chitosan, pluronic F127, and magnesium oxide, revealing promising applications in the biomedical industry. These synthesized nanomaterials effectively eliminated both bacterial infections and cancer cells—specifically cervical cancer HeLa cells—without affecting normal fibroblasts. They function by releasing reactive oxygen species (ROS), which play a crucial role in their antibacterial and anticancer properties. Factors such as particle size, surface area, and ion release contribute to increased ROS production. These nanomaterials exhibited enhanced antibacterial activity compared to magnesium oxide alone and more potent cancer cell inhibition than any of their individual components.

Mickymaray et al. (2023) explored the application of nanotechnology in treating breast cancer, specifically employing chitosan-encapsulated NiO-TiO_2-farnesol hybrid nanomaterials. These nanomaterials were effective in killing MDA-MB-231 breast cancer cells and demonstrated high antibacterial effectiveness against drug-resistant bacterial pathogens. Characterization techniques such as electron microscopy and spectroscopy confirmed their potential as both anticancer and antibacterial agents. Kabay et al. (2023) took a different approach, developing nanoparticle-nanofiber hybrids by incorporating photoresponsive melanin nanoparticles into a poly(ε-caprolactone) nanofiber matrix. These hybrid nanomaterials were further optimized as drug delivery systems by loading them with ampicillin. Drug release tests indicated that the developed nanomaterials improved the drug release profile, making it more sustained and controlled. When subjected to photodynamic therapy using ultraviolet-A irradiation, these nanocarriers resulted in significant melanoma cell death. da Silva Costa et al. (2023) synthesized sodium titanate nanotubes modified by incorporating gallium and cerium ions. This modification increased cytotoxicity against melanoma cells. This targeted approach has the potential to not only enhance the efficacy of cancer treatments but also mitigate some of the debilitating side effects commonly associated with systemic chemotherapy.

Finally, in the field of sonodynamic therapy for cancer treatment, low-intensity ultrasound is utilized in conjunction with specialized hybrid nanomaterials known as sonosensitizers. Figure 3 illustrates how these hybrid nanomaterial carriers target cancer cells and release therapeutic agents upon activation by ultrasound waves. This allows for localized treatment and minimizes damage to surrounding healthy tissues. These materials are engineered to react to low-intensity ultrasound, generating free radicals—highly reactive molecules that are capable of destroying cancer cells. The therapy leverages the mechanical and thermal effects of ultrasound, including cavitation. This phenomenon involves the formation and collapse of small bubbles in a liquid, leading to cellular disruption. This action either directly destroys cancer cells or increases membrane permeability, facilitating the delivery of anticancer medications. Hybrid nanomaterials, serving as sonosensitizers, are essential for amplifying these effects. They are specifically designed to interact with low-intensity ultrasound to maximize free radical production, thereby enhancing the therapy's cell-killing efficiency. Sonosensitizers can also be optimized to target specific types of cancer cells, which increases treatment selectivity and reduces harm to adjacent healthy tissues. Additionally, the mechanical forces generated by low-intensity ultrasound can disrupt the membranes of drug carriers, enabling targeted drug release for more effective treatment. Consequently, the synergistic effects of low-intensity ultrasound and sonosensitizers in sonodynamic therapy present a potent, targeted, and minimally invasive approach for treating deeply situated tumors (Y. Zhu et al., 2023).

2) Antimicrobial applications

Antibacterial applications are another promising avenue for hybrid nanomaterials. By incorporating antibacterial metals such as silver or copper into the nanocomposite structure, these materials can perform dual functions: they can both carry and deliver antibiotics while providing their own inherent antibacterial effects. This multipronged strategy is a powerful tool for fighting drug-resistant bacterial strains. Given the growing prevalence of antibiotic-resistant bacteria, there has never been an urgent need

Figure 3. Schematic representation of ultrasound-triggered cancer therapy utilizing hybrid nanomaterials

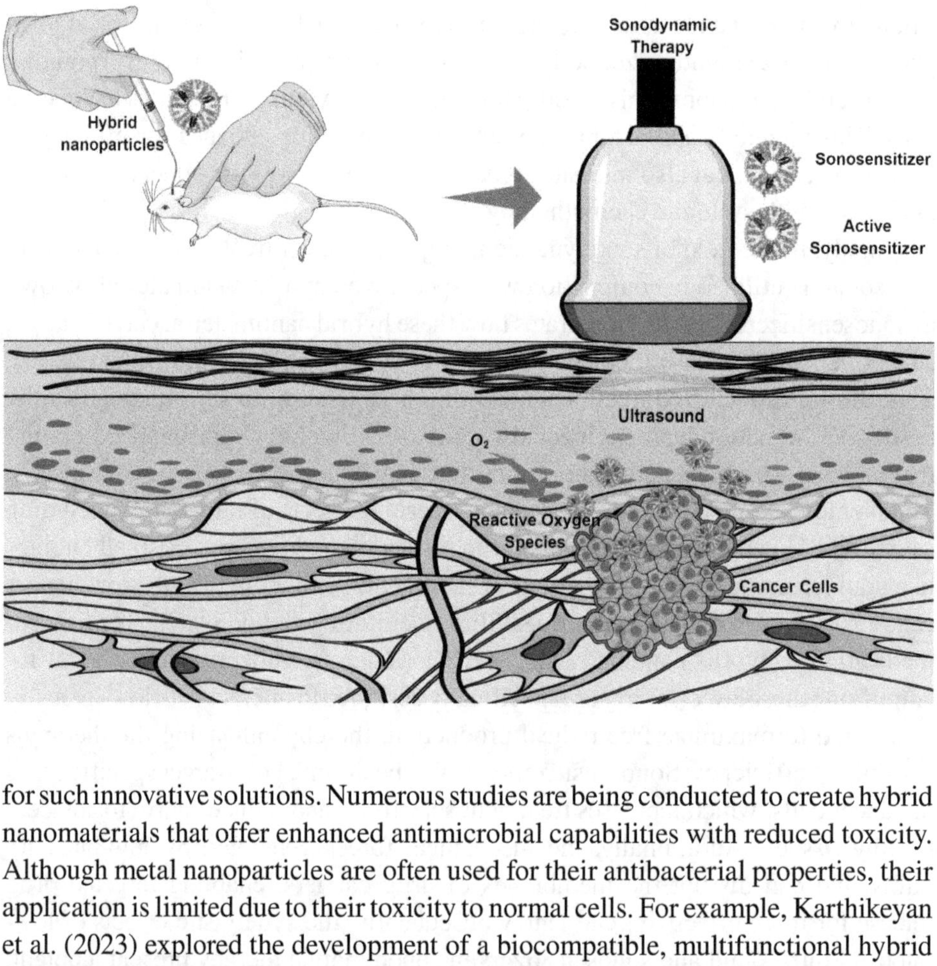

for such innovative solutions. Numerous studies are being conducted to create hybrid nanomaterials that offer enhanced antimicrobial capabilities with reduced toxicity. Although metal nanoparticles are often used for their antibacterial properties, their application is limited due to their toxicity to normal cells. For example, Karthikeyan et al. (2023) explored the development of a biocompatible, multifunctional hybrid nanomaterial aimed at mitigating the common toxicity issues associated with metal nanoparticles. Utilizing a double precipitation method, researchers combined chitosan, curcumin, ZnO, and TiO_2 to formulate these hybrid nanomaterials. The inclusion of the biomolecules chitosan and curcumin was specifically designed to counteract the toxicity of ZnO and TiO_2 while enhancing their antimicrobial and anticancer properties. Tests on human breast cancer cells and fibroblast lines demonstrated that the material has reduced toxicity and effective cytotoxic capabilities. Moreover, the study assessed the antimicrobial effectiveness of the nanomaterial against *Escherichia coli* and *Staphylococcus aureus*, as well as its antioxidant properties. In addition, Thakur et al. (2020) reported the antimicrobial properties of ZnO nanomaterials. Given ZnO's high surface-to-volume ratio, the study focused on creating ultrafine

ZnO nanofibers using a polyvinyl alcohol base through an electrospinning process. The researchers then subjected them to calcination at temperatures of 400 and 500°C. Through Rietveld refinement of X-ray diffraction patterns, the team confirmed that the ZnO nanofibers possessed a hexagonal wurtzite structure, identified by the space group P63mc. Scanning electron microscopy was employed to analyze the fibers, revealing an average diameter ranging between 315 and 292 nm. A key finding of the study was the effectiveness of these electrospun ZnO nanofibers in combating bacteria, specifically *E. coli* and *S. aureus*. The fibers were found to be particularly effective in breaking down the cell walls of these bacteria and adversely affecting their genetic material, thereby demonstrating significant potential for antimicrobial applications. Overall, the results indicate that this novel nanomaterial has significant potential as a biocidal agent in clinical and healthcare settings.

3) Gene therapy

Gene therapy aims to cure or alleviate diseases by modifying gene expression, manipulating gene function, or repairing faulty genes. Hybrid nanomaterials offer a range of beneficial properties that make them especially well suited for gene delivery applications (Yadav et al., 2023). For instance, biocompatible elements such as lipids or biopolymers can be integrated to enhance compatibility with biological systems. In a study by Rennie et al. (2023), researchers focused on treatments for type 1 diabetes mellitus. They specifically addressed the challenges posed by the body's immune response to insulin-secreting cell implants. To address this issue, the team developed a unique core-shell nanostructure for delivering an immunomodulatory agent called MCC950. Composed of upconversion nanoparticles and mesoporous silica, this system can be activated by near-infrared light for localized, controlled delivery of the agent. Tests showed that this nanostructure did not inhibit the growth of insulin-secreting cells. It was also effective in reducing inflammation, preventing fibroblast differentiation, and minimizing monocyte adhesion. These findings suggest that this novel MCC950-loaded nanomedicine system has the potential to extend the functional lifespan of insulin-secreting cell implants, representing a promising avenue for future type 1 diabetes treatments.

Furthermore, metals or magnetic materials can also be incorporated into the composite structure to enable targeted delivery under an external magnetic field. The amalgamation of diverse materials enhances the mechanical stability of the gene delivery vector, ensuring that it remains intact until it reaches its intended destination. Some materials can even facilitate a controlled release mechanism, allowing for the gradual release of the gene therapy agent and thereby providing sustained therapeutic effects (Jia et al., 2022). Various applications underscore the versatility of hybrid nanomaterials in gene therapy. Lipid-polymer hybrid nanoparticles have proven effective in delivering CRISPR/Cas9 systems. The lipid layer enhances biocompatibility and interaction with cell membranes, while the polymer core provides mechanical strength and protects the gene-editing components (Lin et al., 2023). Gold-silica nanoshells have been developed for photothermal gene therapies, where the gold layer absorbs light and generates heat, triggering the release of therapeutic genes encapsulated in the silica layer upon exposure to light (Mandal, Banerjee, & Mandal, 2022).

Finally, magnetic nanocomposites offer another intriguing possibility for targeted gene delivery. These composites generally consist of magnetic nanoparticles, such as magnetite or maghemite, mixed with biocompatible polymers or lipids. The magnetic components guide the composite to specific tissues or cells under an externally applied magnetic field (Mandal et al., 2022). This allows for high-precision targeting and minimizes the risk of off-target effects. The biocompatible component, usually a polymer or lipid, ensures that the gene therapy agent is safely encapsulated and released in a controlled manner at the target site. In summary, magnetic nanocomposites present an innovative approach for the precise delivery of gene therapy agents, opening the door to more effective and safer treatments.

4) Neurological disorder treatments

Neurological disorders such as Alzheimer's disease, Parkinson's disease, epilepsy, and multiple sclerosis present formidable challenges for treatment, largely due to the complexities of the central nervous system (Conklin, Conley, Hou, Chen, & Lee, 2023; Nekoucian, Akhoundian, Wester, & Laurila, 2023). Traditional drug delivery methods often fail to overcome biological barriers such as the blood–brain barrier (BBB)—a highly selective membrane that blocks most substances from entering the brain. As a result, hybrid nanomaterials have emerged as promising platforms for overcoming these hurdles and enhancing treatment efficacy for neurological disorders (Desai & Shinde, 2022; Tan et al., 2023). Figure 4 illustrates the various mechanisms by which engineered hybrid nanoparticles overcome the BBB to deliver therapeutic agents for the treatment of neurodegenerative diseases. These mechanisms include strategies such as passive targeting, active targeting through receptor-mediated transcytosis, controlled drug release, and external guidance methods such as magnetic fields. One approach uses passive targeting, where nanoparticles are designed with specific sizes and surface properties to naturally traverse the BBB. Specifically, nanoparticles smaller than 200 nm and with hydrophilic surfaces are more likely to passively cross this biological barrier (Desai & Shinde, 2022).

Active targeting methods have also been developed, including receptor-mediated and adsorptive-mediated transcytosis. In receptor-mediated transcytosis, nanoparticles are engineered to include ligands that bind specifically to receptors on the BBB, facilitating their uptake and crossing. Various mechanisms are employed for the controlled release of drugs from these hybrid nanoparticles, adding another layer of complexity and effectiveness to their design. These mechanisms include both controlled and responsive release techniques. Controlled release often involves using polymers or other materials that enable a slow, sustained release of therapeutic agents. Conversely, responsive release systems are designed to discharge their drug payload in response to specific triggers such as changes in pH, enzyme action, or external stimuli such as light or heat (Mishra et al., 2023).

Additionally, some nanoparticles can be designed for more complex applications, such as codelivery systems, which carry both a drug and another substance, such as an enzyme that modifies the BBB to facilitate easier drug delivery. These systems may also include synergistic components such as metallic particles for imaging and polymers for drug delivery, serving dual purposes of diagnosis and treatment. Moreover, techniques such as magnetic guidance and sonoporation utilize external magnetic fields and ultrasound, respectively, to temporarily open the BBB for drug delivery (Desai & Shinde, 2022). These innovative methods underscore the multifaceted potential of hybrid nanoparticles in treating neurological disorders.

One critical challenge is effectively delivering therapeutic agents across the BBB (Kour, Panda, Singh, Kumar, & Panda, 2023). Hybrid nanomaterials address this issue by combining lipids, polymers, and sometimes metallic particles to create nanocomposites capable of crossing this barrier. For instance, a study by Elkasabgy, Salama, and Salama (2023) successfully employed polymer/lipid hybrid nanoparticles loaded with quetiapine fumarate to treat schizophrenia in mice. The researchers used poly(vinyl alcohol) and an emulsion/solvent evaporation method to create these nanoparticles, finding an optimal formulation with a small particle size and high entrapment efficiency. This formulation released approximately 95% of its quetiapine fumarate content over 28 days and showed promising results in reducing behavioral symptoms and inflammation in mice. Furthermore, hybrid nanomaterials offer the advantage of controlled drug release, enabling sustained therapeutic effects (Omar, Osman, Mamdouh, Abdel-Bar, & Awad, 2020). Through the use of lipid- and polysaccharide-modified hybrid nanoparticles, it is possible to engineer systems that release neuroprotective agents at a controlled rate. This feature minimizes side effects and provides prolonged therapeutic benefits, making it especially useful for managing conditions such as Alzheimer's disease, where consistent drug levels are essential.

5) Wound healing applications

Wound healing is a complex physiological process that involves multiple phases, including hemostasis, inflammation, proliferation, and tissue remodeling (Navabhatra

Figure 4. Schematic representation of hybrid nanoparticle-mediated brain drug delivery

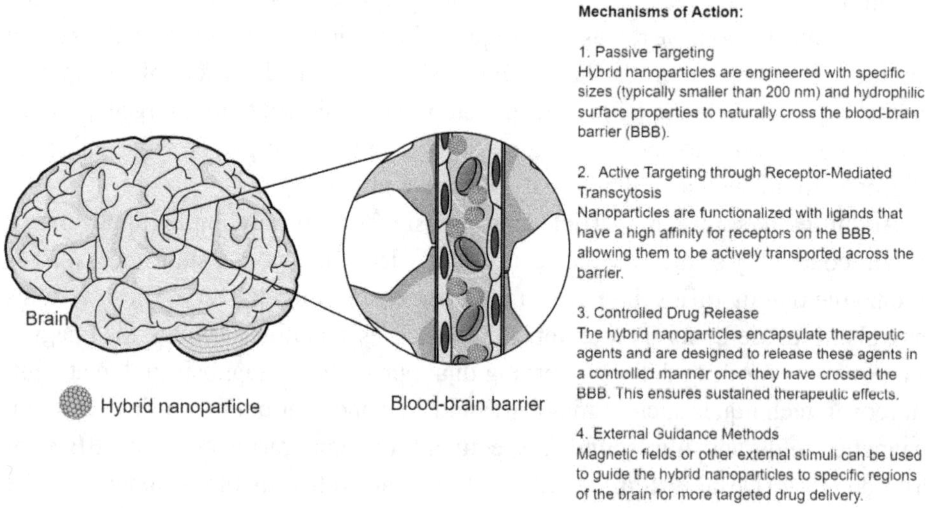

et al., 2022). In recent years, hybrid nanomaterials have emerged as promising agents to accelerate and enhance the wound-healing process. These nanocomposites combine the properties of different materials at the nanoscale to address the multifaceted challenges associated with wound care. One of the primary attributes that make hybrid nanomaterials favorable for wound healing applications is their antibacterial properties. These materials can effectively prevent wound infections by incorporating metal nanoparticles such as silver or copper into the composite structure. Antibacterial metals work to inhibit bacterial growth, thereby reducing the microbial load at the wound site and creating a cleaner environment conducive to tissue repair (Yayehrad, Siraj, Matsabisa, & Birhanu, 2023).

Another essential feature of hybrid nanomaterials in wound healing is their ability to support cell adhesion and proliferation. The inclusion of bioactive materials such as hydrogels and biopolymers creates a favorable environment for cellular activities crucial for tissue regeneration. These components interact with biological systems to facilitate cell adhesion to the wound site, enabling faster tissue growth and, ultimately, quicker healing (Y. Wang et al., 2023). The importance of bioactivity and cell adhesion cannot be overstated, as they serve as foundational elements for successful tissue regeneration. The third critical attribute of hybrid nanomaterials in wound healing is their ability to control drug release. These materials can act as reservoirs for antimicrobial or anti-inflammatory agents, allowing for a controlled, gradual release of therapeutic substances over an extended period. This sustained release ensures continuous care for the wound, contributing to a more effective and efficient healing process. For instance, one study investigated the challenges of treating diabetic wounds, particularly those infected with the common pathogen *Staphylococcus aureus*. Researchers developed a gel-like wound dressing using ferrous sulfide (FeS) nanoparticles in a glycyrrhizic acid solution. This formulation resulted in a biocompatible material with excellent mechanical properties, including injectability. Glycyrrhizic acid served multiple roles: it protected the FeS nanoparticles from oxidation and created a slightly acidic environment that enabled the controlled release of hydrogen sulfide and ferrous ions. These substances demonstrated strong antibacterial activity against both regular and drug-resistant *S. aureus* by disrupting the bacteria's energy metabolism and inducing cell death through a process called ferroptosis. The hydrogel also showed anti-inflammatory properties and promoted fibroblast growth in laboratory tests. When applied to infected diabetic wounds in a model system, the hydrogel efficiently killed bacteria and regulated the local inflammatory response, thereby accelerating wound healing (Z. Xu et al., 2023). This study thus presents a promising new approach to treating bacterial infections in diabetic wounds using multifunctional hydrogels.

Overall, hybrid nanomaterials play important roles in drug delivery systems, and Table 3 provides some examples of recent publications in the field. For instance, in a study conducted by Alipour, Sheikhnejad, Fouani, Bardania, and Hosseinkhani (2023), researchers created a novel nanohybrid particle designed to specifically target and neutralize the BCL-2 oncogene. This gene plays a key role in preventing programmed cell death and is linked to chemoresistance in many types of cancer, including breast cancer. The nanohybrid, composed of a peptide-based carrier and a DNA inhibitor (DNAi), is engineered to induce apoptosis in breast cancer cells. Using recombinant DNA technology and an internalizing RGD (iRGD) targeting motif, the team successfully created particles with an average size of 30 nm that could efficiently penetrate cancer cells. These particles suppressed BCL-2 expression, leading to the activation of apoptotic pathways. The study found that the nanohybrid was highly effective in killing cancer cells while leaving normal cells unharmed, suggesting its potential as a targeted therapy for breast cancer without adverse side effects.

Tissue Engineering and Regeneration

Tissue engineering and regeneration aim to repair or replace damaged biological tissues. Hybrid nanomaterials have emerged as especially promising candidates for advancements in these fields. The use of these nanomaterials offers several advantages. For instance, the synergistic combination of different materials can create composites with enhanced mechanical strength, which is essential for load-bearing tissues such

Table 3. Biomedical applications of some hybrid nanomaterials for drug delivery

Therapy	Hybrid Nanomaterial	Active Pharmaceutical Ingredient	Application	Reference
Breast cancer	Peptide-based nanocarrieres	Single-stranded DNA inhibitor (DNAi)	DNAi-peptide nanohybrid smart particles specifically target and inhibit the BCL-2 oncogene, triggering programmed cell death (apoptosis) in breast cancer cells and offering a potential targeted therapy for breast cancer treatment.	Alipour et al. (2023)
Cervical cancer	Magnetic nanohybrids	Organic/inorganic hybrid nanomaterials combining chitosan -pluronic F127 and magnesium oxide	Targeted delivery of chemotherapeutics to tumor cells by using magnetically guided delivery to tumor sites while minimizing damage to healthy cells.	Beemkumar et al. (2023)
Antibacterial therapy	Metal hybrid nanoparticles	Hybrid nanomaterial composed of chitosan, curcumin, ZnO and TiO_2	The developed nanohybrids exhibited good antibacterial properties against *Escherichia coli* and *Staphyloccus aureus*.	Karthikeyan et al. (2023)
Wound healing therapy	Metal nanoparticle-loaded nanofibrous scaffold	Oxidized sodium alginate and its electrospun biohybrids with zinc oxide nanoparticles	The developed nanomaterials exhibited effective antibacterial and anti-inflammatory properties, along with promoting tissue regeneration and blood vessel formation, making them a promising material for wound healing applications.	W. Wang et al. (2023)

as bone and cartilage. Additionally, bioactive components can be incorporated to boost cell adhesion, proliferation, and differentiation. The following sections will discuss applications in tissue engineering and regenerative medicine.

Applications In Tissue Engineering

1) Bone tissue engineering

One of the most popular hybrid nanomaterials used in bone tissue engineering is hydroxyapatite-polymer composites. Hydroxyapatite is a naturally occurring mineral found in bones and teeth. When combined with polymers such as poly(ε-caprolactone) or poly(lactic-co-glycolic acid) (PLGA), the resulting composite mimics both the structural and functional characteristics of natural bone. This is crucial, as the mechanical properties of bone—including its strength and stiffness—are essential for its load-bearing functions (Cao et al., 2023). In clinical settings, hydroxyapatite-PLGA composites have been used for bone grafts. Here, PLGA provides mechanical durability and tensile strength to the scaffold, while hydroxyapatite enhances its bioactivity. This composite encourages osteoblasts, the bone-forming cells, to adhere, proliferate, and eventually form new bone tissue. As the material is biodegradable, it dissolves over time, leaving behind only the newly regenerated bone (Soleymani & Naghib, 2023).

In addition, graphene—an allotrope of carbon arranged in a single-layer hexagonal lattice—adds another dimension to bone tissue engineering. When integrated into polymer scaffolds such as PLGA, graphene offers both mechanical reinforcement and electrical conductivity. Electrical properties are particularly important in bone tissue engineering because electrical signals are known to influence cellular activities such as proliferation and differentiation. Graphene-polymer composites have found applications beyond bone tissue engineering, notably in neural and muscle tissue regeneration, due to their electrical conductivity. However, within the realm of bone tissue, these composites have shown great promise in promoting osteoblast proliferation. Impressively, they can even influence stem cells to differentiate into bone cells, thereby contributing to more rapid and effective bone tissue regeneration (Amiryaghoubi, Fathi, Barar, Omidian, & Omidi, 2022).

2) Cartilage tissue engineering

Cartilage tissue engineering is an interdisciplinary field that combines principles from biology, engineering, and materials science to regenerate or replace damaged cartilage tissue. Given the avascular nature of cartilage, self-repair mechanisms are limited, necessitating engineered solutions. Hybrid nanomaterials offer unique advantages for cartilage tissue engineering due to their mechanical strength, biocompatibility, and other functional properties. For instance, composites of polymers and carbon nanotubes can provide the needed mechanical reinforcement while mimicking the anisotropic properties of native cartilage (Pathak et al., 2023). Inflammatory conditions such as osteoarthritis often accompany cartilage defects; hybrid nanomaterials can serve as effective carriers for anti-inflammatory drugs or growth factors. Hydrogels embedded with nanoparticles can store and gradually release therapeutic agents, reducing inflammation and enhancing cartilage regeneration (El-Husseiny, Mady, Yoshida, & Tanaka, 2024). The cellular microenvironment is crucial for cartilage tissue engineering; hybrid nanomaterials can be designed to promote chondrocyte adhesion and proliferation. Scaffolds made from hybrid materials that incorporate hydrogels and bioceramics can simulate the natural extracellular matrix, providing anchorage for cells and facilitating tissue growth (H. Li, Wu, Yu, & Zhang, 2023).

3) Neural Tissue Engineering

Neural tissue engineering seeks to restore or replace damaged neural tissues, which are often compromised due to injuries, degenerative diseases, or congenital anomalies. The repair and regeneration of neural tissue present unique challenges because of the complex architecture and function of the nervous system. Hybrid nanomaterials offer promising solutions for overcoming these challenges. One essential requirement for neural tissue engineering is electrical conductivity to facilitate signal transduction. Hybrid nanomaterials, such as composites incorporating carbon nanotubes or metallic nanoparticles into biocompatible polymers, can offer enhanced electrical properties. These materials can mimic the electrical characteristics of neural tissue, promoting efficient signal relay across repaired or regenerated areas (Ma et al., 2023). Neurological conditions often involve deficiencies or excesses of specific neurotransmitters or growth factors. Hybrid nanomaterials can serve as reservoirs for these bioactive molecules; for example, nanoparticles coated with biodegradable polymers can carry neurotrophic factors and release them in a controlled manner (Chircov, Ioniţă, Sîrmon, Neacşu, & Ficai, 2023). Hybrid nanomaterials can also be engineered for excellent biocompatibility and biodegradability. For instance, scaffolds made from a composite of natural materials such as chitosan and synthetic materials such as PLGA can offer both structural support and a natural interface for cell growth. Moreover, hybrid nanomaterials can be designed to have surface properties that promote neuron adhesion and axonal growth, facilitating more natural integration with existing neural networks (Sahin, Ciki, & Karademir-Yilmaz, 2023).

Applications In Regenerative Medicine

For tissue repair, the controlled delivery of growth factors and bioactive agents is crucial for promoting cell proliferation and differentiation. Traditional drug delivery systems have limitations, such as a lack of target specificity or rapid degradation of bioactive molecules. To overcome these challenges, lipid-polymer hybrid nanoparticles have emerged as a sophisticated platform for therapeutic delivery. These hybrid nanoparticles combine the best attributes of liposomes and polymeric nanoparticles: the lipid component enhances biocompatibility and cellular uptake, while the polymer core allows for the encapsulation of both hydrophilic and hydrophobic agents, enabling controlled release kinetics. Among the noteworthy applications of lipid-polymer hybrid nanoparticles is the encapsulation of vascular endothelial growth factor (VEGF). VEGF is a vital signaling protein involved in angiogenesis—the formation of new blood vessels. Sustained and localized delivery of VEGF is essential for various tissue engineering applications, especially where an enhanced blood supply is needed for tissue viability. Lipid-polymer hybrid nanoparticles encapsulating VEGF have been shown to significantly improve angiogenesis, paving the way for better outcomes in therapies requiring vascularization, such as in the treatment of chronic wounds or ischemic conditions (Beheshtizadeh et al., 2023). Similarly, lipid-polymer hybrid nanoparticles have also been effectively used to deliver bone morphogenetic protein-2 (BMP-2), a critical molecule for osteogenic differentiation. In cases where bone regeneration is needed, such as spinal fusions or fracture repairs, the targeted and sustained release of BMP-2 can dramatically improve clinical outcomes. Researchers have successfully incorporated BMP-2 into lipid-polymer hybrid nanoparticles, demonstrating their effectiveness in inducing osteogenic differentiation in stem cells and promoting bone tissue formation (Hallan, Amirian, Brangule, & Bandere, 2022).

Diagnostic Imaging

In the field of targeted imaging, hybrid nanomaterials offer unique advantages, such as high surface area, stability, and multifunctionality. This section explores the various biomedical applications of hybrid nanomaterials specifically designed for targeted imaging techniques, including magnetic resonance imaging (MRI), positron emission tomography (PET), and ultrasound imaging.

Magnetic Resonance Imaging

1) Gadolinium-based hybrid nanomaterials

Traditional MRI contrast agents, such as gadolinium, can be nephrotoxic in high doses (Iyad, S.Ahmad, Alkhatib, & Hjouj, 2023). Hybrid nanomaterials offer a way to minimize this risk by enhancing specificity and reducing the amount of gadolinium needed. Core-shell nanoparticles, which have a gadolinium core and an organic shell, have shown significantly higher contrast-to-noise ratios, enabling more precise imaging. For instance, gadolinium-loaded, surface-modified Nd^{3+}-activated $Gd_{1-x}Nd_x(CO_3)OH@SiO_2$ core–shell nanospheres have been utilized to produce enhanced MRI contrast. A biocompatible polymer coating ensures both stability and minimal toxicity, permitting targeted imaging with a reduced gadolinium load. The major advantage of these hybrid nanomaterials lies in their ability to improve the contrast-to-noise ratio (Junyi Li et al., 2023). Unlike conventional gadolinium chelates, their hybrid nature allows for controlled release and targeted delivery, which means less gadolinium is needed for imaging. This reduces the risk of nephrotoxicity. Studies have demonstrated that when gadolinium-coated mesoporous silica nanoparticles (Gd_2O_3@MSNs) were injected into rats, they offered an approximately threefold improvement in the contrast-to-noise ratio compared to conventional gadolinium agents (Z. Li, Guo, Zhang, Li, & Hao, 2022).

2) Multimodal imaging

The use of iron oxide-gold nanoparticles offers dual-modal MRI and computed tomography (CT) imaging capabilities. These particles facilitate the coregistration of images from both modalities, thereby enhancing the accuracy of disease diagnosis and monitoring. Specifically, iron oxide-gold (Fe_3O_4-Au) nanoparticles are ideal for this purpose due to their unique magnetic properties and the X-ray attenuation characteristics of gold. As a result, multimodal imaging has become an essential tool for comprehensive disease diagnosis and monitoring. While MRI excels at providing excellent soft tissue contrast, CT offers detailed anatomical information. The ability to coregister images using hybrid iron oxide-gold nanoparticles significantly improves the reliability and comprehensiveness of diagnostic data. This is particularly useful in oncology; for instance, a study by (Navyatha & Nara, 2021) demonstrated that Fe_3O_4-Au nanoparticles were effective in accurately localizing tumors in mice. The coregistered MRI-CT images provided a more comprehensive understanding of the tumor's morphology and spread.

Positron Emission Tomography

1) Carbon-based hybrid nanomaterials

Carbon dots combined with heavy isotopes, such as aluminum-fluorine-18 and lutetium-177, have been employed to enhance the efficiency and stability of PET tracers. These hybrid nanomaterials offer advantages in reducing the degradation rate of tracers, thereby providing more reliable and longer-lasting imaging solutions. Specifically, carbon dots serve as stable carriers for ^{18}F, facilitating a longer half-life and increased bioavailability for imaging applications. Compared to conventional PET tracers, carbon-based hybrid nanomaterials offer several benefits, including a larger surface area and chemical stability that ensure minimal degradation over time. According to a study by Laurène et al. (2023), carbon dot Al $^{18}F/^{177}Lu$ complexation hybrids exhibited a longer half-life than conventional tracers. This extended half-life allows these tracers to be useful for long-distance transportation to imaging facilities, offering a new level of convenience and expanding the range of PET applications. Furthermore, carbon dots possess excellent biocompatibility and low toxicity, minimizing risks to patients during the imaging process. The durability of these hybrid nanomaterials makes them particularly suitable for tracking slow metabolic processes or for use in longitudinal studies that require extended periods of observation.

2) Hybrid nanoparticles with antibody conjugation

Antibodies specific to tumor markers can be attached to PET-active hybrid nanoparticles. This allows for targeted imaging of cancerous tissues, increasing diagnostic sensitivity and specificity. A promising example in this category involves gold nanoparticles conjugated with antibodies against a specific cancer marker, such as human epidermal growth factor receptor-2 (HER2), and labeled with a PET-active isotope such as ^{64}Cu. These nanoparticles have shown high efficacy in targeting HER2-positive breast cancer cells (Sitia et al., 2022). Compared to traditional imaging agents, antibody-conjugated hybrid nanoparticles offer unparalleled specificity. According to a study conducted by (Chen et al., 2020) using mouse tumor models, these antibody-conjugated hybrid nanoparticles were able to localize and accurately identify HER2-positive tumors. Impressively, they achieved a 20% greater signal-to-noise ratio compared to conventional PET agents. Such advancements could significantly impact early cancer detection and therapy monitoring. Furthermore, the hybrid nature of these nanoparticles allows for the codelivery of therapeutic agents, opening new avenues for theranostic applications — that is, combining therapy and diagnostics in a single platform.

Ultrasound Imaging

1) Perfluorocarbon-embedded nanomaterials

Perfluorocarbon-filled hybrid nanomaterials have revolutionized enhanced ultrasound imaging by offering superior echogenic properties. This makes it easier to visualize internal structures, particularly in cardiovascular applications. In the ultrasound imaging field, perfluorocarbon-embedded silica nanoparticles have shown improved echogenic properties. These nanomaterials can encapsulate perfluorocarbon gases, which are stable and highly echogenic, thereby improving ultrasound signal intensity. Perfluorocarbon-embedded nanomaterials are especially valuable in cardiovascular applications where accurate visualization of blood flow and heart structures is crucial. These hybrid nanoparticles offer several advantages over traditional contrast agents, including extended circulation time and enhanced stability. According to a study by Alphandéry (2022), perfluorocarbon-embedded silica nanoparticles improved the visibility of myocardial infarctions in rat models compared to traditional ultrasound contrast agents. Such improvements in echogenic properties could substantially enhance the diagnostic capabilities of ultrasound for cardiovascular applications.

2) Acoustic-responsive nanomaterials

Hybrid nanomaterials can be engineered to produce distinct acoustic responses upon ultrasound stimulation. These offer particular advantages for imaging in highly vascular or dense tissues, where traditional methods may be less effective. For example, lipid-coated gold nanorods can be designed to respond specifically to certain ultrasound frequencies, generating a unique and detectable acoustic signal that enhances imaging. These materials offer improved contrast and better visualization in tissues that are challenging to image due to factors such as signal scattering or attenuation. In a review article by Gao et al. (2021), lipid-coated gold nanorods were shown to produce a measurable acoustic response, improving the imaging of liver tumors in mouse models. This research suggests that such hybrid nanomaterials have the potential to significantly advance ultrasound imaging of complex anatomical structures.

Photothermal and Photodynamic Therapy

Photothermal and photodynamic therapies represent an evolving frontier in the application of hybrid nanomaterials for targeted biomedical treatment. These therapies employ light-sensitive nanomaterials that convert light energy into heat or reactive oxygen species, facilitating localized damage to targeted cells or tissues. This section explores three major applications: the treatment of solid tumors, microbial inactivation, and atherosclerotic plaques.

Treatment of Solid Tumors

Solid tumors are abnormal masses of tissue that arise when cells grow and divide uncontrollably. Unlike hematologic cancers, which originate in the blood and blood-forming tissues, solid tumors develop in organs or tissues. These tumors can be either benign (noncancerous) or malignant (cancerous). Common examples of solid tumors include carcinomas of the breast, lung, prostate, and colon, as well as sarcomas, gliomas, and melanomas, among others (Maharaj, Polineni, Abreu, & Gouvea, 2023). Gold nanoparticles coated with photosensitive dyes are an example of hybrid nanomaterials employed in photothermal therapy. When exposed to near-infrared light, these particles generate heat, effectively destroying tumor cells while sparing adjacent healthy tissues. The application of photothermal and photodynamic therapies for solid tumors has shown significant promise in preclinical studies. For instance, a review study by Taylor, Wilson, Amrhein, and Huang (2022) demonstrated that gold nanoparticle-based photothermal therapy effectively shrank breast cancer tumors in mouse models without causing significant side effects. The key advantages of these therapies include minimally invasive treatment options, fewer side effects compared to traditional chemotherapy, and the potential for real-time monitoring during treatment.

Inactivation of Microbes

In recent years, hybrid nanomaterials have emerged as promising agents for the inactivation of microbial pathogens. This is particularly important given the rising prevalence of antibiotic-resistant bacteria and the limitations associated with conventional disinfection methods. Key points highlighting the significance of using hybrid nanomaterials for microbial inactivation are as follows: hybrid nanomaterials can be engineered to possess broad-spectrum antimicrobial activity, enabling them to effectively neutralize a wide array of pathogens, including bacteria, viruses, and fungi. This versatility makes them useful tools in healthcare settings as well as in public sanitation efforts. For instance, the combination of silver nanoparticles with methylene blue produces hybrid nanomaterials capable of generating reactive oxygen species when exposed to visible light. This combination has been effective in inactivating a variety of bacteria and fungi. Photodynamic therapy employing hybrid nanomaterials presents a compelling alternative to traditional antimicrobial treatments, especially in an era marked by increasing antibiotic resistance. Studies such as those conducted by Shirzadi-Ahodashti et al. (2021) have demonstrated the efficacy of silver nanoparticles in eradicating drug-resistant bacteria. In their research, silver nanoparticles (AgNPs) were synthesized using an eco-friendly method involving *Stachys inflata* extract. These nanoparticles were comprehensively characterized and optimized. They exhibited a spherical shape and a size range of 35-45 nm. The study confirmed their antibacterial effectiveness against both gram-positive and gram-negative bacteria, including strains that are resistant to multiple drugs. Minimum inhibitory concentration values were also reported for each bacterial type tested. Additionally, these AgNPs demonstrated the capability to degrade certain hazardous pollutants under both ultraviolet and visible light conditions. The study concludes that these biosynthesized AgNPs hold promise for applications in both the biomedical and environmental fields.

Treatment of Atherosclerotic Plaques

Atherosclerotic plaques are a leading cause of cardiovascular diseases such as heart attacks and strokes, which continue to be the top causes of mortality worldwide (Ambrose & Sharma, 2023; G. Zhang et al., 2023). Traditional treatment methods often come with risks, including invasive surgeries or long-term pharmaceutical regimens that may have potential side effects. Hybrid nanomaterials are emerging as an innovative approach for the diagnosis and treatment of atherosclerotic plaques. Specifically, nanoparticles composed of gold and coated with photosensitizers have been tested for use in the photodynamic treatment of these plaques. These nanoparticles can be directed to plaques using targeting agents such as antibodies. One major challenge in treating atherosclerosis is the targeted removal of plaques without causing damage to the surrounding blood vessels. Photothermal and photodynamic therapies offer highly localized treatment options. A study by M. Xu et al. (2022) demonstrated that targeted photodynamic therapy could effectively destabilize atherosclerotic plaques in animal models, thereby reducing the risk of heart attacks or strokes. This study investigates a novel method for the early detection and treatment of vulnerable atherosclerotic plaques, which are a major cause of acute cardiovascular events. Researchers created specialized nanoparticles from human serum albumin and targeted them with a peptide specific to osteopontin. These nanoparticles were also loaded with a photosensitizer and a hypoxia-activatable compound. When injected into mice with atherosclerosis, these osteopontin-targeted nanoparticles specifically accumulated in atherosclerotic plaques. Utilizing dual-mode imaging, precise laser irradiation was administered to the nanoparticles, generating reactive oxygen species that effectively broke down the plaques while increasing localized hypoxia. This environmental change activated the hypoxia-sensitive compound, which further suppressed plaque-forming cells. After two weeks of treatment with these nanoparticles, both the plaque area and the level of arterial stenosis were significantly reduced. Additionally, the nanoparticles displayed excellent biocompatibility. The study concludes that these specialized nanoparticles offer a targeted and effective treatment for atherosclerosis.

Theragnostic

The term "theragnostic" is a portmanteau of "therapeutic" and "diagnostic", and it refers to technologies or materials that serve both therapeutic and diagnostic functions simultaneously. In the context of biomedical applications, theragnostic nanomaterials can provide real-time monitoring and targeted treatment in a unified system. These materials are particularly promising for personalized medicine, as they enable the precise delivery of treatment while continuously monitoring the disease state or treatment response. This dual functionality allows for more effective and adaptable treatment plans. Hybrid nanomaterials have been increasingly used in the development of theragnostic agents due to their versatile properties, such as optical activity, electrical conductivity, and magnetic behavior. By combining different types of materials, such as organic polymers, metal oxides, or quantum dots, hybrid nanomaterials can be engineered to possess both therapeutic and diagnostic functions. For example, in cancer treatment, hybrid nanomaterials can be designed to carry anticancer drugs (therapeutic function) while being tagged with imaging agents (diagnostic function). This allows healthcare providers to simultaneously track the location of the material in the body using medical imaging techniques and deliver targeted therapy to the tumor site. Real-time monitoring helps minimize side effects and adapt the treatment in response to how the tumor is reacting. In cardiovascular disease management, theragnostic nanoparticles might be used to both dissolve arterial plaques (therapeutic) and release a dye to help visualize the arteries (diagnostic), thereby aiding in real-time monitoring of the disease's progression or regression. Similarly, in the treatment of neurodegenerative disorders, hybrid nanomaterials could be designed to both deliver neuroprotective agents and enable imaging of specific brain regions, thus facilitating a comprehensive and personalized treatment plan.

Biosensors

Biosensors are analytical devices designed to detect specific biological molecules by combining a biological recognition element with a transducer, thereby providing qualitative or quantitative data. Essentially, these devices convert a biological response into an electrical signal. A biosensor comprises three main components: the biological recognition element, the transducer, and the signal processor. The biological recognition element specifically interacts with the target biological molecule and can be an enzyme, an antibody, strands of DNA/RNA, or even whole cells. The transducer translates this biological interaction into a measurable signal, which can be optical, electrochemical, thermometric, piezoelectric, or magnetic in nature. Finally, the signal processor amplifies and interprets this signal to produce a readable output, often displayed on a screen or stored in a data storage unit for further analysis. Recently, hybrid nanomaterials have emerged as a highly versatile platform for developing advanced biosensors. These nanomaterials combine the advantageous properties of both organic and inorganic materials—such as conductivity, mechanical strength, optical properties, and biocompatibility—to enable highly sensitive and specific detection mechanisms (Yang et al., 2023). This section aims to explore the biomedical applications of hybrid nanomaterials in the development of biosensors for various purposes, including disease diagnosis, environmental monitoring, and drug delivery monitoring.

1) Enzyme-based Biosensors

Enzyme-based biosensors are one of the pioneering applications in which hybrid nanomaterials have been employed to enhance both specificity and sensitivity. These biosensors utilize enzymes as biological recognition elements to selectively detect specific substrates such as glucose, lactate, and cholesterol. Enzymes are immobilized onto the surface of the transducer, and their interaction with the substrate is converted into a measurable signal. One of the most promising hybrid nanomaterials for this application is a composite made from carbon nanotubes and polymers. Carbon nanotubes offer remarkable mechanical strength and electrical conductivity, while polymers provide a flexible and biocompatible matrix. The high surface area of these nanocomposites allows ample space for enzyme immobilization, making them an ideal platform for enzyme-based biosensors. Some of the most impactful applications of these biosensors are in healthcare and lifestyle monitoring. For instance, glucose sensors that employ glucose oxidase-immobilized carbon nanotube-polymer composites have shown high sensitivity, allowing real-time and accurate blood glucose monitoring for better diabetes management. Lactate sensors, which employ lactate oxidase on similar platforms, are useful in sports science for assessing muscle fatigue and in medical settings for monitoring sepsis. Furthermore, cholesterol biosensors that utilize cholesterol oxidase have become crucial tools for cardiovascular health, showing promise in point-of-care testing and making it easier for individuals to manage conditions such as hypercholesterolemia.

Table 4 provides some examples of biosensors used in the biomedical field. For instance, Deepapriya et al. (2023) developed a novel enzymatic glucose biosensor using a unique chitosan nanocomposite composed of $Ni_{0.25}Zn_{0.25}Cu_{0.25}Co_{0.25}La_{0.06}Fe_{1.94}O_4$ synthesized via the combustion method. The biosensor was constructed by immobilizing glucose oxidase onto a chitosan-based nanocomposite coated on a fluorine-doped tin oxide glass substrate. Electrochemical analysis using cyclic voltammetry and electrochemical impedance spectroscopy revealed excellent molecular diffusion, superior charge transfer rates, high sensitivity, and quick response times. The biosensor demonstrated a high sensitivity level of 52.76 $\mu AmM^{-1}cm^{-2}$ and exhibited strong catalytic properties in H_2O_2 solution. Moreover, it has a low Michaelis–Menten constant, indicating better affinity for glucose oxidase. This innovation represents a significant advance in the field of noninvasive glucose monitoring systems.

2) Nucleic Acid Sensors

Nucleic acid sensors, particularly DNA biosensors, represent another promising avenue where hybrid nanomaterials have shown considerable potential for biomedical applications. These sensors employ strands of DNA or RNA as biological recognition elements to selectively identify complementary nucleic acid sequences. A significant advancement in this field is the incorporation of gold nanoparticle-polymer hybrids as substrates for immobilizing nucleic acid probes. The unique optical and electronic properties of gold nanoparticles synergize with polymers to amplify the signal-to-noise ratio, substantially enhancing the biosensor's sensitivity and specificity. One groundbreaking application of these DNA biosensors is in the early diagnosis of cancer; they can detect mutations or specific biomarkers in DNA sequences indicative of early-stage malignancy. These sensors are also invaluable for identifying genetic disorders, as they can pinpoint mutations or irregularities in DNA sequences. For instance, they can detect single-nucleotide polymorphisms (SNPs) associated with diseases such as cystic fibrosis or sickle cell anemia, offering critical insights for early intervention and management.

3) Immunological Detection

Immunological detection is another area where hybrid nanomaterials are making a considerable impact, particularly in the field of immunoassays. Immunological biosensors, or immunosensors, employ antibodies or antigens as biological recognition elements, making them incredibly specific and selective for detecting corresponding antigens or antibodies in a sample. A key innovation in immunosensors is the use of quantum dot-polymer hybrids. Quantum dots are semiconductor nanoparticles with unique optical properties, such as broad absorption spectra and narrow emission spectra. When combined with organic polymers, these nanomaterials provide a high degree of functional flexibility. This hybrid material serves as an excellent platform for the immobilization of antibodies or antigens, significantly improving the sensitivity of the biosensor. Quantum dot-polymer hybrid immunosensors are revolutionizing various fields of medicine due to their rapid and sensitive detection capabilities. These sensors have been effectively employed for detecting cancer-related biomarkers such as prostate-specific antigen and carcinoembryonic antigen, offering new avenues for early cancer diagnosis. In the realm of cardiovascular diseases, these immunosensors have shown promise in detecting key biomarkers such as troponin and C-reactive protein, enabling early intervention and better management of these conditions. Furthermore, they have demonstrated efficacy in rapidly identifying antibodies related to infectious diseases such as HIV, hepatitis, and dengue, thereby streamlining the diagnostic process and facilitating timely treatment.

4) Electrochemical Biosensors

Electrochemical biosensors are a rapidly emerging field in which hybrid nanomaterials offer significant advantages. These biosensors utilize a blend of metal oxide semiconductors and conductive polymers to form a versatile platform that hosts both biological recognition elements and transduction mechanisms. The semiconductor properties of metal oxides, combined with the high conductivity of polymers, facilitate rapid electron transfer, leading to fast and accurate readings. These electrochemical biosensors are particularly effective in detecting neurotransmitters such as dopamine and serotonin, providing new possibilities for diagnosing and treating neurological disorders such as depression, Parkinson's disease, and schizophrenia. Moreover, they are skilled at identifying environmental toxins, such as heavy metals and pesticides, making them indispensable for real-time water quality and environmental safety assessments. Their adaptability extends to detecting other small organic molecules, including glucose and lactate, which showcases their broad applicability in both medical and environmental domains. The rapidity and precision of these hybrid electrochemical biosensors underscore the synergistic benefits of combining metal oxide semiconductors with conductive polymers, heralding a new era of highly efficient biosensors.

5) Portable and Wearable Biosensors

The advent of miniaturized biosensors is a groundbreaking achievement, made possible in large part by the integration of hybrid nanomaterials. Among the materials leading the way are hybrid nanofibers, which meld organic and inorganic components to create a remarkably flexible and lightweight platform. These properties make hybrid nanofibers perfectly suited for integration into wearable devices designed to monitor physiological parameters in real time. In the realm of wearable biosensors, hybrid nanofibers play a transformative role by enabling real-time monitoring of key physiological parameters. For instance, these biosensors can continuously track heart rate, offering vital insights for athletes and those with cardiovascular conditions and allowing for immediate intervention in cases of irregularities such as arrhythmia. In addition to heart rate, they provide real-time data on blood pressure levels, revolutionizing early detection and management of hypertension—a leading risk factor for various cardiovascular diseases. Furthermore, the mechanical flexibility and lightweight characteristics of hybrid nanofibers make them well suited for continuous body temperature monitoring, a feature especially crucial during infectious disease outbreaks for facilitating early medical intervention. Overall, hybrid nanofibers are making significant strides in preventive healthcare by making it easier to integrate continuous health monitoring into daily life.

CASE STUDY

Chronic wounds and bacterial infections continue to pose significant challenges in healthcare. Traditional wound dressings often fall short of offering both mechanical support and antibacterial properties. Electrospinning is a versatile technique widely used for fabricating nanofibers from a variety of materials. The underlying principle of electrospinning involves applying an electric field to draw charged threads of polymer solutions or melts, thereby creating fibers with diameters in the nanometer to micrometer range. In the case of silver nanoparticle-loaded nanofiber scaffolds, the electrospinning process is adapted to incorporate silver nanoparticles into the nanofibers. This combination offers both the mechanical advantages of nanofibers

Table 4. Examples of hybrid nanomaterials for biosensor applications

Biosensor	Composition	Application	Reference
Enzymatic-based biosensor	An enzymatic glucose biosensor containing glucose oxidase immobilized on $Ni_{0.25}Zn_{0.25}Cu_{0.25}Co_{0.25}La_{0.06}Fe_{1.94}O_4$ and chitosan nanocomposite	An enzymatic glucose biosensor utilizes an effective immobilization of glucose oxidase on a $Ni_{0.25}Zn_{0.25}Cu_{0.25}Co_{0.25}La_{0.06}Fe_{1.94}O_4$-chitosan nanocomposite, offering high sensitivity and fast response times for glucose monitoring.	Deepapriya et al. (2023)
Immunosensor	Graphene oxide decorated with gold nanoflower structures (GO@Au-NS)	A label-free biosensor made to early detection of colorectal cancer through the identification of miR-223, a specific microRNA biomarker.	Akbari et al. (2023)
Genosensor	Flexible electrochemical genosensors composed of polypyrrole (PPy) and graphene quantum dots (GQDs)	These flexible nanosensors offer a promising tool for the early detection and monitoring of the PML/RARα fusion genes commonly found in pediatric cases of acute promyelocytic leukemia.	L. S. Oliveira et al. (2023)
Nonenzymatic amperometric biosensor	Amperometric biosensor with Ni nanoparticles anchored in a nonenzymatic configuration	The wearable nonenzymatic amperometric biosensor, featuring anchored Ni nanoparticles, provides a reliable method for the quantification of glucose levels in urine.	Nugba, Mousa, Osman, and El-Moneim (2023)

and the antibacterial properties of silver. Thus, this case study aims to explore the potential of electrospun nanofibers loaded with silver nanoparticles as a promising alternative for effective wound management. The authors used poly(vinyl alcohol) (with a degree of alcoholysis between 87% and 89%) and polyethylene oxide (with a molecular weight of 400,000 Da) to create nanofibrous scaffolds by electrospinning. Silver nanoparticles, synthesized separately, were incorporated into the nanofiber matrix during the electrospinning process.

Initially, silver nanoparticles were synthesized using chemical reduction methods, where silver ions (Ag^+) are reduced to form nanoparticles. Briefly, He et al. (2020) described a method for creating silver nanoparticles that are dispersible in water. This process involved the use of polyvinylpyrrolidone as a stabilizing agent and was conducted in a liquid-phase reduction setting. Initially, polyvinylpyrrolidone was dissolved in distilled water in a flask of 250 mL capacity. The addition of tannic acid and aqueous ammonia to this mixture, while stirring at ambient temperature, resulted in the solution changing color to brown. The subsequent gradual introduction of silver nitrate caused the solution to turn black. This reaction was maintained for an hour, followed by the concentration of the solution using a rotary evaporator at 60°C over a period of four hours. To precipitate the nanoparticles, acetone was added to the concentrated mixture, which was then filtered and washed multiple times with distilled water and acetone. The final product, water-dispersible silver nanoparticles, was achieved by vacuum drying at ambient temperature.

These nanoparticles were then stabilized in distilled water to form a colloidal solution (1.2% w/w). Separately, a polymer mixture of poly(vinyl alcohol) (8.4% w/w) and polyethylene oxide (3.6% w/w) was chosen for electrospinning. A polymer solution was prepared by dissolving the polymers in distilled water with continuous stirring at 80°C. These synthesized silver nanoparticles were then added to the polymer solution. This mixture is thoroughly stirred to ensure an even distribution of the nanoparticles throughout the polymer solution. Figure 5 demonstrates the electrospinning device setup used in this case study. The mixed solution is loaded into a syringe fitted with a spinneret. The syringe was placed in a syringe pump, which controlled the flow rate of the solution (0.6 mL/minute). An electric field is applied between the spinneret and a grounded collecting surface (25 kV). As the polymer solution is extruded from the spinneret, the applied electric field induces a charge on the liquid. When the electric force overcomes the surface tension of the liquid droplet at the tip of the spinneret, a jet is ejected. This jet undergoes a whipping process as it travels toward the collecting surface, during which the solvent evaporates, leaving behind a fiber. The silver nanoparticles encapsulated in the polymer matrix are thereby incorporated into the resulting nanofibers. The fibers are collected on a rotating drum, resulting in a nonwoven mat of nanofibers loaded with silver nanoparticles. Finally, the nanofiber mat undergoes postprocessing steps and characterization methods, including scanning electron microscopy (SEM) for morphology, tensile tests for mechanical properties, and antibacterial tests to confirm the successful incorporation and activity of the silver nanoparticles.

SEM analysis revealed a highly porous network of intertwined nanofibers, each with a relatively uniform diameter ranging from 490 to 730 nm (Figure 6). The fibers displayed a smooth surface texture, indicative of a successful electrospinning

Figure 5. Electrospinning device setup used by the research group of the authors to fabricate the silver nanoparticle-loaded nanofiber scaffolds

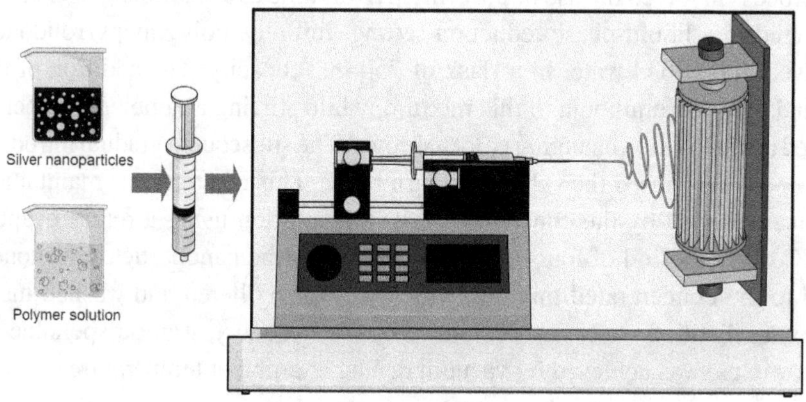

process with consistent material deposition. The silver nanoparticles were evenly distributed throughout the nanofiber matrix, appearing as distinguishable nanosized particles embedded within the fibers. The average diameter of these embedded silver nanoparticles was measured to be approximately 10–20 nm. No agglomeration of silver nanoparticles was observed, suggesting that the mixture and electrospinning processes effectively maintained nanoparticle dispersion within the polymer matrix. The SEM images further showed that the fibers were randomly oriented, forming a nonwoven mesh essential for good mechanical integrity and porosity. The pore sizes were estimated to be between 1 and 3 nm, which are conducive for cell infiltration and nutrient transport—important factors for wound healing applications.

Texture analysis revealed that the silver nanoparticle-loaded nanofiber scaffolds exhibited a tensile strength of approximately 4–6 MPa and an elongation at break of 20–30%. These mechanical properties are comparable to those of native skin, making them crucial for materials intended for use in wound dressings. Such dressings must be both flexible and robust to adapt to the wound environment and provide support. The texture profile indicated a moderate level of surface roughness, which is favorable for cellular attachment and tissue integration. This surface topology could enhance the bioactivity of the wound dressing, aiding in quicker wound closure and healing. In addition, the nanomaterial demonstrated excellent water absorption properties, which are essential for maintaining a moist wound environment. The scaffolds could absorb up to 300% of their weight in water, confirming their ability to facilitate fluid management in the wound area.

The *in vitro* study utilized standardized bacterial cultures, including *Staphylococcus aureus* and *Escherichia coli*, to evaluate antibacterial properties. Three different scaffold types were tested: (1) control (traditional wound dressing materials); (2)

Figure 6. Scanning electron microscopy images showing the silver nanoparticle-loaded nanofiber scaffolds

nanofiber scaffolds (electrospun nanofibers); and (3) silver nanoparticle-loaded nanofiber scaffolds (electrospun nanofibers embedded with silver nanoparticles). The antibacterial efficacy was assessed through agar disk diffusion tests and bacterial colony count measurements over a period of 48 hours. The study yielded unequivocal evidence affirming the antibacterial potency of nanofibrous scaffolds loaded with silver nanoparticles. Utilizing standardized tests, the composite scaffolds consistently outperformed their counterparts in several crucial aspects. When compared with polymer-only nanofibers and conventional wound dressing materials, the silver nanoparticle-loaded scaffolds showed a markedly enhanced zone of inhibition. This larger zone suggests that the antibacterial agents (silver nanoparticles) were effectively released from the scaffold, achieving a broader reach in attacking bacterial cells and reducing their ability to proliferate. Beyond the zones of inhibition, the data revealed a significant and dramatic reduction in the number of bacterial colonies for samples treated with the silver nanoparticle-loaded scaffolds. This result is pivotal, as it directly correlates with a material's ability not only to inhibit bacterial growth but also to effectively neutralize bacterial cells. The drastic drop in bacterial colonies indicates that these composite nanofibers could offer more than just passive protection; they could actively eliminate infection-causing bacteria. Silver nanoparticles interact with bacterial cells in several ways to inhibit their growth or cause cell death. First, they attach to bacterial cell membranes, disrupting their structure and leading to a loss of essential ions and nutrients. Second, after entering the cell, silver nanoparticles interfere with key biological processes by binding to proteins, DNA, and RNA. This effectively disables nutrient metabolism, DNA replication, and protein synthesis. Third, they induce the production of reactive oxygen species, causing oxidative stress and damage to cellular components. Additionally, silver ions released from nanoparticles can inhibit metabolic pathways, such as enzymes in the electron transport chain, disrupting ATP production and thus cellular energy. Importantly, their multitarget approach makes it difficult for bacteria to develop resistance, which offers an advantage over single-target antibiotics.

To gauge the safety of these composite materials, cell viability tests were performed using L929 mouse fibroblasts, as demonstrated in Figure 7. This is crucial because an ideal wound dressing must be effective against bacteria while also being safe for surrounding tissues. The tests confirmed that the silver nanoparticle-loaded nanofibers did not exhibit cytotoxic effects on mammalian cells, thereby assuring their biocompatibility. This provides a preliminary but reassuring indication that the material could be safely employed for *in vivo* applications without eliciting adverse cellular reactions. In summary, the silver nanoparticle-loaded nanofiber scaffolds achieved a trifecta of desired attributes: strong antibacterial capabilities, mechanical robustness, and high biocompatibility. The *in vivo* wound healing properties of the fabricated materials will be studied in the future to substantiate their benefits.

These findings strongly suggest that such composite scaffolds could revolutionize wound care, offering a new paradigm that outperforms traditional wound dressing materials in multiple dimensions.

CHALLENGES IN CLINICAL TRANSLATION

The emerging field of biomedicine has witnessed substantial advancements through the incorporation of hybrid nanomaterials. These materials hold great promise for a range of applications, from targeted drug delivery to innovative diagnostic techniques. However, translating these promising innovations from the laboratory to clinical settings presents a host of technical challenges. One of the primary bottlenecks is scalability; methods effective for small-scale lab production often become costly and inefficient when applied on a larger, clinical scale. Alongside scalability, achieving consistent reproducibility in the nanomaterials' characteristics—such as

Figure 7. Photographs showing the cytotoxic study of silver nanoparticles against L929 mouse fibroblasts using the 3-(4,5-dimethylthiazol-2-yl)-2,5-diphenyltetrazolium bromide (MTT) assay

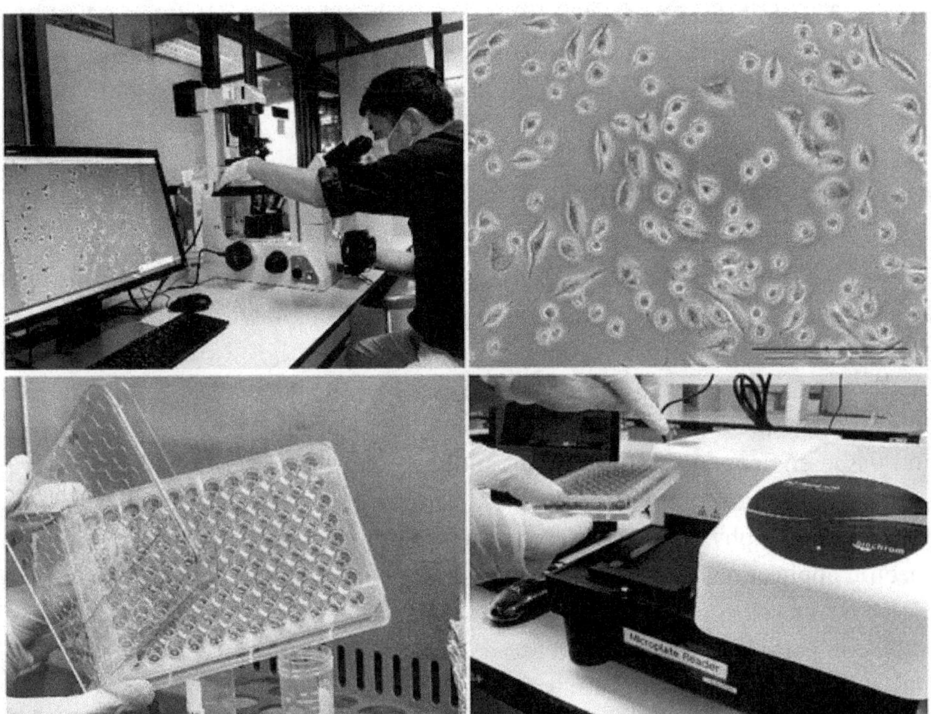

size, shape, and composition—is essential for ensuring predictable pharmacokinetics and toxicity profiles. Additionally, stability concerns arise due to the inherently high surface-to-volume ratios of these nanomaterials, which can lead to issues such as aggregation, sedimentation, and chemical degradation over time. Another set of hurdles pertains to biocompatibility and toxicity. The complex makeup of hybrid nanomaterials, often involving a blend of organic and inorganic components, introduces new variables that require rigorous, resource-intensive testing. These variables can affect the interactions of nanomaterials with biological systems, potentially leading to cytotoxicity, genotoxicity, and immunogenicity. Moreover, even when biocompatibility is confirmed, there is the intricate challenge of engineering the materials for high targeting efficiency while minimizing off-target effects, a task made complex by the biological intricacy of targeted tissues and cells. Beyond these technical issues, regulatory and integration challenges also loom large. Regulatory bodies demand stringent safety and efficacy evaluations, yet standardized protocols for such evaluations are often lacking in this rapidly evolving field. For these materials to be smoothly integrated into existing healthcare settings, they must also be compatible with current biomedical technologies and workflows, including imaging systems, drug delivery methods, and surgical instruments. Despite these substantial challenges, the transformative potential of hybrid nanomaterials in healthcare justifies the intensive scrutiny and interdisciplinary collaboration they require. This calls for a concerted effort from experts in materials science, biology, engineering, and medicine to successfully transition these promising materials from bench to bedside (Sowmiya et al., 2023).

CONCLUSION AND FUTURE SCOPE

The exploration of hybrid nanomaterials and their applications in the biomedical field represents a frontier of research experiencing an unprecedented surge. This chapter aims to illuminate the various applications of these novel materials, revealing their substantial potential to fundamentally reshape the biomedical landscape. The spectrum of applications for hybrid nanomaterials ranges from targeted drug delivery and tissue engineering to bioimaging and theragnostics, offering novel solutions to some of the most pressing challenges in contemporary medicine. Characterized by their unique physicochemical properties and the capacity for customization to specific applications, hybrid nanomaterials serve as a powerful instrument in the biomedical research toolkit. However, this potential is not without obstacles. Issues of biocompatibility, potential toxicity, scalable manufacturing, cost-effectiveness, and regulatory compliance pose significant hurdles. Therefore, it is incumbent upon the scientific community to foster a collaborative, multidisciplinary strategy to fully realize the possibilities offered by hybrid nanomaterials. Despite these challenges, the prognosis for the incorporation of hybrid nanomaterials into biomedicine remains promising. As the exploration of their unique properties continues and the refinement of their synthesis and application progresses, it is anticipated that the influence of hybrid nanomaterials within the biomedical sphere will expand significantly. In conclusion, hybrid nanomaterials are poised to play an instrumental role in shaping the future of biomedicine. Success in overcoming the inherent challenges in their development and application will have a profound impact on healthcare delivery and patient care. As a result, this success will pave an optimistic path forward for the evolution of medical science.

Looking forward, one of the most critical areas to address is the issue of long-term toxicity and biocompatibility. As these nanomaterials become increasingly complex, assessing their long-term effects on biological systems becomes paramount. Future research will likely focus on creating nanocomposites that are not only effective in their intended functions but also pose minimal risks to human health. The field also stands to benefit from more interdisciplinary collaborations, incorporating insights from materials science, biology, engineering, pharmaceuticals, and medicine. For instance, a deep understanding of cellular mechanisms can guide the design of more efficient drug delivery systems, while advances in materials science can contribute to the development of more robust and durable bioactive scaffolds. Another emerging trend is the integration of machine learning and computational methods to predict the properties and behaviors of novel hybrid nanomaterials before their synthesis. This approach can significantly accelerate the development cycle and identify promising candidates for specific biomedical applications, thereby making the research process more efficient and cost-effective. Finally, addressing regulatory hurdles will be crucial for the successful translation of these promising materials from the laboratory to the clinic. Strategies to streamline this process could include establishing standardized testing protocols for safety and efficacy and fostering closer relationships between researchers, medical professionals, and regulatory agencies. In conclusion, the application of hybrid nanomaterials in biomedical sciences holds enormous potential to revolutionize healthcare. By addressing existing challenges and capitalizing on interdisciplinary research, the future looks promising for the development of more effective, safe, and personalized medical treatments.

ABBREVIATIONS

AgNPs Silver nanoparticles

BBB Blood–brain barrier

BMP-2 Bone morphogenetic protein-2

CT Computed tomography
HER-2 Human epidermal growth factor receptor-2
MRI Magnetic resonance imaging
PET Positron emission tomography
PLGA Poly(lactic-co-glycolic acid)
VEGF Vascular endothelial growth factor

REFERENCES

Akbari, A., Hashemzadeh, H., Eshkiki, Z. S., Masoodi, M., Tabaeian, S. P., Naderi-Manesh, H., Zare, A. A., & Agah, S. (2023). Detection of plasma miR-223 by a novel label-free graphene oxide/gold nanocomposite immunosensor in colorectal cancer patients: An electrochemical biosensor approach. *Biosensors & Bioelectronics: X, 14*, 100331. doi:10.1016/j.biosx.2023.100331

Alghamdi, M., Fatima, B., Hussain, Z., Nisar, Z., & Alghamdi, H. A. (2023). Peristaltic pumping of hybrid nanofluids through an inclined asymmetric channel: A biomedical application. *Materials Today. Communications, 35*, 105684. doi:10.1016/j.mtcomm.2023.105684

Alipour, M., Sheikhnejad, R., Fouani, M. H., Bardania, H., & Hosseinkhani, S. (2023). DNAi-peptide nanohybrid smart particles target BCL-2 oncogene and induce apoptosis in breast cancer cells. *Biomedicine and Pharmacotherapy, 166*, 115299. doi:10.1016/j.biopha.2023.115299 PMID:37573657

Alphandéry, E. (2022). Nanomaterials as ultrasound theragnostic tools for heart disease treatment/diagnosis. *International Journal of Molecular Sciences, 23*(3), 1683. doi:10.3390/ijms23031683 PMID:35163604

Alvarez-Lorenzo, C., Vivero-Lopez, M., & Concheiro, A. (2023). Contact lenses that transform gold into nanoparticles for prophylaxis of light-related events and photothermal therapy. *International Journal of Pharmaceutics, 641*, 123048. doi:10.1016/j.ijpharm.2023.123048 PMID:37192704

Ambrose, J. A., & Sharma, A. V. (2023). Identifying and treating vulnerable atherosclerotic plaques. *The American Journal of Cardiology, 205*, 214–222. doi:10.1016/j.amjcard.2023.07.121 PMID:37611413

Amiryaghoubi, N., Fathi, M., Barar, J., Omidian, H., & Omidi, Y. (2022). Recent advances in graphene-based polymer composite scaffolds for bone/cartilage tissue engineering. *Journal of Drug Delivery Science and Technology, 72*, 103360. doi:10.1016/j.jddst.2022.103360

Ashfaq, M., Talreja, N., Chauhan, D., Afreen, S., Sultana, A., & Srituravanich, W. (2022). Two-dimensional (2D) hybrid nanomaterials for diagnosis and treatment of cancer. *Journal of Drug Delivery Science and Technology*, *70*, 103268. doi:10.1016/j.jddst.2022.103268

Bashir, M. A. (2021). Use of dynamic mechanical analysis (dma) for characterizing interfacial interactions in filled polymers. *Solids*, *2*(1), 108–120. doi:10.3390/solids2010006

Beemkumar, N., Kaushik, M., Tripathi, A. D., Sharma, M., Khan, S. A., & Sharma, R. C. (2023). Pluronic F127-chitosan modified magnesium oxide hybrid nanomaterials prepared via a one-pot method: Potential uses in antibacterial and anticancer agents. *Surfaces and Interfaces*, *42*, 103327. doi:10.1016/j.surfin.2023.103327

Beheshtizadeh, N., Gharibshahian, M., Bayati, M., Maleki, R., Strachan, H., Doughty, S., & Tayebi, L. (2023). Vascular endothelial growth factor (VEGF) delivery approaches in regenerative medicine. *Biomedicine and Pharmacotherapy*, *166*, 115301. doi:10.1016/j.biopha.2023.115301 PMID:37562236

Butto-Miranda, N., Cabrera-Barjas, G., Ibáñez, A., & Neira-Carrillo, A. (2023). Preparation of Novel Electrospun Organic–Inorganic Hybrid Nanofibers Based on Polycaprolactone, Glycine Betaine, and Calcium Carbonate (P/G/CaCO3-EPnF). *Crystals*, *13*(4), 611. Advance online publication. doi:10.3390/cryst13040611

Cao, Z., Bian, Y., Hu, T., Yang, Y., Cui, Z., Wang, T., Yang, S., Weng, X., Liang, R., & Tan, C. (2023). Recent advances in two-dimensional nanomaterials for bone tissue engineering. *Journal of Materiomics*, *9*(5), 930–958. doi:10.1016/j.jmat.2023.02.016

Charitos, I., Georgousis, G., Klonos, P. A., Kyritsis, A., Mouzakis, D., Raptis, Y., Kontos, A., & Kontou, E. (2021). The synergistic effect on the thermomechanical and electrical properties of carbonaceous hybrid polymer nanocomposites. *Polymer Testing*, *95*, 107102. doi:10.1016/j.polymertesting.2021.107102

Chen, J. S., Chen, J., Bhattacharjee, S., Cao, Z., Wang, H., Swanson, S. D., Zong, H., Baker, J. R. Jr, & Wang, S. H. (2020). Functionalized nanoparticles with targeted antibody to enhance imaging of breast cancer *in vivo*. *Journal of Nanobiotechnology*, *18*(1), 135. doi:10.1186/s12951-020-00695-2 PMID:32948179

Chircov, C., Ioniţă, D.-A., Sîrmon, A.-M., Neacşu, I. A., & Ficai, A. (2023). Natural, synthetic, and hybrid and composite biomaterials for neural tissue engineering. In O. Gunduz, C. B. Ustundag, & M. Sengor (Eds.), *Biomaterials for Neural Tissue Engineering* (pp. 21–58). Woodhead Publishing. doi:10.1016/B978-0-323-90554-1.00008-2

Conklin, B., Conley, B. M., Hou, Y., Chen, M., & Lee, K.-B. (2023). Advanced theragnostics for the central nervous system (CNS) and neurological disorders using functional inorganic nanomaterials. *Advanced Drug Delivery Reviews*, *192*, 114636. doi:10.1016/j.addr.2022.114636 PMID:36481291

Curl, R. F., Jr., Kroto, H., Smalley, R. E., Curl, R. F., Jr., & Kroto, H. W. (1996). *The Nobel Prize in Chemistry 1996*. Explore Press Release.

da Silva Costa, K. R. B., de Sá, M. L., de Carvalho Oliveira, F., Dittz, D., Leal, B. S., de Castro e Sousa, J. M., Abreu, G. J. P., de Oliveira, T. G., Gusmão, S. B. S., Ghosh, A., Guerra, Y., & Viana, B. C. (2023). Titanate nanotubes modified with gallium and cerium and their cytotoxic activity. *Journal of Nanoparticle Research*, *25*(9), 177. Advance online publication. doi:10.1007/s11051-023-05824-0

Deepapriya, S., Rodney, J. D., Flora John, J., Joshi, S., Udayashankar, N. K., Lakshmi Devi, S., & Jerome Das, S. (2023). A novel effective immobilization of glucose oxidase on Ni0.25Zn0.25Cu0.25Co0.25La0.06Fe1.94O4 – Chitosan nanocomposite as an enzymatic glucose biosensor. *Inorganic Chemistry Communications*, *153*, 110822. doi:10.1016/j.inoche.2023.110822

Desai, H. R., & Shinde, U. A. (2022). 20 - Hybrid nanoparticles to cross the blood–brain barrier. In L. Kumar & Y. Y. Pathak (Eds.), *Nanocarriers for Drug-Targeting Brain Tumors* (pp. 565–586). Elsevier. doi:10.1016/B978-0-323-90773-6.00002-6

Deshmukh, M. A., Thorat, H. N., Shirsat, M. D., & Ramanavicius, A. (2023). 1 - Engineered nanostructures: an introduction. In A. K. Kaushik, S. Kumar, & G. R. Chaudhary (Eds.), *Engineered Nanostructures for Therapeutics and Biomedical Applications* (pp. 1–43). Woodhead Publishing. doi:10.1016/B978-0-12-821240-0.00002-0

El-Husseiny, H. M., Mady, E. A., Yoshida, T., & Tanaka, R. (2024). Stimuli-responsive hydrogels: cutting-edge platforms for cartilage tissue engineering. In A. R. Nochehdehi, F. Nemavhola, S. Thomas, & H. J. Maria (Eds.), *Cartilage Tissue and Knee Joint Biomechanics* (pp. 467–486). Academic Press. doi:10.1016/B978-0-323-90597-8.00009-8

El-Khawaga, A. M., Zidan, A., & El-Mageed, A. I. A. A. (2023). Preparation methods of different nanomaterials for various potential applications: A review. *Journal of Molecular Structure*, *1281*, 135148. doi:10.1016/j.molstruc.2023.135148

Elkasabgy, N. A., Salama, A., & Salama, A. H. (2023). Exploring the effect of intramuscularly injected polymer/lipid hybrid nanoparticles loaded with quetiapine fumarate on the behavioral and neurological changes in cuprizone-induced schizophrenia in mice. *Journal of Drug Delivery Science and Technology*, *79*, 104064. doi:10.1016/j.jddst.2022.104064

Fernandes, N. B., Nayak, Y., Garg, S., & Nayak, U. Y. (2023). Multifunctional engineered mesoporous silica/inorganic material hybrid nanoparticles: Theranostic perspectives. *Coordination Chemistry Reviews*, *478*, 214977. doi:10.1016/j.ccr.2022.214977

Fithri, N. A., Wu, Y., Cowin, G., Akther, F., Tran, H. D. N., Tse, B., van Holthe, N. W., Moonshi, S. S., Peter, K., Wang, X., Troung, N. P., & Ta, H. T. (2023). Gold-iron oxide nanoparticle: A unique multimodal theranostic approach for thrombosis. *Applied Materials Today*, *31*, 101750. doi:10.1016/j.apmt.2023.101750

Flimelová, M., Ryabchikov, Y. V., Behrends, J., & Bulgakova, N. M. (2023). Environmentally Friendly Improvement of Plasmonic Nanostructure Functionality toward Magnetic Resonance Applications. *Nanomaterials (Basel, Switzerland)*, *13*(4), 764. Advance online publication. doi:10.3390/nano13040764 PMID:36839132

Ganguly, S., & Margel, S. (2023). Fluorescent quantum dots-based hydrogels: Synthesis, fabrication and multimodal biosensing. *Talanta Open*, *8*, 100243. doi:10.1016/j.talo.2023.100243

Gao, Q., Zhang, J., Gao, J., Zhang, Z., Zhu, H., & Wang, D. (2021). Gold Nanoparticles in Cancer Theranostics. *Frontiers in Bioengineering and Biotechnology*, *9*, 647905. Advance online publication. doi:10.3389/fbioe.2021.647905 PMID:33928072

Gulati, K., & Adachi, T. (2023). Profiling to Probing: Atomic force microscopy to characterize nanoengineered implants. *Acta Biomaterialia*, *170*, 15–38. Advance online publication. doi:10.1016/j.actbio.2023.08.006 PMID:37562516

Hallan, S. S., Amirian, J., Brangule, A., & Bandere, D. (2022). Lipid-Based Nano-Sized Cargos as a Promising Strategy in Bone Complications: A Review. *Nanomaterials (Basel, Switzerland)*, *12*(7), 1146. Advance online publication. doi:10.3390/nano12071146 PMID:35407263

He, M., Chen, M., Dou, Y., Ding, J., Yue, H., Yin, G., Chen, X., & Cui, Y. (2020). Electrospun silver nanoparticles-embedded feather keratin/poly(vinyl alcohol)/poly(ethylene oxide) antibacterial composite nanofibers. *Polymers*, *12*(2), 305. doi:10.3390/polym12020305 PMID:32028586

Huo, Y., Xiu, S., Meng, L.-Y., & Quan, B. (2023). Solvothermal synthesis and applications of micro/nano carbons: A review. *Chemical Engineering Journal, 451*, 138572. doi:10.1016/j.cej.2022.138572

Iijima, S. (1994). Carbon nanotubes. *MRS Bulletin, 19*(11), 43–49. doi:10.1557/S0883769400048405

Ilyas, S. U., Shamsuddin, R., Kai Xiang, T., Estellé, P., & Pendyala, R. (2023). Rheological profile of graphene-based nanofluids in thermal oil with hybrid additives of carbon nanotubes and nanofibers. *Journal of Molecular Liquids, 376*, 121443. doi:10.1016/j.molliq.2023.121443

Iyad, N. S., Ahmad, M., Alkhatib, S. G., & Hjouj, M. (2023). Gadolinium contrast agents- challenges and opportunities of a multidisciplinary approach: Literature review. *European Journal of Radiology Open, 11*, 100503. doi:10.1016/j.ejro.2023.100503 PMID:37456927

Jeyasubramanian, K., Sakthivel, A., Thangagiri, B., & Raja, J. D. (2023). Trends and developments in point-of-care diagnostics using quantum dots. In K. Yun & S. Govindaraju (Eds.), *Carbon and Graphene Quantum Dots for Biomedical Applications* (pp. 193–209). Woodhead Publishing. doi:10.1016/B978-0-323-98362-4.00015-5

Jia, X., Lv, M., Fei, Y., Dong, Q., Wang, H., Liu, Q., Li, D., Wang, J., & Wang, E. (2022). Facile one-step synthesis of NIR-Responsive siRNA-Inorganic hybrid nanoplatform for imaging-guided photothermal and gene synergistic therapy. *Biomaterials, 282*, 121404. doi:10.1016/j.biomaterials.2022.121404 PMID:35134700

Jiang, Q., Gao, X., Zhang, W., & Chen, Z. (2023). Recent progress in metal-organic framework-based sonosensitizers for sonodynamic tumor therapy. *Biomaterials Science, 11*(13), 4452–4470. doi:10.1039/D3BM00556A PMID:37221901

Kabay, G., Meydan, A. E., Eom, T., Shim, B. S., Mutlu, M., & Kaleli-Can, G. (2023). Stimuli-responsive nanoparticle-nanofiber hybrids for drug delivery and photodynamic therapy. *International Journal of Pharmaceutics, 630*, 122442. Advance online publication. doi:10.1016/j.ijpharm.2022.122442 PMID:36442721

Karri, R. R., Mubarak, N. M., Koduru, J. R., Lingamdinne, L. P., Dehghani, M. H., Jatoi, A. S., & Mazari, S. A. (2023). 1 - Role of hybrid nanomaterials for a sustainable environment. In J. R. Koduru, R. R. Karri, & N. M. Mubarak (Eds.), *Hybrid Nanomaterials for Sustainable Applications* (pp. 1–24). Elsevier. doi:10.1016/B978-0-323-98371-6.00017-3

Karthikeyan, C., Jayaramudu, T., Núñez, D., Jara, N., Opazo-Capurro, A., Varaprasad, K., Kim, K., Yallapu, M. M., & Sadiku, R. (2023). Hybrid nanomaterial composed of chitosan, curcumin, ZnO and TiO_2 for antibacterial therapies. *International Journal of Biological Macromolecules*, *242*, 124814. Advance online publication. doi:10.1016/j.ijbiomac.2023.124814 PMID:37201889

Kharey, P., Goel, M., Husain, Z., Gupta, R., Sharma, D. M. M., Palani, I. A., & Gupta, S. (2023). Green synthesis of biocompatible superparamagnetic iron oxide-gold composite nanoparticles for magnetic resonance imaging, hyperthermia and photothermal therapeutic applications. *Materials Chemistry and Physics*, *293*, 126859. doi:10.1016/j.matchemphys.2022.126859

Kour, A., Panda, H. S., Singh, I. R., Kumar, A., & Panda, J. J. (2023). Peptide-metal nanohybrids (PMN): Promising entities for combating neurological maladies. *Advances in Colloid and Interface Science*, *318*, 102954. doi:10.1016/j.cis.2023.102954 PMID:37487364

Lal, M., Shandilya, M., Sharma, S., & Rai, R. (2017). Smart maerial nanofibers for day to day life. In R. Rai (Ed.), *Smart Materials for Smart Living* (pp. 1–66). Nova Scientific Publishers.

Las-Casas, B., Dias, I. K. R., Yupanqui-Mendoza, S. L., Pereira, B., Costa, G. R., Rojas, O. J., & Arantes, V. (2023). The emergence of hybrid cellulose nanomaterials as promising biomaterials. *International Journal of Biological Macromolecules*, *250*, 126007. doi:10.1016/j.ijbiomac.2023.126007 PMID:37524277

Laurène, W., Raúl, L., Katalin, S., Céline, F., Gilles, K., Antonio, M., Charlotte, C., & Samir, A. (2023). Design and synthesis of a new bifunctional chelating agent: Application for Al 18F/177Lu complexation. *Journal of Inorganic Biochemistry*, *246*, 112267. doi:10.1016/j.jinorgbio.2023.112267 PMID:37329775

Lavrova, D. G., Zvonarev, A. N., Alferov, V. A., Khonina, T. G., Shadrina, E. V., Alferov, S. V., & Ponamoreva, O. N. (2023). Biocompatible silica-polyethylene glycol-based composites for immobilization of microbial cells by sol-gel synthesis. *Polymers*, *15*(2), 458. doi:10.3390/polym15020458 PMID:36679338

Li, H., Wu, C., Yu, X., & Zhang, W. (2023). Recent advances of PVA-based hydrogels in cartilage repair application. *Journal of Materials Research and Technology*, *24*, 2279–2298. doi:10.1016/j.jmrt.2023.03.130

Li, J., Duan, J., Liao, Y., Liu, X., Rong, P., Chen, G., Wan, H., & Ma, R. (2023). Surface modification of Nd3+ activated gadolinium core-shell nanospheres for near-infrared and magnetic resonance dual functional bioimaging system. *Materials & Design*, *232*, 112132. doi:10.1016/j.matdes.2023.112132

Li, J., Lu, W., Yang, Y., Xiang, R., Ling, Y., Yu, C., & Zhou, Y. (2023). Hybrid Nanomaterials for Cancer Immunotherapy. *Advancement of Science*, *10*(6), 2204932. Advance online publication. doi:10.1002/advs.202204932 PMID:36567305

Li, Z., Guo, J., Zhang, M., Li, G., & Hao, L. (2022). Gadolinium-Coated Mesoporous Silica Nanoparticle for Magnetic Resonance Imaging. *Frontiers in Chemistry*, *10*, 837032. doi:10.3389/fchem.2022.837032 PMID:35242742

Lima, D., Ribicki, A., Gonçalves, L., Hacke, A. C. M., Lopes, L. C., Pereira, R. P., Wohnrath, K., Fujiwara, S. T., & Pessôa, C. A. (2022). Nanoconjugates based on a novel organic–inorganic hybrid silsesquioxane and gold nanoparticles as hemocompatible nanomaterials for promising biosensing applications. *Colloids and Surfaces. B, Biointerfaces*, *213*, 112355. doi:10.1016/j.colsurfb.2022.112355 PMID:35158220

Lin, Y.-Q., Feng, K.-K., Lu, J.-Y., Le, J.-Q., Li, W.-L., Zhang, B.-C., Li, C.-L., Song, X., Tong, L.-W., & Shao, J.-W. (2023). CRISPR/Cas9-based application for cancer therapy: Challenges and solutions for nonviral delivery. *Journal of Controlled Release*, *361*, 727–749. doi:10.1016/j.jconrel.2023.08.028 PMID:37591461

Lofgreen, J. E., & Ozin, G. A. (2014). Controlling morphology and porosity to improve performance of molecularly imprinted sol–gel silica. *Chemical Society Reviews*, *43*(3), 911–933. doi:10.1039/C3CS60276A PMID:24247659

Ma, H., Yu, K., Wang, H., Liu, J., Cheng, Y. Y., Kang, Y., Wang, H., Zhang, J., & Song, K. (2023). Fabrication and detection of a novel hybrid conductive scaffold based on alginate/gelatin/carboxylated carbon nanotubes (Alg/Gel/mMWCNTs) for neural tissue engineering. *Tissue & Cell*, *80*, 101995. doi:10.1016/j.tice.2022.101995 PMID:36512950

Maharaj, D., Polineni, V., Abreu, M. M., & Gouvea, J. (2023). Molecular remission using personalized low-dose immunotherapy with minimal toxicities for poor prognosis hematological and solid tumor cancers. In A. Jewett & Y. Fong (Eds.), *NK Cells in Cancer Immunotherapy: Successes and Challenges* (pp. 427–453). Academic Press. doi:10.1016/B978-0-12-822620-9.00002-1

Mallakpour, S., Azadi, E., & Hussain, C. M. (2023). Innovative progress in functionalized carbon nanomaterials, their hybrids, and nanocomposites: Fabrication, antibacterial, biomedical, bioactivity, and biosensor applications. In S. Mallakpour & C. M. Hussain (Eds.), *Functionalized Carbon Nanomaterials for Theranostic Applications* (pp. 439–479). Elsevier. doi:10.1016/B978-0-12-824366-4.00017-0

Mandal, M., Banerjee, I., & Mandal, M. (2022). Nanoparticle-mediated gene therapy as a novel strategy for the treatment of retinoblastoma. *Colloids and Surfaces. B, Biointerfaces*, *220*, 112899. doi:10.1016/j.colsurfb.2022.112899 PMID:36252537

Mathur, V. (2013). *Effects of nanoscale inclusions on impact resistance of Kevlar-epoxy laminate composites* [Doctoral dissertation]. Wichita State University.

Mickymaray, S., Al Aboody, M. S., Eraqi, M. M., Alhoqail, W. A., Alothaim, A. S., Suresh, K., & Arulselvan, P. (2023). Chitosan-encapsulated nickel oxide, tin dioxide, and farnesol nanoparticles: Antimicrobial and anticancer properties in breast cancer cells. *International Journal of Biological Macromolecules*, *248*, 125799. Advance online publication. doi:10.1016/j.ijbiomac.2023.125799 PMID:37451381

Mishra, A., Kumar, R., Mishra, J., Dutta, K., Ahlawat, P., Kumar, A., Dhanasekaran, S., Gupta, A. K., Sinha, S., Bishi, D. K., Gupta, P. K., & Nayak, S. (2023). Strategies facilitating the permeation of nanoparticles through blood–brain barrier: An insight toward the development of brain-targeted drug delivery system. *Journal of Drug Delivery Science and Technology*, *86*, 104694. doi:10.1016/j.jddst.2023.104694

Mukhortova, Y. R., Pryadko, A. S., Chernozem, R. V., Pariy, I. O., Akoulina, E. A., Demianova, I. V., Zharkova, I. I., Ivanov, Y. F., Wagner, D. V., Bonartsev, A. P., Surmenev, R. A., & Surmeneva, M. A. (2022). Fabrication and characterization of a magnetic biocomposite of magnetite nanoparticles and reduced graphene oxide for biomedical applications. *Nano-Structures & Nano-Objects*, *29*, 100843. doi:10.1016/j.nanoso.2022.100843

Nadaf, S., Jena, G. K., Rarokar, N., Gurav, N., Ayyanar, M., Prasad, S., & Gurav, S. (2023). Biogenic and biomimetic functionalized magnetic nanosystem: Synthesis, properties, and biomedical applications. *Hybrid Advances*, *3*, 100038. doi:10.1016/j.hybadv.2023.100038

Navabhatra, A., Maniratanachote, R., & Yingngam, B. (2022). Antiphotoaging properties of Zingiber montanum essential oil isolated by solvent-free microwave extraction against ultraviolet B-irradiated human dermal fibroblasts. *Toxicological Research*, *38*(2), 235–248. doi:10.1007/s43188-021-00107-z PMID:35419276

Navyatha, B., & Nara, S. (2021). Gold nanotheranostics: Future emblem of cancer nanomedicine. *Nanobiomedicine*, *8*, 18495435211053945. doi:10.1177/18495435211053945 PMID:34737836

Nekoueian, K., Akhoundian, M., Wester, N., & Laurila, T. (2023). An ultrasensitive dopamine measurement platform based on molecularly imprinted polymer-carbon hybrid nanomaterials for *in vitro* use. *Electrochimica Acta*, *445*, 142029. doi:10.1016/j.electacta.2023.142029

Nezhad, E. Z., Qu, X., Musharavati, F., Jaber, F., Appleford, M. R., Bae, S., Uzun, K., Struthers, M., Chowdhury, M. E. H., & Khandakar, A. (2021). Effects of titanium and carbon nanotubes on nano/micromechanical properties of HA/TNT/CNT nanocomposites. *Applied Surface Science*, *538*, 148123. doi:10.1016/j.apsusc.2020.148123

Niu, B., Chen, Y., Zhang, L., & Tan, J. (2022). Organic–inorganic hybrid nanomaterials prepared via polymerization-induced self-assembly: Recent developments and future opportunities. *Polymer Chemistry*, *13*(18), 2554–2569. doi:10.1039/D2PY00180B

Nugba, B. E., Mousa, N. O., Osman, A., & El-Moneim, A. A. (2023). Nonenzymatic amperometric biosensor with anchored Ni nanoparticles for urinary glucose quantification. *Diamond and Related Materials*, *137*, 110171. doi:10.1016/j.diamond.2023.110171

Oliveira, F. C., Carvalho, J. O., Magalhães, L. S. S. M., da Silva, J. M., Pereira, S. R., Gomes Júnior, A. L., Soares, L. M., Cariman, L. I. C., da Silva, R. I., Viana, B. C., Silva-Filho, E. C., Afewerki, S., da Cunha, H. N., Vega, M. L., Marciano, F. R., & Lobo, A. O. (2021). Biomineralization inspired engineering of nanobiomaterials promoting bone repair. *Materials Science and Engineering C*, *120*, 111776. doi:10.1016/j.msec.2020.111776 PMID:33545906

Oliveira, L. S., Avelino, K. Y. P. S., Oliveira, S. R. D. E., Lucena-Silva, N., de Oliveira, H. P., Andrade, C. A. S., & Oliveira, M. D. L. (2023). Flexible genosensors based on polypyrrole and graphene quantum dots for PML/RARα fusion gene detection: A study of acute promyelocytic leukemia in children. *Journal of Pharmaceutical and Biomedical Analysis*, *235*, 115606. doi:10.1016/j.jpba.2023.115606 PMID:37544275

Omar, S. H., Osman, R., Mamdouh, W., Abdel-Bar, H. M., & Awad, G. A. S. (2020). Bioinspired lipid-polysaccharide modified hybrid nanoparticles as a brain-targeted highly loaded carrier for a hydrophilic drug. *International Journal of Biological Macromolecules*, *165*, 483–494. doi:10.1016/j.ijbiomac.2020.09.170 PMID:32987085

Palacio-Vergara, M., Álvarez-Gómez, M., Gallego, J., & López, D. (2023). Biomass solvothermal treatment methodologies to obtain carbon quantum dots: A systematic review. *Talanta Open*, *8*, 100244. doi:10.1016/j.talo.2023.100244

Panda, P. K., Verma, A., Saraf, S., Raikwar, S., Da Bidla, P., & Jain, S. K. (2023). Poly(lactic acid) and poly(lactic-co-glycolic acid)-based nanocarrier systems for theranostic applications. In M. S. Hasnain, A. K. Nayak, & T. M. Aminabhavi (Eds.), *Polymeric Nanosystems* (pp. 13–36). Academic Press. doi:10.1016/B978-0-323-85656-0.00017-6

Pathak, R., Punetha, V. D., Bhatt, S., & Punetha, M. (2023). Carbon nanotube-based biocompatible polymer nanocomposites as an emerging tool for biomedical applications. *European Polymer Journal, 196*, 112257. doi:10.1016/j.eurpolymj.2023.112257

Paul, R., Chatterjee, D., Das Ghosh, L., Narayanswamy, V., Singh, M. P., Agarwal, M., Ghosh, D., Radhakrishna, M., Tiwary, C. S., Provazník, I., & Chattopadhyay, K. (2023). Synthesis, characterization and *In vitro* studies of CNT/Gd2O3 hybrid structure. *Carbon Trends, 11*, 100272. doi:10.1016/j.cartre.2023.100272

Potara, M., Suarasan, S., Craciun, A.-M., Focsan, M., Hada, A.-M., & Astilean, S. (2022). Probing polyvinylpyrrolidone-passivated graphene oxide nanoflakes as contrast agents inside tissue-like phantoms via multimodal confocal microscopy. *Talanta, 247*, 123581. doi:10.1016/j.talanta.2022.123581 PMID:35636370

Qi, K., Sun, B., Liu, S.-y., & Zhang, M. (2023). Research progress on carbon materials in tumor photothermal therapy. *Biomedicine and Pharmacotherapy, 165*, 115070. doi:10.1016/j.biopha.2023.115070 PMID:37390711

Qi, X., Xiang, Y., Cai, E., Ge, X., Chen, X., Zhang, W., Li, Z., & Shen, J. (2023). Inorganic–organic hybrid nanomaterials for photothermal antibacterial therapy. *Coordination Chemistry Reviews, 496*, 215426. doi:10.1016/j.ccr.2023.215426

Rahman, A., Chowdhury, M. A., & Hossain, N. (2022). Green synthesis of hybrid nanoparticles for biomedical applications: A review. *Applied Surface Science Advances, 11*, 100296. doi:10.1016/j.apsadv.2022.100296

Rennie, C., Huang, Y., Siwakoti, P., Du, Z., Padula, M., Bao, G., Tuch, B. E., Xu, X., & McClements, L. (2023). *In vitro* evaluation of a hybrid drug-delivery nanosystem for fibrosis prevention in cell therapy for Type 1 diabetes. *Nanomedicine (London), 18*(1), 53–66. doi:10.2217/nnm-2022-0231 PMID:36938861

Revathy, R., Sajini, T., Augustine, C., & Joseph, N. (2023). Iron-based magnetic nanomaterials: Sustainable approaches of synthesis and applications. *Results in Engineering, 18*, 101114. doi:10.1016/j.rineng.2023.101114

Rodríguez-Galván, A., & Contreras-Torres, F. F. (2022). Scanning tunneling microscopy of biological structures: An elusive goal for many years. *Nanomaterials (Basel, Switzerland), 12*(17), 3013. doi:10.3390/nano12173013 PMID:36080050

Sahin, A., Ciki, B., & Karademir-Yilmaz, B. (2023). In vitro evaluation of biomaterials for neural tissue engineering. In O. Gunduz, C. B. Ustundag, & M. Sengor (Eds.), *Biomaterials for Neural Tissue Engineering* (pp. 367–415). Woodhead Publishing. doi:10.1016/B978-0-323-90554-1.00003-3

Sanchis-Gual, R., Coronado-Puchau, M., Mallah, T., & Coronado, E. (2023). Hybrid nanostructures based on gold nanoparticles and functional coordination polymers: Chemistry, physics and applications in biomedicine, catalysis and magnetism. *Coordination Chemistry Reviews*, *480*, 215025. doi:10.1016/j.ccr.2023.215025

Sharmila, S., Gowri, S., Karthikeyan, C., & Faiyazuddin, M. (2023). Multifunctional hybrid nanomaterials and their antimicrobial activity. In C. M. Hussain, K. V. Anand, & S. Mallakpour (Eds.), *Antimicrobial Nanosystems* (pp. 213–225). Elsevier. doi:10.1016/B978-0-323-91156-6.00019-1

Shevtsova, T., Cavallaro, G., Lazzara, G., Milioto, S., Donchak, V., Harhay, K., Korolko, S., Budkowski, A., & Stetsyshyn, Y. (2022). Temperature-responsive hybrid nanomaterials based on modified halloysite nanotubes uploaded with silver nanoparticles. *Colloids and Surfaces. A, Physicochemical and Engineering Aspects*, *641*, 128525. doi:10.1016/j.colsurfa.2022.128525

Shirzadi-Ahodashti, M., Hashemi, Z., Mortazavi, Y., Khormali, K., Mortazavi-Derazkola, S., & Ebrahimzadeh, M. A. (2021). Discovery of high antibacterial and catalytic activities against multidrug resistant clinical bacteria and hazardous pollutants by biosynthesized of silver nanoparticles using Stachys inflata extract (AgNPs@SI). *Colloids and Surfaces. A, Physicochemical and Engineering Aspects*, *617*, 126383. doi:10.1016/j.colsurfa.2021.126383

Sitia, L., Sevieri, M., Signati, L., Bonizzi, A., Chesi, A., Mainini, F., Corsi, F., & Mazzucchelli, S. (2022). HER-2-targeted nanoparticles for breast cancer diagnosis and treatment. *Cancers (Basel)*, *14*(10), 2424. Advance online publication. doi:10.3390/cancers14102424 PMID:35626028

Solangi, N. H., Karri, R. R., Mubarak, N. M., & Mazari, S. A. (2023). *Mechanism of polymer composite-based nanomaterial for biomedical applications*. Advanced Industrial and Engineering Polymer Research. doi:10.1016/j.aiepr.2023.09.002

Soleymani, S., & Naghib, S. M. (2023). 3D and 4D printing hydroxyapatite-based scaffolds for bone tissue engineering and regeneration. *Heliyon*, *9*(9), e19363. doi:10.1016/j.heliyon.2023.e19363 PMID:37662765

Sowmiya, P., Dhas, T. S., Inbakandan, D., Anandakumar, N., Nalini, S., Suganya, K. S. U., Remya, R. R., Karthick, V., & Kumar, C. M. V. (2023). Optically active organic and inorganic nanomaterials for biological imaging applications: A review. *Micron (Oxford, England)*, *172*, 103486. doi:10.1016/j.micron.2023.103486 PMID:37262930

Suh, H., Kim, G., Cho, S., Li, P., Son, D.-H., Koo, D., Lim, J., Choi, C.-S., Seok, S., & Bae, S. (2023). Comparative analysis of the synergistic effects of hybrid nanomaterial reinforcement in cementitious Composites: A perspective for pore refinement and thermal resistance. *Construction & Building Materials*, *401*, 132856. doi:10.1016/j.conbuildmat.2023.132856

Sun, T., Li, Y., Liu, Y., Deng, B., Liao, C., & Zhu, Y. (2023). Advanced scanning electron microscopy and microanalysis: Applications to nanomaterials. In Y. Yin, Y. Lu, & Y. Xia (Eds.), *Encyclopedia of Nanomaterials* (1st ed., pp. 183–209). Elsevier. doi:10.1016/B978-0-12-822425-0.00104-4

Tan, Q., Zhao, S., Xu, T., Wang, Q., Zhang, M., Yan, L., Chen, X., & Lan, M. (2023). Inorganic nanodrug delivery systems for crossing the blood–brain barrier: Advances and challenges. *Coordination Chemistry Reviews*, *494*, 215344. doi:10.1016/j.ccr.2023.215344

Tarasenka, N., Kornev, V., Cai, X., Li, R., Dudchik, N., Emeliyanova, O., & Tarasenko, N. (2023). Nanoceria and hybrid silver–ceria nanoparticles fabricated by liquid-mediated laser ablation as antimicrobial agents. *Nano-Structures & Nano-Objects*, *34*, 100971. doi:10.1016/j.nanoso.2023.100971

Taylor, M. L., Wilson, R. E. Jr, Amrhein, K. D., & Huang, X. (2022). Gold Nanorod-Assisted Photothermal Therapy and Improvement Strategies. *Bioengineering (Basel, Switzerland)*, *9*(5), 200. Advance online publication. doi:10.3390/bioengineering9050200 PMID:35621478

Thakur, S., Kaur, M., Lim, W. F., & Lal, M. (2020). Fabrication and characterization of electrospun ZnO nanofibers; antimicrobial assessement. *Materials Letters*, *264*, 127279. doi:10.1016/j.matlet.2019.127279

Thakur, S., Thakur, S., Shandilya, M., Lal, M., & Rai, R. (2017). Biosynthesis of nanoparticles using plant extracts. In R. Rai (Ed.), *Smart Materials for Smart Living* (pp. 365–382). Nova Scientific Publishers.

Umair Wani, T., Hamid Rather, A., Saleem Khan, R., Macossay, J., Jadhav, A. H., Srinivasappa, P. M., Abdal-hay, A., Rather, S., & Sheikh, F. A. (2023). Titanium dioxide functionalized multi-walled carbon nanotubes, and silver nanoparticles reinforced polyurethane nanofibers as a novel scaffold for tissue engineering applications. *Journal of Industrial and Engineering Chemistry*, *121*, 200–214. doi:10.1016/j.jiec.2023.01.024

Wang, P., Li, N., Li, S., & Zhang, Y. (2022). Strategies for preparing hybrid nanomaterials via polymerization-induced self-assembly. *European Polymer Journal*, *172*, 111234. doi:10.1016/j.eurpolymj.2022.111234

Wang, W., Liu, M., Shafiq, M., Li, H., Hashim, R., El-Newehy, M., EL-Hamshary, H., Morsi, Y., & Mo, X. (2023). Synthesis of oxidized sodium alginate and its electrospun bio-hybrids with zinc oxide nanoparticles to promote wound healing. *International Journal of Biological Macromolecules*, *232*, 123480. doi:10.1016/j.ijbiomac.2023.123480 PMID:36720331

Wang, Y., Qi, W., Mao, Z., Wang, J., Zhao, R. C., & Chen, H. (2023). rPDAs doped antibacterial MOF-hydrogel: Bio-inspired synergistic whole-process wound healing. *Materials Today Nano*, *23*, 100363. doi:10.1016/j.mtnano.2023.100363

Xu, M., Mao, C., Chen, H., Liu, L., Wang, Y., Hussain, A., Li, S., Zhang, X., Tuguntaev, R. G., Liang, X.-J., Guo, W., & Cao, F. (2022). Osteopontin targeted theranostic nanoprobes for laser-induced synergistic regression of vulnerable atherosclerotic plaques. *Acta Pharmaceutica Sinica. B*, *12*(4), 2014–2028. doi:10.1016/j.apsb.2021.12.020 PMID:35847489

Xu, Z., Xu, Z., Gu, J., Zhou, J., Sha, G., Huang, Y., Wang, T., Fan, L., Zhang, Y., & Xi, J. (2023). In situ formation of ferrous sulfide in glycyrrhizic acid hydrogels to promote healing of multi-drug resistant Staphylococcus aureus-infected diabetic wounds. *Journal of Colloid and Interface Science*, *650*, 1918–1929. doi:10.1016/j.jcis.2023.07.141 PMID:37517191

Yadav, K., Sahu, K. K., Sucheta, Gnanakani, S. P. E., Sure, P., Vijayalakshmi, R., Sundar, V. D., Sharma, V., Antil, R., Jha, M., Minz, S., Bagchi, A., & Pradhan, M. (2023). Biomedical applications of nanomaterials in the advancement of nucleic acid therapy: Mechanistic challenges, delivery strategies, and therapeutic applications. *International Journal of Biological Macromolecules*, *241*, 124582. doi:10.1016/j.ijbiomac.2023.124582 PMID:37116843

Yang, G., Liu, F., Zhao, J., Fu, L., Gu, Y., Qu, L., Zhu, C., Zhu, J.-J., & Lin, Y. (2023). MXenes-based nanomaterials for biosensing and biomedicine. *Coordination Chemistry Reviews*, *479*, 215002. doi:10.1016/j.ccr.2022.215002

Yayehrad, A. T., Siraj, E. A., Matsabisa, M., & Birhanu, G. (2023). 3D printed drug loaded nanomaterials for wound healing applications. *Regenerative Therapy*, *24*, 361–376. doi:10.1016/j.reth.2023.08.007 PMID:37692197

Yingngam, B., Chiangsom, A., Pharikarn, P., Vonganakasame, K., Kanoknitthiran, V., Rungseevijitprapa, W., & Prasitpuriprecha, C. (2019). Optimization of menthol-loaded nanocapsules for skin application using the response surface methodology. *Journal of Drug Delivery Science and Technology*, *53*, 101138. doi:10.1016/j.jddst.2019.101138

Yingngam, B., Kacha, W., Rungseevijitprapa, W., Sudta, P., Prasitpuriprecha, C., & Brantner, A. (2019). Response surface optimization of spray-dried citronella oil microcapsules with reduced volatility and irritation for cosmetic textile uses. *Powder Technology*, *355*, 372–385. doi:10.1016/j.powtec.2019.07.065

Zhang, G., Cui, X., Qin, Z., Wang, Z., Lu, Y., Xu, Y., Xu, S., Tang, L., Zhang, L., Liu, G., Wang, X., Zhang, J., & Tang, J. (2023). Atherosclerotic plaque vulnerability quantification system for clinical and biological interpretability. *iScience*, *26*(9), 107587. doi:10.1016/j.isci.2023.107587 PMID:37664595

Zhang, Y., Li, Z., Zhao, Z., Li, Y., Liu, Z., & Sun, S. (2023). Mesoporous carbon in biomedicine: Modification strategies and biocompatibility. *Carbon*, *212*, 118121. doi:10.1016/j.carbon.2023.118121

Zhu, D., Kong, H., Sun, Z., Xu, Y., Han, P., Xi, Y., & Wei, G. (2023). Recent advance in tailoring the structure and functions of self-assembled peptide nanomaterials for biomedical applications. *Coordination Chemistry Reviews*, *494*, 215374. Advance online publication. doi:10.1016/j.ccr.2023.215374

Zhu, Y., Liu, Y., Khan, K., Arkin, G., Tareen, A. K., Xie, Z., He, T., Su, L., Guo, F., Lai, X., Xu, J., & Zhang, J. (2023). Ultrasound combined with nanomaterials for cancer therapy. *Materials Today. Advances*, *17*, 100330. doi:10.1016/j.mtadv.2022.100330

Chapter 8 Exploring the Transformative Potential of Hybrid Nanoparticles in Biomedical Applications:
Relevance of Hybrid Nanoparticles

Arshya Hashim
Dr. D.Y. Patil Arts, Commerce, and Science College, India

Faria Fatima
Integral University, India

ABSTRACT

Hybrid nanoparticles are nanoscale particles that are composed by interaction between the different components, resulting in enhanced properties that can be harnessed for wide range of application across fields like medicine, electronics energy, and more. These nanoparticles are typically in the size range from 1-100 nanometers, which is advantageous. Because at this scale, they often exhibit novel behavior due to their quantum and surface effects. The choice of material and the way they are combined can be tailored to achieve specific goals. For example, in biomedical applications, hybrid nanoparticles can be engineered to have specific targeting abilities such as targeted drug delivery, theranostics, gene therapy, phototherapy, tissue regeneration, vaccines, antibacterial, biomolecules detection, imaging probes, tissue engineering, biosensing, and cancer treatment. They have enhanced qualities with increased target specificity and sensitivity, extended circulation times, and resistance to biological barriers.

DOI: 10.4018/979-8-3693-1261-2.ch008

1. INTRODUCTION

Nanohybrids are composite materials consisting of two or more different nanoscale components, each with distinct properties and functions (Park et al., 2020). The term "hybrid" is derived from the Latin "hybrida" first originated in the English language in the 17th century. Experiments in Plant Hybridization, reported by Gregor Mendel in 1865, ignited a scientific revolution in the field of hybridization. True hybrid nano systems containing inorganic and organic structural components were. reported regarding synthesis of nanoscale silica-polymer hybrid sphere (Bagheri et al., 2018).

The initial groundwork for nanotechnology began during the period of 1970s - 1980s. Scientists explored the synthesis and characterization of nanomaterials, including nanoparticles of various compositions (Shnoudeh et al., 2019). In 1990s, research in nanotechnology gained momentum. The potential of nanoparticles for drug delivery, imaging, and diagnostic applications due to their unique size-dependent properties was explored. Early 2000s nanohybrids started gaining attention as researchers sought to combine different nanomaterials to exploit their synergistic properties (Sahoo et al., 2023). Initial studies focused on creating hybrid structures to enhance drug delivery systems and imaging agents.

From the Mid-2000s, studies demonstrated the advantages of hybrid nanoparticles in biomedical applications. Core-shell structures, combining metals, polymers, and organic materials, emerged as promising candidates for targeted drug delivery and imaging (Wang et al., 2017). Early 2010s scientists began fine-tuning the fabrication methods of nanohybrids to optimize their properties. Surface functionalization and modification techniques were explored to improve biocompatibility, stability, and targeting capabilities. Further, in Mid-2010s advancements in nanotechnology, biomaterials, and surface chemistry led to more sophisticated nanohybrid designs. Responsive nanohybrids capable of stimuli-triggered drug release and targeted therapies began to emerge (Alwattar et al., 2021).

Moreover, research continued to focus on addressing challenges such as scalability, toxicity, and clinical translation. Notable advancements included the development of multifunctional nanohybrids for combination therapies, regenerative medicine, and highly sensitive biosensing applications (Seaberg et al., 2021). Ongoing research efforts aimed at improving the understanding of nanohybrid interactions within biological systems, optimizing their performance, and paving the way for clinical application.

Around the turn of the 21st Century, the development of hybrid nano systems expanded greatly through the hybridization of silica with micelles, liposomes, and polymers along with the introduction of polymer-functionalized metallic nanoparticles (Nik Mohd Adnan, 2018). The combination of these components in a nanohybrid material often results in enhanced properties, improved functionality, and expanded applications. Nanohybrids encompass a diverse range of structures that combine different nanomaterials, offering a wide array of functionalities. Here are some common types of nanohybrids used in various biomedical applications (Bohara, 2019).

[1] **Core-Shell Nanoparticles**: These hybrids consist of a core material encapsulated within another material, forming a shell. For example, a magnetic core (like iron oxide) coated with a biocompatible polymer or lipid layer. Core-shell structures offer protection to the core material and enable functionalization of the surface for targeted drug delivery or imaging (Kumar et al., 2020).

[2] **Organic-Inorganic Nanohybrids**: Combinations of organic and inorganic materials, such as polymers with metallic or ceramic nanoparticles, offer synergistic properties. For instance, polymers like polyethylene glycol (PEG) combined with gold or silver nanoparticles exhibit enhanced stability and biocompatibility for various applications (Kukushkina et al., 2021).

[3] **Polymer-Metal Nanocomposites**: Integration of polymers with metallic nanoparticles, like gold or silver, yields nanocomposites with unique optical, electronic, or catalytic properties. These hybrids find applications in sensing, imaging, and therapeutic delivery systems (Tamayo et al., 2018).

[4] **Lipid-Polymer Hybrids**: Combinations of lipids and polymers form versatile structures for drug delivery. Lipid-polymer hybrids can encapsulate drugs within a lipid bilayer while incorporating polymeric materials for stability and controlled release. Up to now, the FDA has approved only a few types of NPs for clinical trials, among these the most relevant are liposomes, albumin, iron-based, PLGA-based and silica nanoparticles (Sivadasan et al., 2021).

[5] **Carbon-Based Nanohybrids**: Combining carbon-based materials (such as carbon nanotubes or graphene) with other nanomaterials creates hybrids with excellent mechanical, electrical, and thermal properties. These hybrids have potential applications in biosensing, drug delivery, and tissue engineering (Karimi et al., 2018).

[6] **Mesoporous Nanoparticles**: These hybrids possess a porous structure that allows for high drug-loading capacities. They can be composed of various materials like silica, metals, or polymers, providing a large surface area for drug adsorption and controlled release (Park & Ha, 2018).

[7] **Hybrid Nanofibers**: Nanofibers created by combining different nanomaterials or by electrospinning techniques offer a platform for tissue engineering, wound healing, and drug delivery due to their high surface area and structural similarities to the extracellular matrix (Keshvardoostchokami et al., 2020).

Nanohybrids have gained significant attention in various fields due to their unique properties and versatile applications. Given below in Figure 1 is the list of two major type of hybrid system and their applications. This chapter investigates hybrid nano systems broadly into two categories firstly inorganic, which include metallic nanoparticles like gold (Au) and silver (Ag), bimetallic nanoparticles like iron cobalt (Fe–Co) and iron platinum (Fe–Pt) and metal oxides including titanium dioxide (TiO_2), cerium dioxide (CeO_2), silica (SiO_2) and zinc oxide (ZnO) for possible use in biomedical applications. We focus on investigating the potential of metallic (Ag), bimetallic (Fe–Pt) and metal oxide (TiO_2). Secondly, Organic nanohybrids, such as liposomes, micelles and dendrimer for their possible use in drug delivery, hyperthermia, bioimaging, photothermal therapy, cancer therapy and biosensors.

2. INORGANIC BUILDING BLOCKS

Inorganic building blocks in the context of hybrid nanoparticles refer to the inorganic components or structures that make up a significant part of the overall nanoparticle. These inorganic building blocks can be nanoparticles themselves or other inorganic structures, and they are often combined with organic molecules, polymers, or other materials to create hybrid nanoparticles with unique properties and functionalities (Mir et al., 2018). Here are some common types of inorganic building blocks used in hybrid nanoparticles.

2.1 Metal Nanoparticles

Gold (Au) and silver (Ag) nanoparticles are frequently used for their unique optical, electronic, and catalytic properties. Also, nanoparticles of transition metals like iron (Fe), cobalt (Co), and platinum (Pt) have magnetic and catalytic properties (Adamski et al., 2018). Among these metal nanoparticles Gold nanostructures display self-therapeutic properties to fight cancers, including inhibition of tumoral angiogenesis, inhibition of MAPK-signaling to reverse epithelial-mesenchymal transition in cancer, and alteration of non-viral gene delivery uptake pathways to avoid lysosomal degradation.

2.2 Metal Oxide Nanoparticles

Iron oxide nanoparticles, such as magnetite (Fe_3O_4) and maghemite (γ-Fe_2O_3), are commonly used for their magnetic properties and applications in imaging, drug delivery, and hyperthermia treatment (Ajinkya et al., 2020). Semiconductor oxides such as Titanium dioxide (TiO_2), zinc oxide (ZnO), and other semiconductor oxide nanoparticles are used for their photocatalytic and electronic properties (Sazonov et al., 2021) (Table 1). Also, semiconductor nanocrystals known as quantum dots (QDs) exhibit size-dependent optical properties, such as tunable fluorescence, making them useful for imaging and sensing applications and Lanthanide ions (e.g., europium, gadolinium) and their complexes are used for luminescence-based applications, including bioimaging and sensing (Suárez et al., 2019). Small assemblies of metal atoms such as gold-silver alloy, known as nanoclusters, exhibit properties that can differ from both individual atoms and bulk materials. They are used in catalysis and electronics.

2.3 Other Inorganic Structures

Carbon-based materials, such as carbon nanotubes and graphene, are sometimes considered inorganic building blocks due to their unique electronic and mechanical. Silica (SiO_2) nanoparticles are commonly used for their biocompatibility and surface functionalization capabilities.

3. ORGANIC BUILDING BLOCKS

These organic building blocks can be designed to provide specific functions, surface properties, or interactions within the nanohybrid structure. Organic polymers can be used to coat inorganic nanoparticles, enhancing their stability, biocompatibility, and functionalization capabilities (Datta et al., 2022). Further, organic nanoparticles are nano-sized organic materials, such as dendrimers, micelles, and liposomes, that can encapsulate drugs, genes, or other molecules for targeted delivery and release. Liposomes are a type of organic nanohybrid nanoparticle composed of lipid molecules that self-assemble into spherical vesicles. These vesicles consist of one or more lipid bilayers surrounding an aqueous core. Liposomes are widely used in various fields, including medicine, cosmetics, and research, due to their unique properties and ability to encapsulate wide range of substance. AuNP-liposomes can also deliver a combined gene/protein payload such as that required for CRISPR/Cas9 therapy (Seaberg et al., 2021). The delivery of Cas9 protein and sgRNA using a gold/lipid-based vector significantly reduced PIk1 expression in A375 cells *in vitro*. A subsequent investigation revealed that the plasmonic characteristics of AuNPs enabled the targeted release of Cas9-sgPlk-1 plasmids in response to laser light to more precisely regulate gene-editing procedures *in vivo* (Rahimi et al., 2020). As such, AuNP-liposome hybrids appear poised to deliver next-generation gene-editing technologies in upcoming years.

Polymers form the second class of organic building blocks. Cationic polymers such as chitosan, poly(ethylene imine) (PEI), and poly(L-lysine) (PLL) enhance cellular uptake through electrostatic interactions with anionic cell membranes (Zheng et al., 2021). The third organic building blocks are dendrimers, which are branching, radially-symmetric star-shaped biomolecules made up of repetitive units dispersing from an origin. Dendrimers are especially suited for biomedical applications because of their monodispersity, non-toxicity, solubility, self-assembly, and stability (Qi et al., 2023). Polyamidoamine (PAMAM) is the standard dendrimer used in hybrid nano systems and has been useful for gene delivery, chemotherapy, and *in vivo* imaging.

4. BIOORGANIC BUILDING BLOCKS

Natural or engineered biomolecules can be integrated into nanohybrids to create biocompatible and functional materials for applications in medicine and biotechnology. The choice of organic building blocks depends on the desired properties and functions of the nanohybrid structure. These organic components can introduce functionalities like biocompatibility, controlled release, surface modifications, and specific interactions (Sanità et al., 2020). The characterization and optimization of nanohybrids with organic building blocks involve understanding their interactions, structural arrangement, and resulting properties for various applications.

Figure 1. Types of nano-hybrid system and their biomedical application

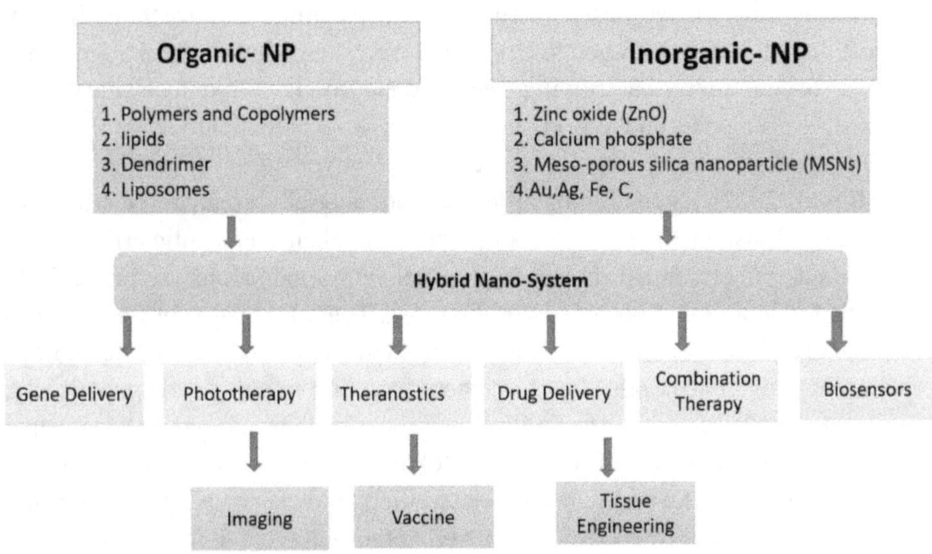

5. BIOMEDICAL APPLICATIONS OF NANOHYBRIDS

5.1. Gene Delivery

Nanotechnology has played a significant role in advancing gene delivery techniques. Nanoparticles and other nanomaterials can be engineered to carry and protect genetic material, improve cellular uptake, and enhance the targeted delivery of genes to specific tissues or cells (Mirza & Karim, 2021). This can increase the efficiency and reduce the potential side effects of gene therapy. Nanoscale gold is unmatched in combination gene-PTT due to its self-therapeutic, biocompatible, and programmable absorption capabilities. In order to build nano systems for the treatment of lung cancer, breast cancer, and gastric cancer, biocompatible polymers have been coupled with AuNPs, AuNRs, AuNSs, and gold nano prisms (Bolaños et al., 2019). Gold nanostructures are frequently used in chemotherapy-phototherapy, a form of advanced cancer treatment, to collect near-infrared radiation (NIR) and transform it into thermal energy. Laser-activated medication release can be combined with PTT using AuNS-coated liposomes containing chemotherapeutics (Kumar & Lim, 2023). Nanohybrids are utilized in gene delivery systems to efficiently transport genetic material (such as DNA or RNA) into cells. The combination of nanoparticles with nucleic acids enables targeted and controlled gene therapy approaches. By modifying the surface properties of nanohybrids, researchers can achieve effective cellular uptake and gene expression (Hasan et al., 2019).

5.2. Drug Delivery

Nanohybrids are extensively used in drug delivery systems to improve the efficiency and targeting of therapeutic agents. Drug delivery systems (DDSs) seek to increase the local concentration of a drug in a specified environment to reduce side effects and improve efficiency, and hybrid nano systems show potential to overcome drug delivery barriers like rapid *in vivo* clearance and selective drug encapsulation/release (Garello et al., 2022). By combining different nanomaterials, researchers can achieve controlled and sustained release of drugs, enhancing their therapeutic efficacy and reducing side effects. Small molecule drug delivery applications are particularly well-suited to inorganic-liposome hybridizations. The lipid bilayer of liposomes is useful for drug loading and can isolate medicines that have been encapsulated from their surroundings. On the other hand, magnetic targeting, imaging, biocompatibility, and photothermal drug release capabilities may be offered by inorganic components (Choi et al., 2021). For use in medication delivery, chemotherapeutic drugs like paclitaxel (PTX) and doxorubicin (DOX) have frequently been encapsulated inside gold-liposome hybrids. Nanoparticles with photothermal agents or photosensitizers, respectively, nanohybrids can selectively target and destroy cancer cells upon exposure to light such as Photothermal Therapy (PTT) and Photodynamic Therapy (PDT): Nanohybrids are employed in PTT and PDT these therapies offer minimally invasive approaches with reduced side effects compared to traditional treatments (Wang et al., 2020).

5.3. Combination Therapy

Hybrid nano-systems excel in combination therapies using several techniques (e.g., combined chemo/phototherapy, gene/drug codelivery, etc.) to more effectively treat malignancies (Seaberg et al., 2021). This is because they combine the advantages of hybridization in gene and drug delivery as well as phototherapy. Moreover, an architectural standpoint, gold-functionalized liposomes combine the loading capacity and transfection of liposomes with the optical and self-therapeutic capabilities of nanoscale gold, making them ideal for chemo-phototherapy and co-delivery of genes and small molecule medicines (Song et al., 2021). Some of the popular combination therapy are gene/drug co delivery is nanoscale gold and liposomes, and these systems have been further modified for magnetic targeting and photo-triggered release of encapsulated therapeutics. Example of gene delivery and photothermal therapy are nanoscale gold combined with tunable absorption properties make it unparalleled in combination gene-PTT. AuNPs and gold nano-prism have been combined with biocompatible polymers to create nano-systems to treat lung cancer, breast cancer, and gastric cancer. Another combination therapy is the Chemo-phototherapy which has recently emerged as a next-generation cancer treatment in which gold nanostructures are regularly employed to harvest NIR radiation and convert it to thermal energy (Gao et al., 2021). AuNPs-coated liposomes containing chemotherapeutics can couple laser-activated drug release with PTT.

5.4. Tissue Regeneration and Remodeling

To create biological replacements for injured tissues, tissue engineering attempts to mimic the properties of native tissue. In the creation of biomimetic tissue structures, hybrid nano systems have significantly advanced the field. Both organic and inorganic building blocks have demonstrated potential by controlling cell division and proliferation as well as modifying the extracellular matrix' (ECM) physical attributes (Gao et al., 2021). These improvements are demonstrated for regeneration of the tissue types such as skin, bone and cartilage, nerve, cardiac, and vascular tissues. One of the most utilized substances for treating tissue burn infections and persistent ulcers is AgNPs.

5.4.1 Wound Management

Hybrid nano systems have facilitated wound healing in different capacities. In particular, metal nanoparticles incorporated with injectable and wound dressing patches can promote wound healing efficacy. This improvement stems from the active tissue regenerative and antimicrobial function of nanoparticle components during the stages of wound healing. AgNPs are one of the most widely used compounds for treating chronic ulcers and tissue burn infections (Singh et al., 2022). The non-specific release of AgNPs which initially raised toxicity concerns was circumvented in the recent silver-based products such as Acticoat Flex 3 wound dressings.

5.4.2 Bone and Cartilage Tissue Engineering

It involves the development of artificial materials and structures to replace or repair damaged or diseased bone and cartilage tissues in the human body. Nano-hybrid materials play a crucial role in enhancing the performance and biocompatibility of these engineered tissues. Here are some examples of bone and cartilage tissue engineering and nano-hybrid materials are Hydroxyapatite (HA)-Based Nanocomposites of which Hydroxyapatite is a biocompatible material found in natural bone (Jiang et al., 2020). Researchers have developed nanocomposites by incorporating HA nanoparticles into various polymer matrices (e.g., poly(lactic-co-glycolic acid) or PLGA) to create scaffolds that mimic the natural bone structure. These nanocomposites promote cell adhesion and proliferation and enhance bone regeneration. Further, Graphene Oxide (GO) and Carbon Nanotubes (CNTs) have been incorporated into tissue-engineered constructs to enhance mechanical properties and electrical conductivity (Shamekhi et al., 2019). In bone tissue engineering, these nanostructures can improve the strength and conductivity of scaffolds, which can be beneficial for bone regeneration and monitoring of tissue growth. Similarly, Chitosan and hyaluronic acid are natural polymers with excellent biocompatibility. Combining them with nanoparticles, such as gold or silver, can enhance their mechanical properties and bioactivity. These nanocomposites have been used in cartilage tissue engineering to support chondrogenesis and maintain the cartilage matrix (Gorain et al., 2018).

5.4.3 Vascular Tissue Engineering

Vascular tissue engineering focuses on developing artificial blood vessels and vascular networks to replace or repair damaged or diseased blood vessels in the human body. Nano-hybrid materials play a crucial role in enhancing the performance and functionality of these engineered vascular tissues (Dhania et al., 2022). Here are some examples of vascular tissue engineering and nano-hybrid materials. Electro spinning techniques are used to create nanofiber scaffolds from biodegradable polymers like poly(lactic-co-glycolic acid) (PLGA) or poly(caprolactone) (PCL). Incorporating nanoscale features into these fibers enhances the mechanical strength and biocompatibility of vascular grafts. Additionally, nanoparticles can be embedded within the fibers to provide controlled drug release for anti-inflammatory or pro-angiogenic purposes (Kargozar et al., 2020). Also, graphene and its derivatives, like graphene oxide (GO), have shown promise in vascular tissue engineering due to their excellent mechanical properties and ability to promote cell adhesion and proliferation. They can be incorporated into vascular scaffolds to enhance their mechanical strength and improve cellular interactions. Also, bioactive ceramics, such as hydroxyapatite nanoparticles or bioactive glass nanoparticles, can be combined with biodegradable polymers to create nanocomposite vascular grafts (Dreanca et al., 2021). These materials improve the graft's mechanical properties and encourage tissue integration by promoting osteogenesis and angiogenesis (Kargozar et al., 2020).

5.5 Antimicrobials

It's critical to understand how various materials interact with various viruses in light of the recent global epidemiological danger posed by SARS-CoV-2. One of the main issues in medicine, for instance, is the decreased effectiveness of antibiotics as a result of rising bacterial resistance. Therefore, it's critical to identify fresh approaches to this problem. Silver has been known as an antimicrobial agent since antiquity. In recent years, the silver nanoparticles (AgNPs) have attracted considerable attention because of their unique antimicrobial activity among other properties (Silva et al., 2017). Poly(N,N′-(dimethylamino)ethyl methacrylate) (PDMAEMA) is a good cationic polymer for stabilizing AgNPs and exhibits biocidal properties by itself (Teper et al., 2020). Another, "Green" synthesis of AgNPs using the *N,N′*-dimethylaminoethyl methacrylate homopolymer has been described in very few researches. Recently, micelles of linear and 4-arm triblock copolymers of DMAEMA, 2-hydroxyethyl methacrylate, and poly(ethylene glycol) monomethyl ether methacrylate (PEGMA) for AgNPs formation and investigated antimicrobial activity against one bacterial strain *Escherichia coli* results showed higher concentration of hybrid AgNPs is used to inhibit the growth of *E. coli*.

5.6. Vaccine Delivery

Vaccines have prevented more than 20 life-threatening diseases, helping people of all ages live longer, healthier lives. Immunization currently prevents 3.5-5 million deaths every year from diseases like diphtheria, tetanus, pertussis, influenza and measles. Despite progress in vaccine production there are still issues with vaccination efficacy and thermostability that can be resolved by hybrid nano systems (Maiti et al., 2023). Nanohybrids are explored for vaccine delivery to improve the immune response. They can enhance antigen presentation, promote targeted delivery to immune cells, and provide controlled release of antigens and adjuvants (Homaeigohar et al., 2021). AuNP-based vaccines have been developed against pathogen ranging from dengue virus to *Burkholderia mallei* and enterohemorrhagic *E. coli*. DNA-based malaria vaccine from IONPs coated in PEI, claiming magnetic targeting and pH-responsive DNA condensation produced. Another field where vaccines are being developed is cancer immunotherapy. A range of nanomaterials and architectural designs have been used to create hybrid nano systems, which are effective cancer vaccines. Hollow MSNs hybridized with PEI have shown improved Th1 antitumor immunity and memory in comparison to hollow MSNs alone. Also, MSNs coated in dextran and encapsulated within cancer cell membranes have shown similar potential for cancer immunotherapy Hybrid nano systems-based vaccines show several benefits over conventional. They provide thermostability, and versatility to treat and prevent a variety of diseases (Park et al., 2020).

5.7. Theranostics

Nanohybrids play a crucial role in theranostics, which integrates diagnosis and therapy into a single platform (Chen et al., 2021). They possess biodegradable and biocompatible leading to the management of disease in a more specific way. By incorporating imaging agents and therapeutic payloads into a single nanohybrid system, clinicians can visualize disease sites and simultaneously deliver treatments (Khot et al., 2020). This approach enables real-time monitoring of treatment response and adjustment if needed. These nanohybrids were found to be particularly promising in cancer treatment, allowing for personalized and precise medicine. Various metallic and metal oxide nanohybrids show theranostics effects such as graphene oxide (GO)-based iron oxide nanoparticles (IONPs) nanohybrids used for diagnostic purposes and as a drug delivery agent. Here, the GO is functionalized with cyclodextrin, improving the loading of doxorubicin (DOX) and also cellular uptake. While for therapy and diagnostic purpose "targeted drug delivery" and "magnetic resonance imaging" (MRI) respectively (Khot et al., 2020). A theranostics approach for NHs having magnetic Fe_3O_4 NPs as an inorganic core for diagnostic purposes in MRI (Magnetic Resonance Imaging) and the tumor treatment magnetic hyperthermia property is being used. $Zn_{0.4}Fe_{2.6}O_4$ has a superparamagnetic feature, the magnetization value of the NHs induces hyperthermia and it activates the release of the drug that was encapsulated.

5.8. Imaging Agents

Nanohybrids are employed as contrast agents in various imaging modalities, such as magnetic resonance imaging (MRI), computed tomography (CT), positron emission tomography (PET), and fluorescence imaging (Luengo Morato et al., 2021). These agents enhance the visibility of specific tissues, cells, or molecular processes, aiding in the early detection and diagnosis of diseases. For example, nanoparticles coated with imaging agents can target cancer cells, providing improved imaging resolution.

5.9. Biosensors

Nanohybrids are integrated into biosensor platforms for the detection of biomolecules, pathogens, and other analytes (Soto & Orozco, 2022). Many different types of biosensors have been developed, such as the cellulose strip sensor. By using core-shell NHs of silver-gold (Ag-Au) NPs with surface-enhanced Raman scattering (SERS). Metal oxides like tungsten oxide (WO_3), titanium oxide (TiO_2), tin oxide (SnO_2), and molybdenum oxide (Mo-Ox) were employed in the fabrication of biosensors NHs in the form of single or coupled (heterostructure or composite form) semiconductors. Carbon-based NPs are also doped with metallic ions. There are many uses for the NPs of TiO_2, whether they are inorganic-inorganic or inorganic-organic NHs, in the development of biosensors. The combination of different nanomaterials enhances sensitivity, selectivity, and response time of biosensors (Naresh & Lee, 2021). Nanohybrid-based biosensors are valuable tools for diagnosing diseases, monitoring biomarkers, and detecting infections.

6. FUTURE DIRECTIONS AND CHALLENGES

The biomedical applications of hybrid nanoparticles hold great promise for revolutionizing the field of medicine (Mottaghitalab et al., 2019). The future

Tabe 1. Inorganic nanohybrid materials and their applications

Iron Nanoparticles	Gold Nanoparticles	Silver Nanoparticles	Carbon Nanoparticles	Titanium Nanoparticles
Antibacterial Tissue Engineering MRI Contrast Targeted Drug delivery (Ezealigo et al., 2021) (Shen et al., 2017)	Non-viral gene delivery Anticancer Inhibit tumoral angiogenesis Chemotherapy-phototherapy (Rani, 2017)	Antimicrobial agent Antiviral Anti-Inflammatory Anti-oxidative (Burduşel et al., 2018)	targeted drug delivery Imaging tissue-engineered Biosensors (Friedrich et al., 2021)	targeted drug delivery photodynamic therapy Biosensors (Vangijzegem et al., 2019)

directions of hybrid nanoparticles hold promising prospects across various fields. In medicine, these nanoparticles are envisioned to revolutionize targeted drug delivery, enabling precise and controlled release of therapeutics while minimizing side effects (Afzal et al., 2022). Their multifunctional nature allows for simultaneous imaging and treatment in diagnostics and therapy. In the realm of renewable energy, hybrid nanoparticles show potential for more efficient solar cells, catalysis, and energy storage systems, contributing to sustainable energy solutions (Banin et al., 2020). The development of novel materials and coatings employing hybrid nanoparticles is anticipated to enhance functionalities in electronics, sensors, and even environmental remediation. With ongoing advancements in nanotechnology, future research will likely focus on refining synthesis methods to achieve better control over nanoparticle properties, addressing challenges of toxicity, biocompatibility, and scalability for broader applications (Chehelgerdi et al., 2023). Furthermore, the exploration of innovative designs and compositions, coupled with a deeper understanding of their behavior within biological systems, will pave the way for safer, more effective, and tailored applications of hybrid nanoparticles, marking a dynamic and transformative path in diverse scientific and industrial domains. Challenges such as regulatory hurdles, manufacturing scalability, and cost-effectiveness remain to be addressed before hybrid nanoparticles can become widely adopted in clinical practice (Younis et al., 2022). Additionally, ongoing research is needed to optimize the design and functionality of hybrid nanoparticles for specific medical applications.

The toxicity of hybrid nanoparticles presents a critical concern across various applications. Comprising diverse materials, these nanoparticles pose potential risks due to their chemical composition, size, surface characteristics, and interactions with biological systems (Egbuna et al., 2021). The materials used can introduce toxic elements or compounds that, when in contact with living organisms, may induce adverse reactions. Moreover, their nano-sized nature enables easy penetration of biological barriers, potentially interacting with cellular structures and triggering harmful effects (Wang et al., 2021). Surface modifications, while intended to improve stability or functionality, might inadvertently increase reactivity, impacting their biological behavior. Understanding how these nanoparticles distribute within the body, potentially accumulating in vital organs, is crucial to evaluate their toxic effects and long-term consequences. Researchers conduct comprehensive studies to assess their impact on cells, tissues, and living organisms, aiming to mitigate these concerns through surface modifications, alternative materials, and encapsulation techniques. Moreover, the potential environmental impact, including effects on ecosystems and bioaccumulation, is a growing area of concern that demands further investigation to ensure the safe and responsible application of hybrid nanoparticles (Corsi et al., 2018).

CONCLUSION

In conclusion, the realm of nanohybrids in biomedical applications represents a promising frontier with remarkable potential for revolutionizing diagnostics, therapeutics, and personalized medicine. These nanoparticles, engineered through the fusion of diverse materials, offer a versatile platform with multifaceted functionalities and unique properties that cater to various biomedical needs. Nanohybrids in biomedical applications continue to evolve through interdisciplinary collaborations, advancements in fabrication techniques, and deeper insights into their interactions within biological systems. Continued research endeavors aim to refine their design, optimize their properties, and unravel new avenues for their utilization in treating diseases, enhancing diagnostics, and advancing regenerative medicine. The journey of nanohybrids in the biomedical realm is poised to shape the future of healthcare by offering tailored, targeted, and efficient solutions that hold immense promise in addressing unmet medical needs and improving patient outcomes. As research progresses, the potential for these nanohybrids to transform the landscape of medicine remains compelling and inspiring.

REFERENCES

Adamski, J., Qadir, M. I., Serna, J. P., Bernardi, F., Baptista, D. L., Salles, B. R., Novak, M. A., Machado, G., & Dupont, J. (2018). Core–shell Fe–Pt nanoparticles in ionic liquids: Magnetic and catalytic properties. *The Journal of Physical Chemistry. C, Nanomaterials and Interfaces*, *122*(8), 4641–4650. doi:10.1021/acs.jpcc.7b12219

Afzal, O., Altamimi, A. S., Nadeem, M. S., Alzarea, S. I., Almalki, W. H., Tariq, A., Mubeen, B., Murtaza, B. N., Iftikhar, S., Riaz, N., & Kazmi, I. (2022). Nanoparticles in Drug Delivery: From History to Therapeutic Applications. *Nanomaterials (Basel, Switzerland)*, *12*(24), 4494. doi:10.3390/nano12244494 PMID:36558344

Ajinkya, N., Yu, X., Kaithal, P., Luo, H., Somani, P., & Ramakrishna, S. (2020). Magnetic iron oxide nanoparticle (IONP) synthesis to applications: Present and future. *Materials (Basel)*, *13*(20), 4644. doi:10.3390/ma13204644 PMID:33080937

Alwattar, J. K., Mneimneh, A. T., Abla, K. K., Mehanna, M. M., & Allam, A. N. (2021). Smart stimuli-responsive liposomal nanohybrid systems: A critical review of theranostic behavior in cancer. *Pharmaceutics*, *13*(3), 355. doi:10.3390/pharmaceutics13030355 PMID:33800292

Bagheri, E., Ansari, L., Abnous, K., Taghdisi, S. M., Charbgoo, F., Ramezani, M., & Alibolandi, M. (2018). Silica based hybrid materials for drug delivery and bioimaging. *Journal of Controlled Release, 277*, 57–76. doi:10.1016/j.jconrel.2018.03.014 PMID:29550399

Banin, U., Waiskopf, N., Hammarström, L., Boschloo, G., Freitag, M., Johansson, E. M., Sá, J., Tian, H., Johnston, M., Herz, L., Milot, R. L., Kanatzidis, M. G., Ke, W., Spanopoulos, I., Kohlstedt, K. L., Schatz, G. C., Lewis, N., Meyer, T., Nozik, A. J., ... Brudvig, G. W. (2020). Nanotechnology for catalysis and solar energy conversion. *Nanotechnology, 32*(4), 042003. doi:10.1088/1361-6528/abbce8 PMID:33155576

Bohara, R. A. (2019). Introduction and types of hybrid nanostructures for medical applications. In *Hybrid Nanostructures for Cancer Theranostics* (pp. 1–16). Elsevier. doi:10.1016/B978-0-12-813906-6.00001-9

Bolaños, K., Kogan, M. J., & Araya, E. (2019). Capping gold nanoparticles with albumin to improve their biomedical properties. *International Journal of Nanomedicine, 14*, 6387–6406. doi:10.2147/IJN.S210992 PMID:31496693

Burduşel, A.-C., Gherasim, O., Grumezescu, A. M., Mogoantă, L., Ficai, A., & Andronescu, E. (2018). Biomedical applications of silver nanoparticles: An up-to-date overview. *Nanomaterials (Basel, Switzerland), 8*(9), 681. doi:10.3390/nano8090681 PMID:30200373

Chehelgerdi, M., Chehelgerdi, M., Allela, O. Q. B., Pecho, R. D. C., Jayasankar, N., Rao, D. P., Thamaraikani, T., Vasanthan, M., Viktor, P., Lakshmaiya, N., Saadh, M. J., Amajd, A., Abo-Zaid, M. A., Castillo-Acobo, R. Y., Ismail, A. H., Amin, A. H., & Akhavan-Sigari, R. (2023). Progressing nanotechnology to improve targeted cancer treatment: Overcoming hurdles in its clinical implementation. *Molecular Cancer, 22*(1), 169. doi:10.1186/s12943-023-01865-0 PMID:37814270

Chen, G., Qian, Y., Zhang, H., Ullah, A., He, X., Zhou, Z., Fenniri, H., & Shen, J. (2021). Advances in cancer theranostics using organic-inorganic hybrid nanotechnology. *Applied Materials Today, 23*, 101003. doi:10.1016/j.apmt.2021.101003

Choi, G., Rejinold, N. S., Piao, H., & Choy, J.-H. (2021). Inorganic–inorganic nanohybrids for drug delivery, imaging and photo-therapy: Recent developments and future scope. *Chemical Science (Cambridge), 12*(14), 5044–5063. doi:10.1039/D0SC06724E PMID:34168768

Corsi, I., Winther-Nielsen, M., Sethi, R., Punta, C., Della Torre, C., Libralato, G., Lofrano, G., Sabatini, L., Aiello, M., Fiordi, L., Cinuzzi, F., Caneschi, A., Pellegrini, D., & Buttino, I. (2018). Ecofriendly nanotechnologies and nanomaterials for environmental applications: Key issue and consensus recommendations for sustainable and ecosafe nanoremediation. *Ecotoxicology and Environmental Safety*, *154*, 237–244. doi:10.1016/j.ecoenv.2018.02.037 PMID:29476973

Datta, D., Das, K. P., Deepak, K., & Das, B. (2022). Candidates of functionalized nanomaterial-based membranes. In *Membranes with Functionalized Nanomaterials* (pp. 81–127). Elsevier. doi:10.1016/B978-0-323-85946-2.00004-7

Dhania, S., Bernela, M., Rani, R., Parsad, M., Grewal, S., Kumari, S., & Thakur, R. (2022). Scaffolds the backbone of tissue engineering: Advancements in use of polyhydroxyalkanoates (PHA). *International Journal of Biological Macromolecules*, *208*, 243–259. doi:10.1016/j.ijbiomac.2022.03.030 PMID:35278518

Dreanca, A., Muresan-Pop, M., Taulescu, M., Tóth, Z.-R., Bogdan, S., Pestean, C., Oren, S., Toma, C., Popescu, A., Páll, E., Sevastre, B., Baia, L., & Magyari, K. (2021). Bioactive glass-biopolymers-gold nanoparticle based composites for tissue engineering applications. *Materials Science and Engineering C*, *123*, 112006. doi:10.1016/j.msec.2021.112006 PMID:33812626

Egbuna, C., Parmar, V. K., Jeevanandam, J., Ezzat, S. M., Patrick-Iwuanyanwu, K. C., Adetunji, C. O., Khan, J., Onyeike, E. N., Uche, C. Z., Akram, M., Ibrahim, M. S., El Mahdy, N. M., Awuchi, C. G., Saravanan, K., Tijjani, H., Odoh, U. E., Messaoudi, M., Ifemeje, J. C., Olisah, M. C., ... Ibeabuchi, C. G. (2021). Toxicity of nanoparticles in biomedical application: Nanotoxicology. *Journal of Toxicology*, *2021*, 1–21. doi:10.1155/2021/9954443 PMID:34422042

Ezealigo, U. S., Ezealigo, B. N., Aisida, S. O., & Ezema, F. I. (2021). Iron oxide nanoparticles in biological systems: Antibacterial and toxicology perspective. *JCIS Open*, *4*, 100027. doi:10.1016/j.jciso.2021.100027

Friedrich, R. P., Cicha, I., & Alexiou, C. (2021). Iron oxide nanoparticles in regenerative medicine and tissue engineering. *Nanomaterials (Basel, Switzerland)*, *11*(9), 2337. doi:10.3390/nano11092337 PMID:34578651

Gao, Q., Zhang, J., Gao, J., Zhang, Z., Zhu, H., & Wang, D. (2021). Gold nanoparticles in cancer theranostics. *Frontiers in Bioengineering and Biotechnology*, *9*, 647905. doi:10.3389/fbioe.2021.647905 PMID:33928072

Garello, F., Svenskaya, Y., Parakhonskiy, B., & Filippi, M. (2022). Micro/nanosystems for magnetic targeted delivery of bioagents. *Pharmaceutics*, *14*(6), 1132. doi:10.3390/pharmaceutics14061132 PMID:35745705

Gorain, B., Choudhury, H., Pandey, M., Kesharwani, P., Abeer, M. M., Tekade, R. K., & Hussain, Z. (2018). Carbon nanotube scaffolds as emerging nanoplatform for myocardial tissue regeneration: A review of recent developments and therapeutic implications. *Biomedicine and Pharmacotherapy*, *104*, 496–508. doi:10.1016/j.biopha.2018.05.066 PMID:29800914

Hasan, M. N., Nafiujjaman, M., & Lee, Y.-K. (2019). 2D nanomaterials for gene delivery. In *Biomedical Applications of Graphene and 2D Nanomaterials* (pp. 87–104). Elsevier. doi:10.1016/B978-0-12-815889-0.00004-0

Homaeigohar, S., Liu, Q., & Kordbacheh, D. (2021). Biomedical applications of antiviral nanohybrid materials relating to the COVID-19 pandemic and other viral crises. *Polymers*, *13*(16), 2833. doi:10.3390/polym13162833 PMID:34451371

Jiang, S., Liu, X., Liu, Y., Liu, J., He, W., & Dong, Y. (2020). Synthesis of silver@hydroxyapatite nanoparticles based biocomposite and their assessment for viability of Osseointegration for rabbit knee joint anterior cruciate ligament rehabilitation. *Journal of Photochemistry and Photobiology. B, Biology*, *202*, 111677. doi:10.1016/j.jphotobiol.2019.111677 PMID:31810037

Kargozar, S., Baino, F., Hamzehlou, S., Hamblin, M. R., & Mozafari, M. (2020). Nanotechnology for angiogenesis: Opportunities and challenges. *Chemical Society Reviews*, *49*(14), 5008–5057. doi:10.1039/C8CS01021H PMID:32538379

Karimi, M., Mansouri, M. R., Rabiee, N., & Hamblin, M. R. (2018). Carbon-based nanomaterials. *Advances in nanomaterials for drug delivery: Polymeric, nanocarbon, and bio-inspired.*

Keshvardoostchokami, M., Majidi, S. S., Huo, P., Ramachandran, R., Chen, M., & Liu, B. (2020). Electrospun nanofibers of natural and synthetic polymers as artificial extracellular matrix for tissue engineering. *Nanomaterials (Basel, Switzerland)*, *11*(1), 21. doi:10.3390/nano11010021 PMID:33374248

Khot, S., Rawal, S. U., & Patel, M. M. (2020). Theranostics Nanoformulations: Merging Diagnostics and Nanotherapeutics. *Nanoformulations in Human Health: Challenges and Approaches*, 73-91.

Kukushkina, E. A., Hossain, S. I., Sportelli, M. C., Ditaranto, N., Picca, R. A., & Cioffi, N. (2021). Ag-based synergistic antimicrobial composites. A critical review. *Nanomaterials (Basel, Switzerland)*, *11*(7), 1687. doi:10.3390/nano11071687 PMID:34199123

Kumar, P. P. P., & Lim, D.-K. (2023). Photothermal Effect of Gold Nanoparticles as a Nanomedicine for Diagnosis and Therapeutics. *Pharmaceutics*, *15*(9), 2349. doi:10.3390/pharmaceutics15092349 PMID:37765317

Kumar, R., Mondal, K., Panda, P. K., Kaushik, A., Abolhassani, R., Ahuja, R., Rubahn, H.-G., & Mishra, Y. K. (2020). Core–shell nanostructures: Perspectives towards drug delivery applications. *Journal of Materials Chemistry. B*, *8*(39), 8992–9027. doi:10.1039/D0TB01559H PMID:32902559

Luengo Morato, Y., Ovejero Paredes, K., Lozano Chamizo, L., Marciello, M., & Filice, M. (2021). Recent advances in multimodal molecular imaging of cancer mediated by hybrid magnetic nanoparticles. *Polymers*, *13*(17), 2989. doi:10.3390/polym13172989 PMID:34503029

Maiti, B., Harshitha, M., Disha, S., Badekila, A. K., Kini, S., & Rai, P. (2023). Nanovaccine. In *Nanotechnological Approaches to the Advancement of Innovations in Aquaculture* (pp. 37–65). Springer. doi:10.1007/978-3-031-15519-2_3

Mir, S. H., Nagahara, L. A., Thundat, T., Mokarian-Tabari, P., Furukawa, H., & Khosla, A. (2018). Organic-inorganic hybrid functional materials: An integrated platform for applied technologies. *Journal of the Electrochemical Society*, *165*(8), B3137–B3156. doi:10.1149/2.0191808jes

Mirza, Z., & Karim, S. (2021). Nanoparticles-based drug delivery and gene therapy for breast cancer: Recent advancements and future challenges. Seminars in cancer biology. *Journal of Controlled Release*, *295*, 250–267.

Naresh, V., & Lee, N. (2021). A review on biosensors and recent development of nanostructured materials-enabled biosensors. *Sensors (Basel)*, *21*(4), 1109. doi:10.3390/s21041109 PMID:33562639

Nik Mohd Adnan, N. (2018). *Hybrid polymer-inorganic nanoparticles for biomedical applications*. UNSW Sydney.

Park, S. S., & Ha, C. S. (2018). Hollow mesoporous functional hybrid materials: Fascinating platforms for advanced applications. *Advanced Functional Materials*, *28*(27), 1703814. doi:10.1002/adfm.201703814

Park, W., Shin, H., Choi, B., Rhim, W.-K., Na, K., & Han, D. K. (2020). Advanced hybrid nanomaterials for biomedical applications. *Progress in Materials Science*, *114*, 100686. doi:10.1016/j.pmatsci.2020.100686

Qi, X., Xiang, Y., Cai, E., Ge, X., Chen, X., Zhang, W., Li, Z., & Shen, J. (2023). Inorganic–organic hybrid nanomaterials for photothermal antibacterial therapy. *Coordination Chemistry Reviews*, *496*, 215426. doi:10.1016/j.ccr.2023.215426

Rahimi, H., Salehiabar, M., Charmi, J., Barsbay, M., Ghaffarlou, M., Razlighi, M. R., Davaran, S., Khalilov, R., Sugiyama, M., & Nosrati, H. (2020). Harnessing nanoparticles for the efficient delivery of the CRISPR/Cas9 system. *Nano Today*, *34*, 100895. doi:10.1016/j.nantod.2020.100895

Rani, K. (2017). Biomedical applications of silver and gold nanoparticles: Effective and safe non-viral delivery vehicles. *Journal of Applied Biotechnology and Bioengineering*, *3*(2). Advance online publication. doi:10.15406/jabb.2017.03.00059

Sahoo, K., Gazi, T. R., Roy, S., & Chakraborty, I. (2023). Nanohybrids of atomically precise metal nanoclusters. *Communications Chemistry*, *6*(1), 157. doi:10.1038/s42004-023-00958-7 PMID:37495665

Sanità, G., Carrese, B., & Lamberti, A. (2020). Nanoparticle surface functionalization: How to improve biocompatibility and cellular internalization. *Frontiers in Molecular Biosciences*, *7*, 587012. doi:10.3389/fmolb.2020.587012 PMID:33324678

Sazonov, R., Kholodnaya, G., Ponomarev, D., Zhuravlev, M., Pyatkov, I., Konusov, F., Lapteva, O., & Gadirov, R. (2021). Physicochemical, optical, and photocatalytic properties of titanium dioxide modified with zinc oxide nanoparticles synthesized by a pulsed plasma chemical method. *Physica B, Condensed Matter*, *619*, 413208. doi:10.1016/j.physb.2021.413208

Seaberg, J., Montazerian, H., Hossen, M. N., Bhattacharya, R., Khademhosseini, A., & Mukherjee, P. (2021). Hybrid nanosystems for biomedical applications. *ACS Nano*, *15*(2), 2099–2142. doi:10.1021/acsnano.0c09382 PMID:33497197

Shamekhi, M. A., Mirzadeh, H., Mahdavi, H., Rabiee, A., Mohebbi-Kalhori, D., & Eslaminejad, M. B. (2019). Graphene oxide containing chitosan scaffolds for cartilage tissue engineering. *International Journal of Biological Macromolecules*, *127*, 396–405. doi:10.1016/j.ijbiomac.2019.01.020 PMID:30625354

Shen, Z., Wu, A., & Chen, X. (2017). Iron oxide nanoparticle based contrast agents for magnetic resonance imaging. *Molecular Pharmaceutics*, *14*(5), 1352–1364. doi:10.1021/acs.molpharmaceut.6b00839 PMID:27776215

Shnoudeh, A. J., Hamad, I., Abdo, R. W., Qadumii, L., Jaber, A. Y., Surchi, H. S., & Alkelany, S. Z. (2019). Synthesis, characterization, and applications of metal nanoparticles. In *Biomaterials and bionanotechnology* (pp. 527–612). Elsevier. doi:10.1016/B978-0-12-814427-5.00015-9

Silva, L. P., Silveira, A. P., Bonatto, C. C., Reis, I. G., & Milreu, P. V. (2017). Silver nanoparticles as antimicrobial agents: Past, present, and future. In *Nanostructures for antimicrobial therapy* (pp. 577–596). Elsevier. doi:10.1016/B978-0-323-46152-8.00026-3

Singh, M., Thakur, V., Kumar, V., Raj, M., Gupta, S., Devi, N., Upadhyay, S. K., Macho, M., Banerjee, A., Ewe, D., & Saurav, K. (2022). Silver nanoparticles and its mechanistic insight for chronic wound healing: Review on recent progress. *Molecules (Basel, Switzerland)*, *27*(17), 5587. doi:10.3390/molecules27175587 PMID:36080353

Sivadasan, D., Sultan, M. H., Madkhali, O., Almoshari, Y., & Thangavel, N. (2021). Polymeric lipid hybrid nanoparticles (plns) as emerging drug delivery platform—A comprehensive review of their properties, preparation methods, and therapeutic applications. *Pharmaceutics*, *13*(8), 1291. doi:10.3390/pharmaceutics13081291 PMID:34452251

Song, M., Liu, C., Chen, S., & Zhang, W. (2021). Nanocarrier-based drug delivery for melanoma therapeutics. *International Journal of Molecular Sciences*, *22*(4), 1873. doi:10.3390/ijms22041873 PMID:33668591

Soto, D., & Orozco, J. (2022). Hybrid Nanobioengineered nanomaterial-based electrochemical biosensors. *Molecules (Basel, Switzerland)*, *27*(12), 3841. doi:10.3390/molecules27123841 PMID:35744967

Suárez, P. L., García-Cortés, M., Fernández-Argüelles, M. T., Encinar, J. R., Valledor, M., Ferrero, F. J., Campo, J. C., & Costa-Fernández, J. M. (2019). Functionalized phosphorescent nanoparticles in (bio) chemical sensing and imaging–a review. *Analytica Chimica Acta*, *1046*, 16–31. doi:10.1016/j.aca.2018.08.018 PMID:30482294

Tamayo, L., Palza, H., Bejarano, J., & Zapata, P. A. (2018). *Polymer Composites With Metal Nanoparticles: Synthesis, Properties, and Applications. In Polymer Composites with Functionalized Nanoparticles: Synthesis, Properties, and Applications*. Elsevier Inc.

Teper, P., Sotirova, A., Mitova, V., Oleszko-Torbus, N., Utrata-Wesołek, A., Koseva, N., Kowalczuk, A., & Mendrek, B. (2020). Antimicrobial activity of hybrid nanomaterials based on star and linear polymers of N, N′-dimethylaminoethyl methacrylate with in situ produced silver nanoparticles. *Materials (Basel)*, *13*(13), 3037. doi:10.3390/ma13133037 PMID:32645957

Vangijzegem, T., Stanicki, D., & Laurent, S. (2019). Magnetic iron oxide nanoparticles for drug delivery: Applications and characteristics. *Expert Opinion on Drug Delivery*, *16*(1), 69–78. doi:10.1080/17425247.2019.1554647 PMID:30496697

Wang, D., Zhou, J., Shi, R., Wu, H., Chen, R., Duan, B., Xia, G., Xu, P., Wang, H., Zhou, S., Wang, C., Wang, H., Guo, Z., & Chen, Q. (2017). Biodegradable core-shell dual-metal-organic-frameworks nanotheranostic agent for multiple imaging guided combination cancer therapy. *Theranostics*, *7*(18), 4605–4617. doi:10.7150/thno.20363 PMID:29158848

Wang, J., Ni, Q., Wang, Y., Zhang, Y., He, H., Gao, D., Ma, X., & Liang, X.-J. (2021). Nanoscale drug delivery systems for controllable drug behaviors by multi-stage barrier penetration. *Journal of Controlled Release*, *331*, 282–295. doi:10.1016/j.jconrel.2020.08.045 PMID:32866590

Wang, J., Wu, X., Shen, P., Wang, J., Shen, Y., Shen, Y., Webster, T. J., & Deng, J. (2020). Applications of inorganic nanomaterials in photothermal therapy based on combinational cancer treatment. *International Journal of Nanomedicine*, *15*, 1903–1914. doi:10.2147/IJN.S239751 PMID:32256067

Younis, M. A., Tawfeek, H. M., Abdellatif, A. A., Abdel-Aleem, J. A., & Harashima, H. (2022). Clinical translation of nanomedicines: Challenges, opportunities, and keys. *Advanced Drug Delivery Reviews*, *181*, 114083. doi:10.1016/j.addr.2021.114083 PMID:34929251

Zheng, M., Pan, M., Zhang, W., Lin, H., Wu, S., Lu, C., Tang, S., Liu, D., & Cai, J. (2021). Poly (α-l-lysine)-based nanomaterials for versatile biomedical applications: Current advances and perspectives. *Bioactive Materials*, *6*(7), 1878–1909. doi:10.1016/j.bioactmat.2020.12.001 PMID:33364529

Chapter 9
Hybrid Nanomaterial Employment in Clinical and Therapeutic Applications

Marriam Zargham
https://orcid.org/0009-0008-2724-8264
University of Health Sciences, Lahore, Pakistan

Muhammad Farhan Khan
University of Health Sciences, Lahore, Pakistan

Anam Munawar
University of Health Sciences, Lahore, Pakistan

ABSTRACT

Hybrid nanomaterials are based on a combination of materials resulting in amalgamation of unique features that can address various challenges in medicine and healthcare. Such materials offer a wide range of functionalities that can revolutionize diagnostics, therapeutics, and medical imaging. In drug delivery, tunable properties of hybrid nanomaterials enable precise targeting of specific cells or tissues, enhancing the therapeutic efficacy while minimizing side effects while their utility in diagnostic techniques facilitate early disease detection with improved sensitivity and selectivity.

INTRODUCTION

Significant developments in the realm of nanotechnology have recently opened up new avenues for healthcare and medicine. The therapeutic application of hybrid

DOI: 10.4018/979-8-3693-1261-2.ch009

nanomaterials, or materials that integrate various nanoscale components to produce multifunctional platforms, is one particularly promising direction. The history of hybrid nanomaterials use in biomedicine shows how materials science, medicine, and nanotechnology have come together, with the ultimate objective of improving patient outcomes and advancing medical treatment technologies. Researchers and clinicians are increasingly investigating the potential of these materials to overcome obstacles in drug delivery, imaging, diagnostics, and regenerative medicine.

This breakthrough has sparked ground-breaking advancements in medicine field providing previously unheard-of chances to improve the accuracy and effectiveness of clinical interventions. This becomes possible because, at the nanoscale level, nanomaterials display unique chemical and physical characteristics. These qualities, which include their large surface area, quantum effects, distinctive optical features, and other physiochemical properties make them appropriate for the employment in many clinical, pharmaceutical, imaging, catalysis and biomedical purposes. These nanometal based hybrids impart antimicrobial, antibacterial and anti-oxidants, antiviral activities and also act as drug delivery agents (Balla & Sumithra, 2023). Particle dimensions of less than one micrometer make nanoscale hybrid systems also referred to as submicrometer systems. Researchers are drawn to them because they offer a variety of therapeutic benefits in a number of domains, such as enhanced therapeutic effect, site specificity, and routes of administration (Aflori, 2021).

Hybrid nanomaterials have transformed imaging methods in the field of diagnostics. The sensitivity and resolution of imaging are greatly increased by the use of nanoparticles into contrast agents. This is especially important for tracking the effectiveness of treatment and early disease identification. Moreover, the advent of theranostic nanoparticles, which can combine therapeutic and diagnostic properties, represents a paradigm change in the direction of personalized medicine. Clinicians can now optimize patient outcomes by customizing treatments based on real-time diagnostic information (Bushara, 2023). Traditional drug delivery methods might have unintended side effects and frequently lack precision. By encapsulating therapeutic chemicals within nanocarriers and delivering them precisely to the region of interest, hybrid nanomaterials enable the creation of focused medication delivery methods that minimize systemic toxicity and increase therapeutic efficacy.

Numerous nanoparticle-based treatment techniques, including gene delivery, medication administration, and photodynamic therapy, have been developed for therapeutic uses. Mesoporous silica nanoparticles are frequently employed for controlled delivery and drug loading., while organic nanoparticles' inherent propensity to produce reactive oxygen species has been used for photodynamic treatment (Nguyen & Zhao, 2015). The field of hybrid nanomaterials aids in the advancement of methods for personalized treatment. Treatments that are more successful and have fewer adverse effects can be developed by customizing nanocarriers to fit the needs of certain patient profiles. This personalization is particularly important for cancer treatments, as different tumor types call for different approaches to treatment.

Another purpose of using nanohybrids for drug delivery purpose is the availability of active chemicals which determines the efficacy of herbal medicine, or medicinal plant species. Since conventional treatment cannot fully meet these conditions, novel carriers should carry the active constituent at a suitable concentration for the treatment and route it toward the chosen target, some nano based carriers are proved to be more effective than other conventional forms (Aflori, 2021).

Hybrid nanomaterials are a testament to the revolutionary power of interdisciplinary collaboration across three different fields in the dynamic landscape of medical innovation. Clinical applications of hybrid nanomaterials are set to change the standard of care as research advances, providing new hope and opportunities for better patient outcomes across a range of medical specialties.

BRIEF HISTORY OF USING HYBRID NANOFORMS IN MEDICINE

The usage of nanotechnological materials in the current pharmaceutical business has emerged since the 19th century and is especially advantageous for the development of delivery methods for different medicines like solid dispersions, liposomes, microcapsules, emulsions, magnetic nanocarriers, and nanocomposites. Medicines benefit from greater therapeutic impact or novel efficacy, improved drug stability, flexible dosage forms, and enhanced drug delivery pathways thanks to nanocarriers (Adel, ElMeligy, Amer, & Elkasabgy, 2023).

The original concept of nanomedicine originated from the visionary hypothesis that sufficiently small nanorobots could be manufactured and inserted into the human body to carry out molecular treatments at the cellular level. Though it was a very radical concept when it was first put forth in 1959, Nobel physicist Richard P. Feynman clearly saw the potential of nanotechnology in medical applications.

From the initial report of the liposomal structure in 1964 to the development of enhanced permeability and retention effect in 1986, the approval of doxorubicin liposomes in 1995, the successful introduction of the first nano-sized antibody drug CabiliviTM (Sanofi, France) in 2018, and the most recent approvals of lipid nanoparticles (LNP)-loaded COVID-19 mRNA vaccines in 2018, the field of nanomedicine has expanded rapidly since then. (J. Zhang et al., 2021).

Here discusses some detail of different applications of nano-based hybrids in biomedicine industry.

Figure 1. Different types of nanocarriers
Source: Reproduced from MDPI (Lôbo et al., 2021)

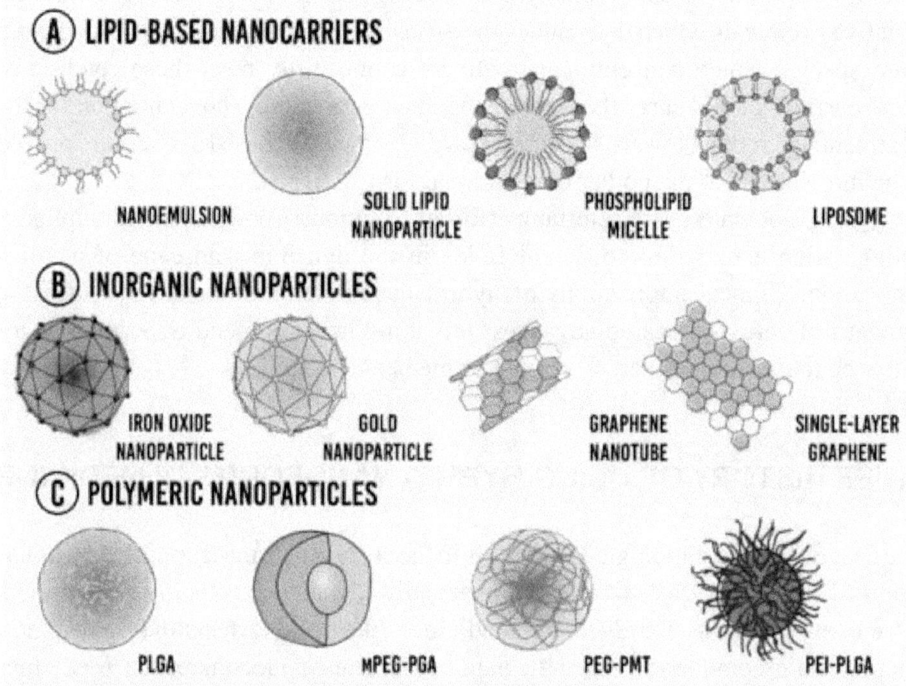

Diagnosis

Early disease identification is linked to diagnosis, after which treatment can start. With the help of appropriate targeting ligands integrated into the nanoparticles, more effective targeted optical imaging should be possible. Nanohybrid with their many marvelous characteristics can be employed in various imaging techniques such as optical imaging, PET (Positron emission tomography), MRI (Magnetic Resonance Imaging, US (Ultrasound) imaging and CT (Computed Tomography) imaging for precise detail regarding molecular, anatomical or physiological phenomena. For clinical purposes, a high standard image contrast is required to visualize the appropriate physical structure, or to make appropriately sensitive for the therapeutic purpose. Utilizing the inherent characteristics of specific nanoparticles, novel imaging methods have been created, including multiphoton excitation and photoacoustic imaging with near-infrared light (Spesivtseva, Gvarzhdets, Autlev, Kruchinin, & Kuznetsov, 2023). While reducing the negative effects on normal tissues, the usage of longer-wavelength excitation provenence enables deeper penetration into the human body for disease diagnosis. Additionally, by mixing various component types in nanoparticles, multimodal imaging techniques have been developed, providing improved spatial perspectives and increased accuracy with the goal of identifying life-threatening disorders before symptoms manifest (Vivero-Escoto & Huang, 2011).

Super magnetic iron oxide (Fe_3O_4) Nanocomplexes with their idiosyncratic magnetic character have been exploited in the biomedical scope for the therapeutic motives and magnetic resonance imaging (MRI) (Comanescu, 2022). Another study exhibit magnetic $ZnFe_2O_4$ Nanoforms are utilized in MRI. The nano based hybrids with the magnitude of 48 nm were contemplated as a supramolecule and it is smoothly piled up and adsorb by the tumor cells or tissue (Ilosvai et al., 2023). The contrasting capacity of magnetic resonance imaging is enhanced by the size variation of hybrids of nanometals (Fe_3O_4). Moreover, under observing the magnetic properties of Fe_3O_4 nanohybrids, the chitosan-Fe_3O_4 were exploit in tumor cells for massive accumulation which occurred due to MRI-guided Photothermal (PTT) with the companionship of the external magnetic field. It is established that manganese-tampered Fe3O4 nanohybrid (Mn-Fe_3O_4-NCs) NPs showed highly sized dependent for the measurement of varied intensities in a dual model MRI the T1-dominated (5 nm), T2-dominated (12 nm), and merged T1-T2 (7 nm) were done by supervising their size (Choi, Rejinold, Piao, & Choy, 2021).

Figure 2. Magnetic nanomaterial in diagnostics
Source: Reproduced from MDPI (Farinha, Coelho, Reis, & Gaspar, 2021)

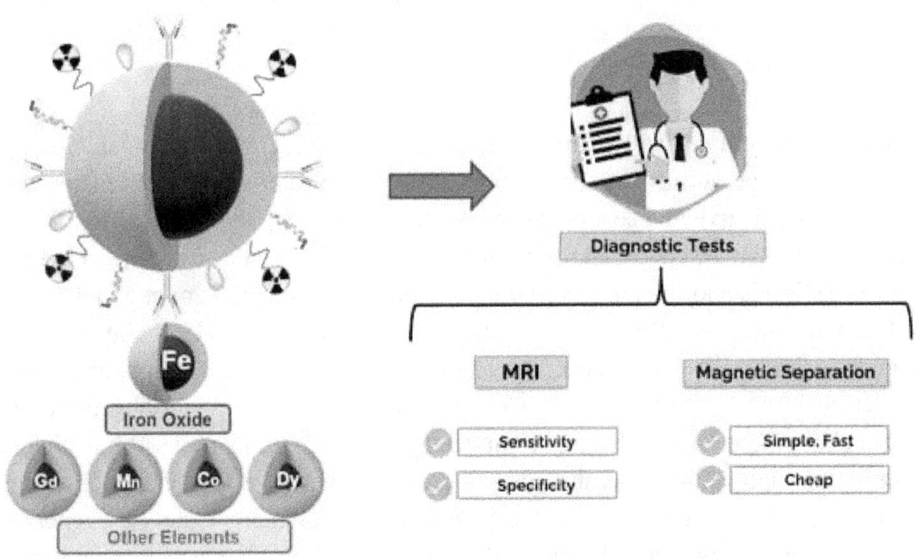

Gold nanohole arrays (Au-NHAs) were used to create a plasmonic biosensor that can detect biomarkers in the pg mL−1 range. Photothermal imaging of cells has been successfully accomplished using in vivo methods based on plasmonic materials (Jin et al., 2022). It has been reported that plasmonic nanobubbles can be used for both in vivo removal of microtumors that remain after surgery in head and neck cancer cells. Therefore, optimizing plasmonic nanomaterials' surface plasmon resonance (SPR) characteristics for biomolecular recognition may help provide superior substitutes for early illness detection (Gurav, Jia, Ye, & Qian, 2019).

Because of their exceptional optical qualities, gold nanoparticles (AuNPs) have been widely exploited in biosensing-related development for decades. As recently discovered functional macromolecules, peptides hold great promise as potential substitutes for substrates, receptors, and antibodies in particular molecular interactions (Khurshid, Amin, Aziz, Ahmed, & Abdel-Haliem, 2023). Peptides and AuNPs may be produced cheaply and with relative ease. They are both strong and versatile. As a result, peptide-AuNP-based bio-nano-technological methods are becoming more and more popular, particularly in the fields of molecular targeting and cell imaging and diagnostics (X. Liu, Zhang, Knoll, Liedberg, & Wang, 2020).

In the last few decades, carbon-based nanomaterials have drawn a lot of interest from a wide range of scientific course of study. In the subject of biomedicine, materials such as carbon quantum dots, carbon nanotubes, graphene and fullerenes can improve diagnostic methods and raise diagnosis accuracy. Although the use of carbon base nanoforms in the diagnosis of various diseases is still in its early stages, bioimaging and nucleic acid amplification techniques establish on CN, depicts novel and more promising diagnostic tools that further contribute in the probable application of CNs in infectious disease and tumor diagnosis. (He et al., 2022).

Gene Delivery System

Gene delivery systems, which are used in gene therapy (GT) to transfer nucleic acid into cells, show potential in treating a range of genetic problems. For the transmission of genes, an effective and secure carrier is necessary. Since they have a significant deal of potential to be candidates for it, the inorganic and organic nano based hybrid materials (ONHs and IONHs) overcome this obstacle (Kesharwani et al., 2018).

Gene transfer is mostly the result of the organic portion of NHs. In the presence of a magnetic field, iron oxide (Fe_3O_4), for instance, possesses a superparamagnetic characteristic that allows for gene transfection and incorporation. According to a study, gold-nickel-silicon (Au-Ni-Si) functionalized polyethyleneimine (PEI) nanospheres can pass through cellular membranes and deliver targeted gene transfection under magnetic guidance. The gene delivery technique uses nanoclusters to generate a nanocomposite, which adheres to the charged pDNA. Bimetallic nanoforms were employed to trace the integration of pDNA into the cells, and the composite nanoparticles are being absorbed by malignant cells, while the hybrid nanoforms are responsible for the luminescence (Messah & Pawitan, 2020) (Wang et al., 2019). Au-Ni nanorods, for example, have been found to be beneficial in the transport or selective transfer of genes (Guo, Zhao, Dai, Zhao, & Xu, 2019).

Drug Delivery System

The need to develop novel, clever drug delivery systems with enhanced potency, efficiency, and specificity has arisen due to the recent advancements in biomolecular treatments. These strategies would guarantee the best possible bioavailability of the encapsulated therapeutic agent and, as a result, the therapeutic efficacy. Nanomaterial engineering and fabrication have gained a lot of attention due to the increased interest in microfluidic technologies that offer quick microscale mixing and precise fluid

Figure 3. Illustration of magnetified-based nanomaterial for gene delivery with the process of DNA pilling into lamellar magnetic hydroxyapatite (MHAp) nanomaterial for transporting

Source: Reproduced from MDPI (Gareev et al., 2022); imaged by Mondal et al. (2017)

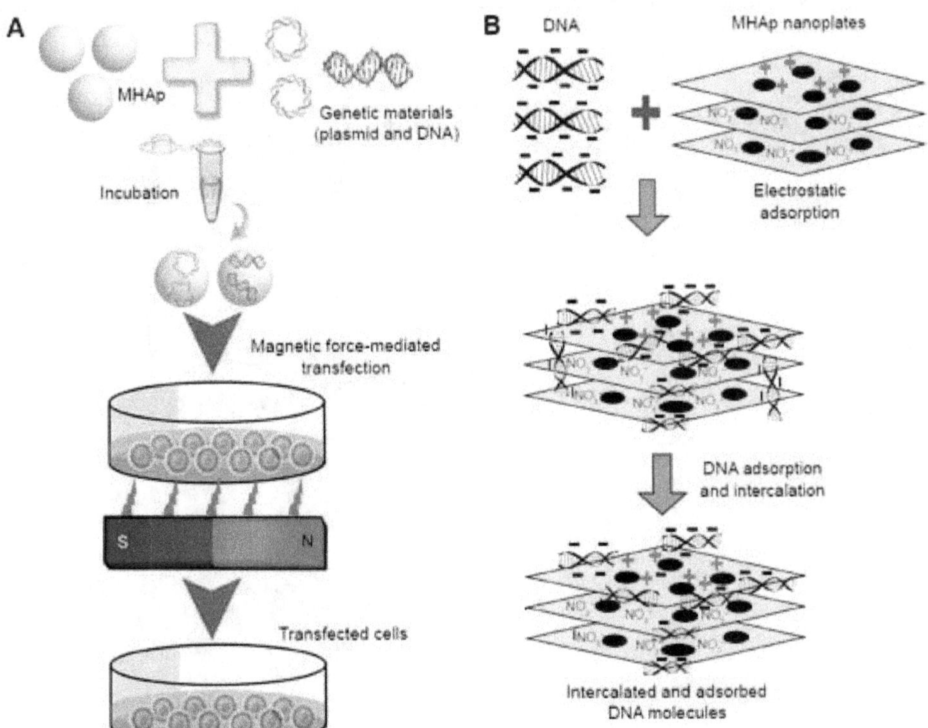

control. In comparison to drug-loaded nanomaterials made using traditional bench techniques, those made using microfluidic techniques can demonstrate improved monodispersity, increased drug encapsulation efficiency, and extended blood circulation times (Zhang, Chen, Ma, & Sun, 2019).

It was demonstrated that lipid hybrid-based polymer nanoforms, which contain a lipid shell and an inner polymeric core, are a promising drug transporting manifesto for nursing of metastatic prostate cancer, breast cancer, and pancreatic cancer (Bushara, 2023). Because these hybrid nanoforms combine the high biocompatibility of lipids with the structural sturdiness offered by polymer nanocomplexes, they can effectively compact both hydrophilic and hydrophobic medicines for improved therapeutic outcomes. According to one study, this mechanism can be dexterously absorbed by cancer cells and evade the reticuloendothelial system's quick clearance (Yao et al., 2020).

Poly (lactic-co-glycolic acid) PLGA-lipid hybrid nanocarriers have been employed in numerous study investigations to deliver chemotherapeutic agents to a specified site/tissue, assessing their biological and therapeutic performances against various cancer cell types, including liver, cervical, prostate, lung, and breast (Ghitman, Biru,

Figure 4. Illustration of NPs in drug delivery mechanism
Source: Reproduced from MDPI (Afzal et al., 2022)

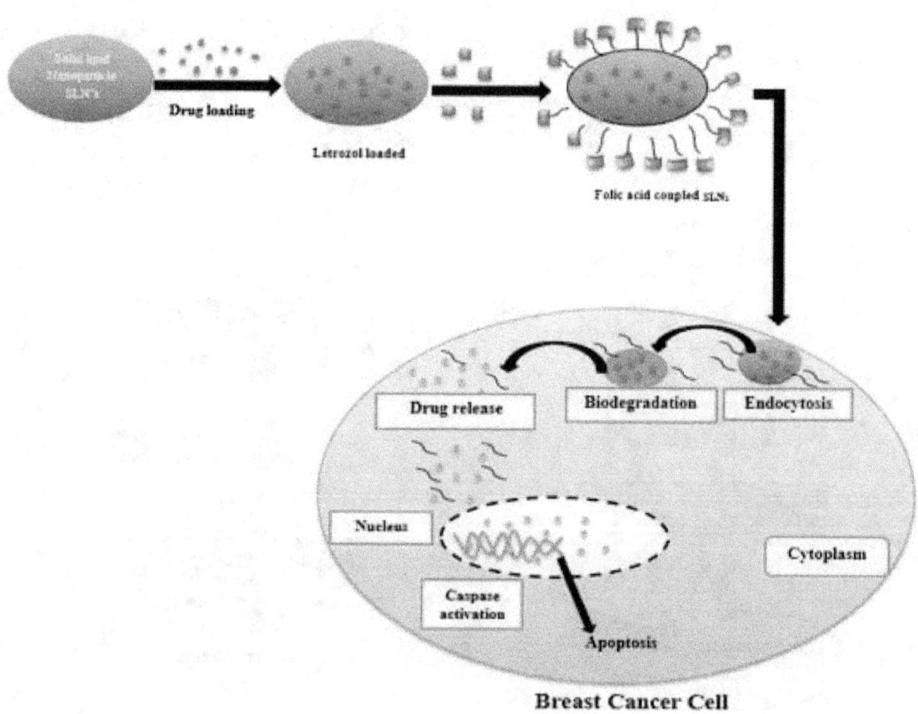

Stan, & Iovu, 2020). The most thoroughly studied Nanohybrids for cancer therapy induced by hyperthermia are iron–gold NHs. Magnetic hyperthermia is the result of changes in the magnetic moment of iron oxide nanoparticles (NHs) caused by noninvasive alternating magnetic field (AMF) (Tarkistani, Komalla, & Kayser, 2021).

Therapeutics

Through the encapsulation or adsorption of medication, nano-particles with their abundance of nano-carriers as inorganic and organic moieties are being employed to Handle a variety of illnesses. Zinc oxide and zinc oxide nanohybrids (ZnO-NHs) are examples of metallic oxide nanohybrids (MONHs) that exhibit antibacterial properties (Sharif, Tunio, & Bano, 2021), exhibit regulated pH-dependent cytotoxicity, and function as drug delivery vehicles against cancer cells (Gul, Memon, Ujjan, Goswami, & Bhatti, 2023). At low pH, the disassembled Zn+ ions form ROS (reactive oxygen species), which stop the proliferation of malignant cells. ZnO-NHs exhibit both antibacterial and anticancer properties against *Staphylococcus aureus* and MCF-7 breast cancer cells, according to a different study that demonstrated the dual therapeutic effects of NHs. Certain nanoparticles, like Cu-Pt, can even undermine the resistance of bacteria. They catalyze H_2O_2 to produce the hydroxy (OH-) radical, which stops the proliferation of bacteria (Khurshid et al., 2023). They exhibit peroxidase-like action. It was shown that the exceptional qualities of gold nanoparticles (Au-NPs), such as their non-toxicity and ability to be handled intracellularly, make them distinctive. When Au (Gold) has gained a negative charge, it can interact with biomolecules such as DNA, proteins, and antibodies. Because

Figure 5. Hybrid Nanomaterial involve in melanoma therapy
Source: Reproduced from MDPI (Cruz et al., 2020)

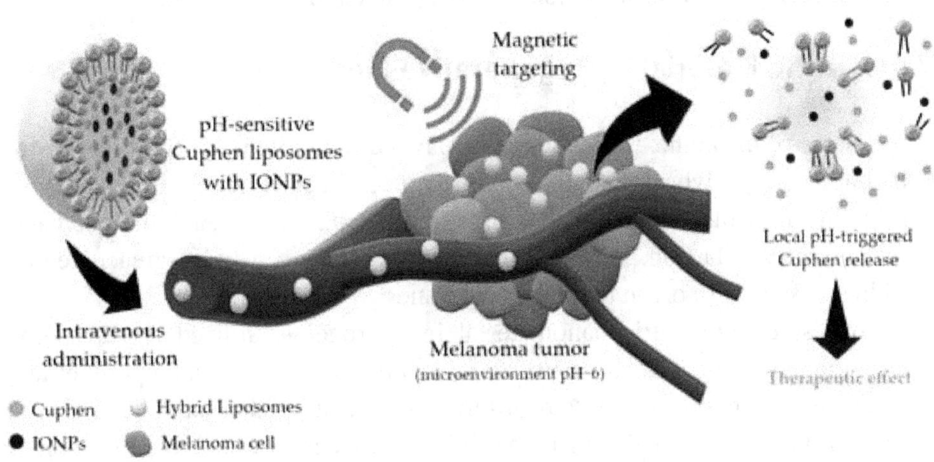

Au-NCs are radiosensitizers and can transform light energy into heat, which stops the propagation of malignant cells, it aids in treatment of cancer.

A few degrees Celsius rise can cause enormous, abrupt variations in the temperature of living things and cells; for example, going from 37 to 42 degrees Celsius is fatal. Because of this, a safe treatment known as hyperthermia is used to provide precise heating that kills malignant cells. Certain magnetic nanoparticles (NPs) have the ability to generate heat or deliver drugs to destroy cancer cells when heated in a regulated magnetic field. Tumor cells typically heat up to 40–45 degrees Celsius, which triggers certain pathways that cause apoptosis(Obaidat, Narayanaswamy, Alaabed, Sambasivam, & Muralee Gopi, 2019).

A separate method claims that the Ag-Au (3:1) Hybrid NPs, which were produced by a chemical reduction method, instigate the cancer cell lines 4T1 (extremely metastatic breast cancer cells), HCT116 (colon cancer cells), and HUH7 (hepatocyte-derived carcinoma tissue line) to undergo apoptosis. Many signaling mechanisms, including the p53, Caspase 3, and Bax/Bcl-2 pathways, caused this apoptosis. A liver cancer therapy was conducted using nanoforms of PEGylated Janus Fe3O4-SiO2. The medicine was encapsulated using iron oxide/magnetite (Fe3O4)'s magnetic property and silica oxide's (SiO2) distinctive feature, which served to target the drug. To target tumor cells, a nanohybrid of Janus PS/Fe_3O_4, SiO_2 was created. Folic acid ($C_{19}H_{19}N_7O_6$) was attached to the PS cast, and the anticancer medication doxorubicin (DOX) was bonded to the silica surface. The release of the drug was linked to a change in pH liable bonding. In an in vivo investigation on the "calvarium" defect model using a 3D-printed scaffold approach, a recent study demonstrates that chitosan-inorganic core nano-based hybrid (Chit-IOC-NCs) demonstrates the therapeutic impact nanohybrids and it inure the new bone production. The 3D tissue regenerating scaffold is now possible thanks to the special qualities of NHs, including their nanohybrid topography, cellular activation, high resilience, and variety of therapeutic transportation systems (Khurshid et al., 2023).

Nanohybrid Material in Regenerative Medicine

In regenerative medicine, these materials play a crtical role in scaffold design for tissue engineering. Hybrid nanomaterials promote the regeneration of injured tissues by imitating the natural extracellular matrix and acomplishing an environment that is favorable for cell adhesion, proliferation, and differentiation. This has great potential to address issues with organ transplantation and wound healing.

In one study polymeric nanomaterial in hydrogel is utilized for jaw bone regeneration. The medication rosuvastatin was put into an alginate hydrogel and ensnared in silica nanoparticles to be inserted into the injured tissue. The improved bone tissue production obtained with AL-RSNP was demonstrated by the outcomes

Figure 6. Polymeric nanomaterial-based therapy action under internal and external stimuli
Source: Reproduced from MDPI (Aflori, 2021); imaged by Cheng, Meng, Deng, Klok, and Zhong (2013)

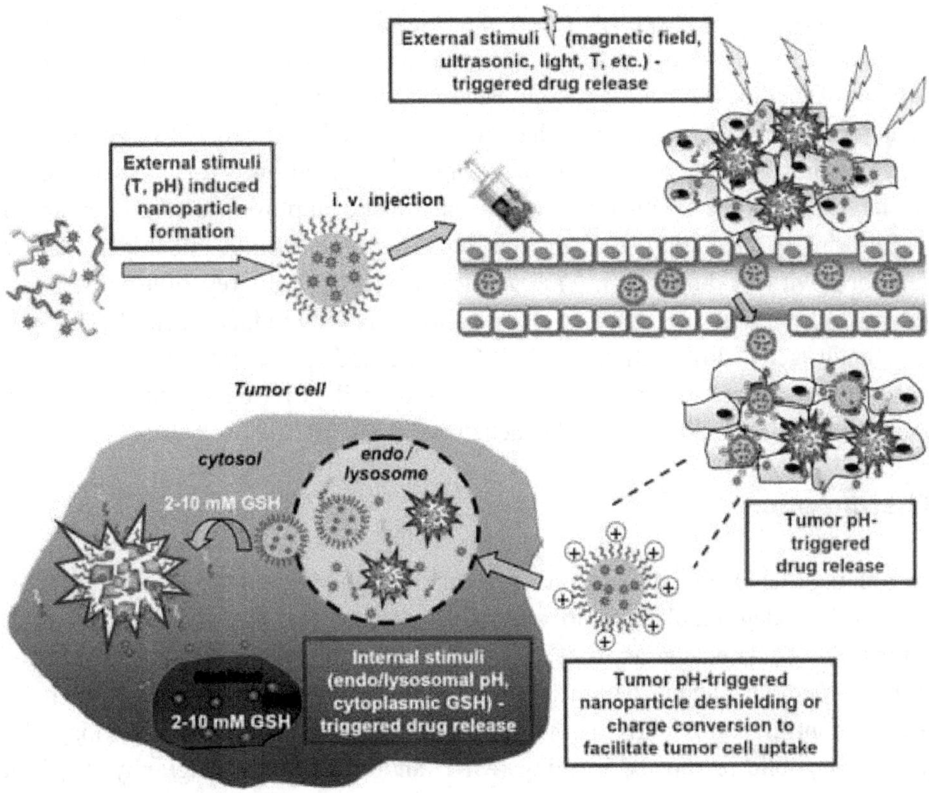

of this in-vivo testing. The newly generated tissue resembled the original tissue in structure and had a uniform hue (Adel et al., 2023).

In 1969, Larry Hench's group created bioactive glass ceramic nanoparticles (n-BGC) having SiO2-CaO–P2O5-Na2O core structures. Silicone, salt, magnesium, potassium, calcium, phosphorous and oxygen are just a few of the materials that can be used to create bioglasses that the cells can absorb. Because of their exceptional bioactivity, antibacterial, and angiogenic qualities, nBGs (nanobioglass) are a promising option for dentin regeneration applications. Via increases in cellular essence, intracellular calcium deposition (ICD), and ALP (alkaline phosphatase) enterprises, the incorporation of boron-modified nBGs into cellulose acetate/oxidized pullulan/gelatin-based formulation has demonstrated encouraging results for dentin regeneration while maintaining boron ion release below toxic levels (Morones et al., 2005). Other uses of nBGs for bone tissue engineering have been researched, such as the use of nBGs containing copper in gelatin-coated scaffolds (Zheng et al., 2018).

Because 3D printing technologies offer great domination over the design and fabrication of 3D scaffolds, they have lately appealed lot of engrossment in the TERM (tissue engineering and regenerative medicine) sector. The primary substance used in 3D bioprinting, known as "bioink," is a solution of one or more biomaterials that may be formed into a hydrogel that encapsulates the necessary cell types. By a number of methods, it can be crosslinked or stabilized during or after 3D printing (Gungor-Ozkerim, Inci, Zhang, Khademhosseini, & Dokmeci, 2018). Because of their customizable chemical structures, high biocompatibility, and advantageous mechanical qualities, synthetic polyurethanes can be employed as a feeder material for 3D printing to create customized scaffolds. A synthetic biodegradable PU elastic nanoform with hyaluronan (HA) as a viscosity and cell aggregation booster and Y27632 inhibitor as a replacement for transforming growth factor beta-3 (TGF-β3) for chondrogenic differentiation of MSCs is needed to generate feeder materials (ink) for water-based 3D printing for customized cartilage tissue engineering (Hung, Tseng, Dai, & Hsu, 2016; Fathi-Achachelouei et al., 2019).

Theranostics

Combining therapy and diagnosis is called theranostics. Theranostics NPs' capacity to be both biodegradable and biocompatible enables more targeted illness control. Prior to 2005, when theranostics was still in its infancy, these delivery vehicles carrying medicinal agents would simply have an imaging function added to them by sticking contrast compounds for imaging tests such as positron emission tomography (PET), magnetic resonance imaging (MRI) and computed tomography (Li, Liu, Sarpong, & Gu). After that theratonics designing enhances as the cancer treating technologies advancement continued (Torchilin, 2011). A theranostic nanoparticle can be divided into at least three parts using Ferrari's classification system, based on their functions and physical locations: surface modification, carrier, and biological payload. Biomedical "payloads" encompass both therapeutic and imaging agents, such as proteins, small interfering RNA [siRNA], anticancer prescription drugs, DNA, hyperthermia-inducing nanoparticles, ROS-generators, CT contrast faculties, organic dyes, upconversion particles [UCNPs], quantum dots [QDs] and MRI contrast faculties. During delivery to the intended target place, the carrier must offer the biological payloads enough physical protection under physiological conditions. Numerous carriers, both inorganic and organic, have been created. To improve imaging or treatment efficacies, the carrier may be engineered to disintegrate precisely in malignancy. Ultimately, the carrier's surface is modified. These modifications are anticipated to give theranostic nanoparticles new capabilities like extended circulation times, barrier-penetrating abilities, and target-specific binding abilities (Yezhelyev et al., 2006; Zhou, Jing, & Chen, 2017).

Graphene oxide (Đorđević et al.)-based iron oxide nanoparticles (IONPs) nanohybrids, which are exploited as drug delivery agents and for diagnostic purposes, are one type of metallic and metal oxide nanohybrid that exhibits theranostics effects. In this case, cyclodextrin functionalizes the GO, enhancing both cellular uptake and doxorubicin (DOX) loading.

Carbon-coated based IONPs offer "targeted drug delivery" and "magnetic resonance imaging" (MRI) capabilities for therapeutic and diagnostic purposes, respectively. Using the tumor therapy magnetic hyperthermia property, a theranostics technique was created for NHs with magnetic Fe3O4 Nanofabricates as an inorganic core for MRI (Magnetic Resonance Imaging) diagnostics. Superparamagnetic $ZnO.4Fe_2.6O_4$ ($ZnFe_2O_4$) is characterized by the magnetism value of the hybrid nanofabricates, which causes hyperthermia and triggers the release of the encapsulated medication.

Tumor selection is made possible by the use of pH-sensitive MnO@Mn3O4, an activatable imaging probe. The design of theranostic devices is made much simpler by imaging probes with drug carrier capabilities, such as hollow iron oxide or manganese oxide nanocontainers (Lim et al., 2021).

Nanoelectronics and Biosensing

Electronic components with smaller dimensions and altered power consumption, displays, and research are the results of nanotechnological involvement in electronics. The application of nanosensors has expanded beyond the home to the field as a result of their development. Materials created utilizing nanoelectronics can be applied to many biological domains and aid in the identification of disease-causing substances and biomarkers. These days, biosensors are extensively employed in a variety of biomedical fields, counting drug development, food control, environmental monitoring, disease detection, monitoring treatments, and biomedical research. Numerous biosensors have been created, including the cellulose strip sensor, which uses surface-enhanced Raman scattering (SERS) and NHs of silver-gold (Ag-Au) core-shell NPs (Laraib, Sargazi, Rahdar, Khatami, & Pandey, 2022).

A biosensor's main component, its bioreceptors, determine its classification as an enzymatic biosensor, an immunosensor, an aptamer biosensor, or a microbiological biosensor. Second, they are divided into five categories: optical biosensors, electronic biosensors, electrochemical biosensors and thermal biosensor (Bhattarai & Hameed, 2020). This is dependent on the type of transducer used. They are further divided into nano-biosensors, surface plasma resonance biosensors, biosensors-on-chip, and electrometers according to the detecting technology. In the coming years, biosensors will achieve the highest levels of wearability, biocompatibility, selectivity, sensitivity and LOD (level of design), thanks to the amazing advancements made in the field of nanotechnology. The following sections go into further detail on a handful of the uses for biosensors (Ramesh, Janani, Deepa, & Rajeshkumar, 2022).

Field-effect transistor (FET)-based devices have garnered significant interest among various electrical biosensor architectures due to their ability to convert biological synergid between target molecules and FET surface into electrical signals that can be read. The use of silicon nanowire field-effect transistors (SiNW-FET)s in biosensor design has recently increased significantly due to their high levels of sensitivity, selectivity, and real-time, label-free detection of proteins, DNA sequences, small molecules, viruses, and cancer biomarkers (Prasad, 2019) .

In the process of creating biosensors, semiconductors such as metal oxides like tin oxide (SnO_2), tungsten oxide (WO_3), titanium oxide (TiO_2) and molybdenum oxide (MoO_x) as well as carbon-based NPs doped with metallic ions were employed as NHs in single or coupled (heterostructure or composite form) (Jadhav, Annaldewar, & Jadhav, 2023). There are numerous uses for TiO2-NPs in the construction of biosensors, regardless of whether they are inorganic-inorganic or inorganic-organic NHs. In the case of iron-gold hybrid nanocomplexes (Fe-Au-NCs), NFs have functioned as both an ionization matrix and a capturing probe. In laser desorption/ionization mass spectrometry (LDI-MS), Au-nanofabricates control cellular uptake, while cationic moieties perform surface functionalization (J. Liu, Zhou, Yang, & Ariga, 2020). According the findings of one study, Au-nanometals are stable in the intracellular environment without the need of a matrix, and these cells can be further employed as a probe for enrichment, treatment, drug delivery, and protein capture (M. Zhang et al., 2021).

CHALLENGES OF USING HYBRID NANOTECHNOLOGY IN MEDICAL SYSTEM

In utilizing Hybrid nanoforms for clinical and therapeutics application also brought some challenges and hurdles in a way which need to be addressed for their safe utilization.

- Toxicological problems may develop from the components used in the hybrid nanomaterials; thus, it is important to ensure their biocompatibility. More research is needed to determine the potential long-term impacts of the hybrid nanomaterials. It is also crucial to comprehend the distribution and clearance rates of hybrid nanomaterials in order to determine the ideal dosage and minimize the risk of accumulation in critical organs.
- It is highly difficult to precisely target hybrid nanomaterials to particular cells or tissues without creating off-target effects, and getting beyond biological

Figure 7. Schematic representation of metallic nanohybrids biosensors
Source: Reproduced from MDPI (Soto & Orozco, 2022); imaged by Li et al. (2019), Paul and Srivastava (2018), and Tian et al. (2020)

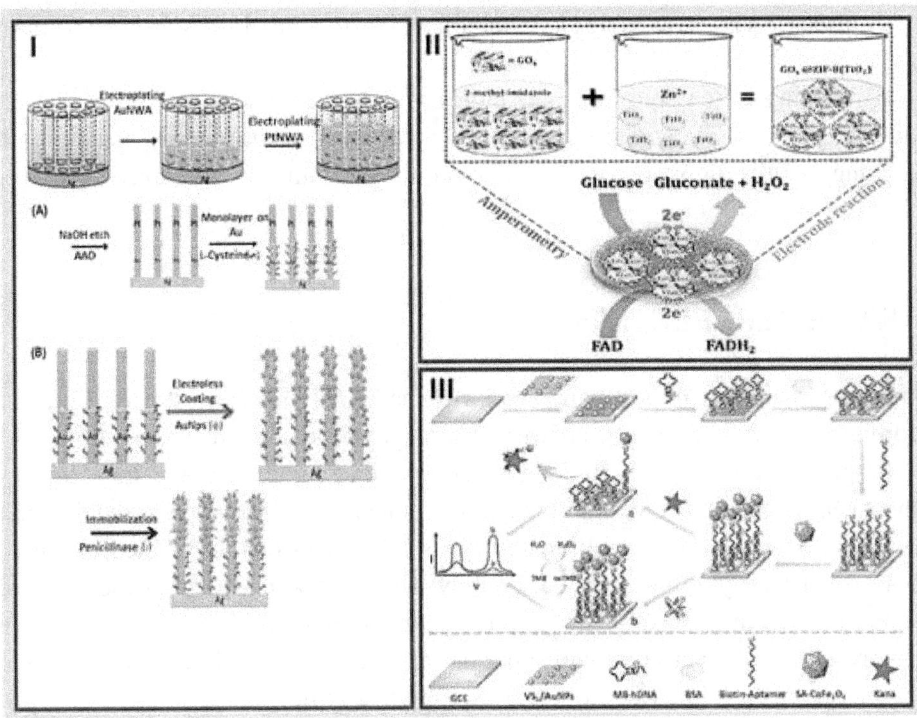

barriers like the blood-brain barrier is necessary for successful delivery to specific therapeutic areas.

- It is important to take into account how the immune system will react to hybrid nanomaterials. Its safety and effectiveness may be impacted by immunogenicity. It is necessary to investigate methods for regulating the immune response and averting undesirable effects.
- For hybrid nanomaterials to have long-lasting therapeutic effects, they must preserve both their structural and functional integrity. Environmental influences and biological component interactions might give rise to stability problems.
- It's critical to address moral issues raised by the application of nanomaterials in human anatomy. The effective adoption of these technologies is also influenced by societal attitudes and acceptance of them.
- Complying with clinical use regulations is a difficult task. Approval depends on standardizing production processes and making sure rules are followed.

- Some drawbacks which have been noted by clinical researchers, including high costs, challenges in scaling up procedures, and the simple inhalability of nanoparticles, which can cause harmful for lungs and frequently lead to further illnesses that may cause modifications to homeostasis, or even demise.

By tackling these issues, safer and more useful hybrid nanomaterials for medical and therapeutic applications will be developed. Overcoming these obstacles and realizing the full promise of nanomedicine requires ongoing research and cooperative initiatives.

FUTURE PERSPECTIVE

Future prospects for hybrid nanomaterials' medicinal and clinical uses are quite promising, providing creative answers to a range of medical problems.

In spite of the absence of clear regulatory guidelines, more than 50 nanomedicines have been commercialized, and this number is increasing fast. Due to the difficult tumor landscape and the need for hazardous substances that are difficult to treat, these are primarily found in cancer therapy. The most well-known of them are, to mention a few, the liposomal preparations Doxil® and AmBisome®, which have more recently shown promise with albumin-drug nanoparticles like Abraxane® and polymeric micelles like Eligard® (Foulkes et al., 2020).

While there have been some initiatives to establish National Characterization Laboratories by government agencies and academic groups, more precise and strict guidelines are required from the major regulatory bodies, such as the FDA and MHRA. A worldly collaboration for the supervision of nanomaterials should be formed to advance these programs and provide official guidelines to the research communities, as many diseases do not discriminate based on factors such as race or geography. Over the past 20 years, billions of dollars have been invested in the development of nanomedicine; however, in the absence of clear leadership and supervision from regulatory agencies, these efforts will not yield goods for the market, and future investment will be directed elsewhere (Đorđević et al., 2022).

- Targeted drug delivery is made possible by hybrid nanomaterials, opening the door to individualized and accurate treatment plans. This could reduce negative effects while increasing treatment efficacy. Future research might concentrate on creating biodegradable and sustainable hybrid nanomaterials, reducing their negative effects on the environment, and addressing long-term exposure issues.

- By targeting several elements of a disease at once and enhancing treatment outcomes, combination treatments made possible by the integration of various therapeutic agents within hybrid nanomaterials may be possible (Foulkes et al., 2020).
- Immunotherapies can be made more successful by customizing hybrid nanomaterials. They might facilitate the passage through immunosuppressive conditions and enhance the focused administration of immunomodulatory drugs.
- Reactive components included into hybrid nanomaterials may allow for regulated drug release and remote sensing, facilitating site-specific and on-demand therapeutic interventions (Foulkes et al., 2020).
- To get hybrid nanomaterials from the lab to clinical applications, researchers, doctors, biologists, and engineers must continue to cooperate. Multidisciplinary methods have the potential to spur innovation.
- Hybrid nanomaterials that incorporate smart materials and technology may be able to adapt to changing physiological conditions, increasing the overall effectiveness of therapeutic interventions.
- As research advances, confirming the safety and effectiveness of hybrid nanomaterials for clinical application will require managing regulatory procedures and carrying out carefully planned clinical trials.

Hybrid nanomaterials have a vibrant future in clinical and therapeutic applications, and they have the ability to completely transform healthcare by offering cutting-edge, practical solutions in a variety of medical specialties. Future developments in technology, cooperative initiatives, and ongoing study will determine how these novel ideas develop.

CONCLUSION

Advancement in the field of nanotechnology also revolutionized the industry of biomedicine and therapeutics. Different nano-based hybrids are synthesized for different clinical applications. These nano hybrids have large surface area, quantum effects and many other physiochemical and optical properties which enable them to generate antibacterial, antiviral, optical and antioxidant etc. activities which are helpful in diagnostics and medicine. Different type of nanohybrids like metallic, metallic oxide, carbon-based polymers are now a days used as drug or gene carriers in therapeutics and many different nanoforms combine to provide theratoincs based treatment for different disease. But with the development of nanohybrids utilization in clinical applications many challenges also arise like ethical issues, high costs and the simple inhalability of nanoparticles, which can cause harmful for lungs and frequently lead to further illnesses that may cause modifications to homeostasis, or even demise to tackle these challenges some collaborative efforts and required, optimal check and balance is need to be establish and requisite of safe research methodologies need to develop under recognized organizations.

REFERENCES

Adel, I. M., ElMeligy, M. F., Amer, M. S., & Elkasabgy, N. A. (2023). Polymeric nanocomposite hydrogel scaffold for jawbone regeneration: The role of rosuvastatin calcium-loaded silica nanoparticles. *International Journal of Pharmaceutics: X*, *6*, 100213. doi:10.1016/j.ijpx.2023.100213 PMID:37927584

Aflori, M. (2021). Smart nanomaterials for biomedical applications—A review. *Nanomaterials (Basel, Switzerland)*, *11*(2), 396. doi:10.3390/nano11020396 PMID:33557177

Afzal, O., Altamimi, A. S., Nadeem, M. S., Alzarea, S. I., Almalki, W. H., Tariq, A., ... Riaz, N. (2022). Nanoparticles in Drug Delivery: From History to Therapeutic Applications. *Nanomaterials (Basel, Switzerland)*, *12*(24), 4494. doi:10.3390/nano12244494 PMID:36558344

Balla, S., & Sumithra, M. (2023). Platinum Nano Particles Role In Cancer Treatment: A Review. *Latin American Journal of Pharmacy*, *42*(3), 961–971.

Bhattarai, P., & Hameed, S. (2020). Basics of biosensors and nanobiosensors. *Nanobiosensors: From Design to Applications*, 1-22.

Bushara, M. O. E. (2023). Characteristics of Patients with Breast Cancer Attending the Breast Cancer Center in Sudan. *Advancements in Life Sciences*, *9*(4), 603–606.

Cheng, R., Meng, F., Deng, C., Klok, H.-A., & Zhong, Z. (2013). Dual and multi-stimuli responsive polymeric nanoparticles for programmed site-specific drug delivery. *Biomaterials*, *34*(14), 3647–3657. doi:10.1016/j.biomaterials.2013.01.084 PMID:23415642

Choi, G., Rejinold, N. S., Piao, H., & Choy, J.-H. (2021). Inorganic–inorganic nanohybrids for drug delivery, imaging and photo-therapy: Recent developments and future scope. *Chemical Science (Cambridge)*, *12*(14), 5044–5063. doi:10.1039/D0SC06724E PMID:34168768

Comanescu, C. (2022). Magnetic Nanoparticles: Current Advances in Nanomedicine, Drug Delivery and MRI. *Chemistry (Basel, Switzerland)*, *4*(3), 872–930. doi:10.3390/chemistry4030063

Đorđević, S., Gonzalez, M. M., Conejos-Sánchez, I., Carreira, B., Pozzi, S., Acúrcio, R. C., ... Vicent, M. J. (2022). Current hurdles to the translation of nanomedicines from bench to the clinic. *Drug Delivery and Translational Research*, 1–26. PMID:34302274

Fathi-Achachelouei, M., Knopf-Marques, H., Ribeiro da Silva, C. E., Barthès, J., Bat, E., Tezcaner, A., & Vrana, N. E. (2019). Use of nanoparticles in tissue engineering and regenerative medicine. *Frontiers in Bioengineering and Biotechnology*, *7*, 113. doi:10.3389/fbioe.2019.00113 PMID:31179276

Foulkes, R., Man, E., Thind, J., Yeung, S., Joy, A., & Hoskins, C. (2020). The regulation of nanomaterials and nanomedicines for clinical application: Current and future perspectives. *Biomaterials Science*, *8*(17), 4653–4664. doi:10.1039/D0BM00558D PMID:32672255

Gareev, K. G., Grouzdev, D. S., Koziaeva, V. V., Sitkov, N. O., Gao, H., Zimina, T. M., & Shevtsov, M. (2022). Biomimetic nanomaterials: Diversity, technology, and biomedical applications. *Nanomaterials (Basel, Switzerland)*, *12*(14), 2485. doi:10.3390/nano12142485 PMID:35889709

Ghitman, J., Biru, E. I., Stan, R., & Iovu, H. (2020). Review of hybrid PLGA nanoparticles: Future of smart drug delivery and theranostics medicine. *Materials & Design*, *193*, 108805. doi:10.1016/j.matdes.2020.108805

Gul, F., Memon, S., Ujjan, I., Goswami, P., & Bhatti, K. A. (2023). The effects of 4β-hydroxy withanolide E extracted from Physalis Peruviana on Complete Blood Count of Dimethylbenz (a) anthracene-induced Breast Cancer in Albino Rats. *Advancements in Life Sciences*, *10*(3), 434–438.

Gungor-Ozkerim, P. S., Inci, I., Zhang, Y. S., Khademhosseini, A., & Dokmeci, M. R. (2018). Bioinks for 3D bioprinting: An overview. *Biomaterials Science*, *6*(5), 915–946. doi:10.1039/C7BM00765E PMID:29492503

Guo, K., Zhao, X., Dai, X., Zhao, N., & Xu, F. J. (2019). Organic/inorganic nanohybrids as multifunctional gene delivery systems. *The Journal of Gene Medicine*, *21*(5), e3084. doi:10.1002/jgm.3084 PMID:30850992

Gurav, D. D., Jia, Y. A., Ye, J., & Qian, K. (2019). Design of plasmonic nanomaterials for diagnostic spectrometry. *Nanoscale Advances*, *1*(2), 459–469. doi:10.1039/C8NA00319J PMID:36132258

He, Y., Hu, C., Li, Z., Wu, C., Zeng, Y., & Peng, C. (2022). Multifunctional carbon nanomaterials for diagnostic applications in infectious diseases and tumors. *Materials Today. Bio*, *14*, 100231. doi:10.1016/j.mtbio.2022.100231 PMID:35280329

Hung, K.-C., Tseng, C.-S., Dai, L.-G., & Hsu, S. (2016). Water-based polyurethane 3D printed scaffolds with controlled release function for customized cartilage tissue engineering. *Biomaterials*, *83*, 156–168. doi:10.1016/j.biomaterials.2016.01.019 PMID:26774563

Ilosvai, Á. M., Forgách, L., Kovács, N., Heydari, F., Szigeti, K., Máthé, D., ... Viskolcz, B. (2023). Development of Polymer-Encapsulated, Amine-Functionalized Zinc Ferrite Nanoparticles as MRI Contrast Agents. *International Journal of Molecular Sciences*, *24*(22), 16203. doi:10.3390/ijms242216203 PMID:38003394

Jadhav, A. C., Annaldewar, B. N., & Jadhav, N. C. (2023). 2 A current perspective on nanocomposite and nanohybrid material: Developments and trends. *Nanocomposite and Nanohybrid Materials: Processing and Applications*, *17*, 29–54. doi:10.1515/9783111137902-002

Jin, C., Wu, Z., Molinski, J. H., Zhou, J., Ren, Y., & Zhang, J. X. (2022). Plasmonic nanosensors for point-of-care biomarker detection. *Materials Today. Bio*, *14*, 100263. doi:10.1016/j.mtbio.2022.100263 PMID:35514435

Kesharwani, P., Gothwal, A., Iyer, A. K., Jain, K., Chourasia, M. K., & Gupta, U. (2018). Dendrimer nanohybrid carrier systems: An expanding horizon for targeted drug and gene delivery. *Drug Discovery Today*, *23*(2), 300–314. doi:10.1016/j.drudis.2017.06.009 PMID:28697371

Khurshid, A., Amin, R., Aziz, A., Ahmed, O., & Abdel-Haliem, M. E. (2023). Deciphering the Potential Therapeutic Intervention Points of 2019-nCoV: A Pharmacological Perspective. *Advancements in Life Sciences*, *9*(4), 412–428.

Laraib, U., Sargazi, S., Rahdar, A., Khatami, M., & Pandey, S. (2022). Nanotechnology-based approaches for effective detection of tumor markers: A comprehensive state-of-the-art review. *International Journal of Biological Macromolecules*, *195*, 356–383. doi:10.1016/j.ijbiomac.2021.12.052 PMID:34920057

Li, Z., Liu, C., Sarpong, V., & Gu, Z. (2019). Multisegment nanowire/nanoparticle hybrid arrays as electrochemical biosensors for simultaneous detection of antibiotics. *Biosensors & Bioelectronics*, *126*, 632–639. doi:10.1016/j.bios.2018.10.025 PMID:30513482

Lim, E.-K., Kim, T., Paik, S., Haam, S., Huh, Y.-M., & Leec, K. (2021). Nanomaterials for Theranostics: Recent Advances and Future Challenges. *Nanomaterials and Neoplasms*, 587-775.

Liu, J., Zhou, H., Yang, W., & Ariga, K. (2020). Soft nanoarchitectonics for enantioselective biosensing. *Accounts of Chemical Research*, *53*(3), 644–653. doi:10.1021/acs.accounts.9b00612 PMID:32073816

Liu, X., Zhang, Q., Knoll, W., Liedberg, B., & Wang, Y. (2020). Rational design of functional peptide–gold hybrid nanomaterials for molecular interactions. *Advanced Materials*, *32*(37), 2000866. doi:10.1002/adma.202000866 PMID:32743897

Messah, A. D. V., & Pawitan, J. A. (2020). Next generation sequencing as rapid diagnosis of multidrug resistance tuberculosis. *Advancements in Life Sciences*, *8*(1), 30–37.

Mondal, S., Manivasagan, P., Bharathiraja, S., Santha Moorthy, M., Kim, H. H., Seo, H., Lee, K. D., & Oh, J. (2017). Magnetic hydroxyapatite: A promising multifunctional platform for nanomedicine application. *International Journal of Nanomedicine*, *12*, 8389–8410. doi:10.2147/IJN.S147355 PMID:29200851

Morones, J. R., Elechiguerra, J. L., Camacho, A., Holt, K., Kouri, J. B., Ramírez, J. T., & Yacaman, M. J. (2005). The bactericidal effect of silver nanoparticles. *Nanotechnology*, *16*(10), 2346–2353. doi:10.1088/0957-4484/16/10/059 PMID:20818017

Nguyen, K. T., & Zhao, Y. (2015). Engineered hybrid nanoparticles for on-demand diagnostics and therapeutics. *Accounts of Chemical Research*, *48*(12), 3016–3025. doi:10.1021/acs.accounts.5b00316 PMID:26605438

Obaidat, I. M., Narayanaswamy, V., Alaabed, S., Sambasivam, S., & Muralee Gopi, C. V. (2019). Principles of magnetic hyperthermia: A focus on using multifunctional hybrid magnetic nanoparticles. *Magnetochemistry (Basel, Switzerland)*, *5*(4), 67. doi:10.3390/magnetochemistry5040067

Paul, A., & Srivastava, D. N. (2018). Amperometric glucose sensing at nanomolar level using MOF-encapsulated TiO2 platform. *ACS Omega*, *3*(11), 14634–14640. doi:10.1021/acsomega.8b01968 PMID:30555983

Prasad, R. (2019). Nanotechnology in the life science. *Springer Nature*.

Ramesh, M., Janani, R., Deepa, C., & Rajeshkumar, L. (2022). Nanotechnology-enabled biosensors: A review of fundamentals, design principles, materials, and applications. *Biosensors (Basel)*, *13*(1), 40. doi:10.3390/bios13010040 PMID:36671875

Sharif, M., Tunio, S. A., & Bano, S. (2021). Synergistic effects of Zinc oxide nanoparticles and conventional antibiotics against methicillin resistant Staphylococcus aureus. *Advancements in Life Sciences*, *8*(2), 167–171.

Spesivtseva, A., Gvarzhdets, N., Autlev, K., Kruchinin, E., & Kuznetsov, I. (2023). Diseases of the Neuroacanthocytosis Group: A Systematic Review of Clinical Cases and Difficulties in their Diagnosis. *Advancements in Life Sciences*, *10*(3), 335–340.

Tarkistani, M. A. M., Komalla, V., & Kayser, V. (2021). Recent advances in the use of iron–gold hybrid nanoparticles for biomedical applications. *Nanomaterials (Basel, Switzerland)*, *11*(5), 1227. doi:10.3390/nano11051227 PMID:34066549

Tian, L., Zhang, Y., Wang, L., Geng, Q., Liu, D., Duan, L., Wang, Y., & Cui, J. (2020). Ratiometric dual signal-enhancing-based electrochemical biosensor for ultrasensitive kanamycin detection. *ACS Applied Materials & Interfaces*, *12*(47), 52713–52720. doi:10.1021/acsami.0c15898 PMID:33170623

Torchilin, V. (2011). Tumor delivery of macromolecular drugs based on the EPR effect. *Advanced Drug Delivery Reviews*, *63*(3), 131–135. doi:10.1016/j.addr.2010.03.011 PMID:20304019

Vivero-Escoto, J. L., & Huang, Y.-T. (2011). Inorganic-organic hybrid nanomaterials for therapeutic and diagnostic imaging applications. *International Journal of Molecular Sciences*, *12*(6), 3888–3927. doi:10.3390/ijms12063888 PMID:21747714

Wang, Y., Wang, X., Ma, X., Chen, Q., He, H., Nau, W. M., & Huang, F. (2019). Coassembly of gold nanoclusters with nucleic acids: Sensing, bioimaging, and gene transfection. *Particle & Particle Systems Characterization*, *36*(10), 1900281. doi:10.1002/ppsc.201900281

Yao, Y., Zhou, Y., Liu, L., Xu, Y., Chen, Q., Wang, Y., Wu, S., Deng, Y., Zhang, J., & Shao, A. (2020). Nanoparticle-based drug delivery in cancer therapy and its role in overcoming drug resistance. *Frontiers in Molecular Biosciences*, *7*, 193. doi:10.3389/fmolb.2020.00193 PMID:32974385

Yezhelyev, M. V., Gao, X., Xing, Y., Al-Hajj, A., Nie, S., & O'Regan, R. M. (2006). Emerging use of nanoparticles in diagnosis and treatment of breast cancer. *The Lancet. Oncology*, *7*(8), 657–667. doi:10.1016/S1470-2045(06)70793-8 PMID:16887483

Zhang, J., Hu, K., Di, L., Wang, P., Liu, Z., Zhang, J., ... Chen, T. (2021). Traditional herbal medicine and nanomedicine: Converging disciplines to improve therapeutic efficacy and human health. *Advanced Drug Delivery Reviews*, *178*, 113964. doi:10.1016/j.addr.2021.113964 PMID:34499982

Zhang, L., Chen, Q., Ma, Y., & Sun, J. (2019). Microfluidic methods for fabrication and engineering of nanoparticle drug delivery systems. *ACS Applied Bio Materials*, *3*(1), 107–120. doi:10.1021/acsabm.9b00853 PMID:35019430

Zhang, M., Shao, S., Yue, H., Wang, X., Zhang, W., Chen, F., Zheng, L., Xing, J., & Qin, Y. (2021). High stability au nps: From design to application in nanomedicine. *International Journal of Nanomedicine*, *16*, 6067–6094. doi:10.2147/IJN.S322900 PMID:34511906

Zheng, K., Wu, J., Li, W., Dippold, D., Wan, Y., & Boccaccini, A. R. (2018). Incorporation of Cu-containing bioactive glass nanoparticles in gelatin-coated scaffolds enhances bioactivity and osteogenic activity. *ACS Biomaterials Science & Engineering*, *4*(5), 1546–1557. doi:10.1021/acsbiomaterials.8b00051 PMID:33445312

Zhou, Y., Jing, X., & Chen, Y. (2017). Material chemistry of graphene oxide-based nanocomposites for theranostic nanomedicine. *Journal of Materials Chemistry. B*, *5*(32), 6451–6470. doi:10.1039/C7TB00680B PMID:32264411

Chapter 10
A Review on the Chemical Vapor Deposition Synthesis of 2D Materials and Their Applications

Sumit Kumar
https://orcid.org/0000-0002-6255-571X
Department of Physics, Patna University, India

ABSTRACT

This chapter explains the principle and synthesis method involved in the chemical vapor deposition (CVD) process. A smart method for making layered 2D atomic crystal nanomaterials on a compatible substrate through bottom-up approach is explored. CVD is a vacuum deposition method used to produce high-quality, high-performance solid materials. CVD is a scalable method to fabricate one layer to multi-layer thin films to bulk layer hybrid materials and their hetero-structure composites by changing the precursor compound with combination of an appropriate element having affinity toward the precursor complex cations. The process stacks two or more atomic layers of different compounds in the rich variety of 2D library to explore novel and collective quantum phenomena at the interfaces. Recently, 2D-SnXSY semiconductor materials have been synthesized by using CVD. These hybrid materials find various applications in sensors, photovoltaics, and quantum phenomena. Latest trends on fabricating hetero-structures of SnS and SnS2 with other hybrid composites by CVD have also been discussed.

DOI: 10.4018/979-8-3693-1261-2.ch010

1. INTRODUCTION

Chemical Vapor Deposition (CVD) is a bottom-up method for the synthesis of 2D-hybrid materials nanoparticles. Two-dimensional (2D) hybrid materials are a class of nanomaterials that have two dimensions (XY-Plane) outside of the nanometric size range and atomic-scale thicknesses (Z-Dimension). The CVD method of synthesis was first used by (Nickl et al., 1972). Later (Goto et al., 1987) used similar method to make monolithic Ti_3SiC_2. The first well-known 2D material was Graphene consisting of a single layer of carbon atoms arranged in a hexagonal lattice. Graphene by having excellent electronic, optical and mechanical properties lacks an electronic bandgap. Thus, ideal Graphene behaves like metallic materials in contrast to the semi-conductive behavior, this is most undesirable for an electronic application. This drawback has stimulated the scientists, researchers and engineers to open up the electronic band of Graphene or search the new 2D atomic crystals nanomaterials with semiconducting nature. The Transition Metal Chalcogenides (TMCs) or Dichalcogenides (TMDs) have been proven a better semiconducting character regarding 2D hybrid material. These 2D crystals exhibit high crystal quality, macroscopic continuity and better surface morphologies which find exhaustive exploitations in catalysis, optoelectronics and sensors based devices. The search for new layered materials does not stop at TMCs and TMDs only moreover, it has open the channel for further new hybrid materials. For instance, a new family of layered 2D hybrid materials has been found, known as MXenes. A family of 2D crystals typically composed of a transition metal paired with a chalcogen in a similar lamella structure resembling that of graphite. Some examples of the typical transition metals are Molybdenum (Mo) and Tungsten (W) whereas the chalcogens are typically Sulphur (S), Selenium (Se) and Tellurium (Te). Among reported TMDs, MoS_2 has a bandgap of 1.8eV, direct bandgap for single layer and 1.2eV, indirect bandgap for bulk layer. The weak interaction in the adjacent layers of MoS_2 helps to obtain its single-layer sheets through exfoliation. Other TMDs like $MoTe_2$ and WSe_2 being recent are unexplored found to have a high level of Valleytronics control and bandgap opening can be prepared by salt-assisted method using CVD.

CVD principle involves the adsorption of precursor molecules from the vapor phase onto a pre-heated substrate at ambient temperature where they break down to form a desired material by chemical reaction with the help of an inert or carrier gas flow like Argon or N_2. Gases like CH_4, H_2 or H_2S are also used with the carrier gases as elemental ingredients. These precursor species may have already begun thermal breakdown in the vapor phase. The precursor molecules are often volatile and are delivered to the heated substrate by an inert carrier gas. However, reactive gases such as oxygen or ammonia can also be used if oxidizing or reducing conditions are required. Precursors can either be single source, where all the elements of the desired material are contained within one molecule or multi source where different precursor molecules react together to produce the required material. Volatile byproducts matter produced during the reaction by the decomposing precursors are removed by vacuum or carrier gas with appropriate flow rate. Scalability is an advantage of CVD. CVD technique is a powerful bottom-up chemical approach to prepare high quality, uniform thickness films and modified heterostructures with faster growth rate.

Chemical Vapor Deposition is the prominent van der Waal heterostructure growth method and can produce high quality TMDs for photovoltaic and energy storage applications. However, uncontrolled nucleation and the lateral width of the individual TMD domains have been the limiting factors to obtain multi-junction planar heterostructures which require the exchange of multiple sources and air exposure at each step. Stacking two or more atomic layers of different compounds in the rich variety of 2D library allows researchers to explore novel and collective quantum phenomena at the interfaces of various types of layers available for the combinatorial stackings. Vertically Stacked (VS) heterostructures have shown very promising performance as tunneling transistors, light emitting devices, photodetectors and photovoltaic cells. The most straight forward technique to fabricate and study vertical heterostructures of 2D-materials is perhaps the mechanical transfer of exfoliated materials. Although this method is not without its limitations like scalability, twist-control and bond-orientation. The limitations of transfer process can be overcome through scalable and more controllable techniques like epitaxial or CVD growth methods.

The CVD method can achieve both lateral as well as vertical epitaxial growth of TMDs. The controlled and homogeneous integration of poly metal-based 2D-materials with tuned opto-properties is an another step which could widen the scope of possible material combinations for the new design of spectral-selective two-dimensional heterogeneous materials for nanophotonics and optoelectronic applications. Controlled CVD technique has enabled the synthesis of large area and uniform thickness 2D layer of metal and insulating surfaces for the large scale device fabrication including electronic and flexible optoelectronic devices. CVD is the primary method used for the fabrication of GaAs, silicon nitride and silicon dioxide thin-films. Each material produced via CVD requires a suitable recipe consisting of a substrate, the best combination of input gases to enhance a surface kinetics, a suitable temperature for the reaction to occur and a pressure at which the chemical reaction occurs at. All these are done in a relatively low vacuum tube in the furnace powered by a vacuum pump. CVD is a widely used growth technique to fabricate thin solid films on the various substrates from the vapor species through chemical reactions. In a conventional CVD reaction, gas precursors are utilized to decompose at relatively higher temperatures and deposited on to transition metal catalysts. In

Figure 1. Elementary transport steps in a CVD process

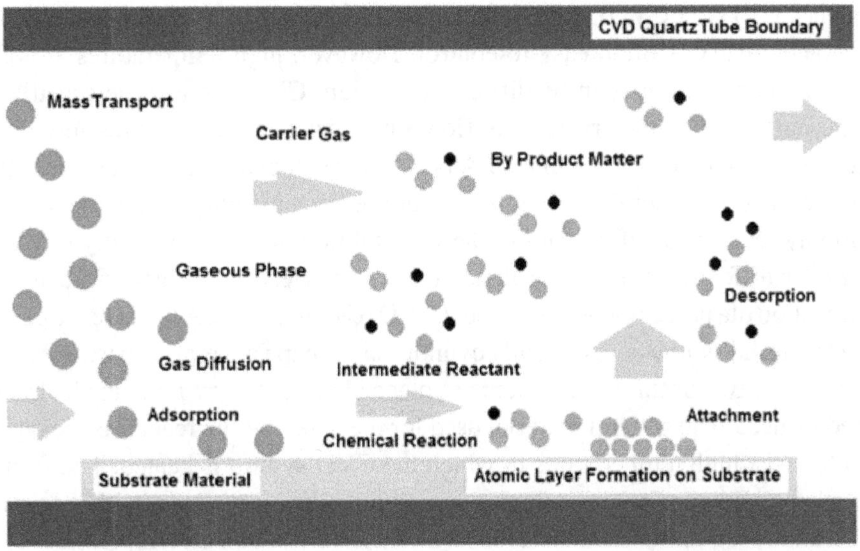

most of the Physical Vapor Deposition (PVD) processes, the source materials will be solid powder and vaporized on to the substrates by transporting through the inert and ambient carrier gases or vacuum. Controlled Chemical Vapor Deposition, Atomic Layer Deposition (ALD) and Ion Beam (IB) induction are nowadays the most used processes to realize almost all kind of electronic devices with the current used materials, it would be highly desirable to develop TMD synthesis processes that share the same deposition equipment used for other materials, since it could open the possibility to directly integrate TMC and TMD with other materials.

Most of the proposed processes utilize one or two of the TMD constituent elements (metal and sulfur/selenium) in vapor form, transporting the selected species over the substrate with a carrier gas generally Argon. This permit, in principle, to achieve a good control over the gas composition and through flow dynamic optimizations obtained with chamber or tube design to deliver the precursors in a homogeneous way over all the substrate area, large area scalability, uniformity and thickness control are routinely achieved for thin films deposited by CVD and can potentially be obtained also for monolayer TMD. During a typical CVD process, a carrier gas such as Argon is used to carry the volatile reactants to the substrate kept at a desired temperature. The byproducts formed during the reaction process get removed by the flow of the carrier gas. On the basis of different features of the CVD, it has been classified into several different methods.

Some of them include plasma-enhanced CVD (PECVD), hot-wall CVD, thermal CVD, low-pressure CVD (LPCVD), ultrahigh-vacuum CVD (UHCVD) and radio-frequency CVD (RFCVD). Additionally thin films are easy to produce. The purity and stoichiometry is often easy to control. However, high temperatures are often required and precursors can be difficult to design. CVD can also be sensitive to a wide range of parameters such as flow rates, gas purities and pressure. These parameters can also vary locally, so depositions even over length scales less than 1cm can vary considerably. This multi parameter sensitivity means depositions are highly reactor specific, making the repeatability and comparability of results reported in different publications difficult. This often leads to the need for extensive optimization ofa process whenever a new CVD reactor is built. A CVD reactors used for 2D material synthesis is typically cylindrical in shape generally quartz tube with a diameter of two or three inches and are placed horizontally within an electrically heated furnace. The general material used for the maker of the reactor is the quartz, so as to resist high temperatures to which the CVD system is heated. The furnace is insulated and consists of a single, two or three heating zones, with a separate temperature programmer available for each zone. Even though the temperatures of the 3-zones can be set differently, there is heat transfer between them such that the final temperatures obtained may be different from the programmed set points (Menzel et al., 2012). The growth mechanics depends on temperature inside the furnace tube, which can be low, moderate or high and is determined by heat transfer inside the reactor. In terms of the pressure used, although atmospheric pressure CVD growth is common (Han et al., 2013), several researchers have investigated low-pressure or vacuum CVD growth of materials (Li et al., 2011; Singhal et al., 2021), with the low pressure maintained using a vacuum pump. The chemical growth in a vacuum tube furnace is also termed as vapor transport when a condensed phase, typically a solid material is used to get volatized in the gaseous reactant called the transport agent and the final product get deposited elsewhere.

Mechanical and chemical exfoliation are the widely used methods for the preparation of TMDs. In the exfoliation, bulk crystals are used as the base materials and either applied force or atomic bond intercalation is employed for the peeling of the adjacent layers. Due to the weak vdW force between them, it is possible to achieve single to some layers of the 2D-materials. Similarly, liquid exfoliation is being widely used method for the preparation of VX_2 nanosheets. CVD is a scalable route to produce large-area 2D-materials (Cai et al., 2018), required in applications such as optoelectronic devices (Cheng et al., 2019), sensors (Tyagi et al., 2020) and membrane separation modules (Hyun et al., 2019). CVD has been used to grow 2D materials such as graphene (Ani et al., 2018; Zhang et al., 2013), transition metal dichalcogenides (TMDs) (Jiang et al., 2019; Kumar et al., 2015; Okada et al.,

Figure 2. A typical two zone split furnace CVD system internal structure

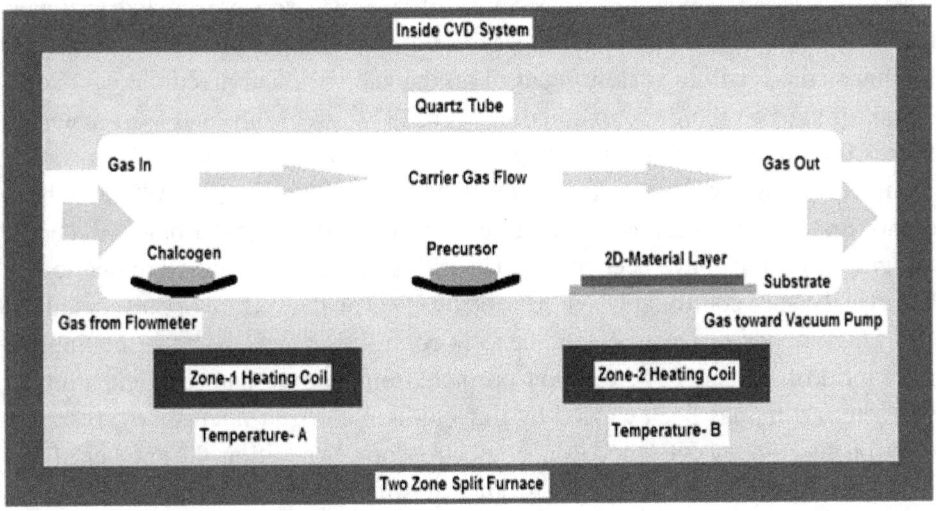

2019). Among TMDs, the prominent examples of materials obtained via CVD are molybdenum disulfide (MoS_2) and tungsten diselenide (WSe_2).

Nowadays, CVD has shown promise to grow high-quality TMDs with a controllable thickness density, scalability and best electronic properties. CVD is a high temperature chemical synthesis process by which a desired material is deposited on chosen substrates. CVD processes have been extensively studied for synthesizing thin film coatings of a wide range of materials, including metals, hybrid materials semiconductors, quasi halide perovskites and topological insulators. The precursor vapor is integrated from other system or made in the reactor furnace. The main advantages of CVD synthesis process are represented by the accesses to high quality, high purity 2D nanomaterials with controlled properties, which allow one to control the surface morphology, crystallinity and defects of 2D nanostructures by tuning the process parameters. Due to these advantages, some 2D nanomaterials have been prepared by this way conveniently, such as graphene, MoS_2, h-BN. CVD method is also another annealing strategy for the growth of high-quality single crystalline 2D sheet on substrates. Even so, CVD suffers from the requirement of high temperature, high vacuum and specific substrates.

Chemical Vapor Transport (CVT) process allows avoiding the mechanical exfoliation process for obtaining monolayer MoS_2 after bulk MoS_2 crystal growth with CVT. CVT provides high quality TMDs and Rare Earth Oxides (REOs) nanosheet samples that are comparable to that of a mechanical exfoliated sample through thermal decomposition. Metal-Organic (MOCVD) growth of TMDs, a direct vapor phase reaction of transition metal oxide also called as organometallics and

Sulfur (S), Selenium (Se), Tellurium (Te) has been widely adopted for the growth of layers such as MoS_2, WS_2, $MoSe_2$, WSe_2 and $MoTe_2$. The shapes were like triangles, hexagons, truncated-triangles, 3-point stars, 6-point stars. Synthesis of TMDs, made by this method with the careful control on the substrate and precursor gas. Some demonstration on the direct method of MOCVD to the MoS_2 and WS_2 materials makes this method highly appealing compared to other known CVD processes. Semiconducting TMD mono layers such as MoS_2, $MoSe_2$, WSe_2 and WS_2 have been demonstrated to be feasible for future electronics and opto-electronics. Recently, the graphene films have been synthesize with thickness that lies from one to few layers and can be controllably synthesized by CVD on the surface of polycrystalline Cu. The number of layers of graphene is controlled precisely by regulating the flow ratio of H_2 and CH_4, reaction pressure, temperature and reaction time. Its characterization has been carried by transmission electron microscopy, scanning electron microscopy, XRD and Raman spectroscopy. In addition, the graphene films which has been transferred from Cu to other substrates are found to have a fine optical transmittance that makes them available for exploitation in conductive materials. For the measurement of electrical conductivity, graphene films were subsequently transferred to silica glass substrate and examined by a 4-point probe technique. There have been attempts to produce MoS_2 layers via simple sulfurization of Mo metal. In the report, Mo film was deposited on a SiO_2/Si wafer and followed by thermal annealing in sulfur vapors to produce MoS_2 film. The direct sulfurization of Mo metal thin film provides a quick and easy way to access atomically thin MoS_2 layers on insulating substrates. The obtained MoS_2 layers show a metallic transport property with a low on/off current ratio, which could be due to the presence of unreacted metal impurities. Similarly, The report reported the vertically aligned MoS_2 and molybdenum diselenide ($MoSe_2$) layers can be produced by a rapid sulfurization/selenization process at 550°C.

1.1 Growth Parameters of CVD

The figure below explains the growth parameters in CVD chamber. To produce large-area, uniform and high-quality TMD sheets with good electrical performance, the growth parameters plays an important role in the whole process. It basically consists of five major physical conditional parameters namely precursor, temperature, substrate, carrier gas and pressure. Collectively they are responsible for fine development of 2D hybrid materials. Their inter-relation explanations are

A. A. **Temperature - Precursor:** Collectively both are responsible for concentration of reactants. Generally a chemical reaction triggers when the temperature is raised with introducing reactants. Concentration of reactants is directly proportional to the above combinations. Principle of chemical thermodynamics are involved in this process.

$$\Delta G^{\circ} = \Delta H^{\circ} - T\Delta S^{\circ}$$

B. **Precursor - Pressure:**Here principle of chemical equilibrium is involved. Law of partial pressure are applied. It also increases the feeding rate and thickness control.

C. **Pressure - Carrier Gas:** Introducing a carrier gas prevent spontaneity of chemical reaction and facilitates the forward reaction with selective evaporation.

D. D. **Carrier Gas - Substrate:** Molecular fusion and adatom flux increases towards the substrate. As the substrate is also heating up the molecular fusion and affinity of the vapors for substrate increases. Variation of reaction speed with temperature is given by Arrhenius's equation as,

$$k = Ae^{-E/RT}$$

where k is rate constant, E is activation energy and A is frequency factor.

E. **Substrate - Temperature:** Deposition and Nucleation sites are directly proportional to the substrate temperature. More the temperature more the number of nucleation sites per unit area.

To produce large-area, uniform and high-quality TMD sheets with good electrical performance, the growth parameters plays an important role in the whole process. It basically consists of five major physical conditional parameters namely precursor, temperature, substrate, carrier gas and pressure. Collectively they are responsible for fine development of 2D hybrid materials. Efforts have been devoted to the thermolysis of the precursor containing Mo and S atoms. A two-step CVD thermolysis process of ammonium thiomolybdate which was able to produce highly crystalline and large-area MoS_2 thin sheets on a variety of insulating substrates. The thermolysis of the single precursor ammonium thiomolybdate $(NH_4)_2MoS_4$ dissociated it into MoS_2, NH_3 and S vapors at above 800°C. In a two-step annealing process high-quality MoS_2 thin film was grown, firstly prepared by the dip-coating method where, $(NH_4)_2MoS_4$ thin film was deposited and further was taken into a quartz tube chamber with the Ar or H_2 flow at temperature 500°C. For the second step, temperature was raised to 1000°C and the sulfur vapors were introduced. In another process, MoO_3 and sulfur as the reactants for MoS_2 growth were used. It is a planned method for large scale MoS_2 single crystal with solid MoO_3 powder and S as precursor.

Figure 3. CVD growth parameters and their inter-effects

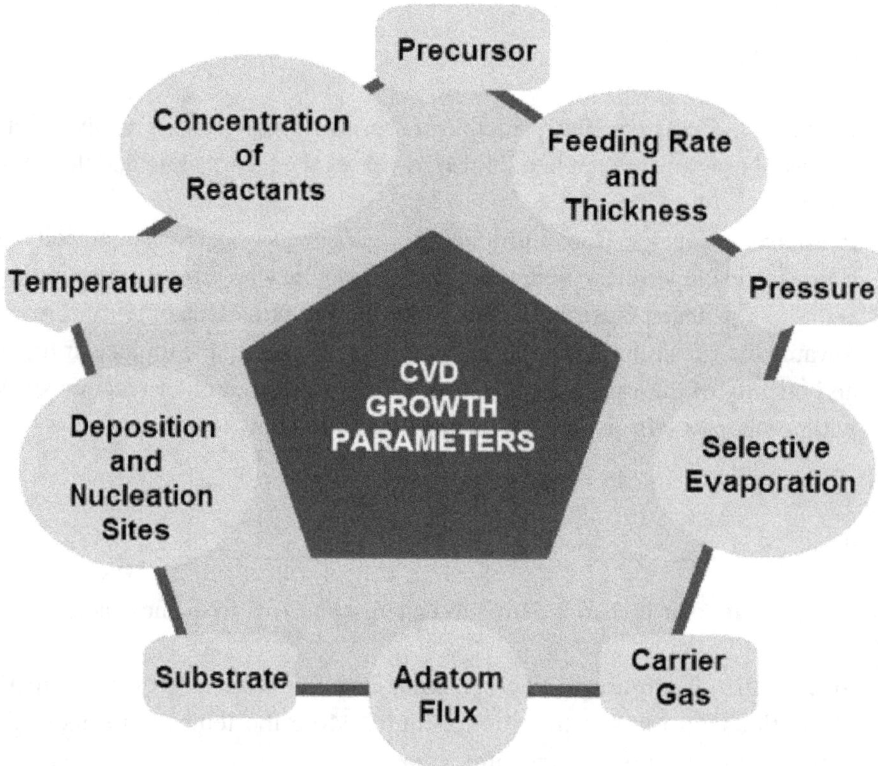

The photoluminescence (PL) polarization obtained from the MoS_2 prepared by VS growth is substantial with a maximum of 35% at 1.92eV, even at room temperature. Centimeter-scale uniform monolayer MoS_2 was synthesized on mica through a low-pressure CVD (LPCVD) process. It was concluded that the MoS_2 growth on mica follows an analogous layer-plus-island (Stranski–Krastanov) growth mode. The various properties of transition metal dichalcogenides (TMDs) can be characterized by utilizing different kinds of instruments and spectroscopic techniques. PL and Raman spectroscopy are the two most commonly used characterization tools. PL spectroscopy can be used to study the optical properties of the material after absorption of photons from a graph of intensity versus wavelength because the interlayer coupling between the bilayers changes the band structure and therefore alters the PL emission.

Raman spectroscopy makes use of the molecule mechanical vibration to provide information on the change of phonon vibrations. Raman spectroscopy has also been applied to investigate the interlayer coupling of WS_2 bilayers. In addition to bilayers made of the same material, Raman spectroscopy can also be used to reveal the interlayer coupling of bilayer heterostructures. Apart from them, optical microscopy (OM), atomic force microscopy (AFM), scanning electron microscopy (SEM), transmission electron microscopy (TEM), density functional theory (DFT) calculations, X-ray diffraction (XRD), micro-structural phases, surface morphologies, topological superconductivity phases, quantum spin hall effect, magneto-resistive effect, etc. have also been used to assist in the identification of twisted bilayers and have also been applied to identify and study the TMD bilayers and multilayer hybrid 2D materials.

These studies now have been used to integrate them in enormous potential applications. CVD synthesis have great significance in the generation of C-based nano-particles. In this method, a thin-film is formed on the substrate via the chemical kinetics of vapor precursors. A precursor is considered appropriate for CVD if it has surplus volatility, good chemical purity, stability at the time of evaporation, economical, a non-toxic nature and a longer life. Also, its decomposition should not results in impurities. For example, in the generation of CNTs via CVD, a substrate is placed in a chamber and heated to high temperatures. Subsequently, a carbon containing such as organic gas is gradually introduced to the system as a precursor. At high temperatures, the decomposition of the gas releases C-atoms, which recombine to form CNTs on substrate. However, the choice of catalyst plays a significant role in the morphology and type of nanomaterial obtained. In the preparation of graphene, Ni/Co catalysts provide few layer graphene, whereas a Copper catalyst provides one layer graphene. Overall, CVD is an excellent method for producing high-quality nanomaterials and it is well-known for the mass production of hybrid 2D nanomaterials.

1.2 Advantages of CVD

Some of the major advantages of CVD over other method includes,

1. Various metallic, glasses, hybrid materials, ceramics, alloys, polymers and semiconducting materials can be used.
2. The surface integrated over other substrate can be precise and controlled.
3. The temperature range can bear extreme variations and controls.
4. Vertical heterostructures are possible on compatible substrates in which molecular tuning can be done with thickness controllability.
5. Cost effectiveness, high purity, wear resistance, chemical inertness, high lubricity, corrosion resistance, fouling resistance are possible in CVD synthesis.

6. Different combinations of precursor gas can be integrated in CVD synthesis to convert metal-organic framework.
7. Due to inert vacuum environment it is comparably stable process.
8. Surface modification using different ionic combinations can be used by integrating one or more ionic precursor gases.
9. Comparably less by-products gases are formed and they can be neutralized by suitable external bypassing methods in CVD systems.

2. SYNTHESIS OF 2D-Sn_XS_Y BY CVD

Although powder vaporization approach can prepare high quality SnS_2 and SnS (using SnO_2 and sulfur as precursors) for fundamental study, it is not suitable for wafer-scale synthesis due to limited control of the precursor concentration. Chemical vapor deposition is able to grow high quality 2D materials with monolayer thickness control over wafer-scale. The CVD of tin sulfides using different precursors has been reported, for instance, $SnCl_4$ combined with H_2S. In this cases, SnS_2 is formed at low temperature (300-500)°C while SnS is formed at high temperature (500-550)°C. Note that both SnS_2 and SnS are stable phases in the temperature range between 300°C and 600°C according to the phase diagram of the Sn-S system.

(Liu et al., 2019) have developed a CVD method involving the use of SnO and S powders in separate boats with Argon as carrier gas with the furnace heated at 650°C. The S-crucible was shifted to the centre of the heated zone when the temperature was static. The SnS nanosheets produced were square faceted, horizontally aligned nanosheets with 25nm thickness with lateral sizes of 2nm, these nanosheets were used in photodetection. Nanosheets have been produced via a similar method by (Mutlu et al., 2018) SnO_2 and S powders were evaporated onto a B-doped Si substrate at 550°C under nitrogen within the tube furnace. The furnace temperature was maintained for 40 min. The formation mechanism involved the initial evaporation of SnO_2 at 400°C and the formation of SnO_2 clusters, which acted as SnS nucleation sites. SnS was formed at 550°C with SnS_2 formed at 425°C. The SnS growth was square in design and was aligned horizontally. At lower and higher temperatures, the following reaction occurs:

$$SnO_2\ (s) + 1.5\ S_2\ (g) \rightarrow SnS_2\ (s) + SO_2\ (g).$$

$$SnO_2\ (s) + S_2\ (g) \rightarrow SnS\ (s) + SO_2\ (g).$$

The CVD methods are very challenging to regenerate, with variability between layer grown in different reactors using the same parameters and conditions (Narayanan et al., 2019). It is also quite difficult to optimize the same reaction conditions, such as the temperature, pressure, flow rate and vapor-phase combinations as required for the synthesis of greater area hybrid materials with the desired density of defects. To assist with optimizing efforts, this processes used for 2D-materials can be classified into two types, based on whether the vapor-phase precursors enter the reactor from outside or are generated inside, with the latter more difficult to model theoretically and control experimentally. A novel synthetic route was developed for the growth of 2D layered materials, which involves vapor transport and a synthesized precursor that has the same chemical composition as the final 2D crystals. SnO_2 powder was transformed into a SnS_2 precursor powder by thermal sulfurization. Then, 2D SnS_2 nanocrystals were successfully synthesized from the precursor powder by vapor transport method.

The control of the different phases in the CVD process and how this is affected by the process conditions, is therefore not yet understood. Moreover, the SnS_2 and SnS layers were micrometer level with random grain orientations, resulting in insulating rather than semiconducting properties. For applications as a 2D semiconductor, the thickness needs to be scaled down to nanometer level, but so far there are only few studies focusing on few-layer thin layers. For example, the nucleation mechanism for CVD of 2D SnS_2 using H_2S and $SnCl_4$ has been investigated recently. SnS_2 shows a growth mode of initial 3D growth followed by 2D lateral growth. The thickness of the 2D SnS_2 crystals is determined by the height of the initial SnS_2 islands, which grow vertically and laterally due to a delayed crystallization mechanism. However, further investigation of the structure and composition of the initial nuclei are required in order to get better control on the height of initial nuclei, so as to obtain monolayer-thin SnS_2. Thus, the mechanism of the $SnCl_4/H_2S$ CVD process and how to control the Sn_xS_y phase from the process conditions remain to be explored. In addition, the thickness needs to be scaled down to nanometer level so that the semiconducting properties of 2D SnS_2 and SnS can be further explored. We study the dependence of the Sn_xS_y phase on both temperature and $H_2S/SnCl_4$ concentration ratio. A mechanism that explains how the phases (SnS_2 versus SnS) are determined by the deposition temperature and $H_2S/SnCl_4$ concentration ratio is possible. Based on the proposed mechanism for the CVD process, nanometer level 2D (SnS_2 and SnS) with well-controlled phase and grain orientation can be synthesized in wide process windows.

SnS_2 has a hexagonal structure and SnS has an orthorhombic structure. To investigate their structural properties, SnS_2 and SnS grown at 400°C with a deposition time of 30 min are characterized by atomic force microscopy (AFM), transmission electron microscopy (TEM) and X-ray diffraction (XRD). With a $H_2S/SnCl_4$ concentration ratio of 10, a nearly continuous SnS_2 film of ~14nm thick (2.1×10^{16} Sn atoms/cm^2, determined by RBS) with spiral growth characteristics is obtained. AFM height profile shows that the thickness of the SnS_2 layer is around 15nm, consistent with that determined by RBS. The grain size of SnS_2 is in the range of 100nm to 200nm. With a $H_2S/SnCl_4$ concentration ratio of 20, isolated square SnS domains rather than continuous film are obtained. The SnS domains are ~8nm thick as shown by the AFM height profile, with a lateral size of several μm. Electron diffraction study verifies the hexagonal structure of SnS_2 and orthorhombic structure of SnS. Only spots of the [0001] zone of SnS_2 with a combination of different sets of well-defined hexagonal patterns with relative rotations are observed. This indicates the good crystallinity of the SnS_2 film yet without preferred in-plane orientation for different domains. In contrast, only one set of well-defined rhombic pattern of the [010] zone is obtained from a single SnS domain, suggesting the good crystallinity of the isolated SnS domains. Moreover, XRD (θ-2θscan) is used to characterize the crystallinity of SnS2 $_a$nd SnS. (001) and (002) peaks of SnS2 $_a$nd (040) peak of SnS are detected, respectively, indicating that the basal planes of SnS2 $_a$nd SnS are mainly parallel to the substrates.

SnS has already been synthesised in mono-crystal, nano-particle and thin-film. Single crystals can either be synthesised by melt growth (also called the Bridgeman method) or melt mixing. Melt growth involves the transfer of the desired material vapor from a hot zone to a cool zone where it condenses to form a single crystal. Melt mixing involves heating the material above its meting point in a vacuum sealed ampoule which on cooling affords single crystals. SnS single crystals have previously been produced by both techniques, which can then be exfoliated to allow characterisation. SnS nanoparticles can be synthesised relatively simply from a tin salt, (often SnCl2$_)$ and a sulphur source (such as Na2$_S$), capping agents such as oleyl amine are used to stop agglomeration. The size of the nanoparticles can be tuned by the reaction temperature and precursor injection rates.

Reid et al. (2007) have used LP-CVD of a tin thio ether, $SnCl_4(O\text{-}C_6H_4(CH_2SMe)_2)$, with a furnace temperature of 650°C to produce SnS or SnS_2, depending on the substrate position within the furnace, however due to difficulties in controlling the oxygen content studies using LP-CVD are limited. The domain list of precursors for AP-CVD of SnS is also limited due to spontaneity and volatility. Therefore precursors that are used in wet chemical techniques and LP-CVD cannot often be used in AP-CVD processes. SnS has already been made from $SnCl_4$ and toxic hydrogen sulphide gas at atmospheric pressure by Price et al. (1999) onto glass substrates at 545°C with good stoichiometric control. Physical settlements techniques have also been used to prepare thin films of few layer SnS. Bulk SnS has been evaporated by resistive evaporation, electron beam evaporation and sputtering. High quality SnS films can be achieved at temperatures as low as room temperature by this technique, however they require access to high vacuum systems, which is a disadvantage when considering scale up.

In a similar manner to SnS, single crystals suitable for exfoliation of SnS_2 can be synthesised by melt mixing or melt growth. Additionally SnS_2 nanoparticles can be synthesised in a similar fashion to SnS nanoparticles using a tin salt and a sulphur source, however, the salt is often $SnCl_4$ rather than $SnCl_2$. (Kumar et al., 2022) have mounted SnS_2 nanoparticles onto graphene nanosheets by heating $SnCl_4$and thiourea in Graphene Oxide (GO) then subsequently reducing the GO. As with SnS, the synthesis of SnS_2 thin films by LP-CVD has been reported by Reid et al. (2007) who have deposited SnS and SnS_2 from a tin thioether $SnCl_4(O\text{-}C_6H_4(CH_2SMe)_2)$, at 650°C. Again as with SnS, the most probable reason for the lack of reports of LP-CVD of SnS_2 is the difficulty of excluding oxygen from the system. Luo et al. (2022) have reported the deposition of SnS_2 nanosheets on carbon cloth at low pressures by the initial depositionof SnO_2 and subsequent sulfurisation. Although the quality of the nanosheets was high, they were produced at temperatures of 850°C and the sheets were < 10μm across. Price et al. (1998) have used AA-CVD of the single source precursor tin(IV) thiolate $(PhS)_4Sn$ in the presence of H_2S to produce SnS_2 films. Liu et al. (2013) avoided the use of toxic H_2S by evaporating sulphur, $Na_2S_2O_3$ and $SnCl_4$ in a PECVD process to produce SnS_2 at 550°C. Amalraj et al. (2002) have also demonstrated SnS_2 films could be deposited by AA-CVD at the low temperature of 125°C using $SnCl_2$ and thiourea dissolved in H_2O and isopropyl alcohol (IPA) respectively by wet method.

3. APPLICATIONS OF 2D-MATERIALS

The 2D materials can be utilized in extensive applications such as catalysis,2D photodetectors, optoelectronic UV detectors, 2D superconductors, supercapacitors, power electronics FETs, organic FETs, 2D electromagnetic super-lens, 2D digital and nonlinear photonic communications, 2D crystal waveguide, contact engineering, 2D Schottky barriers (SBs), 2D nonlinear photonic crystal and smart 2D gas sensors. 2D semiconductors are promising materials for application in nanoelectronics, optoelectronics, catalysis and sensors.

3.1 Digital Sensors

Apart from these applications, these 2D materials are fulfilling the requirement of high light absorption and are therefore suitable for thin-film photovoltaic, quantum dot solar cells and other related hi-tech digital sensor applications. These 2D hybrid materials may be a good substitute for silicon-based techniques such as high-speed digital Internet of Things (IoT) devices, microwave sensors, photoluminescence, biosensors, electromagnetic antennas, conductometric sensors, etc.

Figure 4. Some important industrial applications of 2D-materials

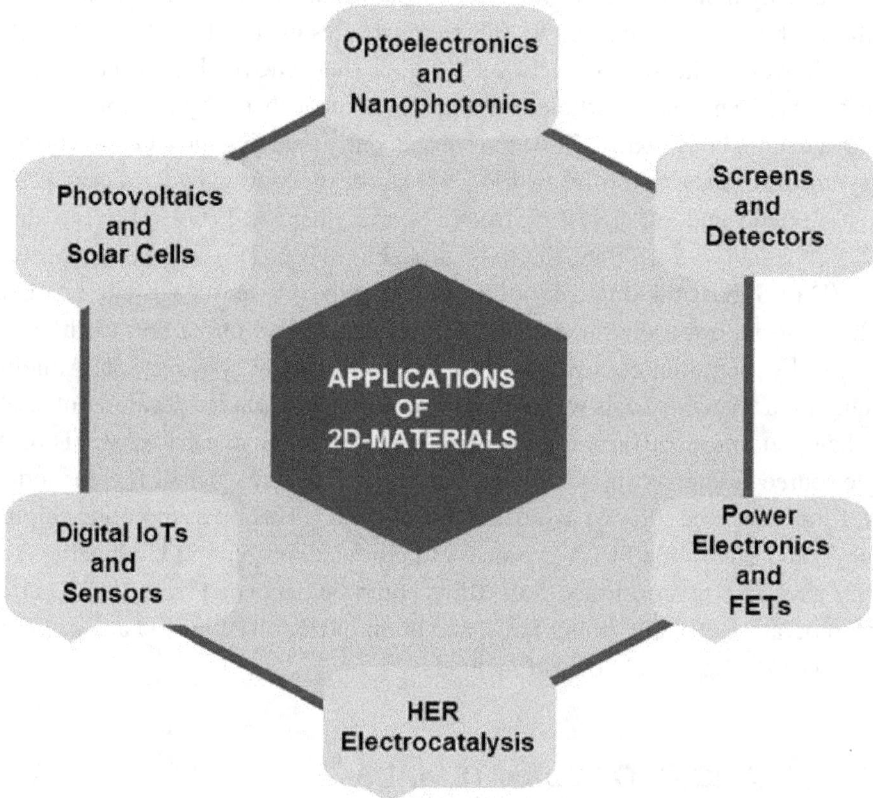

3.2 Power Electronics and FETs

Tin (II) monosulfide (SnS) and Tin (IV) disulfide (SnS_2) are interesting 2D semiconductors which exhibit a different type of conduction. SnS is a p-type semiconductor with orthorhombic crystal structure while SnS_2 is an n-type semiconductor with hexagonal crystal structure. One such material is SnS. With a direct bandgap, it has a high solar absorption coefficient and orthorhombic SnS (α-SnS) has a near ideal direct bandgap of 1.3eV. Tin and sulphur are also earth abundant elements and thus SnS thin films have the received an increasing amount of attention over the last 5 years as a solar cell absorber layer. SnS_2 has a monolayer thickness of 0.59nm and SnS has a monolayer thickness of 0.56 nm. 3nm thick SnS_2 has been predicted to have an electron mobility of 306 $cm^2/V{\bullet}s$. A thin-film dependent hole mobility has been verified for few layer SnS. The monolayer SnS was predicted to have a hole mobility as high as 4300 $cm^2/V \cdot s$, much higher than that of the bulk SnS (~500 $cm^2/V \cdot s$).

The much higher mobility of monolayer SnS than that of the bulk is attributed to the smaller effective mass of holes in monolayer SnS. Both 2D (SnS_2 and SnS) are attractive for application in ultra scaled metal oxide semiconductor field effect transistors (MOSFETs). Based on desired staggered band alignment with other 2D materials, they are promising candidates for tunnel FETs.

3.3 Optoelectronics and Nanophotonics

Also, 2D (SnS_2 and SnS) are interesting materials for ultrathin photovoltaic because of their optically active bandgap, whereby SnS_2 has a bandgap of 2.2eV and SnS has a bandgap of 1.1eV. Like SnS, SnS_2 also has a direct bandgap however it is larger at 2.2eV. In the bulk form it appears yellow and is air stable. Unlike SnS, SnS_2 displays n-type behavior and has potential to be used as a window layer in solar cells, a photo detector and a layered material for super batteries. In addition, 2D (SnS_2 and SnS) show good performance in photo sensing application. To enable the applications of 2D-SnS_2 and 2D-SnS, it is essential to synthesize them with a careful control of the phase, as this defines the semiconducting properties. Also, the deposition needs to be controlled at nanometer level with well-controlled grain orientation over wafer-scale. Due to its electrical and optical properties in addition to it being environmentally stable and composed of earth abundant elements, the major proposed application of SnS is as an absorber layer in a solar cell.

3.4 Detectors

However, SnS has also received attention as an infra-red detector, battery material and in photo electrochemical cells. So far the record efficiency of a solar cell incorporating SnS is 4.4%, with a short circuit current density of 20.2 mA cm^{-2} and an open-circuit voltage of 0.372 V. This was achieved by Park et al. (2015) who used a SnS film synthesized by atomic layer deposition method. They attribute the record efficiency to a few factors, which include H_2S annealing of the SnS film to increase grain size to increase carrier mobility and the fine tuning of the ZnO and ZnS mixed layer to align the conduction bands. The addition of a few mono layers of SnO_2 between SnS and Zn(O,S) reduced recombination at the junction.

3.5 Electrocatalysis

Hydrogen Evolution Reaction (HER) and Oxygen Evolution Reaction (OER) are the major chemical catalysis application where some 2D-materials are used. This

HER is further utilized as hydrogen storage process in energy applications. There are four main applications where SnS_2 is being explored, photocatalysis, solar cells, photodetection and energy storage hybrid materials. Prior SnS_2 photo catalytic abilities have been reviewed. Wei et al. (2014) exfoliated mono layers of SnS_2 mounted onto ITO possessed a light conversion efficiency of 38.7% at 420nm, which is one of the highest efficiencies recorded to date.

3.6 Photovoltaic and Solar Cells

Reddy et al. (2010) have reported a solar cell that has a SnS absorber layer and a SnS_2 window layer grown by chemical bath deposition (CBD) with an overall efficiency of 0.51%. Li et al. (2009) have shown it can be used as a replacement to the Pt-counter electrode in a dye sensitized solar cell, achieving an efficiency of 6.30% compared to 6.67% for Pt. Wang ct al. (2023) continued this work by demonstrating Ag doping of the SnS_2 provided an improvement on the Pt-counter electrode with an efficiency of 8.70%. SnS_2 has also been used as colloidal quantum dots in organic-inorganic hybrid solar cells with efficiencies of 0.31%. SnS_2 is additionally being explored as a photo detector as it has a very fast current response time. Mahajan et al. (2015) show single crystal SnS_2 grown by CVD has a response time of 5μs, an order of magnitude faster than any other 2D material measured. Due to its layered structure, SnS_2 can intercalate ions for energy storage applications and contact engineering. This is particularly useful as a replacement for graphite electrodes in Na-ion batteries due to graphite's poor performance. It has been shown by Meng et al. (2015) that Na can alloy with SnS_2 as the anode to produce Na_2S_2 in a reversible reaction. They report a good performance of 500 $mAhg^{-1}$ for 400 cycles when SnS_2 is deposited by CVD onto rGO. Again using rGO, Jiao et al. (2017) use exfoliated SnS_2 to report a performance of 610 $mAhg^{-1}$ up to 300 cycles.

ACKNOWLEDGMENT

The author would like to acknowledge help and support from the following departments and universities.

1. Advanced Materials Laboratory, Department of Physics, Patna University, India 800005.

2. Physics Laboratory, Government Engineering College, Nawada, Bihar Engineering University, Bihar, India 805111.

3. Department of Science and Technology, Government of Bihar, Patna, India 800001.

Corresponding Author Details

Email: sumitkumar@pup.ac.in
Designation: Assistant Professor, GEC Nawada, Bihar, India 805111.

REFERENCES

Amalraj, L., Sanjeeviraja, C., & Jayachandran, M. (2002). Spray pyrolysised tin disulphide thin film and characterisation. *Journal of Crystal Growth*, *234*(4), 683–689. doi:10.1016/S0022-0248(01)01756-0

Anasori, B., Lukatskaya, M. R., & Gogotsi, Y. (2017). 2D metal carbides and nitrides (MXenes) for energy storage. *Nature Reviews. Materials*, *2*(2), 1–17. doi:10.1038/natrevmats.2016.98

Ani, M. H., Kamarudin, M. A., Ramlan, A. H., Ismail, E., Sirat, M. S., Mohamed, M. A., & Azam, M. A. (2018). A critical review on the contributions of chemical and physical factors toward the nucleation and growth of large-area graphene. *Journal of Materials Science*, *53*(10), 7095–7111. doi:10.1007/s10853-018-1994-0

Cai, Z., Liu, B., Zou, X., & Cheng, H. M. (2018). Chemical vapor deposition growth and applications of two-dimensional materials and their heterostructures. *Chemical Reviews*, *118*(13), 6091–6133. doi:10.1021/acs.chemrev.7b00536 PMID:29384374

Cheng, Z., Xia, M., Liu, S., Hu, R., Liang, G., & Zhang, S. (2019). Role of rough substrate on the growth of large single-crystal MoS2 by chemical vapor deposition. *Applied Surface Science*, *476*, 1008–1015. doi:10.1016/j.apsusc.2019.01.211

Chhowalla, M., Shin, H. S., Eda, G., Li, L. J., Loh, K. P., & Zhang, H. (2013). The chemistry of two-dimensional layered transition metal dichalcogenide nanosheets. *Nature Chemistry*, *5*(4), 263–275. doi:10.1038/nchem.1589 PMID:23511414

Deng, S., & Berry, V. (2016). Wrinkled, rippled and crumpled graphene: An overview of formation mechanism, electronic properties, and applications. *Materials Today*, *19*(4), 197–212. doi:10.1016/j.mattod.2015.10.002

Donarelli, M., & Ottaviano, L. (2018). 2D materials for gas sensing applications: A review on graphene oxide. *Sensors*, *18*(11), 3638. doi:10.3390/s18113638

Garg, M., Gupta, A., Sharma, A. L., & Singh, S. (2021). Advancements in 2D materials based biosensors for oxidative stress biomarkers. *ACS Applied Bio Materials*, *4*(8), 5944–5960. doi:10.1021/acsabm.1c00625 PMID:35006886

Goto, T., & Hirai, T. (1987). Chemically vapor deposited Ti3SiC2. *Materials Research Bulletin*, *22*(9), 1195–1201. doi:10.1016/0025-5408(87)90128-0

Gupta, A., Sakthivel, T., & Seal, S. (2015). Recent development in 2D materials beyond graphene. *Progress in Materials Science*, *73*, 44–126. doi:10.1016/j.pmatsci.2015.02.002

Han, G. H., Rodríguez-Manzo, J. A., Lee, C. W., Kybert, N. J., Lerner, M. B., Qi, Z. J., Dattoli, E. N., Rappe, A. M., Drndic, M., & Johnson, A. T. (2013). Continuous growth of hexagonal graphene and boron nitride in-plane heterostructures by atmospheric pressure chemical vapor deposition. *ACS Nano*, *7*(11), 10129–10138. doi:10.1021/nn404331f PMID:24182310

Hyun, C. M., Choi, J. H., Lee, S. W., Seo, S. Y., Lee, M. J., Kwon, S. H., & Ahn, J. H. (2019). Synthesis of Bi 2 Te 3 Single Crystals with Lateral Size up to Tens of Micrometers by Vapor Transport and Its Potential for Thermoelectric Applications. *Crystal Growth & Design*, *19*(4), 2024–2029. doi:10.1021/acs.cgd.8b01931

Jariwala, D., Sangwan, V. K., Lauhon, L. J., Marks, T. J., & Hersam, M. C. (2014). Emerging device applications for semiconducting two-dimensional transition metal dichalcogenides. *ACS Nano*, *8*(2), 1102–1120. doi:10.1021/nn500064s PMID:24476095

Ji, Q., Zhang, Y., Gao, T., Zhang, Y., Ma, D., Liu, M., Chen, Y., Qiao, X., Tan, P. H., Kan, M., Feng, J., Sun, Q., & Liu, Z. (2013). Epitaxial monolayer MoS2 on mica with novel photoluminescence. *Nano Letters*, *13*(8), 3870–3877. doi:10.1021/nl401938t PMID:23899342

Jiang, L., Liu, J., Tian, A., Ren, X., Huang, S., Zhou, W., Zhang, L., Li, D., Zhang, S., Ikeda, M., & Yang, H. (2019). Influence of substrate misorientation on carbon impurity incorporation and electrical properties of p-GaN grown by metalorganic chemical vapor deposition. *Applied Physics Express*, *12*(5), 055503. doi:10.7567/1882-0786/ab0da2

Jiao, X., Li, X., Jin, X., Sun, Y., Xu, J., Liang, L., Ju, H., Zhu, J., Pan, Y., Yan, W., Lin, Y., & Xie, Y. (2017). Partially oxidized SnS2 atomic layers achieving efficient visible-light-driven CO2 reduction. *Journal of the American Chemical Society*, *139*(49), 18044–18051. doi:10.1021/jacs.7b10287 PMID:29144744

Keum, D. H., Cho, S., Kim, J. H., Choe, D. H., Sung, H. J., Kan, M., Kang, H., Hwang, J., Kim, S. W., Yang, H., Chang, K. J., & Lee, Y. H. (2015). Bandgap opening in few-layered monoclinic $MoTe_2$. *Nature Physics*, *11*(6), 482–486. doi:10.1038/nphys3314

Kong, D., Wang, H., Cha, J. J., Pasta, M., Koski, K. J., Yao, J., & Cui, Y. (2013). Synthesis of MoS_2 and $MoSe_2$ films with vertically aligned layers. *Nano Letters*, *13*(3), 1341–1347. doi:10.1021/nl400258t PMID:23387444

Kumar, D. A. K., Valanarasu, S., Tamilnayagam, V., & Amalraj, L. (2017). Structural, morphological and optical properties of SnS_2 thin films by nebulized spray pyrolysis technique. *Journal of Materials Science Materials in Electronics, 28*(19), 14209–14216. doi:10.1007/s10854-017-7278-7

Kumar, R., Goel, N., Jarwal, D. K., Hu, Y., Zhang, J., & Kumar, M. (2023). Strategic review on chemical vapor deposition technology-derived 2D material nanostructures for room-temperature gas sensors. *Journal of Materials Chemistry. C, Materials for Optical and Electronic Devices, 11*(3), 774–801. doi:10.1039/D2TC04188J

Kumar, S., Pavelyev, V., Mishra, P., Tripathi, N., Sharma, P., & Calle, F. (2020). A review on 2D transition metal di-chalcogenides and metal oxide nanostructures based NO_2. *Materials Science in Semiconductor Processing, 107*, 104865. doi:10.1016/j.mssp.2019.104865

Kumar, S., Tessarek, C., Sarau, G., Christiansen, S., & Singh, R. (2015). Self-Catalytic Growth of β-Ga_2O_3 Nanostructures by Chemical Vapor Deposition. *Advanced Engineering Materials, 17*(5), 709–715. doi:10.1002/adem.201400289

Late, D. J., Rout, C. S., Chakravarty, D., & Ratha, S. (2015). Emerging energy applications of two-dimensional layered materials. *Canadian Chemical Transactions, 3*, 118–157. doi:10.13179/canchemtrans.2015.03.02.0174

Li, C., Cao, Q., Wang, F., Xiao, Y., Li, Y., Delaunay, J. J., & Zhu, H. (2018). Engineering graphene and TMDs based van der Waals heterostructures for photovoltaic and photoelectrochemical solar energy conversion. *Chemical Society Reviews, 47*(13), 4981–5037. doi:10.1039/C8CS00067K PMID:29736528

Li, P., Wu, J., Lin, J., Huang, M., Huang, Y., & Li, Q. (2009). High-performance and low platinum loading Pt/carbon black counter electrode for dye-sensitized solar cells. *Solar Energy, 83*(6), 845–849. doi:10.1016/j.solener.2008.11.012

Li, X., Magnuson, C. W., Venugopal, A., Tromp, R. M., Hannon, J. B., Vogel, E. M., Colombo, L., & Ruoff, R. S. (2011). Large-area graphene single crystals grown by low-pressure chemical vapor deposition of methane on copper. *Journal of the American Chemical Society, 133*(9), 2816–2819. doi:10.1021/ja109793s PMID:21309560

Liu, X., Cao, L., Guo, Z., Li, Y., Gao, W., & Zhou, L. (2019). A review of perovskite photovoltaic materials' synthesis and applications via chemical vapor deposition method. *Materials (Basel), 12*(20), 3304. doi:10.3390/ma12203304 PMID:31614476

Liu, X., Kim, J., Yeluri, R., Lal, S., Li, H., Lu, J., Keller, S., Mazumder, B., Speck, J. S., & Mishra, U. K. (2013). Fixed charge and trap states of in situ Al_2O_3 on Ga-face GaN metal-oxide-semiconductor capacitors grown by metalorganic chemical vapor deposition. *Journal of Applied Physics*, *114*(16), 164507. Advance online publication. doi:10.1063/1.4827201

Liu, Y., Weiss, N. O., Duan, X., Cheng, H. C., Huang, Y., & Duan, X. (2016). Van der Waals heterostructures and devices. *Nature Reviews. Materials*, *1*(9), 1–17. doi:10.1038/natrevmats.2016.42

Luo, T., Lindner, L., Langer, J., Cimalla, V., Vidal, X., Hahl, F., Schreyvogel, C., Onoda, S., Ishii, S., Ohshima, T., Wang, D., Simpson, D. A., Johnson, B. C., Capelli, M., Blinder, R., & Jeske, J. (2022). Creation of nitrogen-vacancy centers in chemical vapor deposition diamond for sensing applications. *New Journal of Physics*, *24*(3), 033030. doi:10.1088/1367-2630/ac58b6

Mahajan, A., Hyun, W. J., Walker, S. B., Lewis, J. A., Francis, L. F., & Frisbie, C. D. (2015). High-resolution, high–aspect ratio conductive wires embedded in plastic substrates. *ACS Applied Materials & Interfaces*, *7*(3), 1841–1847. doi:10.1021/am507539a PMID:25594812

Mak, K. F., & Shan, J. (2016). Photonics and optoelectronics of 2D semiconductor transition metal dichalcogenides. *Nature Photonics*, *10*(4), 216–226. doi:10.1038/nphoton.2015.282

Mannix, A. J., Kiraly, B., Hersam, M. C., & Guisinger, N. P. (2017). Synthesis and chemistry of elemental 2D materials. *Nature Reviews. Chemistry*, *1*(2), 0014. Advance online publication. doi:10.1038/s41570-016-0014

Manzeli, S., Ovchinnikov, D., Pasquier, D., Yazyev, O. V., & Kis, A. (2017). 2D transition metal dichalcogenides. *Nature Reviews. Materials*, *2*(8), 1–15. doi:10.1038/natrevmats.2017.33

Meng, J., Suo, Y., Li, J., Zheng, G., Liu, Y., Zhang, J., & Zheng, X. (2015). Nitrogen-doped graphene aerogels as anode materials for lithium-ion battery: Assembly and electrochemical properties. *Materials Letters*, *160*, 392–396. doi:10.1016/j.matlet.2015.08.024

Menzel, A., Subannajui, K., Bakhda, R., Wang, Y., Thomann, R., & Zacharias, M. (2012). Tuning the growth mechanism of ZnO nanowires by controlled carrier and reaction gas modulation in thermal CVD. *The Journal of Physical Chemistry Letters*, *3*(19), 2815–2821. doi:10.1021/jz301103s

Mutlu, Z., Debnath, B., Su, S., Li, C., Ozkan, M., Bozhilov, K. N., Lake, R. K., & Ozkan, C. S. (2018). Chemical vapor deposition and phase stability of pyrite on SiO_2. *Journal of Materials Chemistry. C, Materials for Optical and Electronic Devices*, *6*(17), 4753–4759. doi:10.1039/C8TC00584B

Narayanan, V., M A, G., & Rahman, A. (2019). How to 'train' your CVD to grow large-area 2D materials. *Materials Research Express*, *6*(12), 125002. doi:10.1088/2053-1591/ab5383

Nickl, J. J., Schweitzer, K. K., & Luxenberg, P. (1972). Gasphasenabscheidung im system TiSiC. *Journal of the Less Common Metals*, *26*(3), 335–353. doi:10.1016/0022-5088(72)90083-5

Nicks, J., Sasitharan, K., Prasad, R. R. R., Ashworth, D. J., & Foster, J. A. (2021). Metal–organic framework nanosheets: Programmable 2D materials for catalysis, sensing, electronics, and separation applications. *Advanced Functional Materials*, *31*(42), 2103723. doi:10.1002/adfm.202103723

Novoselov, K. S., Geim, A. K., Morozov, S. V., Jiang, D. E., Zhang, Y., Dubonos, S. V., Grigorieva, I. V., & Firsov, A. A. (2004). Electric field effect in atomically thin carbon films. *Science*, *306*(5696), 666–669. doi:10.1126/science.1102896 PMID:15499015

Novoselov, K. S., Mishchenko, A., Carvalho, A., & Castro Neto, A. H. (2016). 2D materials and van der Waals heterostructures. *Science*, *353*(6298), aac9439. Advance online publication. doi:10.1126/science.aac9439 PMID:27471306

Okada, M., Okada, N., Chang, W. H., Endo, T., Ando, A., Shimizu, T., Kubo, T., Miyata, Y., & Irisawa, T. (2019). Gas-source CVD growth of atomic layered WS_2 from WF_6 and H_2S precursors with high grain size uniformity. *Scientific Reports*, *9*(1), 17678. doi:10.1038/s41598-019-54049-6 PMID:31776373

Park, J., Choudhary, N., Smith, J., Lee, G., Kim, M., & Choi, W. (2015). Thickness modulated MoS_2 grown by chemical vapor deposition for transparent and flexible electronic devices. *Applied Physics Letters*, *106*(1), 012104. Advance online publication. doi:10.1063/1.4905476

Price, L. S., Parkin, I. P., Hardy, A. M. E., Clark, R. J. H., Hibbert, T. G., & Molloy, K. C. (1999). Atmospheric Pressure Chemical Vapor Deposition of Tin Sulfides (SnS, Sn_2S_3, and SnS_2) on Glass. *Chemistry of Materials*, *11*(7), 1792–1799. doi:10.1021/cm990005z

Price, L. S., Parkin, I. P., Hibbert, T. G., & Molloy, K. C. (1998). Atmospheric Pressure CVD of SnS and SnS_2 on Glass. *Chemical Vapor Deposition*, *4*(6), 222–225. doi:10.1002/(SICI)1521-3862(199812)04:06<222::AID-CVDE222>3.0.CO;2-6

Qin, B., Ma, H., Hossain, M., Zhong, M., Xia, Q., Li, B., & Duan, X. (2020). Substrates in the synthesis of two-dimensional materials via chemical vapor deposition. *Chemistry of Materials*, *32*(24), 10321–10347. doi:10.1021/acs.chemmater.0c03549

Ramakrishna Reddy, K. T., Koteswara Reddy, N., & Miles, R. W. (2006). Photovoltaic properties of SnS based solar cells. *Solar Energy Materials and Solar Cells*, *90*(18–19), 3041–3046. doi:10.1016/j.solmat.2006.06.012

Reddy, A. L. M., Shaijumon, M. M., Gowda, S. R., & Ajayan, P. M. (2010). Multisegmented Au-MnO 2 /Carbon Nanotube Hybrid Coaxial Arrays for High-Power Supercapacitor Applications. *The Journal of Physical Chemistry. C, Nanomaterials and Interfaces*, *114*(1), 658–663. doi:10.1021/jp908739q

Reid, S. D., Hector, A. L., Levason, W., Reid, G., Waller, B. J., & Webster, M. (2007). Thio- and seleno-ether complexes with Group 4 tetrahalides and tin tetrachloride: Preparation and use in CVD for metal chalcogenide films. *Dalton Transactions (Cambridge, England)*, *42*(42), 4769–4777. doi:10.1039/b708809d PMID:17955129

Rout, C. S., Late, D. J., & Morgan, H. (Eds.). (2019). *Fundamentals and Sensing Applications of 2D Materials*. Woodhead Publishing.

Schaibley, J. R., Yu, H., Clark, G., Rivera, P., Ross, J. S., Seyler, K. L., Yao, W., & Xu, X. (2016). Valleytronics in 2D materials. *Nature Reviews. Materials*, *1*(11), 1–15. doi:10.1038/natrevmats.2016.55

Schulman, D. S., Arnold, A. J., & Das, S. (2018). Contact engineering for 2D materials and devices. *Chemical Society Reviews*, *47*(9), 3037–3058. doi:10.1039/C7CS00828G PMID:29498729

Shinde, P., & Rout, C. S. (2021). Advances in synthesis, properties and emerging applications of tin sulfides and its heterostructures. *Materials Chemistry Frontiers*, *5*(2), 516–556. doi:10.1039/D0QM00470G

Singhal, R., Echeverria, E., McIlroy, D. N., & Singh, R. N. (2021). Synthesis of hexagonal boron nitride films on silicon and sapphire substrates by low-pressure chemical vapor deposition. *Thin Solid Films*, *733*, 138812. doi:10.1016/j.tsf.2021.138812

Smith, A. J., Meek, P. E., & Liang, W. Y. (1977). Raman scattering studies of SnS2 and SnSe2. *Journal of Physics. Journal of Physics. C. Solid State Physics*, *10*(8), 1321–1323. doi:10.1088/0022-3719/10/8/035

Song, J. G., Park, K., Park, J., & Kim, H. (2015). Vapor deposition techniques for synthesis of two-dimensional transition metal dichalcogenides. *Applied Microscopy*, *45*(3), 119–125. doi:10.9729/AM.2015.45.3.119

Taffelli, A., Dirè, S., Quaranta, A., & Pancheri, L. (2021). MoS2 based photodetectors: A review. *Sensors (Basel)*, *21*(8), 2758. doi:10.3390/s21082758 PMID:33919731

Tang, L., Tan, J., Nong, H., Liu, B., & Cheng, H. M. (2021). Chemical vapor deposition growth of two-dimensional compound materials: Controllability, material quality, and growth mechanism. *Accounts of Materials Research*, *2*(1), 36–47. doi:10.1021/accountsmr.0c00063

Tyagi, D., Wang, H., Huang, W., Hu, L., Tang, Y., Guo, Z., Ouyang, Z., & Zhang, H. (2020). Recent advances in two-dimensional-material-based sensing technology toward health and environmental monitoring applications. *Nanoscale*, *12*(6), 3535–3559. doi:10.1039/C9NR10178K PMID:32003390

Varghese, S. S., Varghese, S. H., Swaminathan, S., Singh, K. K., & Mittal, V. (2015). Two-dimensional materials for sensing: Graphene and beyond. *Electronics (Basel)*, *4*(3), 651–687. doi:10.3390/electronics4030651

Wang, Y., Zhang, S. T., Li, L., Yang, X., Lu, L., & Li, D. (2023). Dopant-free passivating contacts for crystalline silicon solar cells: Progress and prospects. *EcoMat*, *5*(2), e12292. doi:10.1002/eom2.12292

Wei, R., Hu, J., Zhou, T., Zhou, X., Liu, J., & Li, J. (2014). Ultrathin SnS_2 nanosheets with exposed {001} facets and enhanced photocatalytic properties. *Acta Materialia*, *66*, 163–171. doi:10.1016/j.actamat.2013.11.076

Xia, F., Wang, H., Xiao, D., Dubey, M., & Ramasubramaniam, A. (2014). Two-dimensional material nanophotonics. *Nature Photonics*, *8*(12), 899–907. doi:10.1038/nphoton.2014.271

Yan, J., Qiu, W., Wu, G., Heremans, P., & Chen, H. (2018). Recent progress in 2D/quasi-2D layered metal halide perovskites for solar cells. *Journal of Materials Chemistry. A, Materials for Energy and Sustainability*, *6*(24), 11063–11077. doi:10.1039/C8TA02288G

Ye, G., Gong, Y., Lei, S., He, Y., Li, B., Zhang, X., Jin, Z., Dong, L., Lou, J., Vajtai, R., Zhou, W., & Ajayan, P. M. (2017). Synthesis of large-scale atomic-layer SnS_2 through chemical vapor deposition. *Nano Research, 10*(7), 2386–2394. doi:10.1007/s12274-017-1436-3

Yu, J., Li, J., Zhang, W., & Chang, H. (2015). Synthesis of high quality two-dimensional materials via chemical vapor deposition. *Chemical Science (Cambridge), 6*(12), 6705–6716. doi:10.1039/C5SC01941A PMID:29861920

Zhan, Y., Liu, Z., Najmaei, S., Ajayan, P. M., & Lou, J. (2012). Large-area vapor-phase growth and characterization of MoS_2 atomic layers on a SiO_2 substrate. *Small, 8*(7), 966–971. doi:10.1002/smll.201102654 PMID:22334392

Zhang, J., Wang, F., Shenoy, V. B., Tang, M., & Lou, J. (2020). Towards controlled synthesis of 2D crystals by chemical vapor deposition (CVD). *Materials Today, 40*, 132–139. doi:10.1016/j.mattod.2020.06.012

Zhang, W., Zhang, Y., Qiu, J., Zhao, Z., & Liu, N. (2021). Topological structures of transition metal dichalcogenides: A review on fabrication, effects, applications, and potential. *InfoMat, 3*(2), 133–154. doi:10.1002/inf2.12156

Zhang, Y., Tang, T. T., Girit, C., Hao, Z., Martin, M. C., Zettl, A., Crommie, M. F., Shen, Y. R., & Wang, F. (2009). Direct observation of a widely tunable bandgap in bilayer graphene. *Nature, 459*(7248), 820–823. doi:10.1038/nature08105 PMID:19516337

Zhang, Y. I., Zhang, L., & Zhou, C. (2013). Review of chemical vapor deposition of graphene and related applications. *Accounts of Chemical Research, 46*(10), 2329–2339. doi:10.1021/ar300203n PMID:23480816

Compilation of References

. Miller, M. A., Gadde, S., Pfirschke, C., Engblom, C., Sprachman, M. M., Kohler, R. H., ... Weissleder, R. (2015). Predicting therapeutic nanomedicine efficacy using a companion magnetic resonance imaging nanoparticle. *Science Translational Medicine, 7*(314), 314ra183-314ra183.

Abdelrahman, E. A., Hegazey, R. M., Kotp, Y. H., & Alharbi, A. (2019). Facile synthesis of Fe2O3 nanoparticles from Egyptian insecticide cans for efficient photocatalytic degradation of methylene blue and crystal violet dyes. *Spectrochimica Acta. Part A: Molecular and Biomolecular Spectroscopy, 222*, 117195. doi:10.1016/j.saa.2019.117195 PMID:31176151

Abdelsattar, A. S., Gouda, S. M., Hassana, Y. Y., Farouka, W. M., Makky, S., Nasr, A., Hakim, T. A., & El-Shibiny, A. (2022). In vitro bacteriophage-mediated synthesis of silver nanoparticles for antibacterial applications and heavy metal detection. *Materials Letters, 318*, 132184. doi:10.1016/j.matlet.2022.132184

Abel, S., Jule, L. T., & Belay, F. (2021). Application of titanium dioxide nanoparticles synthesized by sol-gel methods in wastewater treatment. *Journal of Nanomaterials, 2021*, 1–6.

Abidin, N. M. Z., Sultan, M. T. H., Shah, A. U. M., & Safri, S. N. A. (2019). Charpy and Izod impact properties of natural fibre composites. *IOP Conference Series. Materials Science and Engineering, 670*(1), 012031. Advance online publication. doi:10.1088/1757-899X/670/1/012031

Abou El-Mal, H. S. S., Sherbini, A. S., & Sallam, H. E. M. (2015). Mode II Fracture Toughness of Hybrid FRCs. *International Journal of Concrete Structures and Materials, 9*(4), 475–486. doi:10.1007/s40069-015-0117-4

Abu-Hani, A. F. S. (2017). Design, fabrication, and characterization of low-power gas sensors based on organic-inorganic nano-composite. *Organic Electronics C, 42*, 284–292. https://www. infona.pl//resource/bwmeta1.element.elsevier-eb8437f5-ad53-3bb2-83ec-852938746948

Acuña, R., Cristóbal, M. J., Abreu, C. M., & Cabeza, M. (2019). Microstructure and Wear Properties of Surface Composite Layer Produced by Friction Stir Processing (FSP) in AA2024-T351 Aluminum Alloy. *Metallurgical and Materials Transactions. A, Physical Metallurgy and Materials Science, 50*(6), 2860–2874. Advance online publication. doi:10.1007/s11661-019-05172-6

Adams, B. D., & Chen, A. (2011). The role of palladium in a hydrogen economy. *Materials Today, 14*(6), 282–289. doi:10.1016/S1369-7021(11)70143-2

Adamski, J., Qadir, M. I., Serna, J. P., Bernardi, F., Baptista, D. L., Salles, B. R., Novak, M. A., Machado, G., & Dupont, J. (2018). Core–shell Fe–Pt nanoparticles in ionic liquids: Magnetic and catalytic properties. *The Journal of Physical Chemistry. C, Nanomaterials and Interfaces, 122*(8), 4641–4650. doi:10.1021/acs.jpcc.7b12219

Adel, I. M., ElMeligy, M. F., Amer, M. S., & Elkasabgy, N. A. (2023). Polymeric nanocomposite hydrogel scaffold for jawbone regeneration: The role of rosuvastatin calcium-loaded silica nanoparticles. *International Journal of Pharmaceutics: X, 6*, 100213. doi:10.1016/j.ijpx.2023.100213 PMID:37927584

Adnan, M. M., Dalod, A. R. M., Balcı, M. H., Glaum, J., & Einarsrud, M. (2018). In situ synthesis of hybrid Inorganic–polymer nanocomposites. *Polymers, 10*(10), 1129. doi:10.3390/polym10101129 PMID:30961054

Aflori, M. (2021). Smart nanomaterials for biomedical applications—A review. *Nanomaterials (Basel, Switzerland), 11*(2), 396. doi:10.3390/nano11020396 PMID:33557177

Afzal, O., Altamimi, A. S., Nadeem, M. S., Alzarea, S. I., Almalki, W. H., Tariq, A., Mubeen, B., Murtaza, B. N., Iftikhar, S., Riaz, N., & Kazmi, I. (2022). Nanoparticles in Drug Delivery: From History to Therapeutic Applications. *Nanomaterials (Basel, Switzerland), 12*(24), 4494. doi:10.3390/nano12244494 PMID:36558344

Agarwal, N., Nair, M. S., Mazumder, A., & Poluri, K. M. (2017). Characterization of Nanomaterials Using Nuclear Magnetic Resonance Spectroscopy. *Characterization of Nanomaterials*, 61-102. doi:10.1016/B978-0-08-101973-3.00003-1

Ahearne, J. (2004). Storage of electric energy. In *Report on research and development of energy technologies* (p. 22). International Union of Pure and Applied Physics. https://www.iupap.org/wg/energy/report-a.pdf

Ahmed, K., Shahidi, T. R., & Irfanul Alam, S. Md., & Momen, S. (2019). Rice Leaf Disease Detection Using Machine Learning Techniques. *2019 International Conference on Sustainable Technologies for Industry 4.0 (STI)*, 1–5. 10.1109/STI47673.2019.9068096

Ahmed, A., He, P., He, P., Wu, Y., He, Y., & Munir, S. (2023). Environmental effect of agriculture-related manufactured nano-objects on soil microbial communities. In *Environment International* (Vol. 173). Elsevier Ltd. doi:10.1016/j.envint.2023.107819

Ajinkya, N., Yu, X., Kaithal, P., Luo, H., Somani, P., & Ramakrishna, S. (2020). Magnetic iron oxide nanoparticle (IONP) synthesis to applications: Present and future. *Materials (Basel), 13*(20), 4644. doi:10.3390/ma13204644 PMID:33080937

Akbari, A., Hashemzadeh, H., Eshkiki, Z. S., Masoodi, M., Tabaeian, S. P., Naderi-Manesh, H., Zare, A. A., & Agah, S. (2023). Detection of plasma miR-223 by a novel label-free graphene oxide/gold nanocomposite immunosensor in colorectal cancer patients: An electrochemical biosensor approach. *Biosensors & Bioelectronics: X*, *14*, 100331. doi:10.1016/j.biosx.2023.100331

Al-Ahmed, A., Sarı, A., Mazumder, M. J., Salhi, B., Hekimoğlu, G., Al-Sulaiman, F. A., & Inamuddin. (2020). Thermal energy storage and thermal conductivity properties of fatty acid/fatty acid-grafted-CNTs and fatty acid/CNTs as novel composite phase change materials. *Scientific Reports*, *10*(1), 15388. Advance online publication. doi:10.1038/s41598-020-71891-1 PMID:32958838

Albalawi, F., Hussein, M. Z., Fakurazi, S., & Masarudin, M. J. (2021). Engineered Nanomaterials: The Challenges and Opportunities for Nanomedicines. *International Journal of Nanomedicine*, *16*, 161–184. doi:10.2147/IJN.S288236 PMID:33447033

Alberti, G., Marmottini, F., Murcia-Mascarós, S., & Vivani, R. (1994). Preparation and Preliminary Characterization of a Covalently Pillared Zirconium Phosphate-Diphosphonate with Interlayer Microporosity. *Angewandte Chemie International Edition in English*, *33*(15-16), 1594–1597. doi:10.1002/anie.199415941

Alghamdi, M., Fatima, B., Hussain, Z., Nisar, Z., & Alghamdi, H. A. (2023). Peristaltic pumping of hybrid nanofluids through an inclined asymmetric channel: A biomedical application. *Materials Today. Communications*, *35*, 105684. doi:10.1016/j.mtcomm.2023.105684

Alipour, M., Sheikhnejad, R., Fouani, M. H., Bardania, H., & Hosseinkhani, S. (2023). DNAi-peptide nanohybrid smart particles target BCL-2 oncogene and induce apoptosis in breast cancer cells. *Biomedicine and Pharmacotherapy*, *166*, 115299. doi:10.1016/j.biopha.2023.115299 PMID:37573657

Alivisatos, P. (2004). The use of nanocrystals in biological detection. *Nature Biotechnology*, *22*(1), 47–52. doi:10.1038/nbt927 PMID:14704706

Alphandéry, E. (2022). Nanomaterials as ultrasound theragnostic tools for heart disease treatment/diagnosis. *International Journal of Molecular Sciences*, *23*(3), 1683. doi:10.3390/ijms23031683 PMID:35163604

Altamura, D., Sibillano, T., Siliqi, D., Caro, L. D., & Giannini, C. (2012). *Assembled Nanostructured Architectures Studied by Grazing Incidence X-Ray Scattering*. Nanomaterials and Nanotechnology. doi:10.5772/55777

Alvarez-Lorenzo, C., Vivero-Lopez, M., & Concheiro, A. (2023). Contact lenses that transform gold into nanoparticles for prophylaxis of light-related events and photothermal therapy. *International Journal of Pharmaceutics*, *641*, 123048. doi:10.1016/j.ijpharm.2023.123048 PMID:37192704

Alwattar, J. K., Mneimneh, A. T., Abla, K. K., Mehanna, M. M., & Allam, A. N. (2021). Smart stimuli-responsive liposomal nanohybrid systems: A critical review of theranostic behavior in cancer. *Pharmaceutics*, *13*(3), 355. doi:10.3390/pharmaceutics13030355 PMID:33800292

Alzate, M., Gamba, O., Daza, C., Santamaria, A., & Gallego, J. (2022). Iron/multiwalled carbon nanotube (Fe/MWCNT) hybrid materials characterization: Thermogravimetric analysis as a powerful characterization technique. *Journal of Thermal Analysis and Calorimetry*, *147*(22), 12355–12363. doi:10.1007/s10973-022-11446-w

Amalraj, L., Sanjeeviraja, C., & Jayachandran, M. (2002). Spray pyrolysised tin disulphide thin film and characterisation. *Journal of Crystal Growth*, *234*(4), 683–689. doi:10.1016/S0022-0248(01)01756-0

Ambrose, J. A., & Sharma, A. V. (2023). Identifying and treating vulnerable atherosclerotic plaques. *The American Journal of Cardiology*, *205*, 214–222. doi:10.1016/j.amjcard.2023.07.121 PMID:37611413

Amiryaghoubi, N., Fathi, M., Barar, J., Omidian, H., & Omidi, Y. (2022). Recent advances in graphene-based polymer composite scaffolds for bone/cartilage tissue engineering. *Journal of Drug Delivery Science and Technology*, *72*, 103360. doi:10.1016/j.jddst.2022.103360

Ananikov, V. P. (2019). Organic–inorganic hybrid nanomaterials. *Nanomaterials (Basel, Switzerland)*, *9*(9), 1197. doi:10.3390/nano9091197 PMID:31454924

Anasori, B., Lukatskaya, M. R., & Gogotsi, Y. (2017). 2D metal carbides and nitrides (MXenes) for energy storage. *Nature Reviews. Materials*, *2*(2), 1–17. doi:10.1038/natrevmats.2016.98

Anderson, P., Wilson, J. D., & Hiller, F. C. (1990). Respiratory Tract Deposition of Ultrafine Particles in Subjects with Obstructive or Restrictive Lung Disease. *Chest*, *97*(5), 1115–1120. doi:10.1378/chest.97.5.1115 PMID:2331906

Andrew, R. M. (2021). Towards near real-time, monthly fossil CO2 emissions estimates for the European Union with current-year projections. *Atmospheric Pollution Research*, *12*(12), 101229. doi:10.1016/j.apr.2021.101229 PMID:34938143

Ani, M. H., Kamarudin, M. A., Ramlan, A. H., Ismail, E., Sirat, M. S., Mohamed, M. A., & Azam, M. A. (2018). A critical review on the contributions of chemical and physical factors toward the nucleation and growth of large-area graphene. *Journal of Materials Science*, *53*(10), 7095–7111. doi:10.1007/s10853-018-1994-0

An, K. H., Jeon, K. K., Heo, J. K., Lim, S. C., Bae, D., & Lee, Y. H. (2002). High-Capacitance supercapacitor using a nanocomposite electrode of Single-Walled carbon nanotube and polypyrrole. *Journal of the Electrochemical Society*, *149*(8), A1058. doi:10.1149/1.1491235

Arava, L. M. R., Gowda, S. R., Shaijumon, M. M., & Ajayan, P. M. (2012). Hybrid Nanostructures for Energy Storage Applications. *Advanced Materials*, *24*(37), 5045–5064. Advance online publication. doi:10.1002/adma.201104502 PMID:22740354

Asharani, P. V., Wu, Y., Gong, Z., & Valiyaveettil, S. (2008). Toxicity of silver nanoparticles in zebrafish models. *Nanotechnology*, *19*(25), 255102. doi:10.1088/0957-4484/19/25/255102 PMID:21828644

Ashfaq, M., Talreja, N., Chauhan, D., Afreen, S., Sultana, A., & Srituravanich, W. (2022). Two-dimensional (2D) hybrid nanomaterials for diagnosis and treatment of cancer. *Journal of Drug Delivery Science and Technology*, *70*, 103268. doi:10.1016/j.jddst.2022.103268

Avid, A., Jafari, S. H., Khonakdar, H. A., Ghaffari, M., Krause, B., & Pötschke, P. (2019). Surface modification of MWCNT and its influence on properties of paraffin/MWCNT nanocomposites as phase change material. *Journal of Applied Polymer Science*, *137*(9), 48428. Advance online publication. doi:10.1002/app.48428

Ayoub, H. A., Khairy, M., Elsaid, S., Rashwan, F. A., & Abdel-Hafez, H. F. (2018). Pesticidal activity of nanostructured metal oxides for generation of alternative pesticide formulations. *Journal of Agricultural and Food Chemistry*, *66*(22), 5491–5498. doi:10.1021/acs.jafc.8b01600 PMID:29757642

Backov, R. (2006). Combining soft matter and soft chemistry: Integrative chemistry towards designing novel and complex multiscale architectures. *Soft Matter*, *2*(6), 452. doi:10.1039/b602579j PMID:32680243

Bagheri, E., Ansari, L., Abnous, K., Taghdisi, S. M., Charbgoo, F., Ramezani, M., & Alibolandi, M. (2018). Silica based hybrid materials for drug delivery and bioimaging. *Journal of Controlled Release*, *277*, 57–76. doi:10.1016/j.jconrel.2018.03.014 PMID:29550399

Bahadur, Singh, Li, Singh, Srivastava, Sakamoto, Chang, Murugavel, & Vinu. (2024). Hybrid nanoarchitectonics of ordered mesoporous C_{60}–BCN with high surface area for supercapacitors and lithium-ion batteries. *Carbon, 216*(5), 118568. doi:10.1016/j.carbon.2023.118568

Bailey, R. E., & Nie, S. (2003). Alloyed semiconductor quantum dots: Tuning the optical properties without changing the particle size. *Journal of the American Chemical Society*, *125*(23), 7100–7106. doi:10.1021/ja035000o PMID:12783563

Bala, M., & Mehan, V. (2021). Identification of Rice Plant Diseases Using Image Processing, Machine Learning & Deep Learning: A Review. *CEUR Workshop Proceedings, 3058*, 0–3.

Bala, M., Kumar Bansal, S., & Fatima, F. (2023). Nanotechnology: A boon for agriculture. *Materials Today: Proceedings*, *73*(October), 267–270. doi:10.1016/j.matpr.2022.09.498

Bala, M., & Mehan, V. (2022). Metaheuristic Techniques for Classification Used in Identification of Plant Diseases. *ECS Transactions*, *107*(1), 13473–13480. doi:10.1149/10701.13473ecst

Balla, S., & Sumithra, M. (2023). Platinum Nano Particles Role In Cancer Treatment: A Review. *Latin American Journal of Pharmacy*, *42*(3), 961–971.

Balogh, L., Swanson, D. R., Tomalia, D. A., Hagnauer, G. L., & McManus, A. T. (2001). Dendrimer– silver complexes and nanocomposites as antimicrobial agents. *Nano Letters*, *1*(1), 18–21. doi:10.1021/nl005502p

Banin, U., Waiskopf, N., Hammarström, L., Boschloo, G., Freitag, M., Johansson, E. M., Sá, J., Tian, H., Johnston, M., Herz, L., Milot, R. L., Kanatzidis, M. G., Ke, W., Spanopoulos, I., Kohlstedt, K. L., Schatz, G. C., Lewis, N., Meyer, T., Nozik, A. J., ... Brudvig, G. W. (2020). Nanotechnology for catalysis and solar energy conversion. *Nanotechnology*, *32*(4), 042003. doi:10.1088/1361-6528/abbce8 PMID:33155576

Bao, J., Chen, W., Liu, T., Zhu, Y., Jin, P., Wang, L., Liu, J., Wei, Y., & Li, Y. (2007). Bifunctional Au-Fe3O4 nanoparticles for protein separation. *ACS Nano*, *1*(4), 293–298. doi:10.1021/nn700189h PMID:19206679

Bao, Q. Y., Zhang, N., Geng, D. D., Xue, J. W., Merritt, M., Zhang, C., & Ding, Y. (2014). The enhanced longevity and liver targetability of Paclitaxel by hybrid liposomes encapsulating Paclitaxel-conjugated gold nanoparticles. *International Journal of Pharmaceutics*, *477*(1-2), 408–415. doi:10.1016/j.ijpharm.2014.10.040 PMID:25455782

Baranowska-Wójcik, E., Szwajgier, D., Oleszczuk, P., & Winiarska-Mieczan, A. (2020). Effects of titanium dioxide nanoparticles exposure on human health—A review. *Biological Trace Element Research*, *193*(1), 118–129. doi:10.1007/s12011-019-01706-6 PMID:30982201

Basheer, A. A. (2023). Graphene materials for fabrication of robots. In *Materials Chemistry and Physics* (Vol. 302). Elsevier Ltd. doi:10.1016/j.matchemphys.2023.127781

Bashir, M. A. (2021). Use of Dynamic Mechanical Analysis (DMA) for Characterizing Interfacial Interactions in Filled Polymers. In Solids (Vol. 2, Issue 1, pp. 108–120). doi:10.3390/solids2010006

Baysal, A., Kuznek, C., & Ozcan, M. (2018). Starch coated titanium dioxide nanoparticles as a challenging sorbent to separate and preconcentrate some heavy metals using graphite furnace atomic absorption spectrometry. *International Journal of Environmental Analytical Chemistry*, *98*(1), 45–55. doi:10.1080/03067319.2018.1427741

Beemkumar, N., Kaushik, M., Tripathi, A. D., Sharma, M., Khan, S. A., & Sharma, R. C. (2023). Pluronic F127-chitosan modified magnesium oxide hybrid nanomaterials prepared via a one-pot method: Potential uses in antibacterial and anticancer agents. *Surfaces and Interfaces*, *42*, 103327. doi:10.1016/j.surfin.2023.103327

Beheshtizadeh, N., Gharibshahian, M., Bayati, M., Maleki, R., Strachan, H., Doughty, S., & Tayebi, L. (2023). Vascular endothelial growth factor (VEGF) delivery approaches in regenerative medicine. *Biomedicine and Pharmacotherapy*, *166*, 115301. doi:10.1016/j.biopha.2023.115301 PMID:37562236

Beigpour, R., Shokrollahi, H., & Khalili, S. M. R. (2021). Experimental and numerical analysis of a biodegradable hybrid composite under tensile and three-point bending tests. *Composite Structures*, *273*, 114255. doi:10.1016/j.compstruct.2021.114255

Bekbölet, M., & Araz, C. V. (1996). Inactivation of Escherichia coli by photocatalytic oxidation. *Chemosphere*, *32*(5), 959–965. doi:10.1016/0045-6535(95)00359-2 PMID:8867143

Bhagat, S. K., Tung, T. M., & Yaseen, Z. M. (2020). Development of artificial intelligence for modeling wastewater heavy metal removal: State of the art, application assessment and possible future research. In *Journal of Cleaner Production* (Vol. 250). Elsevier Ltd. doi:10.1016/j.jclepro.2019.119473

Bhattarai, P., & Hameed, S. (2020). Basics of biosensors and nanobiosensors. *Nanobiosensors: From Design to Applications*, 1-22.

Bibi, I., Nazar, N., Ata, S., Sultan, M., Ali, A., Abbas, A., Jilani, K., Kamal, S., Sarim, F. M., Khan, M. I., Jalal, F., & Iqbal, M. (2019). Green synthesis of iron oxide nanoparticles using pomegranate seeds extract and photocatalytic activity evaluation for the degradation of textile dye. *Journal of Materials Research and Technology*, *8*(6), 6115–6124. doi:10.1016/j.jmrt.2019.10.006

Bohara, R. A. (2019). Introduction and types of hybrid nanostructures for medical applications. In *Hybrid Nanostructures for Cancer Theranostics* (pp. 1–16). Elsevier. doi:10.1016/B978-0-12-813906-6.00001-9

Bolaños, K., Kogan, M. J., & Araya, E. (2019). Capping gold nanoparticles with albumin to improve their biomedical properties. *International Journal of Nanomedicine*, *14*, 6387–6406. doi:10.2147/IJN.S210992 PMID:31496693

Bookbinder, D. C., & Wrighton, M. S. (1983). Electrochromic polymers covalently anchored to electrode surfaces. Optical and electrochemical properties of a viologen-based polymer. *Journal of the Electrochemical Society*, *130*(5), 1080–1087. doi:10.1149/1.2119888

Bor, B., Giuntini, D., Domènech, B., Swain, M. V., & Schneider, G. A. (2019). Nanoindentation-based study of the mechanical behavior of bulk supercrystalline ceramic-organic nanocomposites. *Journal of the European Ceramic Society*, *39*(10), 3247–3256. doi:10.1016/j.jeurceramsoc.2019.03.053

Bosetti, M., Masse, A., Tobin, E., & Cannas, M. (2002). Silver coated materials for external fixation devices: In vitro biocompatibility and genotoxicity. *Biomaterials*, *23*(3), 887–892. doi:10.1016/S0142-9612(01)00198-3 PMID:11771707

Botes, M., & Eugene Cloete, T. (2010). The potential of nanofibers and nanobiocides in water purification. *Critical Reviews in Microbiology*, *36*(1), 68–81. doi:10.3109/10408410903397332 PMID:20088684

Bottino, A., Capannelli, G., & Comite, A. (2002). Preparation and characterization of novel porous PVDF-ZrO2 composite membranes. *Desalination*, *146*(1-3), 35–40. doi:10.1016/S0011-9164(02)00469-1

Brigger, I., Dubernet, C., & Couvreur, P. (2012). Nanoparticles in cancer therapy and diagnosis. *Advanced Drug Delivery Reviews*, *64*, 24–36. doi:10.1016/j.addr.2012.09.006 PMID:12204596

Burduşel, A.-C., Gherasim, O., Grumezescu, A. M., Mogoantă, L., Ficai, A., & Andronescu, E. (2018). Biomedical applications of silver nanoparticles: An up-to-date overview. *Nanomaterials (Basel, Switzerland)*, *8*(9), 681. doi:10.3390/nano8090681 PMID:30200373

Bushara, M. O. E. (2023). Characteristics of Patients with Breast Cancer Attending the Breast Cancer Center in Sudan. *Advancements in Life Sciences*, *9*(4), 603–606.

Butto-Miranda, N., Cabrera-Barjas, G., Ibáñez, A., & Neira-Carrillo, A. (2023). Preparation of Novel Electrospun Organic–Inorganic Hybrid Nanofibers Based on Polycaprolactone, Glycine Betaine, and Calcium Carbonate (P/G/CaCO3-EPnF). *Crystals*, *13*(4), 611. Advance online publication. doi:10.3390/cryst13040611

Byrne, J. D., Betancourt, T., & Brannon-Peppas, L. (2008). Active targeting schemes for nanoparticle systems in cancer therapeutics. *Advanced Drug Delivery Reviews*, *60*(15), 1615–1626. doi:10.1016/j.addr.2008.08.005 PMID:18840489

Cai, Z., Liu, B., Zou, X., & Cheng, H. M. (2018). Chemical vapor deposition growth and applications of two-dimensional materials and their heterostructures. *Chemical Reviews*, *118*(13), 6091–6133. doi:10.1021/acs.chemrev.7b00536 PMID:29384374

Cao, J., Yan, C., Chai, Z., Wang, Z., Du, M., Li, G., Wang, H., & Deng, H. (2024). Laser-induced transient conversion of rhodochrosite/polyimide into multifunctional MnO_2/graphene electrodes for energy storage applications. *Journal of Colloid and Interface Science*, *653*, 606–616. doi:10.1016/j.jcis.2023.09.083 PMID:37738933

Cao, Z., Bian, Y., Hu, T., Yang, Y., Cui, Z., Wang, T., Yang, S., Weng, X., Liang, R., & Tan, C. (2023). Recent advances in two-dimensional nanomaterials for bone tissue engineering. *Journal of Materiomics*, *9*(5), 930–958. doi:10.1016/j.jmat.2023.02.016

Catauro, M., & Ciprioti, S. V. (2021). Characterization of Hybrid Materials Prepared by Sol-Gel Method for Biomedical Implementations. A Critical Review. *Materials (Basel)*, *14*(7), 1788. doi:10.3390/ma14071788 PMID:33916333

Cecchin, I., Reddy, K. R., Thome, A., Tessaro, E. F., & Schnaid, F. (2017). Nanobioremediation: Integration of nanoparticles and bioremediation for sustainable remediation of chlorinated organic contaminants in soil. *International Biodeterioration & Biodegradation*, *119*, 419–428. doi:10.1016/j.ibiod.2016.09.027

Chacko, R. T., Ventura, J., Zhuang, J., & Thayumanavan, S. (2012). Polymer nanogels: A versatile nanoscopic drug delivery platform. *Advanced Drug Delivery Reviews*, *64*(9), 836–851. doi:10.1016/j.addr.2012.02.002 PMID:22342438

Chae, H. K., Siberio-Pérez, D. Y., Kim, J., Go, Y., Eddaoudi, M., Matzger, A. J., O'Keeffe, M., & Yaghi, O. M. (2004). A route to high surface area, porosity, and inclusion of large molecules in crystals. *Nature*, *427*(6974), 523–527. doi:10.1038/nature02311 PMID:14765190

Chakravarty, P., Deka, H., & Chowdhury, D. (2023). Anthracene removal potential of green synthesized titanium dioxide nanoparticles (TiO2-NPs) and Alcaligenes faecalis HP8 from contaminated soil. *Chemosphere*, *321*, 138102. doi:10.1016/j.chemosphere.2023.138102 PMID:36764617

Chan, C. K., Peng, H., Liu, G., McIlwrath, K., Zhang, X. F., Huggins, R. A., & Cui, Y. (2007). High-performance lithium battery anodes using silicon nanowires. *Nature Nanotechnology*, *3*(1), 31–35. doi:10.1038/nnano.2007.411 PMID:18654447

Chang, H.-Y., Lin, W.-C., Chu, P.-C., Wang, Y.-K., Sogo, M., Iida, S., Peng, C.-J., & Miyayama, T. (2022). X-ray Photoelectron Spectroscopy Equipped with Gas Cluster Ion Beams for Evaluation of the Sputtering Behavior of Various Nanomaterials. *ACS Applied Nano Materials*, *5*(3), 4260–4268. doi:10.1021/acsanm.2c00202

Charitos, I., Georgousis, G., Klonos, P. A., Kyritsis, A., Mouzakis, D., Raptis, Y., Kontos, A., & Kontou, E. (2021). The synergistic effect on the thermomechanical and electrical properties of carbonaceous hybrid polymer nanocomposites. *Polymer Testing*, *95*, 107102. doi:10.1016/j.polymertesting.2021.107102

Chehelgerdi, M., Chehelgerdi, M., Allela, O. Q. B., Pecho, R. D. C., Jayasankar, N., Rao, D. P., Thamaraikani, T., Vasanthan, M., Viktor, P., Lakshmaiya, N., Saadh, M. J., Amajd, A., Abo-Zaid, M. A., Castillo-Acobo, R. Y., Ismail, A. H., Amin, A. H., & Akhavan-Sigari, R. (2023). Progressing nanotechnology to improve targeted cancer treatment: Overcoming hurdles in its clinical implementation. *Molecular Cancer*, *22*(1), 169. doi:10.1186/s12943-023-01865-0 PMID:37814270

Chen, H. (2006). *WO2007096656A1 - A method of storing energy and a cryogenic energy storage system -Google Patents.* https://patents.google.com/patent/WO2007096656A1/en

Chen, H., Cong, T. N., Yang, W., Tan, C., Li, Y., & Ding, Y. (2009). Progress in electrical energy storage system: A critical review. *Progress in Natural Science, 19*(3), 291-312. doi:10.1016/j.pnsc.2008.07.014

Chen, C., Zhang, G., Dai, Z., Xiang, Y., Liu, B., Bian, P., Zheng, K., Wu, Z., & Cai, D. (2018). Fabrication of light-responsively controlled-release herbicide using a nanocomposite. *Chemical Engineering Journal*, *349*, 101–110. doi:10.1016/j.cej.2018.05.079

Chen, D., Wang, D., Yang, Y., Huang, Q., Zhu, S., & Zheng, Z. (2017). Self-Healing Materials for Next-Generation Energy Harvesting and Storage Devices. *Advanced Energy Materials*, *7*(23), 1700890. doi:10.1002/aenm.201700890

Chen, G., Qian, Y., Zhang, H., Ullah, A., He, X., Zhou, Z., Fenniri, H., & Shen, J. (2021). Advances in cancer theranostics using organic-inorganic hybrid nanotechnology. *Applied Materials Today*, *23*, 101003. doi:10.1016/j.apmt.2021.101003

Chen, G., Wang, Z., & Xia, D. (2008). One-Pot Synthesis of Carbon Nanotube@SnO_2–Au Coaxial Nanocable for Lithium-Ion Batteries with High Rate Capability. *Chemistry of Materials*, *20*(22), 6951–6956. doi:10.1021/cm801853c

Cheng, L., Yang, K., Li, Y., Zeng, X., Shao, M., Lee, S. T., & Liu, Z. (2012). Multifunctional nanoparticles for upconversion luminescence/MR multimodal imaging and magnetically targeted photothermal therapy. *Biomaterials*, *33*(7), 2215–2222. doi:10.1016/j.biomaterials.2011.11.069 PMID:22169825

Cheng, R., Meng, F., Deng, C., Klok, H.-A., & Zhong, Z. (2013). Dual and multi-stimuli responsive polymeric nanoparticles for programmed site-specific drug delivery. *Biomaterials*, *34*(14), 3647–3657. doi:10.1016/j.biomaterials.2013.01.084 PMID:23415642

Cheng, Z., Xia, M., Liu, S., Hu, R., Liang, G., & Zhang, S. (2019). Role of rough substrate on the growth of large single-crystal MoS2 by chemical vapor deposition. *Applied Surface Science*, *476*, 1008–1015. doi:10.1016/j.apsusc.2019.01.211

Chen, J. S., Chen, J., Bhattacharjee, S., Cao, Z., Wang, H., Swanson, S. D., Zong, H., Baker, J. R. Jr, & Wang, S. H. (2020). Functionalized nanoparticles with targeted antibody to enhance imaging of breast cancer *in vivo*. *Journal of Nanobiotechnology*, *18*(1), 135. doi:10.1186/s12951-020-00695-2 PMID:32948179

Chen, Y., Wang, L., Jiang, S., & Yu, H. J. (2003). Study on novel antibacterial polymer materials (I) preparation of zeolite antibacterial agents and antibacterial polymer composite and their antibacterial properties. *Journal of Polymer Materials*, *20*(3), 279–284.

Chen, Y., Zhu, C., Xue, X., Shi, X., & Cao, M. (2008). High capacity and excellent cycling stability of single-walled carbon nanotube/SnO2 core-shell structures as Li-insertion materials. *Applied Physics Letters*, *92*(22), 223101. Advance online publication. doi:10.1063/1.2937839

Chhowalla, M., Shin, H. S., Eda, G., Li, L. J., Loh, K. P., & Zhang, H. (2013). The chemistry of two-dimensional layered transition metal dichalcogenide nanosheets. *Nature Chemistry*, *5*(4), 263–275. doi:10.1038/nchem.1589 PMID:23511414

Chircov, C., Ioniță, D.-A., Sîrmon, A.-M., Neacşu, I. A., & Ficai, A. (2023). Natural, synthetic, and hybrid and composite biomaterials for neural tissue engineering. In O. Gunduz, C. B. Ustundag, & M. Sengor (Eds.), *Biomaterials for Neural Tissue Engineering* (pp. 21–58). Woodhead Publishing. doi:10.1016/B978-0-323-90554-1.00008-2

Choi, G., Rejinold, N. S., Piao, H., & Choy, J.-H. (2021). Inorganic–inorganic nanohybrids for drug delivery, imaging and photo-therapy: Recent developments and future scope. *Chemical Science (Cambridge)*, *12*(14), 5044–5063. doi:10.1039/D0SC06724E PMID:34168768

Choi, J., Zhang, Q., Reipa, V., Wang, N. S., Stratmeyer, M. E., Hitchins, V. M., & Goering, P. L. (2009). Comparison of cytotoxic and inflammatory responses of photoluminescent silicon nanoparticles with silicon micron-sized particles in RAW 264.7 macrophages. *Journal of Applied Toxicology*, *29*(1), 52–60. doi:10.1002/jat.1382 PMID:18785685

Cho, K., Wang, X. U., Nie, S., Chen, Z., & Shin, D. M. (2008). Therapeutic nanoparticles for drug delivery in cancer. *Clinical Cancer Research*, *14*(5), 1310–1316. doi:10.1158/1078-0432.CCR-07-1441 PMID:18316549

Cho, M., Chung, H., Choi, W., & Yoon, J. (2004). Linear correlation between inactivation of E. coli and OH radical concentration in TiO2 photocatalytic disinfection. *Water Research*, *38*(4), 1069–1077. doi:10.1016/j.watres.2003.10.029 PMID:14769428

Cho, M., Chung, H., Choi, W., & Yoon, J. (2005). Different inactivation behaviors of MS-2 phage and Escherichia coli in TiO2 photocatalytic disinfection. *Applied and Environmental Microbiology*, *71*(1), 270–275. doi:10.1128/AEM.71.1.270-275.2005 PMID:15640197

Chou, K.-S., Lu, Y.-C., & Lee, H.-H. (2005). Effect of alkaline ion on the mechanism and kinetics of chemical reduction of silver. *Materials Chemistry and Physics*, *94*(2-3), 429–433. doi:10.1016/j.matchemphys.2005.05.029

Christoforidis, K. C., Montini, T., Bontempi, E., Zafeiratos, S., Jaén, J. J. D., & Fornasiero, P. (2016). Synthesis and photocatalytic application of visible-light active β -Fe2O3/g-C3N4 hybrid nanocomposites. *Applied Catalysis B: Environmental*, *187*, 171–180. doi:10.1016/j.apcatb.2016.01.013

Chrysochoou, M., Johnston, C. P., & Dahal, G. (2012). A comparative evaluation of hexavalent chromium treatment in contaminated soil by calcium polysulfide and green-tea nanoscale zero-valent iron. *Journal of Hazardous Materials*, *201*, 33–42. doi:10.1016/j.jhazmat.2011.11.003 PMID:22169240

Chun, C. K., Sofer, Z., Simek, P., Jankovsky, O., Klimova, K., Bakardjieva, S., Kuckova, S. H., & Pumera, M. (2015). Synthesis of Strongly Fluorescent Graphene Quantum Dots by Cage-Opening Buckminsterfullerene. *ACS Nano*, *9*(3), 2548–2555. doi:10.1021/nn505639q PMID:25761306

Coll, J. L. (2011). Cancer optical imaging using fluorescent nanoparticles. *Nanomedicine (London)*, *6*(1), 7–10. doi:10.2217/nnm.10.144 PMID:21182412

Comanescu, C. (2022). Magnetic Nanoparticles: Current Advances in Nanomedicine, Drug Delivery and MRI. *Chemistry (Basel, Switzerland)*, *4*(3), 872–930. doi:10.3390/chemistry4030063

Conklin, B., Conley, B. M., Hou, Y., Chen, M., & Lee, K.-B. (2023). Advanced theragnostics for the central nervous system (CNS) and neurological disorders using functional inorganic nanomaterials. *Advanced Drug Delivery Reviews*, *192*, 114636. doi:10.1016/j.addr.2022.114636 PMID:36481291

Corsi, I., Winther-Nielsen, M., Sethi, R., Punta, C., Della Torre, C., Libralato, G., Lofrano, G., Sabatini, L., Aiello, M., Fiordi, L., Cinuzzi, F., Caneschi, A., Pellegrini, D., & Buttino, I. (2018). Ecofriendly nanotechnologies and nanomaterials for environmental applications: Key issue and consensus recommendations for sustainable and ecosafe nanoremediation. *Ecotoxicology and Environmental Safety*, *154*, 237–244. doi:10.1016/j.ecoenv.2018.02.037 PMID:29476973

Coşkun, Z., Çelik, T., & Kisioglu, Y. (2021). Design and Manufacture of the Torque Test Setup For Small and Shapeless Materials. *Turkish Journal of Engineering*, *6*(1), 81–86. Advance online publication. doi:10.31127/tuje.843320

Crutzen, P. J., Mosier, A. R., Smith, K. A., & Winiwarter, W. (2008b). N_2O release from agro-biofuel production negates global warming reduction by replacing fossil fuels. *Atmospheric Chemistry and Physics*, *8*(2), 389–395. doi:10.5194/acp-8-389-2008

Cui, S. (2013). Indium-doped SNO2 nanoparticle-graphene nanohybrids: simple one-pot synthesis and their selective detection of NO2. *J Mater Chem A, 1*(14), 4462–4467. www.rsc. org/MaterialsA

Curl, R. F., Jr., Kroto, H., Smalley, R. E., Curl, R. F., Jr., & Kroto, H. W. (1996). *The Nobel Prize in Chemistry 1996.* Explore Press Release.

Cushing, B. L., Kolesnichenko, V. L., & O'Connor, C. J. (2004). Recent advances in the liquid-phase syntheses of inorganic nanoparticles. *Chemical Reviews, 104*(9), 3893–3946. doi:10.1021/cr030027b PMID:15352782

da Silva Costa, K. R. B., de Sá, M. L., de Carvalho Oliveira, F., Dittz, D., Leal, B. S., de Castro e Sousa, J. M., Abreu, G. J. P., de Oliveira, T. G., Gusmão, S. B. S., Ghosh, A., Guerra, Y., & Viana, B. C. (2023). Titanate nanotubes modified with gallium and cerium and their cytotoxic activity. *Journal of Nanoparticle Research, 25*(9), 177. Advance online publication. doi:10.1007/s11051-023-05824-0

Danai-Varsou, D., Zhang, P., Afantitis, A., Guo, Z., Lynch, I., & Melagraki, G. (2023). Nanoinformatics and artificial intelligence for nano-enabled sustainable agriculture. In *Nano-Enabled Sustainable and Precision Agriculture* (pp. 503–531). Elsevier. doi:10.1016/B978-0-323-91233-4.00015-6

Daniel, S., Rao, T. P., Rao, K. S., Rani, S. U., Naidu, G. R. K., Lee, H. Y., & Kawai, T. (2007). A review of DNA functionalized/grafted carbon nanotubes and their characterization. *Sensors and Actuators. B, Chemical, 122*(2), 672–682. doi:10.1016/j.snb.2006.06.014

Danilczuk, M., Lund, A., Sadlo, J., Yamada, H., & Michalik, J. (2006). Conduction electron spin resonance of small silver particles. *Spectrochimica Acta. Part A: Molecular and Biomolecular Spectroscopy, 63*(1), 189–191. doi:10.1016/j.saa.2005.05.002 PMID:15978868

Darkrim, F., Malbrunot, P., & Tartaglia, G. (2002). Review of hydrogen storage by adsorption in carbon nanotubes. *International Journal of Hydrogen Energy, 27*(2), 193–202. doi:10.1016/S0360-3199(01)00103-3

Dasgupta, N., Ranjan, S., Mundekkad, D., Ramalingam, C., Shanker, R., & Kumar, A. (2015). Nanotechnology in agro-food: From field to plate. *Food Research International, 69*, 381–400. doi:10.1016/j.foodres.2015.01.005

Das, S., Sen, B., & Debnath, N. (2015). Recent trends in nanomaterials applications in environmental monitoring and remediation. *Environmental Science and Pollution Research International, 22*(23), 18333–18344. doi:10.1007/s11356-015-5491-6 PMID:26490920

Datta, D., Das, K. P., Deepak, K., & Das, B. (2022). Candidates of functionalized nanomaterial-based membranes. In *Membranes with Functionalized Nanomaterials* (pp. 81–127). Elsevier. doi:10.1016/B978-0-323-85946-2.00004-7

Deepagan, V. G., You, D. G., Um, W., Ko, H., Kwon, S., Choi, K. Y., Yi, G.-R., Lee, J. Y., Lee, D. S., Kim, K., Kwon, I. C., & Park, J. H. (2016). Long-circulating Au-TiO2 nanocomposite as a sonosensitizer for ROS-mediated eradication of cancer. *Nano Letters*, *16*(10), 6257–6264. doi:10.1021/acs.nanolett.6b02547 PMID:27643533

Deepapriya, S., Rodney, J. D., Flora John, J., Joshi, S., Udayashankar, N. K., Lakshmi Devi, S., & Jerome Das, S. (2023). A novel effective immobilization of glucose oxidase on Ni0.25Zn0.25Cu0.25Co0.25La0.06Fe1.94O4 – Chitosan nanocomposite as an enzymatic glucose biosensor. *Inorganic Chemistry Communications*, *153*, 110822. doi:10.1016/j.inoche.2023.110822

Dembinska, B., & Kulesza, P. J. (2009). Multi-walled carbon nanotube-supported tungsten oxide-containing multifunctional hybrid electrocatalytic system for oxygen reduction in acid medium. *Electrochimica Acta*, *54*(20), 4682–4687. doi:10.1016/j.electacta.2009.03.043

Deng, S., & Berry, V. (2016). Wrinkled, rippled and crumpled graphene: An overview of formation mechanism, electronic properties, and applications. *Materials Today*, *19*(4), 197–212. doi:10.1016/j.mattod.2015.10.002

Derfus, A. M., Chan, W. C., & Bhatia, S. N. (2004). Probing the cytotoxicity of semiconductor quantum dots. *Nano Letters*, *4*(1), 11–18. doi:10.1021/nl0347334 PMID:28890669

Desai, H. R., & Shinde, U. A. (2022). 20 - Hybrid nanoparticles to cross the blood–brain barrier. In L. Kumar & Y. Y. Pathak (Eds.), *Nanocarriers for Drug-Targeting Brain Tumors* (pp. 565–586). Elsevier. doi:10.1016/B978-0-323-90773-6.00002-6

Deshmukh, M. A., Thorat, H. N., Shirsat, M. D., & Ramanavicius, A. (2023). 1 - Engineered nanostructures: an introduction. In A. K. Kaushik, S. Kumar, & G. R. Chaudhary (Eds.), *Engineered Nanostructures for Therapeutics and Biomedical Applications* (pp. 1–43). Woodhead Publishing. doi:10.1016/B978-0-12-821240-0.00002-0

Devarakonda, S. B., Myers, M. R., Lanier, M., Dumoulin, C., & Banerjee, R. K. (2017). Assessment of gold nanoparticle-mediated-enhanced hyperthermia using MR-guided high-intensity focused ultrasound ablation procedure. *Nano Letters*, *17*(4), 2532–2538. doi:10.1021/acs.nanolett.7b00272 PMID:28287747

Dhania, S., Bernela, M., Rani, R., Parsad, M., Grewal, S., Kumari, S., & Thakur, R. (2022). Scaffolds the backbone of tissue engineering: Advancements in use of polyhydroxyalkanoates (PHA). *International Journal of Biological Macromolecules*, *208*, 243–259. doi:10.1016/j.ijbiomac.2022.03.030 PMID:35278518

Dillon, A. C., Mahan, A. H., Deshpande, R., Parilla, P. A., Jones, K. M., & Lee, S. (2008). Metal oxide nano-particles for improved electrochromic and lithium-ion battery technologies. *Thin Solid Films*, *516*(5), 794–797. doi:10.1016/j.tsf.2007.06.177

Diltemiz, S. E., Say, R., Büyüktiryaki, S., Hür, D., Denizli, A., & Ersöz, A. (2008). Quantum dot nanocrystals having guanosine imprinted nanoshell for DNA recognition. *Talanta*, *75*(4), 890–896. doi:10.1016/j.talanta.2007.12.036 PMID:18585161

Divya, S., Panda, S., Hajra, S., Jeyaraj, R., Paul, A., Park, S. H., Kim, H. J., & Oh, T. H. (2023). Smart data processing for energy harvesting systems using artificial intelligence. In *Nano Energy* (Vol. 106). Elsevier Ltd. doi:10.1016/j.nanoen.2022.108084

Donarelli, M., & Ottaviano, L. (2018). 2D materials for gas sensing applications: A review on graphene oxide. *Sensors, 18*(11), 3638. doi:10.3390/s18113638

Dong, P., Wang, Y., Li, H., Li, H., Ma, X., & Han, L. (2013). Shape-controllable synthesis and morphology-dependent photocatalytic properties of Ag3PO4 crystals. *Journal of Materials Chemistry. A, Materials for Energy and Sustainability, 1*(15), 4651. doi:10.1039/c3ta00130j

Đorđević, S., Gonzalez, M. M., Conejos-Sánchez, I., Carreira, B., Pozzi, S., Acúrcio, R. C., ... Vicent, M. J. (2022). Current hurdles to the translation of nanomedicines from bench to the clinic. *Drug Delivery and Translational Research*, 1–26. PMID:34302274

Dragoman, D. (2023). Nanomaterials for Energy Harvesting. *Nanomaterials (Basel, Switzerland), 13*(7), 1154. doi:10.3390/nano13071154 PMID:37049248

Dreanca, A., Muresan-Pop, M., Taulescu, M., Tóth, Z.-R., Bogdan, S., Pestean, C., Oren, S., Toma, C., Popescu, A., Páll, E., Sevastre, B., Baia, L., & Magyari, K. (2021). Bioactive glass-biopolymers-gold nanoparticle based composites for tissue engineering applications. *Materials Science and Engineering C, 123*, 112006. doi:10.1016/j.msec.2021.112006 PMID:33812626

Duhan, J. S., Kumar, R., Kumar, N., Kaur, P., Nehra, K., & Duhan, S. (2017). Nanotechnology: The new perspective in precision agriculture. In Biotechnology Reports (Vol. 15, pp. 11–23). Elsevier B.V. doi:10.1016/j.btre.2017.03.002

Duoss, E. B., Twardowski, M., & Lewis, J. A. (2007). Sol-gel inks for direct-write assembly of functional oxides. *Advanced Materials, 19*(21), 3485–3489. doi:10.1002/adma.200701372

Durán, N., Marcato, P. D., De Souza, G. I., Alves, O. L., & Esposito, E. (2007). Antibacterial effect of silver nanoparticles produced by fungal process on textile fabrics and their effluent treatment. *Journal of Biomedical Nanotechnology, 3*(2), 203–208. doi:10.1166/jbn.2007.022

Durgalakshmi, D., Rajendran, S., & Naushad, M. (2019). Current role of nanomaterials in environmental remediation. In N. Mu, S. Rajendran, & F. Gracia (Eds.), *Adv. Nanostructured Mater. Environ. Remediat* (pp. 1–20). Springer International Publishing. doi:10.1007/978-3-030-04477-0_1

Dutta, D., & Das, B. M. (2021). Scope of green nanotechnology towards amalgamation of green chemistry for cleaner environment: A review on synthesis and applications of green nanoparticles. *Environmental Nanotechnology, Monitoring & Management, 15*, 100418. doi:10.1016/j.enmm.2020.100418

Dutta, V., Verma, R., Gopalkrishnan, C., Yuan, M.-H., Batoo, K. M., Jayavel, R., Chauhan, A., Lin, K.-Y. A., Balasubramani, R., & Ghotekar, S. (2022). Bio-Inspired Synthesis of Carbon-Based Nanomaterials and Their Potential Environmental Applications: A State-of-the-Art Review. *Inorganics, 10*(10), 169. doi:10.3390/inorganics10100169

Ebadollahi, A., Valizadeh, B., Panahandeh, S., Mirhosseini, H., Zolfaghari, M., & Changbunjong, T. (2022). Nanoencapsulation of Acetamiprid by Sodium Alginate and Polyethylene Glycol Enhanced Its Insecticidal Efficiency. *Nanomaterials (Basel, Switzerland)*, *12*(17), 2971. doi:10.3390/nano12172971 PMID:36080008

Egbuna, C., Parmar, V. K., Jeevanandam, J., Ezzat, S. M., Patrick-Iwuanyanwu, K. C., Adetunji, C. O., Khan, J., Onyeike, E. N., Uche, C. Z., Akram, M., Ibrahim, M. S., El Mahdy, N. M., Awuchi, C. G., Saravanan, K., Tijjani, H., Odoh, U. E., Messaoudi, M., Ifemeje, J. C., Olisah, M. C., ... Ibeabuchi, C. G. (2021). Toxicity of nanoparticles in biomedical application: Nanotoxicology. *Journal of Toxicology*, *2021*, 1–21. doi:10.1155/2021/9954443 PMID:34422042

Egerton, T. A., Kosa, S. A., & Christensen, P. A. (2006). Photoelectrocatalytic disinfection of E. coli suspensions by iron doped TiO 2. *Physical Chemistry Chemical Physics*, *8*(3), 398–406. doi:10.1039/B507516E PMID:16482283

Einaga, H., Futamura, S., & Ibusuki, T. (1999). Photocatalytic decomposition of benzene over TiO2 in a humidified airstream. *Physical Chemistry Chemical Physics*, *1*(20), 4903–4908. doi:10.1039/a906214i

El-Husseiny, H. M., Mady, E. A., Yoshida, T., & Tanaka, R. (2024). Stimuli-responsive hydrogels: cutting-edge platforms for cartilage tissue engineering. In A. R. Nochehdehi, F. Nemavhola, S. Thomas, & H. J. Maria (Eds.), *Cartilage Tissue and Knee Joint Biomechanics* (pp. 467–486). Academic Press. doi:10.1016/B978-0-323-90597-8.00009-8

Elkasabgy, N. A., Salama, A., & Salama, A. H. (2023). Exploring the effect of intramuscularly injected polymer/lipid hybrid nanoparticles loaded with quetiapine fumarate on the behavioral and neurological changes in cuprizone-induced schizophrenia in mice. *Journal of Drug Delivery Science and Technology*, *79*, 104064. doi:10.1016/j.jddst.2022.104064

El-Khawaga, A. M., Zidan, A., & El-Mageed, A. I. A. A. (2023). Preparation methods of different nanomaterials for various potential applications: A review. *Journal of Molecular Structure*, *1281*, 135148. doi:10.1016/j.molstruc.2023.135148

Ellis, A. V., Vijayamohanan, K., Goswami, R., Chakrapani, N., Ramanathan, L. S., Ajayan, P. M., & Ramanath, G. (2003). Hydrophobic anchoring of monolayer-protected gold nanoclusters to carbon nanotubes. *Nano Letters*, *3*(3), 279–282. doi:10.1021/nl025824o

El-Sayed, M. A. (2001). Some interesting properties of metals confined in time and nanometer space of different shapes. *Accounts of Chemical Research*, *34*(4), 257–264. doi:10.1021/ar960016n PMID:11308299

El-Temsah, Y. S., & Joner, E. J. (2013). Effects of nano-sized zero-valent iron (nZVI) on DDT degradation in soil and its toxicity to collembola and ostracods. *Chemosphere*, *92*(1), 131–137. doi:10.1016/j.chemosphere.2013.02.039 PMID:23522781

El-Temsah, Y. S., Sevcu, A., Bobcikova, K., Cernik, M., & Joner, E. J. (2016). DDT degradation efficiency and exotoxicological effects of two types of nano-sized zero-valent iron (nZVI) in water and soil. *Chemosphere*, *144*, 2221–2228. doi:10.1016/j.chemosphere.2015.10.122 PMID:26598990

Ennis, P. (2016). Creep and Stress Rupture Testing and Evaluation of Data. In Reference Module in Materials Science and Materials Engineering. doi:10.1016/B978-0-12-803581-8.03415-9

Esmaeili, A., & Beni, A. A. (2014). A novel fixed-bed reactor design incorporating an electrospun PVA/chitosan nanofiber membrane. *Journal of Hazardous Materials*, *280*, 788–796. doi:10.1016/j.jhazmat.2014.08.048 PMID:25240648

Ezealigo, U. S., Ezealigo, B. N., Aisida, S. O., & Ezema, F. I. (2021). Iron oxide nanoparticles in biological systems: Antibacterial and toxicology perspective. *JCIS Open*, *4*, 100027. doi:10.1016/j.jciso.2021.100027

Ezendam, S., Herran, M., Nan, L., Gruber, C., Kang, Y., Gröbmeyer, F., Lin, R., Gargiulo, J., Sousa-Castillo, A., & Cortés, E. (2022). Hybrid Plasmonic Nanomaterials for Hydrogen Generation and Carbon Dioxide Reduction. *ACS Energy Letters*, *7*(2), 778–815. doi:10.1021/acsenergylett.1c02241 PMID:35178471

Faghihi-Zarandi, A., Shirkhanloo, H., & Jamshidzadeh, C. (2019). A new method for removal of hazardous toluene vapor from air based on ionic liquid-phase adsorbent. *International Journal of Environmental Science and Technology*, *16*(6), 2797–2808. doi:10.1007/s13762-018-1975-5

Fahad, N., Kadhim, A., Al-Khayat, R., & Al-Waily, M. (2021). Effect of SiO2 and Al2O3 Hybrid Nano Materials on Fatigue Behavior for Laminated Composite Materials Used to Manufacture Artificial Socket Prostheses. *Materials Science Forum*, *1039*, 493–509. . doi:10.4028/www.scientific.net/MSF.1039.493

Fajardo, C., Ortiz, L. T., Rodríguez-Membibre, M. L., Nande, M., Lobo, M. C., & Martin, M. (2012). Assessing the impact of zero-valent iron (ZVI) nanotechnology on soil microbial structure and functionality: A molecular approach. *Chemosphere*, *86*(8), 802–808. doi:10.1016/j.chemosphere.2011.11.041 PMID:22169206

Fajardo, C., Sánchez-Fortún, S., Costa, G., Nande, M., Botias, P., García-Cantalejo, J., Mengs, G., & Martin, M. (2020). Evaluation of nanoremediation strategy in a Pb, Zn and Cd contaminated soil. *The Science of the Total Environment*, *706*, 136041. doi:10.1016/j.scitotenv.2019.136041 PMID:31855644

Fási, A., Pálinkó, I., Seo, J. W., Kónya, Z., Hernádi, K., & Kiricsi, I. (2003). Sonication assisted gold deposition on multiwall carbon nanotubes. *Chemical Physics Letters*, *372*(5-6), 848–852. doi:10.1016/S0009-2614(03)00527-X

Fasiolo, D. T., Scalera, L., Maset, E., & Gasparetto, A. (2023). Towards autonomous mapping in agriculture: A review of supportive technologies for ground robotics. *Robotics and Autonomous Systems*, *169*, 104514. Advance online publication. doi:10.1016/j.robot.2023.104514

Fathi-Achachelouei, M., Knopf-Marques, H., Ribeiro da Silva, C. E., Barthès, J., Bat, E., Tezcaner, A., & Vrana, N. E. (2019). Use of nanoparticles in tissue engineering and regenerative medicine. *Frontiers in Bioengineering and Biotechnology*, *7*, 113. doi:10.3389/fbioe.2019.00113 PMID:31179276

Fayaz, M., Lai, W., Li, J., Chen, W., Luo, X., Wang, Z., Chen, Y., De Li, S. M. A., & Chen, Y. (2024). Prussian blue analogues and their derived materials for electrochemical energy storage: Promises and Challenges. *Materials Research Bulletin, 112593*. Advance online publication. doi:10.1016/j.materresbull.2023.112593

Felber, A. E., Dufresne, M. H., & Leroux, J. C. (2012). pH-sensitive vesicles, polymeric micelles, and nanospheres prepared with polycarboxylates. *Advanced Drug Delivery Reviews*, *64*(11), 979–992. doi:10.1016/j.addr.2011.09.006 PMID:21996056

Feng, Q. L., Wu, J., Chen, G. Q., Cui, F. Z., Kim, T. N., & Kim, J. O. (2000). A mechanistic study of the antibacterial effect of silver ions on Escherichia coli and Staphylococcus aureus. *Journal of Biomedical Materials Research*, *52*(4), 662–668. doi:10.1002/1097-4636(20001215)52:4<662::AID-JBM10>3.0.CO;2-3 PMID:11033548

Fernandes, N. B., Nayak, Y., Garg, S., & Nayak, U. Y. (2023). Multifunctional engineered mesoporous silica/inorganic material hybrid nanoparticles: Theranostic perspectives. *Coordination Chemistry Reviews*, *478*, 214977. doi:10.1016/j.ccr.2022.214977

Ferri, M., Dash, M., Cometa, S., De Giglio, E., Sabbatini, L., & Chiellini, F. (2014). Preparation and characterization of hybrid nanoparticles based on chitosan and poly(methacryloylglycylglyc ine). *Journal of Nanoparticle Research*, *16*(5), 2422. doi:10.1007/s11051-014-2422-2

Fithri, N. A., Wu, Y., Cowin, G., Akther, F., Tran, H. D. N., Tse, B., van Holthe, N. W., Moonshi, S. S., Peter, K., Wang, X., Troung, N. P., & Ta, H. T. (2023). Gold-iron oxide nanoparticle: A unique multimodal theranostic approach for thrombosis. *Applied Materials Today*, *31*, 101750. doi:10.1016/j.apmt.2023.101750

Flimelová, M., Ryabchikov, Y. V., Behrends, J., & Bulgakova, N. M. (2023). Environmentally Friendly Improvement of Plasmonic Nanostructure Functionality toward Magnetic Resonance Applications. *Nanomaterials (Basel, Switzerland)*, *13*(4), 764. Advance online publication. doi:10.3390/nano13040764 PMID:36839132

Flores, K., Turley, R. S., Valdes, C., Ye, Y., Cantu, J., Hernandez-Viezcas, J. A., Parsons, J. G., & Gardea-Torresdey, J. L. (2021). Environmental applications and recent innovations in single particle inductively coupled plasma mass spectrometry (SP-ICP-MS). *Applied Spectroscopy Reviews*, *56*(1), 1–26. doi:10.1080/05704928.2019.1694937

Fontaras, G., Zacharof, N. G., & Ciuffo, B. (2017). Fuel consumption and CO2 emissions from passenger cars in Europe – Laboratory versus real-world emissions. In *Progress in Energy and Combustion Science* (Vol. 60, pp. 97–131). doi:10.1016/j.pecs.2016.12.004

Ford, W. E., & Rodgers, M. A. (1994). Interfacial electron transfer in colloidal SnO2 hydrosols photosensitized by electrostatically and covalently attached ruthenium (II) polypyridine complexes. *Journal of Physical Chemistry*, *98*(14), 3822–3831. doi:10.1021/j100065a044

Fotoohi Nezhad Khales, M., Sajjadi, S. A., & Kamyabi-Gol, A. (2020). Multipass Friction Stir Processing of Steel/SiC Nanocomposite: Assessment of Microstructure and Tribological Properties. *Journal of Materials Engineering and Performance*, *29*(7), 4241–4250. doi:10.1007/s11665-020-04947-y

Foulkes, R., Man, E., Thind, J., Yeung, S., Joy, A., & Hoskins, C. (2020). The regulation of nanomaterials and nanomedicines for clinical application: Current and future perspectives. *Biomaterials Science*, *8*(17), 4653–4664. doi:10.1039/D0BM00558D PMID:32672255

Frąckowiak, E., & Béguin, F. (2002). Electrochemical storage of energy in carbon nanotubes and nanostructured carbons. *Carbon*, *40*(10), 1775–1787. doi:10.1016/S0008-6223(02)00045-3

Frąckowiak, E., Méténier, K., Bertagna, V., & Béguin, F. (2000). Supercapacitor electrodes from multiwalled carbon nanotubes. *Applied Physics Letters*, *77*(15), 2421–2423. doi:10.1063/1.1290146

Franco, D. V., Da Silva, L. M., & Jardim, W. F. (2008). Reduction of hexavalent chromium in soil and groundwater using zero-valent iron under batch and semi-batch conditions. *Water, Air, and Soil Pollution*, *197*(1-4), 49–60. doi:10.1007/s11270-008-9790-0

Franklin, N. M., Rogers, N. J., Apte, S. C., Batley, G. E., Gadd, G. M., & Casey, P. (2007). Comparative Toxicity of Nanoparticulate ZnO, Bulk ZnO, and $ZnCl_2$ to a Freshwater Microalga (Pseudokirchneriella subcapitata): The Importance of Particle Solubility. *Environmental Science & Technology*, *41*(24), 8484–8490. doi:10.1021/es071445r PMID:18200883

Friedrich, R. P., Cicha, I., & Alexiou, C. (2021). Iron oxide nanoparticles in regenerative medicine and tissue engineering. *Nanomaterials (Basel, Switzerland)*, *11*(9), 2337. doi:10.3390/nano11092337 PMID:34578651

Fu, X., Zheng, Z., Sha, Z., Cao, H., Yuan, Q., Yu, H., & Li, Q. (2022). Biorefining waste into nanobiotechnologies can revolutionize sustainable agriculture. In Trends in Biotechnology (Vol. 40, Issue 12, pp. 1503–1518). Elsevier Ltd. doi:10.1016/j.tibtech.2022.09.013

Fujishima, A., Rao, T. N., & Tryk, D. A. (2000). Titanium dioxide photocatalysis. *Journal of Photochemistry and Photobiology C, Photochemistry Reviews*, *1*(1), 1–21. doi:10.1016/S1389-5567(00)00002-2

Fu, L., Liu, H., Li, C., Wu, Y., Rahm, E., Holze, R., & Wu, H. (2006). Surface modifications of electrode materials for lithium ion batteries. *Solid State Sciences*, *8*(2), 113–128. doi:10.1016/j.solidstatesciences.2005.10.019

Galdames, A., Mendoza, A., Orueta, M., de Soto García, I. S., Sánchez, M., Virto, I., & Vilas, J. L. (2017). Development of new remediation technologies for contaminated soils based on the application of zero-valent iron nanoparticles and bioremediation with compost. *Resource-Efficient Technologies*, *3*(2), 166–176. doi:10.1016/j.reffit.2017.03.008

Gamage, A., Punniamoorthy, T., & Madhujith, T. (2021). *Starch-based hybrid nanomaterials for environmental remediation*. Starch-Evolution and Recent Advances.

Ganguly, S., & Margel, S. (2023). Fluorescent quantum dots-based hydrogels: Synthesis, fabrication and multimodal biosensing. *Talanta Open*, *8*, 100243. doi:10.1016/j.talo.2023.100243

Gao, S. (2019). Ionic liquid functionalized 3D mesoporous FDU-12 for effective SO2 capture. *ACS Sustain Chem Eng, 8*(1), 586–593. doi:10.1021/acssuschemeng.9b06129

Gao, Q., Zhang, J., Gao, J., Zhang, Z., Zhu, H., & Wang, D. (2021). Gold Nanoparticles in Cancer Theranostics. *Frontiers in Bioengineering and Biotechnology*, *9*, 647905. Advance online publication. doi:10.3389/fbioe.2021.647905 PMID:33928072

Gareev, K. G., Grouzdev, D. S., Koziaeva, V. V., Sitkov, N. O., Gao, H., Zimina, T. M., & Shevtsov, M. (2022). Biomimetic nanomaterials: Diversity, technology, and biomedical applications. *Nanomaterials (Basel, Switzerland)*, *12*(14), 2485. doi:10.3390/nano12142485 PMID:35889709

Garello, F., Svenskaya, Y., Parakhonskiy, B., & Filippi, M. (2022). Micro/nanosystems for magnetic targeted delivery of bioagents. *Pharmaceutics*, *14*(6), 1132. doi:10.3390/pharmaceutics14061132 PMID:35745705

Garg, M., Gupta, A., Sharma, A. L., & Singh, S. (2021). Advancements in 2D materials based biosensors for oxidative stress biomarkers. *ACS Applied Bio Materials*, *4*(8), 5944–5960. doi:10.1021/acsabm.1c00625 PMID:35006886

Garrido-Ramírez, E. G., Theng, B. K. G., & Mora, M. L. (2010). Clays and oxide minerals as catalysts and nanocatalysts in Fenton-like reactions—A review. *Applied Clay Science*, *47*(3-4), 182–192. doi:10.1016/j.clay.2009.11.044

Gerion, D., Pinaud, F., Williams, S. C., Parak, W. J., Zanchet, D., Weiss, S., & Alivisatos, A. P. (2001). Synthesis and properties of biocompatible water-soluble silica-coated CdSe/ZnS semiconductor quantum dots. *The Journal of Physical Chemistry B*, *105*(37), 8861–8871. doi:10.1021/jp0105488

Ghitman, J., Biru, E. I., Stan, R., & Iovu, H. (2020). Review of hybrid PLGA nanoparticles: Future of smart drug delivery and theranostics medicine. *Materials & Design*, *193*, 108805. doi:10.1016/j.matdes.2020.108805

Ghosh, P., & Spiro, T. G. (1980). Photoelectrochemistry of tris (bipyridyl) ruthenium (II) covalently attached to n-type tin (IV) oxide. *Journal of the American Chemical Society*, *102*(17), 5543–5549. doi:10.1021/ja00537a021

Gobin, A. M., Lee, M. H., Halas, N. J., James, W. D., Drezek, R. A., & West, J. L. (2007). Near-infrared resonant nanoshells for combined optical imaging and photothermal cancer therapy. *Nano Letters*, *7*(7), 1929–1934. doi:10.1021/nl070610y PMID:17550297

Golodnitsky, D., Nathan, M., Yufit, V., Strauss, E., Freedman, K., Burstein, L., Gladkich, A., & Peled, E. (2006). Progress in three-dimensional (3D) Li-ion microbatteries. *Solid State Ionics, 177*(26–32), 2811–2819. doi:10.1016/j.ssi.2006.02.048

Gómez-Pastora, J., Dominguez, S., Bringas, E., Rivero, M. J., Ortiz, I., & Dionysiou, D. D. (2017). Review and perspectives on the use of magnetic nanophotocatalysts (MNPCs) in water treatment. *Chemical Engineering Journal*, *310*, 407–427. doi:10.1016/j.cej.2016.04.140

Goodman, C. M., & Rotello, V. M. (2004). Biomacromolecule surface recognition using nanoparticles. *Mini-Reviews in Organic Chemistry*, *1*(1), 103–114. doi:10.2174/1570193043488935

Gorain, B., Choudhury, H., Pandey, M., Kesharwani, P., Abeer, M. M., Tekade, R. K., & Hussain, Z. (2018). Carbon nanotube scaffolds as emerging nanoplatform for myocardial tissue regeneration: A review of recent developments and therapeutic implications. *Biomedicine and Pharmacotherapy*, *104*, 496–508. doi:10.1016/j.biopha.2018.05.066 PMID:29800914

Goto, T., & Hirai, T. (1987). Chemically vapor deposited Ti3SiC2. *Materials Research Bulletin*, *22*(9), 1195–1201. doi:10.1016/0025-5408(87)90128-0

Gowda, S. R., Reddy, A. L. M., Zhan, X., Jafry, H. R., & Ajayan, P. M. (2012). 3D nanoporous nanowire current collectors for thin film microbatteries. *Nano Letters*, *12*(3), 1198–1202. doi:10.1021/nl2034464 PMID:22313375

Green, M., & Howman, E. (2005). Semiconductor quantum dots and free radical induced DNA nicking. *Chemical Communications*, (1), 121–123. doi:10.1039/b413175d PMID:15614393

Griffitt, R. J., Luo, J., Gao, J., Bonzongo, J. J., & Barber, D. S. (2008). Effects of particle composition and species on toxicity of metallic nanomaterials in aquatic organisms. *Environmental Toxicology and Chemistry*, *27*(9), 1972–1978. doi:10.1897/08-002.1 PMID:18690762

Guan, G., & Han, M. (2019). Functionalized hybridization of 2D nanomaterials. *Advancement of Science*, *6*(23), 1901837. Advance online publication. doi:10.1002/advs.201901837 PMID:31832321

Guerra, F. D., Attia, M. F., Whitehead, D. C., & Alexis, F. (2018). Nanotechnology for environmental remediation: Materials and applications. *Molecules (Basel, Switzerland)*, *23*(7), 1760. doi:10.3390/molecules23071760 PMID:30021974

Gulati, K., & Adachi, T. (2023). Profiling to Probing: Atomic force microscopy to characterize nanoengineered implants. *Acta Biomaterialia*, *170*, 15–38. Advance online publication. doi:10.1016/j.actbio.2023.08.006 PMID:37562516

Gul, F., Memon, S., Ujjan, I., Goswami, P., & Bhatti, K. A. (2023). The effects of 4β-hydroxy withanolide E extracted from Physalis Peruviana on Complete Blood Count of Dimethylbenz (a) anthracene-induced Breast Cancer in Albino Rats. *Advancements in Life Sciences*, *10*(3), 434–438.

Gungor-Ozkerim, P. S., Inci, I., Zhang, Y. S., Khademhosseini, A., & Dokmeci, M. R. (2018). Bioinks for 3D bioprinting: An overview. *Biomaterials Science*, *6*(5), 915–946. doi:10.1039/C7BM00765E PMID:29492503

Guo, K., Zhao, X., Dai, X., Zhao, N., & Xu, F. J. (2019). Organic/inorganic nanohybrids as multifunctional gene delivery systems. *The Journal of Gene Medicine*, *21*(5), e3084. doi:10.1002/jgm.3084 PMID:30850992

Gupta, A., Sakthivel, T., & Seal, S. (2015). Recent development in 2D materials beyond graphene. *Progress in Materials Science*, *73*, 44–126. doi:10.1016/j.pmatsci.2015.02.002

Gupta, A., & Silver, S. (1998). Molecular genetics: Silver as a biocide: Will resistance become a problem. *Nature Biotechnology*, *16*(10), 888. doi:10.1038/nbt1098-888 PMID:9788326

Gupta, R., Chauhan, H., Garg, V. K., & Kataria, N. (2022). Chemical and physical properties of nanoparticles and hybrid materials. In *Sustainable Nanotechnology for Environmental Remediation* (pp. 199–220). Elsevier. doi:10.1016/B978-0-12-824547-7.00024-2

Gurav, D. D., Jia, Y. A., Ye, J., & Qian, K. (2019). Design of plasmonic nanomaterials for diagnostic spectrometry. *Nanoscale Advances*, *1*(2), 459–469. doi:10.1039/C8NA00319J PMID:36132258

Gurlhosur, S. H., & Sreekanth, B. (2018). Synthesis, characterization of iron oxide (α-Fe2O3) nanoparticles and its application in photocatalytic reduction of chromium (VI). *Rasayan Journal of Chemistry*, *11*, 1678–1685. doi:10.31788/RJC.2018.1145051

Gutés, A. (2012) Graphene decoration with metal nanoparticles: towards easy integration for sensing applications. *Nanoscale*, *4*(2), 438–440. https://pubs.rsc.org/en/content/articlehtml/2012/nr/c1nr115 37e

Hadei, M., Mesdaghinia, A., Nabizadeh, R., Mahvi, A. H., Rabbani, S., & Naddafi, K. (2021). A comprehensive systematic review of photocatalytic degradation of pesticides using nano TiO2. *Environmental Science and Pollution Research International*, *28*(11), 13055–13071. doi:10.1007/s11356-021-12576-8 PMID:33483929

Häggström, F., & Delsing, J. (2018). IoT Energy Storage - A Forecast. *Energy Harvesting and Systems*, *5*(3–4), 43–51. doi:10.1515/ehs-2018-0010

Ha, H. K., Kim, J. W., Lee, M. R., Jun, W., & Lee, W. J. (2015). Cellular uptake and cytotoxicity of β-lactoglobulin nanoparticles: The effects of particle size and surface charge. *Asian-Australasian Journal of Animal Sciences*, *28*(3), 420–427. doi:10.5713/ajas.14.0761 PMID:25656189

Hajkova, P., Spatenka, P., Horsky, J., Horska, I., & Kolouch, A. (2007). Photocatalytic effect of TiO2 films on viruses and bacteria. *Plasma Processes and Polymers*, *4*(S1), S397–S401. doi:10.1002/ppap.200731007

Hallan, S. S., Amirian, J., Brangule, A., & Bandere, D. (2022). Lipid-Based Nano-Sized Cargos as a Promising Strategy in Bone Complications: A Review. *Nanomaterials (Basel, Switzerland)*, *12*(7), 1146. Advance online publication. doi:10.3390/nano12071146 PMID:35407263

Han, G. H., Rodríguez-Manzo, J. A., Lee, C. W., Kybert, N. J., Lerner, M. B., Qi, Z. J., Dattoli, E. N., Rappe, A. M., Drndic, M., & Johnson, A. T. (2013). Continuous growth of hexagonal graphene and boron nitride in-plane heterostructures by atmospheric pressure chemical vapor deposition. *ACS Nano*, *7*(11), 10129–10138. doi:10.1021/nn404331f PMID:24182310

Han, M., Zhu, S. J., Lu, S. Y., Song, Y. B., Feng, T. L., Tao, S. Y., Liu, J. J., & Yang, B. (2018). Recent progress on the photocatalysis of carbon dots: Classification, mechanism and applications. *Nano Today*, *19*, 201–218. doi:10.1016/j.nantod.2018.02.008

Hans, M. L., & Lowman, A. M. (2002). Biodegradable nanoparticles for drug delivery and targeting. *Current Opinion in Solid State and Materials Science*, *6*(4), 319–327. doi:10.1016/S1359-0286(02)00117-1

Haris, N. I. N., Hassan, M. Z., Ilyas, R. A., Suhot, M. A., Sapuan, S. M., Dolah, R., Mohammad, R., & Asyraf, M. R. M. (2022). Dynamic mechanical properties of natural fiber reinforced hybrid polymer composites: A review. *Journal of Materials Research and Technology*, *19*, 167–182. doi:10.1016/j.jmrt.2022.04.155

Hasanin, M. S., Hashem, A. H., Al-Askar, A. A., Haponiuk, J., & Saied, E. (2023). A novel nanocomposite based on mycosynthesized bimetallic zinc-copperoxide nanoparticles, nanocellulose and chitosan: Characterization, antimicrobial and photocatalytic activities. *Electronic Journal of Biotechnology*, *65*, 45–55. doi:10.1016/j.ejbt.2023.05.001

Hasan, M. N., Nafiujjaman, M., & Lee, Y.-K. (2019). 2D nanomaterials for gene delivery. In *Biomedical Applications of Graphene and 2D Nanomaterials* (pp. 87–104). Elsevier. doi:10.1016/B978-0-12-815889-0.00004-0

Hegde, M. N., Hegde, P., Bhandary, S., & Deepika, K. (2011). An evalution of compressive strength of newer nanocomposite: An in vitro study. *Journal of Conservative Dentistry*, *14*(1), 36–39. doi:10.4103/0972-0707.80734 PMID:21691503

Heimer, T. A., D'Arcangelis, S. T., Farzad, F., Stipkala, J. M., & Meyer, G. J. (1996). An acetylacetonate-based semiconductor– sensitizer linkage. *Inorganic Chemistry*, *35*(18), 5319–5324. doi:10.1021/ic960419j

He, L., Wang, L., Zhu, H., Wang, Z., Zhang, L., Yang, L., Dai, Y., Mo, H., Zhang, J., & Shen, J. (2021). A reusable Fe_3O_4/GO-COOH nanoadsorbent for Ca^{2+} and Cu^{2+} removal from oilfield wastewater. *Chemical Engineering Research & Design*, *166*, 248–258. doi:10.1016/j.cherd.2020.12.019

He, M., Chen, M., Dou, Y., Ding, J., Yue, H., Yin, G., Chen, X., & Cui, Y. (2020). Electrospun silver nanoparticles-embedded feather keratin/poly(vinyl alcohol)/poly(ethylene oxide) antibacterial composite nanofibers. *Polymers*, *12*(2), 305. doi:10.3390/polym12020305 PMID:32028586

Hernandez-Viezcas, J. A., Castillo-Michel, H., Andrews, J. C., Cotte, M., Rico, C., Peralta-Videa, J. R., Ge, Y., Priester, J. H., Holden, P. A., & Gardea-Torresdey, J. L. (2013). In situ synchrotron X-ray fluorescence mapping and speciation of CeO2 and ZnO nanoparticles in soil cultivated soybean (Glycine max). *ACS Nano*, *7*(2), 1415–1423. doi:10.1021/nn305196q PMID:23320560

Herrmann, J.-M. (1999). Heterogeneous photocatalysis: Fundamentals and applications to the removal of various types of aqueous pollutants. *Catalysis Today*, *53*(1), 115–129. doi:10.1016/S0920-5861(99)00107-8

He, Y., Hu, C., Li, Z., Wu, C., Zeng, Y., & Peng, C. (2022). Multifunctional carbon nanomaterials for diagnostic applications in infectious diseases and tumors. *Materials Today. Bio*, *14*, 100231. doi:10.1016/j.mtbio.2022.100231 PMID:35280329

Hlongwane, G. N., Sekoai, P. T., Meyyappan, M., & Moothi, K. (2019). Simultaneous removal of pollutants from water using nanoparticles: A shift from single pollutant control to multiple pollutant control. *The Science of the Total Environment*, *656*, 808–833. doi:10.1016/j.scitotenv.2018.11.257 PMID:30530150

Homaeigohar, S., Liu, Q., & Kordbacheh, D. (2021). Biomedical applications of antiviral nanohybrid materials relating to the COVID-19 pandemic and other viral crises. *Polymers*, *13*(16), 2833. doi:10.3390/polym13162833 PMID:34451371

Hot, J., Topalov, J., Ringot, E., & Bertron, A. (2017). Investigation on Parameters Affecting the Effectiveness of Photocatalytic Functional Coatings to Degrade NO: TiO_2 Amount on Surface, Illumination, and Substrate Roughness. *International Journal of Photoenergy*, *2017*, 1–14. doi:10.1155/2017/6241615

Hou, L., Wang, L., Royer, S., & Zhang, H. (2016). Ultrasound-assisted heterogeneous Fenton-like degradation of tetracycline over a magnetite catalyst. *Journal of Hazardous Materials*, *302*, 458–467. doi:10.1016/j.jhazmat.2015.09.033 PMID:26521091

Hou, S., Wu, B., Peng, D., Wang, Z., Wang, Y., & Xu, H. (2019). Remediation performance and mechanism of hexavalent chromium in alkaline soil using multi-layer loaded nano-zero-valent iron. *Environmental Pollution*, *252*, 553–561. doi:10.1016/j.envpol.2019.05.083 PMID:31181500

Hsueh, Y. H., Lin, K. S., Ke, W. J., Hsieh, C. T., Chiang, C. L., Tzou, D. Y., & Liu, S. T. (2015). The antimicrobial properties of silver nanoparticles in bacillus subtilis are mediated by released Ag+ ions. *PLoS One*, *10*(12), e0144306. doi:10.1371/journal.pone.0144306 PMID:26669836

Huang, C.-C., & Chang, H.-T. (2007). Parameters for selective colorimetric sensing of mercury(II) in aqueous solutions using mercaptopropionic acid-modified gold nanoparticles. *Chemical Communications (Cambridge)*, (12), 1215–1217. doi:10.1039/B615383F PMID:17356760

Huang, H. C., Barua, S., Sharma, G., Dey, S. K., & Rege, K. (2011). Inorganic nanoparticles for cancer imaging and therapy. *Journal of Controlled Release*, *155*(3), 344–357. doi:10.1016/j.jconrel.2011.06.004 PMID:21723891

Huang, S. H., & Juang, R. S. (2011). Biochemical and biomedical applications of multifunctional magnetic nanoparticles: A review. *Journal of Nanoparticle Research*, *13*(10), 4411–4430. doi:10.1007/s11051-011-0551-4

Huang, X., El-Sayed, I. H., Qian, W., & El-Sayed, M. A. (2006). Cancer cell imaging and photothermal therapy in the near-infrared region by using gold nanorods. *Journal of the American Chemical Society*, *128*(6), 2115–2120. doi:10.1021/ja057254a PMID:16464114

Hufenbach, W., Ibraim, F., Langkamp, A., Böhm, R., & Hornig, A. (2008). Charpy impact tests on composite structures – An experimental and numerical investigation. *Composites Science and Technology*, *68*(12), 2391–2400. doi:10.1016/j.compscitech.2007.10.008

Hu, N., Wang, Y., Chai, J., Gao, R., Yang, Z., Kong, E. S.-W., & Zhang, Y. (2012). Gas sensor based on P-Phenylenediamine reduced graphene oxide. *Sensors and Actuators. B, Chemical*, *163*(1), 107–114. doi:10.1016/j.snb.2012.01.016

Hung, K.-C., Tseng, C.-S., Dai, L.-G., & Hsu, S. (2016). Water-based polyurethane 3D printed scaffolds with controlled release function for customized cartilage tissue engineering. *Biomaterials*, *83*, 156–168. doi:10.1016/j.biomaterials.2016.01.019 PMID:26774563

Huo, Q. (2007). A perspective on bioconjugated nanoparticles and quantum dots. *Colloids and Surfaces. B, Biointerfaces*, *59*(1), 1–10. doi:10.1016/j.colsurfb.2007.04.019 PMID:17544637

Huo, Y., Xiu, S., Meng, L.-Y., & Quan, B. (2023). Solvothermal synthesis and applications of micro/nano carbons: A review. *Chemical Engineering Journal*, *451*, 138572. doi:10.1016/j.cej.2022.138572

Hyun, C. M., Choi, J. H., Lee, S. W., Seo, S. Y., Lee, M. J., Kwon, S. H., & Ahn, J. H. (2019). Synthesis of Bi 2 Te 3 Single Crystals with Lateral Size up to Tens of Micrometers by Vapor Transport and Its Potential for Thermoelectric Applications. *Crystal Growth & Design*, *19*(4), 2024–2029. doi:10.1021/acs.cgd.8b01931

Ibáñez, J. A., Litter, M. I., & Pizarro, R. A. (2003). Photocatalytic bactericidal effect of TiO2 on Enterobacter cloacae: Comparative study with other Gram (−) bacteria. *Journal of Photochemistry and Photobiology A Chemistry*, *157*(1), 81–85. doi:10.1016/S1010-6030(03)00074-1

Ibrahim, I. D., Sadiku, E. R., Jamiru, T., Hamam, Y., Alayli, Y., & Eze, A. A. (2020). Prospects of nanostructured composite materials for energy harvesting and storage. *Journal of King Saud University. Science*, *32*(1), 758–764. doi:10.1016/j.jksus.2019.01.006

Ibrahim, R. K., Hayyan, M., AlSaadi, M. A., Hayyan, A., & Ibrahim, S. (2016). Environmental application of nanotechnology: Air, soil, and water. *Environmental Science and Pollution Research International*, *23*(14), 13754–13788. doi:10.1007/s11356-016-6457-z PMID:27074929

Iijima, S. (1994). Carbon nanotubes. *MRS Bulletin*, *19*(11), 43–49. doi:10.1557/S0883769400048405

Ilosvai, Á. M., Forgách, L., Kovács, N., Heydari, F., Szigeti, K., Máthé, D., ... Viskolcz, B. (2023). Development of Polymer-Encapsulated, Amine-Functionalized Zinc Ferrite Nanoparticles as MRI Contrast Agents. *International Journal of Molecular Sciences*, *24*(22), 16203. doi:10.3390/ijms242216203 PMID:38003394

Ilyas, S. U., Shamsuddin, R., Kai Xiang, T., Estellé, P., & Pendyala, R. (2023). Rheological profile of graphene-based nanofluids in thermal oil with hybrid additives of carbon nanotubes and nanofibers. *Journal of Molecular Liquids*, *376*, 121443. doi:10.1016/j.molliq.2023.121443

Imada, K., Sakai, S., Kajihara, H., Tanaka, S., & Ito, S. (2016). Magnesium oxide nanoparticles induce systemic resistance in tomato against bacterial wilt disease. *Plant Pathology*, *65*(4), 551–560. doi:10.1111/ppa.12443

Ireland, J. C., Klostermann, P., Rice, E. W., & Clark, R. M. (1993). Inactivation of Escherichia coli by titanium dioxide photocatalytic oxidation. *Applied and Environmental Microbiology*, *59*(5), 1668–1670. doi:10.1128/aem.59.5.1668-1670.1993 PMID:8390819

Iyad, N. S., Ahmad, M., Alkhatib, S. G., & Hjouj, M. (2023). Gadolinium contrast agents-challenges and opportunities of a multidisciplinary approach: Literature review. *European Journal of Radiology Open*, *11*, 100503. doi:10.1016/j.ejro.2023.100503 PMID:37456927

Jadhav, A. C., Annaldewar, B. N., & Jadhav, N. C. (2023). 2 A current perspective on nanocomposite and nanohybrid material: Developments and trends. *Nanocomposite and Nanohybrid Materials: Processing and Applications*, *17*, 29–54. doi:10.1515/9783111137902-002

Jamieson, T., Bakhshi, R., Petrova, D., Pocock, R., Imani, M., & Seifalian, A. M. (2007). Biological applications of quantum dots. *Biomaterials*, *28*(31), 4717–4732. doi:10.1016/j.biomaterials.2007.07.014 PMID:17686516

Jana, N. R., Earhart, C., & Ying, J. Y. (2007). Synthesis of water-soluble and functionalized nanoparticles by silica coating. *Chemistry of Materials*, *19*(21), 5074–5082. doi:10.1021/cm071368z

Jariwala, D., Sangwan, V. K., Lauhon, L. J., Marks, T. J., & Hersam, M. C. (2014). Emerging device applications for semiconducting two-dimensional transition metal dichalcogenides. *ACS Nano*, *8*(2), 1102–1120. doi:10.1021/nn500064s PMID:24476095

Jeon, G., Ha, D., Park, Y., & Jeong, C. (2022). Three-Point Bending Properties of Hybrid Multi-Materials Using Adhesive Bonding Dependent on Strength Difference between Steel and Aluminum. In Materials (Vol. 15, Issue 9). doi:10.3390/ma15093328

Jerin, V. M., & Remya, R. (2019). Investigation on the removal of toxic chromium ion from waste water using Fe2O3 nanoparticles. *Materials Today: Proceedings*, *9*, 27–31. doi:10.1016/j.matpr.2019.02.032

Jeyasubramanian, K., Sakthivel, A., Thangagiri, B., & Raja, J. D. (2023). Trends and developments in point-of-care diagnostics using quantum dots. In K. Yun & S. Govindaraju (Eds.), *Carbon and Graphene Quantum Dots for Biomedical Applications* (pp. 193–209). Woodhead Publishing. doi:10.1016/B978-0-323-98362-4.00015-5

Jia, F., Liu, X., Li, L., Mallapragada, S., Narasimhan, B., & Wang, Q. (2013). Multifunctional nanoparticles for targeted delivery of immune activating and cancer therapeutic agents. *Journal of Controlled Release*, *172*(3), 1020–1034. doi:10.1016/j.jconrel.2013.10.012 PMID:24140748

Jiang, C., Hu, D., & Lu, Y. (2016). Digital Micromirror Device (DMD)-Based High-Cycle Torsional Fatigue Testing Micromachine for 1D Nanomaterials. *Micromachines*, *7*(3), 49. Advance online publication. doi:10.3390/mi7030049 PMID:30407421

Jiang, K., Eitan, A., Schadler, L. S., Ajayan, P. M., Siegel, R. W., Grobert, N., Mayne, M., Reyes-Reyes, M., Terrones, H., & Terrones, M. (2003). Selective attachment of gold nanoparticles to nitrogen-doped carbon nanotubes. *Nano Letters*, *3*(3), 275–277. doi:10.1021/nl025914t

Jiang, L., & Gao, L. (2003). Modified carbon nanotubes: An effective way to selective attachment of gold nanoparticles. *Carbon, 41*(15), 2923–2929. doi:10.1016/S0008-6223(03)00339-7

Jiang, L., Liu, J., Tian, A., Ren, X., Huang, S., Zhou, W., Zhang, L., Li, D., Zhang, S., Ikeda, M., & Yang, H. (2019). Influence of substrate misorientation on carbon impurity incorporation and electrical properties of p-GaN grown by metalorganic chemical vapor deposition. *Applied Physics Express, 12*(5), 055503. doi:10.7567/1882-0786/ab0da2

Jiang, M., Qi, Y., Liu, H., & Chen, Y. (2018). The Role of Nanomaterials and Nanotechnologies in Wastewater Treatment: A Bibliometric Analysis. *Nanoscale Research Letters, 13*(1), 233. doi:10.1186/s11671-018-2649-4 PMID:30097816

Jiang, Q., Gao, X., Zhang, W., & Chen, Z. (2023). Recent progress in metal-organic framework-based sonosensitizers for sonodynamic tumor therapy. *Biomaterials Science, 11*(13), 4452–4470. doi:10.1039/D3BM00556A PMID:37221901

Jiang, Q., Xie, Y., Peng, M., Wang, Z., Li, T., Yin, M., Shen, J., & Yan, S. (2022). A nanocarrier pesticide delivery system with promising benefits in the case of dinotefuran: Strikingly enhanced bioactivity and reduced pesticide residue. *Environmental Science. Nano, 9*(3), 988–999. doi:10.1039/D1EN00752A

Jiang, S., Liu, X., Liu, Y., Liu, J., He, W., & Dong, Y. (2020). Synthesis of silver@ hydroxyapatite nanoparticles based biocomposite and their assessment for viability of Osseointegration for rabbit knee joint anterior cruciate ligament rehabilitation. *Journal of Photochemistry and Photobiology. B, Biology, 202*, 111677. doi:10.1016/j.jphotobiol.2019.111677 PMID:31810037

Jiao, X., Li, X., Jin, X., Sun, Y., Xu, J., Liang, L., Ju, H., Zhu, J., Pan, Y., Yan, W., Lin, Y., & Xie, Y. (2017). Partially oxidized SnS2 atomic layers achieving efficient visible-light-driven CO2 reduction. *Journal of the American Chemical Society, 139*(49), 18044–18051. doi:10.1021/jacs.7b10287 PMID:29144744

Jia, X., Lv, M., Fei, Y., Dong, Q., Wang, H., Liu, Q., Li, D., Wang, J., & Wang, E. (2022). Facile one-step synthesis of NIR-Responsive siRNA-Inorganic hybrid nanoplatform for imaging-guided photothermal and gene synergistic therapy. *Biomaterials, 282*, 121404. doi:10.1016/j.biomaterials.2022.121404 PMID:35134700

Ji, M., AL-Huqail, A. A., Majdi, A., Liu, H., Ali, I., & Marzouki, R. (2022). Remediation of mine polluted soil with nano-enhanced materials: Development of extreme learning machine approaches. *Chemosphere, 307*, 135772. Advance online publication. doi:10.1016/j.chemosphere.2022.135772 PMID:35931267

Jimenez-Cadena, G., Riu, J., & Xavier Rius, F. (2007). Gas sensors based on nanostructured materials. *Analyst, 132*(11), 1083–1099. doi:10.1039/b704562j PMID:17955141

Jin, C., Wu, Z., Molinski, J. H., Zhou, J., Ren, Y., & Zhang, J. X. (2022). Plasmonic nanosensors for point-of-care biomarker detection. *Materials Today. Bio, 14*, 100263. doi:10.1016/j.mtbio.2022.100263 PMID:35514435

Jing, Q., Xie, Y., Zhu, G., Han, R. P. S., & Wang, Z. L. (2015). Self-powered thin-film motion vector sensor. *Nature Communications*, *6*(1), 8031. doi:10.1038/ncomms9031 PMID:26271603

Jin, J., Xiao, T., Zhang, Y., Han, Z., Wang, H., Wang, R., Gong, Y., He, B., Liu, X., & Zhou, K. (2021). Hierarchical MXene/transition metal chalcogenide heterostructures for electrochemical energy storage and conversion. *Nanoscale*, *13*(47), 19740–19770. doi:10.1039/D1NR05799E PMID:34821248

Jin, Y., Kannan, S., Wu, M., & Zhao, J. X. (2007). Toxicity of luminescent silica nanoparticles to living cells. *Chemical Research in Toxicology*, *20*(8), 1126–1133. doi:10.1021/tx7001959 PMID:17630705

Ji, Q., Zhang, Y., Gao, T., Zhang, Y., Ma, D., Liu, M., Chen, Y., Qiao, X., Tan, P. H., Kan, M., Feng, J., Sun, Q., & Liu, Z. (2013). Epitaxial monolayer MoS2 on mica with novel photoluminescence. *Nano Letters*, *13*(8), 3870–3877. doi:10.1021/nl401938t PMID:23899342

Kabay, G., Meydan, A. E., Eom, T., Shim, B. S., Mutlu, M., & Kaleli-Can, G. (2023). Stimuli-responsive nanoparticle-nanofiber hybrids for drug delivery and photodynamic therapy. *International Journal of Pharmaceutics*, *630*, 122442. Advance online publication. doi:10.1016/j.ijpharm.2022.122442 PMID:36442721

Kabra, K., Chaudhary, R., & Sawhney, R. L. (2004). Treatment of hazardous organic and inorganic compounds through aqueous-phase photocatalysis: A review. *Industrial & Engineering Chemistry Research*, *43*(24), 7683–7696. doi:10.1021/ie0498551

Kah, M., & Hofmann, T. (2014). Nanopesticide research: Current trends and future priorities. *Environment International*, *63*, 224–235. doi:10.1016/j.envint.2013.11.015 PMID:24333990

Kalidasan, B., Pandey, A. K., Saidur, R., Yadav, A., Samykano, M., & Tyagi, V. (2022). Graphene–Silver Hybrid Nanoparticle-based Organic Phase Change Materials for Enhanced Thermal Energy Storage. *Sustainability (Basel)*, *14*(20), 13240. doi:10.3390/su142013240

Kang, B., & Ceder, G. (2009). Battery materials for ultrafast charging and discharging. *Nature*, *458*(7235), 190–193. doi:10.1038/nature07853 PMID:19279634

Kanjana, D. (2020). Foliar application of magnesium oxide nanoparticles on nutrient element concentrations, growth, physiological, and yield parameters of cotton. *Journal of Plant Nutrition*, *43*(20), 3035–3049. doi:10.1080/01904167.2020.1799001

Kargozar, S., Baino, F., Hamzehlou, S., Hamblin, M. R., & Mozafari, M. (2020). Nanotechnology for angiogenesis: Opportunities and challenges. *Chemical Society Reviews*, *49*(14), 5008–5057. doi:10.1039/C8CS01021H PMID:32538379

Karimi, M., Mansouri, M. R., Rabiee, N., & Hamblin, M. R. (2018). Carbon-based nanomaterials. *Advances in nanomaterials for drug delivery: Polymeric, nanocarbon, and bio-inspired.*

Karlsson, H., Cronholm, P., Gustafsson, J., & Möller, L. (2008). Copper Oxide Nanoparticles Are Highly Toxic: A Comparison between Metal Oxide Nanoparticles and Carbon Nanotubes. *Chemical Research in Toxicology*, *21*(9), 1726–1732. doi:10.1021/tx800064j PMID:18710264

Karri, R. R., Mubarak, N. M., Koduru, J. R., Lingamdinne, L. P., Dehghani, M. H., Jatoi, A. S., & Mazari, S. A. (2023). 1 - Role of hybrid nanomaterials for a sustainable environment. In J. R. Koduru, R. R. Karri, & N. M. Mubarak (Eds.), *Hybrid Nanomaterials for Sustainable Applications* (pp. 1–24). Elsevier. doi:10.1016/B978-0-323-98371-6.00017-3

Karthikeyan, C., Jayaramudu, T., Núñez, D., Jara, N., Opazo-Capurro, A., Varaprasad, K., Kim, K., Yallapu, M. M., & Sadiku, R. (2023). Hybrid nanomaterial composed of chitosan, curcumin, ZnO and TiO_2 for antibacterial therapies. *International Journal of Biological Macromolecules, 242*, 124814. Advance online publication. doi:10.1016/j.ijbiomac.2023.124814 PMID:37201889

Kaskhedikar, N., & Maier, J. (2009). Lithium storage in carbon nanostructures. *Advanced Materials, 21*(25–26), 2664–2680. doi:10.1002/adma.200901079 PMID:36751065

Katz, E., & Willner, I. (2004). Integrated Nanoparticle–Biomolecule Hybrid Systems: Synthesis, Properties, and Applications. *Angewandte Chemie International Edition, 43*(45), 6042–6108. doi:10.1002/anie.200400651 PMID:15538757

Kaur, J., Pathak, T., Singh, A., & Kumar, K. (2017). Application of nanotechnology in the environment biotechnology. In R. Kumar, A. K. Sharma, & S. S. Ahluwalia (Eds.), *Adv. Environ. Biotechnol* (pp. 155–165). Springer. doi:10.1007/978-981-10-4041-2_9

Kawajiri, K., Kobayashi, M., & Sakamoto, K. (2020). Lightweight materials equal lightweight greenhouse gas emissions?: A historical analysis of greenhouse gases of vehicle material substitution. *Journal of Cleaner Production, 253*, 119805. doi:10.1016/j.jclepro.2019.119805

Kesharwani, P., Gothwal, A., Iyer, A. K., Jain, K., Chourasia, M. K., & Gupta, U. (2018). Dendrimer nanohybrid carrier systems: An expanding horizon for targeted drug and gene delivery. *Drug Discovery Today, 23*(2), 300–314. doi:10.1016/j.drudis.2017.06.009 PMID:28697371

Keshvardoostchokami, M., Majidi, S. S., Huo, P., Ramachandran, R., Chen, M., & Liu, B. (2020). Electrospun nanofibers of natural and synthetic polymers as artificial extracellular matrix for tissue engineering. *Nanomaterials (Basel, Switzerland), 11*(1), 21. doi:10.3390/nano11010021 PMID:33374248

Keum, D. H., Cho, S., Kim, J. H., Choe, D. H., Sung, H. J., Kan, M., Kang, H., Hwang, J., Kim, S. W., Yang, H., Chang, K. J., & Lee, Y. H. (2015). Bandgap opening in few-layered monoclinic $MoTe_2$. *Nature Physics, 11*(6), 482–486. doi:10.1038/nphys3314

Khalid, Z. B., Nasrullah, M., Nayeem, A., Wahid, Z. A., Singh, L., & Krishnan, S. (2020). Application of 2D Graphene-Based Nanomaterials for Pollutant Removal from Advanced Water and Wastewater Treatment Processes. In *ACS Symposium Series* (*Vol. 1353*, pp. 191–217). American Chemical Society. 10.1021/bk-2020-1353.ch009

Khan, Y., Sadia, H., Ali Shah, S. Z., Khan, M. N., Shah, A. A., Ullah, N., Ullah, M. F., Bibi, H., Bafakeeh, O. T., Khedher, N. B., Eldin, S. M., Fadhl, B. M., & Khan, M. I. (2022). Classification, Synthetic, and Characterization Approaches to Nanoparticles, and Their Applications in Various Fields of Nanotechnology: A Review. In Catalysts (Vol. 12, Issue 11). doi:10.3390/catal12111386

Khare, P., Yadav, A., Ramkumar, J., & Verma, N. (2016). Microchannel-embedded metal–carbon–polymer nanocomposite as a novel support for chitosan for efficient removal of hexavalent chromium from water under dynamic conditions. *Chemical Engineering Journal*, *293*, 44–54. doi:10.1016/j.cej.2016.02.049

Kharey, P., Goel, M., Husain, Z., Gupta, R., Sharma, D. M. M., Palani, I. A., & Gupta, S. (2023). Green synthesis of biocompatible superparamagnetic iron oxide-gold composite nanoparticles for magnetic resonance imaging, hyperthermia and photothermal therapeutic applications. *Materials Chemistry and Physics*, *293*, 126859. doi:10.1016/j.matchemphys.2022.126859

Kharisov, B. I., Rasika Dias, H. V., Kharissova, O. V., Manuel Jiménez-Pérez, V., Olvera Pérez, B., & Muñoz Flores, B. (2012). Iron-containing nanomaterials: Synthesis, properties, and environmental applications. *RSC Advances*, *2*(25), 9325–9358. doi:10.1039/c2ra20812a

Kharwar, S., & Singh, S. (2021). First-principles investigation of zigzag graphene nanoribbons based nanosensor for heavy metal detector. *Materials Today: Proceedings*, *47*, 2227–2231. Advance online publication. doi:10.1016/j.matpr.2021.04.183

Khin, M. M., Nair, A. S., Babu, V. J., Murugan, R., & Ramakrishna, S. (2012). A review on nanomaterials for environmental remediation. *Energy & Environmental Science*, *5*(8), 8075–8109. doi:10.1039/c2ee21818f

Khot, S., Rawal, S. U., & Patel, M. M. (2020). Theranostics Nanoformulations: Merging Diagnostics and Nanotherapeutics. *Nanoformulations in Human Health: Challenges and Approaches*, 73-91.

Khurmi, R. S. (2014). *Material Science*. Chand Publishers.

Khurshid, A., Amin, R., Aziz, A., Ahmed, O., & Abdel-Haliem, M. E. (2023). Deciphering the Potential Therapeutic Intervention Points of 2019-nCoV: A Pharmacological Perspective. *Advancements in Life Sciences*, *9*(4), 412–428.

Kievit, F. M., & Zhang, M. (2011). Surface engineering of iron oxide nanoparticles for targeted cancer therapy. *Accounts of Chemical Research*, *44*(10), 853–862. doi:10.1021/ar2000277 PMID:21528865

Kim, D. W., Hwang, I. S., Kwon, S., Kang, H. Y., Park, K. S., Choi, Y. J., Choi, K. J., & Park, J. (2007). Highly Conductive Coaxial SnO_2–In_2O_3 Heterostructured Nanowires for Li Ion Battery Electrodes. *Nano Letters*, *7*(10), 3041–3045. doi:10.1021/nl0715037 PMID:17760477

Kim, D., & Jon, S. (2012). Gold nanoparticles in image-guided cancer therapy. *Inorganica Chimica Acta*, *393*, 154–164. doi:10.1016/j.ica.2012.07.001

Kim, D., Shin, K., Kwon, S. G., & Hyeon, T. (2018). Synthesis and biomedical applications of multifunctional nanoparticles. *Advanced Materials*, *30*(49), 1802309. doi:10.1002/adma.201802309 PMID:30133009

Kim, H., Achermann, M., Balet, L. P., Hollingsworth, J. A., & Klimov, V. I. (2005). Synthesis and characterization of Co/CdSe core/shell nanocomposites: Bifunctional magnetic-optical nanocrystals. *Journal of the American Chemical Society*, *127*(2), 544–546. doi:10.1021/ja047107x PMID:15643877

Kim, J. S., Kuk, E., Yu, K. N., Kim, J. H., Park, S. J., Lee, H. J., Kim, S. H., Park, Y. K., Park, Y. H., Hwang, C.-Y., Kim, Y.-K., Lee, Y.-S., Jeong, D. H., & Cho, M. H. (2007). Antimicrobial effects of silver nanoparticles. *Nanomedicine; Nanotechnology, Biology, and Medicine*, *3*(1), 95–101. doi:10.1016/j.nano.2006.12.001 PMID:17379174

Kim, S. H., Kwak, S. Y., Sohn, B. H., & Park, T. H. (2003). Design of TiO2 nanoparticle self-assembled aromatic polyamide thin-film-composite (TFC) membrane as an approach to solve biofouling problem. *Journal of Membrane Science*, *211*(1), 157–165. doi:10.1016/S0376-7388(02)00418-0

Kim, S. I., Lee, K. H., Mun, H. A., Kim, H. S., Hwang, S. W., Roh, J. W., Yang, D. J., Shin, W. H., Li, X. S., Lee, Y. H., Snyder, G. J., & Kim, S. W. (2015). Dense dislocation arrays embedded in grain boundaries for high-performance bulk thermoelectrics. *Science*, *348*(6230), 109–114. doi:10.1126/science.aaa4166 PMID:25838382

Kim, T., Song, W., Son, D.-Y., Ono, L. K., & Qi, Y. (2019). Lithium-ion batteries: Outlook on present, future, and hybridized technologies. *Journal of Materials Chemistry. A, Materials for Energy and Sustainability*, *7*(7), 2942–2964. doi:10.1039/C8TA10513H

Kim, Y., Ahn, H., Nam, S. H., Lee, S. H., Shim, H., & Kim, W. (2008). Honeycomb pattern array of vertically standing core-shell nanorods: Its application to Li energy electrodes. *Applied Physics Letters*, *93*(10), 103104. Advance online publication. doi:10.1063/1.2977862

Kim, Y., Johnson, R. C., & Hupp, J. T. (2001). Gold nanoparticle-based sensing of "spectroscopically silent" heavy metal ions. *Nano Letters*, *1*(4), 165–167. doi:10.1021/nl0100116

Kinter-Meyer, M., Colella, W., Elizondo, M., Nguyen, T., Viswanathan, V., & Zhang, Y. (2012). *Book National Assessment of Energy Storage for Grid Balancing and Arbitrage: Phase 1, WECC* (M. B. P. Kinter-Meyer, Ed.). Pacific Northwest National Laboratory. doi:10.2172/1131386

Kiranmai, B., Venu Vasantha, S., & Rama Krishna, S. (2021). Techniques for Rice Leaf Disease Detection using Machine Learning Algorithms. *International Journal of Engineering and Technical Research, 9*(8), 162–166. www.ijert.org

Kircher, M. F., De La Zerda, A., Jokerst, J. V., Zavaleta, C. L., Kempen, P. J., Mittra, E., Pitter, K., Huang, R., Campos, C., Habte, F., Sinclair, R., Brennan, C. W., Mellinghoff, I. K., Holland, E. C., & Gambhir, S. S. (2012). A brain tumor molecular imaging strategy using a new triple-modality MRI-photoacoustic-Raman nanoparticle. *Nature Medicine*, *18*(5), 829–834. doi:10.1038/nm.2721 PMID:22504484

Kirchner, C., Liedl, T., Kudera, S., Pellegrino, T., Muñoz Javier, A., Gaub, H. E., Stölzle, S., Fertig, N., & Parak, W. J. (2005). Cytotoxicity of colloidal CdSe and CdSe/ZnS nanoparticles. *Nano Letters*, *5*(2), 331–338. doi:10.1021/nl047996m PMID:15794621

Kneipp, J., Kneipp, H., & Kneipp, K. (2008). SERS—A single-molecule and nanoscale tool for bioanalytics. *Chemical Society Reviews*, *37*(5), 1052–1060. doi:10.1039/b708459p PMID:18443689

Kojima, C., Kono, K., Maruyama, K., & Takagishi, T. (2000). Synthesis of polyamidoamine dendrimers having poly (ethylene glycol) grafts and their ability to encapsulate anticancer drugs. *Bioconjugate Chemistry*, *11*(6), 910–917. doi:10.1021/bc0000583 PMID:11087341

Kong, D., Wang, H., Cha, J. J., Pasta, M., Koski, K. J., Yao, J., & Cui, Y. (2013). Synthesis of MoS_2 and $MoSe_2$ films with vertically aligned layers. *Nano Letters*, *13*(3), 1341–1347. doi:10.1021/nl400258t PMID:23387444

Kothandam, G., Singh, G., Guan, X., Lee, J. M., Ramadass, K., Joseph, S., Benzigar, M., Karakoti, A., Yi, J., Kumar, P., & Vinu, A. (2023). Recent Advances in Carbon-Based Electrodes for Energy Storage and Conversion. *Advancement of Science*, *10*(18), 2301045. doi:10.1002/advs.202301045 PMID:37096838

Kour, A., Panda, H. S., Singh, I. R., Kumar, A., & Panda, J. J. (2023). Peptide-metal nanohybrids (PMN): Promising entities for combating neurological maladies. *Advances in Colloid and Interface Science*, *318*, 102954. doi:10.1016/j.cis.2023.102954 PMID:37487364

Kreyling, W. G., Semmler-Behnke, M., & Möller, W. (2006). Ultrafine Particle–Lung interactions: Does size matter? *Journal of Aerosol Medicine*, *19*(1), 74–83. doi:10.1089/jam.2006.19.74 PMID:16551218

Kukushkina, E. A., Hossain, S. I., Sportelli, M. C., Ditaranto, N., Picca, R. A., & Cioffi, N. (2021). Ag-based synergistic antimicrobial composites. A critical review. *Nanomaterials (Basel, Switzerland)*, *11*(7), 1687. doi:10.3390/nano11071687 PMID:34199123

Kulurumotlakatla, D. K., Ramachandran, T., Kumar, Y. A., Mohammed, A. A. A., & Kang, M. C. (2024). Hierarchically fabricated nano flakes-rod-like CoMoO–S supported Ni-foam for high-performance supercapacitor electrode material. *Journal of Physics and Chemistry of Solids*, *185*, 111735. doi:10.1016/j.jpcs.2023.111735

Kumar, Hietala, & Oksman. (2019). Lignin-Based Electrospun Carbon Nanofibers. *Front. Mater.*, *6*, 62.

Kumar, A., Gupta, G., Bapna, K., & Shivagan, D. D. (2023). Semiconductor-metal-oxide-based nano-composites for humidity sensing applications. In *Materials Research Bulletin* (Vol. 158). Elsevier Ltd. doi:10.1016/j.materresbull.2022.112053

Kumar, D. A. K., Valanarasu, S., Tamilnayagam, V., & Amalraj, L. (2017). Structural, morphological and optical properties of SnS_2 thin films by nebulized spray pyrolysis technique. *Journal of Materials Science Materials in Electronics*, *28*(19), 14209–14216. doi:10.1007/s10854-017-7278-7

Kumari, M., Pandey, S., Bhattacharya, A., Mishra, A., & Nautiyal, C. S. (2017). Protective role of biosynthesized silver nanoparticles against early blight disease in Solanum lycopersicum. *Plant Physiology and Biochemistry*, *121*, 216–225. doi:10.1016/j.plaphy.2017.11.004 PMID:29149700

Kumar, P. P. P., & Lim, D.-K. (2023). Photothermal Effect of Gold Nanoparticles as a Nanomedicine for Diagnosis and Therapeutics. *Pharmaceutics, 15*(9), 2349. doi:10.3390/pharmaceutics15092349 PMID:37765317

Kumar, R., Goel, N., Jarwal, D. K., Hu, Y., Zhang, J., & Kumar, M. (2023). Strategic review on chemical vapor deposition technology-derived 2D material nanostructures for room-temperature gas sensors. *Journal of Materials Chemistry. C, Materials for Optical and Electronic Devices, 11*(3), 774–801. doi:10.1039/D2TC04188J

Kumar, R., Mondal, K., Panda, P. K., Kaushik, A., Abolhassani, R., Ahuja, R., Rubahn, H.-G., & Mishra, Y. K. (2020). Core–shell nanostructures: Perspectives towards drug delivery applications. *Journal of Materials Chemistry. B, 8*(39), 8992–9027. doi:10.1039/D0TB01559H PMID:32902559

Kumar, S., Nehra, M., Dilbaghi, N., Marrazza, G., Hassan, A. A., & Kim, K. H. (2019). Nano-based smart pesticide formulations: Emerging opportunities for agriculture. *Journal of Controlled Release, 294*, 131–153. doi:10.1016/j.jconrel.2018.12.012 PMID:30552953

Kumar, S., Pavelyev, V., Mishra, P., Tripathi, N., Sharma, P., & Calle, F. (2020). A review on 2D transition metal di-chalcogenides and metal oxide nanostructures based NO_2. *Materials Science in Semiconductor Processing, 107*, 104865. doi:10.1016/j.mssp.2019.104865

Kumar, S., Tessarek, C., Sarau, G., Christiansen, S., & Singh, R. (2015). Self-Catalytic Growth of β-Ga_2O_3 Nanostructures by Chemical Vapor Deposition. *Advanced Engineering Materials, 17*(5), 709–715. doi:10.1002/adem.201400289

Kumar, V. S., Nagaraja, B. M., Shashikala, V., Padmasri, A. H., Madhavendra, S. S., Raju, B. D., & Rao, K. R. (2004). Highly efficient Ag/C catalyst prepared by electro-chemical deposition method in controlling microorganisms in water. *Journal of Molecular Catalysis A Chemical, 223*(1-2), 313–319. doi:10.1016/j.molcata.2003.09.047

Kuwata, N., Kawamura, J., Toribami, K., Hattori, T., & Sata, N. (2004). Thin-film lithium-ion battery with amorphous solid electrolyte fabricated by pulsed laser deposition. *Electrochemistry Communications, 6*(4), 417–421. doi:10.1016/j.elecom.2004.02.010

Lakouraj, M. M., Mojerlou, F., & Zare, E. N. (2014). Nanogel and superparamagnetic nanocomposite based on sodium alginate for sorption of heavy metal ions. *Carbohydrate Polymers, 106*, 34–41. doi:10.1016/j.carbpol.2014.01.092 PMID:24721048

Lal, M., Shandilya, M., Sharma, S., & Rai, R. (2017). Smart maerial nanofibers for day to day life. In R. Rai (Ed.), *Smart Materials for Smart Living* (pp. 1–66). Nova Scientific Publishers.

Laraib, U., Sargazi, S., Rahdar, A., Khatami, M., & Pandey, S. (2022). Nanotechnology-based approaches for effective detection of tumor markers: A comprehensive state-of-the-art review. *International Journal of Biological Macromolecules, 195*, 356–383. doi:10.1016/j.ijbiomac.2021.12.052 PMID:34920057

Larminie, J. (2003). *Fuel Cell Systems Explained* (2nd ed.). SAE International. doi:10.1002/9781118878330

Las-Casas, B., Dias, I. K. R., Yupanqui-Mendoza, S. L., Pereira, B., Costa, G. R., Rojas, O. J., & Arantes, V. (2023). The emergence of hybrid cellulose nanomaterials as promising biomaterials. *International Journal of Biological Macromolecules*, *250*, 126007. doi:10.1016/j.ijbiomac.2023.126007 PMID:37524277

Late, D. J., Rout, C. S., Chakravarty, D., & Ratha, S. (2015). Emerging energy applications of two-dimensional layered materials. *Canadian Chemical Transactions*, *3*, 118–157. doi:10.13179/canchemtrans.2015.03.02.0174

Launey, M. E., & Ritchie, R. O. (2009). On the fracture toughness of advanced materials. *Advanced Materials*, *21*(20), 2103–2110. doi:10.1002/adma.200803322

Laurène, W., Raúl, L., Katalin, S., Céline, F., Gilles, K., Antonio, M., Charlotte, C., & Samir, A. (2023). Design and synthesis of a new bifunctional chelating agent: Application for Al 18F/177Lu complexation. *Journal of Inorganic Biochemistry*, *246*, 112267. doi:10.1016/j.jinorgbio.2023.112267 PMID:37329775

Lavrova, D. G., Zvonarev, A. N., Alferov, V. A., Khonina, T. G., Shadrina, E. V., Alferov, S. V., & Ponamoreva, O. N. (2023). Biocompatible silica-polyethylene glycol-based composites for immobilization of microbial cells by sol-gel synthesis. *Polymers*, *15*(2), 458. doi:10.3390/polym15020458 PMID:36679338

Lee, N. A., Gilligan, G. E., & Rochford, J. (2018). Solar energy conversion. In Elsevier eBooks (pp. 881–918). doi:10.1016/B978-0-12-809270-5.00030-3

Lee, H., Lee, Y., Song, C., Cho, H. R., Ghaffari, R., Choi, T. K., Kim, K. H., Lee, Y. B., Ling, D., Lee, H., Yu, S. J., Choi, S. H., Hyeon, T., & Kim, D. H. (2015). An endoscope with integrated transparent bioelectronics and theranostic nanoparticles for colon cancer treatment. *Nature Communications*, *6*(1), 10059. doi:10.1038/ncomms10059 PMID:26616435

Lee, J. H., Huh, Y. M., Jun, Y. W., Seo, J. W., Jang, J. T., Song, H. T., Kim, S., Cho, E.-J., Yoon, H.-G., Suh, J.-S., & Cheon, J. (2007). Artificially engineered magnetic nanoparticles for ultrasensitive molecular imaging. *Nature Medicine*, *13*(1), 95–99. doi:10.1038/nm1467 PMID:17187073

Lee, N., Choi, S. H., & Hyeon, T. (2013). Nano-sized CT contrast agents. *Advanced Materials*, *25*(19), 2641–2660. doi:10.1002/adma.201300081 PMID:23553799

Liang, W., Dai, C., Zhou, X., & Zhang, Y. (2014). Application of zero-valent iron nanoparticles for the removal of aqueous zinc ions under various experimental conditions. *PLoS One*, *9*(1). https:// pubmed.ncbi.nlm.nih.gov/24416439/

Liang, C. H., Wang, C. C., Lin, Y. C., Chen, C. H., Wong, C. H., & Wu, C. Y. (2009). Iron oxide/gold core/shell nanoparticles for ultrasensitive detection of carbohydrate– protein interactions. *Analytical Chemistry*, *81*(18), 7750–7756. doi:10.1021/ac9012286 PMID:19689135

Liang, P. C. (2017). Adsorption of Radionuclide Uranium onto Carbon-Based Nanomaterials from Aqueous Systems. *Huaxue Jinzhan*, *29*(9), 1062–1071.

Liang, R., Wang, J., Wu, X., Dong, L., Deng, R., Wang, K., Sullivan, M., Liu, S., Wu, M., Tao, J., Yang, X., & Zhu, J. (2013). Multifunctional biodegradable polymer nanoparticles with uniform sizes: Generation and in vitro anti-melanoma activity. *Nanotechnology*, *24*(45), 455302. doi:10.1088/0957-4484/24/45/455302 PMID:24145641

Liau, S. Y., Read, D. C., Pugh, W. J., Furr, J. R., & Russell, A. D. (1997). Interaction of silver nitrate with readily identifiable groups: Relationship to the antibacterialaction of silver ions. *Letters in Applied Microbiology*, *25*(4), 279–283. doi:10.1046/j.1472-765X.1997.00219.x PMID:9351278

Li, C., Cao, Q., Wang, F., Xiao, Y., Li, Y., Delaunay, J. J., & Zhu, H. (2018). Engineering graphene and TMDs based van der Waals heterostructures for photovoltaic and photoelectrochemical solar energy conversion. *Chemical Society Reviews*, *47*(13), 4981–5037. doi:10.1039/C8CS00067K PMID:29736528

Li, C., Guo, Y., Li, B., Wang, C., Li, W., & Chen, B. (2005). Template synthesis of SC@C82(I) nanowires and nanotubes at room temperature. *Advanced Materials*, *17*(1), 71–73. doi:10.1002/adma.200400155

Li, H., Wu, C., Yu, X., & Zhang, W. (2023). Recent advances of PVA-based hydrogels in cartilage repair application. *Journal of Materials Research and Technology*, *24*, 2279–2298. doi:10.1016/j.jmrt.2023.03.130

Li, J. (2013). Addressing the grand challenges in energy storage. *Advanced Functional Materials*, *23*(8), 924–928. doi:10.1002/adfm.201203058

Li, J. H., Xu, Y. Y., Zhu, L. P., Wang, J. H., & Du, C. H. (2009). Fabrication and characterization of a novel TiO2 nanoparticle self-assembly membrane with improved fouling resistance. *Journal of Membrane Science*, *326*(2), 659–666. doi:10.1016/j.memsci.2008.10.049

Li, J., Duan, J., Liao, Y., Liu, X., Rong, P., Chen, G., Wan, H., & Ma, R. (2023). Surface modification of Nd3+ activated gadolinium core-shell nanospheres for near-infrared and magnetic resonance dual functional bioimaging system. *Materials & Design*, *232*, 112132. doi:10.1016/j.matdes.2023.112132

Li, J., Lu, W., Yang, Y., Xiang, R., Ling, Y., Yu, C., & Zhou, Y. (2023). Hybrid Nanomaterials for Cancer Immunotherapy. *Advancement of Science*, *10*(6), 2204932. Advance online publication. doi:10.1002/advs.202204932 PMID:36567305

Lim, E.-K., Kim, T., Paik, S., Haam, S., Huh, Y.-M., & Leec, K. (2021). Nanomaterials for Theranostics: Recent Advances and Future Challenges. *Nanomaterials and Neoplasms*, 587-775.

Lima, D., Ribicki, A., Gonçalves, L., Hacke, A. C. M., Lopes, L. C., Pereira, R. P., Wohnrath, K., Fujiwara, S. T., & Pessôa, C. A. (2022). Nanoconjugates based on a novel organic–inorganic hybrid silsesquioxane and gold nanoparticles as hemocompatible nanomaterials for promising biosensing applications. *Colloids and Surfaces. B, Biointerfaces*, *213*, 112355. doi:10.1016/j.colsurfb.2022.112355 PMID:35158220

Li, N., Martin, C. R., & Scrosati, B. (1999). A High-Rate, High-Capacity, nanostructured tin oxide electrode. *Electrochemical and Solid-State Letters*, *3*(7), 316. doi:10.1149/1.1391134

Linko, V., Zhang, H., Nonappa, N., Kostiainen, M. A., & Ikkala, O. (2022). From precision colloidal hybrid materials to advanced functional assemblies. *Accounts of Chemical Research*, *55*(13), 1785–1795. doi:10.1021/acs.accounts.2c00093 PMID:35647700

Lin, X. M., & Samia, A. C. (2006). Synthesis, assembly and physical properties of magnetic nanoparticles. *Journal of Magnetism and Magnetic Materials*, *305*(1), 100–109. doi:10.1016/j.jmmm.2005.11.042

Lin, Y. Y., Kao, H. W., Li, J. J., Hwang, J. J., Tseng, Y. L., Lin, W. J., Lin, M.-H., Ting, G., & Wang, H. E. (2013). Tumor burden talks in cancer treatment with PEGylated liposomal drugs. *PLoS One*, *8*(5), e63078. doi:10.1371/journal.pone.0063078 PMID:23675454

Lin, Y.-Q., Feng, K.-K., Lu, J.-Y., Le, J.-Q., Li, W.-L., Zhang, B.-C., Li, C.-L., Song, X., Tong, L.-W., & Shao, J.-W. (2023). CRISPR/Cas9-based application for cancer therapy: Challenges and solutions for nonviral delivery. *Journal of Controlled Release*, *361*, 727–749. doi:10.1016/j.jconrel.2023.08.028 PMID:37591461

Li, P., Wu, J., Lin, J., Huang, M., Huang, Y., & Li, Q. (2009). High-performance and low platinum loading Pt/carbon black counter electrode for dye-sensitized solar cells. *Solar Energy*, *83*(6), 845–849. doi:10.1016/j.solener.2008.11.012

Li, Q., Mahendra, S., Lyon, D. Y., Brunet, L., Liga, M. V., Li, D., & Alvarez, P. J. (2008). Antimicrobial nanomaterials for water disinfection and microbial control: Potential applications and implications. *Water Research*, *42*(18), 4591–4602. doi:10.1016/j.watres.2008.08.015 PMID:18804836

Lisha, K. P., Anshup, & Pradeep, T. (2009). Towards a practical solution for removing inorganic mercury from drinking water using gold nanoparticles. *Gold Bulletin*, *42*(2), 144–152. doi:10.1007/BF03214924

Liu, X. (2012). A survey on gas sensing technology. *Sensors, 12*(7), 9635–9665. http://www.mdpi.com/1424-8220/12/7/9635

Liu, C., Li, F., Ma, L., & Cheng, H. (2010). Advanced materials for energy storage. *Advanced Materials*, *22*(8), E28–E62. doi:10.1002/adma.200903328 PMID:20217798

Liu, F., Zhu, S., Li, D., Chen, G., & Ho, S. H. (2020). Detecting ferric iron by microalgal residuederived fluorescent nanosensor with an advanced kinetic model. *iScience*, *23*(6), 101174. doi:10.1016/j.isci.2020.101174 PMID:32498017

Liu, H. L., & Yang, T. C. K. (2003). Photocatalytic inactivation of Escherichia coli and Lactobacillus helveticus by ZnO and TiO2 activated with ultraviolet light. *Process Biochemistry (Barking, London, England)*, *39*(4), 475–481. doi:10.1016/S0032-9592(03)00084-0

Liu, H., & Li, Y. (2020). Modified carbon nanotubes for hydrogen storage at moderate pressure and room temperature. *Fullerenes, Nanotubes, and Carbon Nanostructures*, *28*(8), 663–670. doi:10.1080/1536383X.2020.1738396

Liu, J., Wang, T., Uchida, T., & Kumar, S. (2005). Carbon nanotube core-polymer shell nanofibers. *Journal of Applied Polymer Science*, *96*(5), 1992–1995. doi:10.1002/app.21662

Liu, J., Zhou, H., Yang, W., & Ariga, K. (2020). Soft nanoarchitectonics for enantioselective biosensing. *Accounts of Chemical Research*, *53*(3), 644–653. doi:10.1021/acs.accounts.9b00612 PMID:32073816

Liu, R., Duay, J., Lane, T., & Lee, S. B. (2010). Synthesis and characterization of RuO2/poly(3,4-ethylenedioxythiophene) composite nanotubes for supercapacitors. *Physical Chemistry Chemical Physics*, *12*(17), 4309. doi:10.1039/b918589p PMID:20407700

Liu, X., Cao, L., Guo, Z., Li, Y., Gao, W., & Zhou, L. (2019). A review of perovskite photovoltaic materials' synthesis and applications via chemical vapor deposition method. *Materials (Basel)*, *12*(20), 3304. doi:10.3390/ma12203304 PMID:31614476

Liu, X., Kim, J., Yeluri, R., Lal, S., Li, H., Lu, J., Keller, S., Mazumder, B., Speck, J. S., & Mishra, U. K. (2013). Fixed charge and trap states of in situ Al_2O_3 on Ga-face GaN metal-oxide-semiconductor capacitors grown by metalorganic chemical vapor deposition. *Journal of Applied Physics*, *114*(16), 164507. Advance online publication. doi:10.1063/1.4827201

Liu, X., Sheng, G., Zhong, M., & Zhou, X. (2017). Hybrid nanowires and nanoparticles of WO_3 in carbon aerogel for supercapacitor applications. *Nanoscale*, *00*, 1–9. doi:10.1039/C7NR07191D PMID:29350741

Liu, X., Zhang, Q., Knoll, W., Liedberg, B., & Wang, Y. (2020). Rational design of functional peptide–gold hybrid nanomaterials for molecular interactions. *Advanced Materials*, *32*(37), 2000866. doi:10.1002/adma.202000866 PMID:32743897

Liu, Y., Li, P., Jiang, Q., Jiang, F., Liu, J., Liu, G., Liu, C., Du, Y., & Xu, J. (2021). Organic/inorganic hybrid for flexible thermoelectric fibers. *Chemical Engineering Journal*, *405*, 126510. doi:10.1016/j.cej.2020.126510

Liu, Y., Weiss, N. O., Duan, X., Cheng, H. C., Huang, Y., & Duan, X. (2016). Van der Waals heterostructures and devices. *Nature Reviews. Materials*, *1*(9), 1–17. doi:10.1038/natrevmats.2016.42

Liu, Y., Zhang, Y., Xin, X., Xu, X., Wang, G., Gao, S., Qiao, L., Yin, S., Liu, H., Jia, C., Shen, W., Xu, L., Ji, Y., & Zhou, C. (2022). Design and Preparation of Avermectin Nanopesticide for Control and Prevention of Pine Wilt Disease. *Nanomaterials (Basel, Switzerland)*, *12*(11), 1863. doi:10.3390/nano12111863 PMID:35683719

Livani, M. J., Ghorbani, M., & Mehdipour, H. (2018). Preparation of an activated carbon from hazelnut shells and its composites with magnetic NiFe2O4 nanoparticles. *Xinxing Tan Cailiao. New Carbon Materials*, *33*(6), 578–586. doi:10.1016/S1872-5805(18)60358-0

Li, X. Q., & Zhang, W. X. (2006). Iron nanoparticles: The core–shell structure and unique properties for Ni(II) sequestration. *Langmuir*, *22*(10), 4638–4642. doi:10.1021/la060057k PMID:16649775

Li, X., Magnuson, C. W., Venugopal, A., Tromp, R. M., Hannon, J. B., Vogel, E. M., Colombo, L., & Ruoff, R. S. (2011). Large-area graphene single crystals grown by low-pressure chemical vapor deposition of methane on copper. *Journal of the American Chemical Society*, *133*(9), 2816–2819. doi:10.1021/ja109793s PMID:21309560

Li, Y., Bando, Y., & Golberg, D. (2004). Indium-Assisted growth of aligned Ultra-Long Silica nanotubes. *Advanced Materials*, *16*(1), 37–40. doi:10.1002/adma.200305732

Li, Y., Kröger, M., & Liu, W. K. (2015). Shape effect in cellular uptake of PEGylated nanoparticles: Comparison between sphere, rod, cube and disk. *Nanoscale*, *7*(40), 16631–16646. doi:10.1039/C5NR02970H PMID:26204104

Li, Z., Guo, J., Zhang, M., Li, G., & Hao, L. (2022). Gadolinium-Coated Mesoporous Silica Nanoparticle for Magnetic Resonance Imaging. *Frontiers in Chemistry*, *10*, 837032. doi:10.3389/fchem.2022.837032 PMID:35242742

Li, Z., Liu, C., Sarpong, V., & Gu, Z. (2019). Multisegment nanowire/nanoparticle hybrid arrays as electrochemical biosensors for simultaneous detection of antibiotics. *Biosensors & Bioelectronics*, *126*, 632–639. doi:10.1016/j.bios.2018.10.025 PMID:30513482

Liz-Marzán, L. M., Giersig, M., & Mulvaney, P. (1996). Synthesis of nanosized gold– silica core– shell particles. *Langmuir*, *12*(18), 4329–4335. doi:10.1021/la9601871

Lofgreen, J. E., & Ozin, G. A. (2014). Controlling morphology and porosity to improve performance of molecularly imprinted sol–gel silica. *Chemical Society Reviews*, *43*(3), 911–933. doi:10.1039/C3CS60276A PMID:24247659

Long, Q., Zhang, Z., Qi, G., Wang, Z., Chen, Y., & Liu, Z.-Q. (2020). Fabrication of Chitosan Nanofiltration Membranes by the Film Casting Strategy for Effective Removal of Dyes/Salts in Textile Wastewater. *ACS Sustainable Chemistry & Engineering*, *8*(6), 2512–2522. doi:10.1021/acssuschemeng.9b07026

Louie, A. (2010). Multimodality imaging probes: Design and challenges. *Chemical Reviews*, *110*(5), 3146–3195. doi:10.1021/cr9003538 PMID:20225900

Lovrić, J., Bazzi, H. S., Cuie, Y., Fortin, G. R., Winnik, F. M., & Maysinger, D. (2005). Differences in subcellular distribution and toxicity of green and red emitting CdTe quantum dots. *Journal of Molecular Medicine*, *83*(5), 377–385. doi:10.1007/s00109-004-0629-x PMID:15688234

Lucky, S. S., Soo, K. C., & Zhang, Y. (2015). Nanoparticles in photodynamic therapy. *Chemical Reviews*, *115*(4), 1990–2042. doi:10.1021/cr5004198 PMID:25602130

Luengo Morato, Y., Ovejero Paredes, K., Lozano Chamizo, L., Marciello, M., & Filice, M. (2021). Recent advances in multimodal molecular imaging of cancer mediated by hybrid magnetic nanoparticles. *Polymers*, *13*(17), 2989. doi:10.3390/polym13172989 PMID:34503029

Lu, M., Zhang, Z., Qiao, W., Guan, Y., Xiao, M., & Peng, C. (2010). Removal of residual contaminants in petroleum-contaminated soil by Fenton-like oxidation. *Journal of Hazardous Materials, 179*(1-3), 604–611. doi:10.1016/j.jhazmat.2010.03.046 PMID:20378246

Luo, J., & Zhu, J. (2006). Arrays of one-dimensional metal/silicon and metal/carbon nanotube heterojunctions. *Nanotechnology, 17*(11), S262–S270. doi:10.1088/0957-4484/17/11/S06

Luo, T., Lindner, L., Langer, J., Cimalla, V., Vidal, X., Hahl, F., Schreyvogel, C., Onoda, S., Ishii, S., Ohshima, T., Wang, D., Simpson, D. A., Johnson, B. C., Capelli, M., Blinder, R., & Jeske, J. (2022). Creation of nitrogen-vacancy centers in chemical vapor deposition diamond for sensing applications. *New Journal of Physics, 24*(3), 033030. doi:10.1088/1367-2630/ac58b6

Lutsey, N. (2010). *Review of technical literature and trends related to automobile mass-reduction technology.* Institute of Transportation Studies, UC Davis, Institute of Transportation Studies, Working Paper Series.

Lutz, É., & Coradi, P. C. (2022). Applications of new technologies for monitoring and predicting grains quality stored: Sensors, Internet of Things, and Artificial Intelligence. In *Measurement: Journal of the International Measurement Confederation* (Vol. 188). Elsevier B.V. doi:10.1016/j.measurement.2021.110609

Lykogianni, M., Bempelou, E., Karamaouna, F., & Aliferis, K. A. (2021). Do pesticides promote or hinder sustainability in agriculture? The challenge of sustainable use of pesticides in modern agriculture. *The Science of the Total Environment, 795*, 148625. doi:10.1016/j.scitotenv.2021.148625 PMID:34247073

Machado, T. O., Grabow, J., Sayer, C., de Araújo, P. H. H., Ehrenhard, M. L., & Wurm, F. R. (2022). Biopolymer-based nanocarriers for sustained release of agrochemicals: A review on materials and social science perspectives for a sustainable future of agri- and horticulture. In *Advances in Colloid and Interface Science* (Vol. 303). Elsevier B.V. doi:10.1016/j.cis.2022.102645

Maensiri, S., Laokul, P., & Klinkaewnarong, J. (2008). Indium oxide (in 2O3) nanoparticles using aloe vera plant extract: Synthesis and optical properties. *Journal of Optoelectronics and Advanced Materials, 10*, 161–165.

Ma, H., Yu, K., Wang, H., Liu, J., Cheng, Y. Y., Kang, Y., Wang, H., Zhang, J., & Song, K. (2023). Fabrication and detection of a novel hybrid conductive scaffold based on alginate/gelatin/carboxylated carbon nanotubes (Alg/Gel/mMWCNTs) for neural tissue engineering. *Tissue & Cell, 80*, 101995. doi:10.1016/j.tice.2022.101995 PMID:36512950

Mahajan, A., Hyun, W. J., Walker, S. B., Lewis, J. A., Francis, L. F., & Frisbie, C. D. (2015). High-resolution, high–aspect ratio conductive wires embedded in plastic substrates. *ACS Applied Materials & Interfaces, 7*(3), 1841–1847. doi:10.1021/am507539a PMID:25594812

Maharaj, D., Polineni, V., Abreu, M. M., & Gouvea, J. (2023). Molecular remission using personalized low-dose immunotherapy with minimal toxicities for poor prognosis hematological and solid tumor cancers. In A. Jewett & Y. Fong (Eds.), *NK Cells in Cancer Immunotherapy: Successes and Challenges* (pp. 427–453). Academic Press. doi:10.1016/B978-0-12-822620-9.00002-1

Mahmudul Hasan, Hasan, Reza, Islam, & Susan. (2021). Carbon dots as nano-modules for energy conversion and storage. *Materials Today. Communications, 29.*

Maiti, B., Harshitha, M., Disha, S., Badekila, A. K., Kini, S., & Rai, P. (2023). Nanovaccine. In *Nanotechnological Approaches to the Advancement of Innovations in Aquaculture* (pp. 37–65). Springer. doi:10.1007/978-3-031-15519-2_3

Mak, K. F., & Shan, J. (2016). Photonics and optoelectronics of 2D semiconductor transition metal dichalcogenides. *Nature Photonics, 10*(4), 216–226. doi:10.1038/nphoton.2015.282

Mali, S. C., Raj, S., & Trivedi, R. (2020). Nanotechnology a novel approach to enhance crop productivity. *Biochemistry and Biophysics Reports, 24*, 100821. doi:10.1016/j.bbrep.2020.100821 PMID:33015378

Mallakpour, S., Azadi, E., & Hussain, C. M. (2023). Innovative progress in functionalized carbon nanomaterials, their hybrids, and nanocomposites: Fabrication, antibacterial, biomedical, bioactivity, and biosensor applications. In S. Mallakpour & C. M. Hussain (Eds.), *Functionalized Carbon Nanomaterials for Theranostic Applications* (pp. 439–479). Elsevier. doi:10.1016/B978-0-12-824366-4.00017-0

Mallikarjuna, H. M., Ramesh, C. S., Koppad, P. G., Keshavamurthy, R., & Sethuram, D. (2017). Nanoindentation and wear behaviour of copper based hybrid composites reinforced with SiC and MWCNTs synthesized by spark plasma sintering. *Vacuum, 145*, 320–333. doi:10.1016/j.vacuum.2017.09.016

Mandal, M., Banerjee, I., & Mandal, M. (2022). Nanoparticle-mediated gene therapy as a novel strategy for the treatment of retinoblastoma. *Colloids and Surfaces. B, Biointerfaces, 220*, 112899. doi:10.1016/j.colsurfb.2022.112899 PMID:36252537

Maness, P. C., Smolinski, S., Blake, D. M., Huang, Z., Wolfrum, E. J., & Jacoby, W. A. (1999). Bactericidal activity of photocatalytic TiO2 reaction: Toward an understanding of its killing mechanism. *Applied and Environmental Microbiology, 65*(9), 4094–4098. doi:10.1128/AEM.65.9.4094-4098.1999 PMID:10473421

Mannix, A. J., Kiraly, B., Hersam, M. C., & Guisinger, N. P. (2017). Synthesis and chemistry of elemental 2D materials. *Nature Reviews. Chemistry, 1*(2), 0014. Advance online publication. doi:10.1038/s41570-016-0014

Manoharan, Y., Hosseini, S. E., Butler, B., Alzhahrani, H., Senior, B. T. F., Ashuri, T., & Krohn, J. (2019). Hydrogen Fuel Cell Vehicles; Current Status and Future Prospect. *Applied Sciences (Basel, Switzerland), 9*(11), 2296. doi:10.3390/app9112296

Mansas, C., Atfane-Karfane, L., Petit, E., Mendret, J., Brosillon, S., & Ayral, A. (2020). Functionalized ceramic nanofilter for wastewater treatment by coupling membrane separation and catalytic ozonation. *Journal of Environmental Chemical Engineering*, *8*(4), 104043. doi:10.1016/j.jece.2020.104043

Manzeli, S., Ovchinnikov, D., Pasquier, D., Yazyev, O. V., & Kis, A. (2017). 2D transition metal dichalcogenides. *Nature Reviews. Materials*, *2*(8), 1–15. doi:10.1038/natrevmats.2017.33

Maschmann, M. R., Franklin, A. D., Amama, P. B., Zakharov, D. N., Stach, E. A., Sands, T. D., & Fisher, T. S. (2006). Vertical single- and double-walled carbon nanotubes grown from modified porous anodic alumina templates. *Nanotechnology*, *17*(15), 3925–3929. doi:10.1088/0957-4484/17/15/052

Mathew, J., Joy, J., & George, S. C. (2019). Potential applications of nanotechnology in transportation: A review. *Journal of King Saud University. Science*, *31*(4), 586–594. doi:10.1016/j.jksus.2018.03.015

Mathur, V. (2013). *Effects of nanoscale inclusions on impact resistance of Kevlar-epoxy laminate composites* [Doctoral dissertation]. Wichita State University.

Mattia, D., Korneva, G., Sabur, A., Friedman, G., & Gogotsi, Y. (2007). Multifunctional carbon nanotubes with nanoparticles embedded in their walls. *Nanotechnology*, *18*(15), 155305. doi:10.1088/0957-4484/18/15/155305

Mazumder, S., Dey, R., Mitra, M. K., Mukherjee, S., & Das, G. C. (2009). Biofunctionalized quantum dots in biology and medicine. *Journal of Nanomaterials*, *2009*, 2009. doi:10.1155/2009/815734

McLain, A. A. (2019). *Photocatalytic Properties of Zinc Oxide and Graphene Nanocomposites.* Public Knowledge Project. doi:10.17307/wsc.v1i1.253

Meireles, M., Robaina, M., & Magueta, D. (2021). The Effectiveness of Environmental Taxes in Reducing CO2 Emissions in Passenger Vehicles: The Case of Mediterranean Countries. In International Journal of Environmental Research and Public Health (Vol. 18, Issue 10). doi:10.3390/ijerph18105442

Meng, J., Suo, Y., Li, J., Zheng, G., Liu, Y., Zhang, J., & Zheng, X. (2015). Nitrogen-doped graphene aerogels as anode materials for lithium-ion battery: Assembly and electrochemical properties. *Materials Letters*, *160*, 392–396. doi:10.1016/j.matlet.2015.08.024

Menzel, A., Subannajui, K., Bakhda, R., Wang, Y., Thomann, R., & Zacharias, M. (2012). Tuning the growth mechanism of ZnO nanowires by controlled carrier and reaction gas modulation in thermal CVD. *The Journal of Physical Chemistry Letters*, *3*(19), 2815–2821. doi:10.1021/jz301103s

Merino, D., Tomadoni, B., Salcedo, M. F., Mansilla, A. Y., Casalongué, C. A., & Alvarez, V. A. (2020). Nanoclay as Carriers of Bioactive Molecules Applied to Agriculture. In *Handbook of Nanomaterials and Nanocomposites for Energy and Environmental Applications* (pp. 1–22). Springer International Publishing. doi:10.1007/978-3-030-11155-7_62-1

Messah, A. D. V., & Pawitan, J. A. (2020). Next generation sequencing as rapid diagnosis of multidrug resistance tuberculosis. *Advancements in Life Sciences*, *8*(1), 30–37.

Mez-Romero, P. G. O., & Cuentas-Gallegos, K. (2005). Hybrid nanocomposite materials for energy storage and conversion applications. *Journal of Materials Science*, *4*(0), 1423–1428. doi:10.1007/s10853-005-0578-y

Mićić, O. I., & Nozik, A. J. (1996). Synthesis and characterization of binary and ternary III–V quantum dots. *Journal of Luminescence*, *70*(1-6), 95–107. doi:10.1016/0022-2313(96)00047-6

Mickymaray, S., Al Aboody, M. S., Eraqi, M. M., Alhoqail, W. A., Alothaim, A. S., Suresh, K., & Arulselvan, P. (2023). Chitosan-encapsulated nickel oxide, tin dioxide, and farnesol nanoparticles: Antimicrobial and anticancer properties in breast cancer cells. *International Journal of Biological Macromolecules*, *248*, 125799. Advance online publication. doi:10.1016/j.ijbiomac.2023.125799 PMID:37451381

Milano, A., Nasti, G., Iaffaioli, R. V., & Caponigro, F. (2007). First line targeted therapies in breast cancer: Focus on bevacizumab. *Biologics*, *1*(1), 3–10. PMID:19707343

Mir, S. H., Nagahara, L. A., Thundat, T., Mokarian-Tabari, P., Furukawa, H., & Khosla, A. (2018). Review—Organic-Inorganic Hybrid Functional Materials: An Integrated Platform for Applied Technologies. *Journal of the Electrochemical Society*, *165*(8), B3137–B3156. doi:10.1149/2.0191808jes

Mirza, Z., & Karim, S. (2021). Nanoparticles-based drug delivery and gene therapy for breast cancer: Recent advancements and future challenges. Seminars in cancer biology. *Journal of Controlled Release*, *295*, 250–267.

Mishra, A., Kumar, R., Mishra, J., Dutta, K., Ahlawat, P., Kumar, A., Dhanasekaran, S., Gupta, A. K., Sinha, S., Bishi, D. K., Gupta, P. K., & Nayak, S. (2023). Strategies facilitating the permeation of nanoparticles through blood–brain barrier: An insight toward the development of brain-targeted drug delivery system. *Journal of Drug Delivery Science and Technology*, *86*, 104694. doi:10.1016/j.jddst.2023.104694

Miyako, Pichon, Menard-Moyon, Vacchi, Lefevre, Begin-Colin, & Bianco. (2016). Design, synthesis, characterization and properties of magnetic nanoparticle nanocarbon hybrids. *Carbon*, *96*, 49-56. doi:10.1016/j.carbon.2015.09.045

Moghimi, S. M., & Szebeni, J. (2003). Stealth liposomes and long circulating nanoparticles: Critical issues in pharmacokinetics, opsonization and protein-binding properties. *Progress in Lipid Research*, *42*(6), 463–478. doi:10.1016/S0163-7827(03)00033-X PMID:14559067

Mohamed, E.F., & Do, T.-O. (2020). Synthesis of new hollow nanocomposite photocatalysts: sunlight applications for removal of gaseous organic pollutants. *J. Taiwan Ins. Chem. E. 6*(10).

Mohamed, E. F., & Awad, G. (2020). Photodegradation of gaseous toluene and disinfection of airborne microorganisms from polluted air using immobilized TiO_2 nanoparticle photocatalyst–based filter. *Environmental Science and Pollution Research International*, *27*(19), 24507–24517. doi:10.1007/s11356-020-08779-0 PMID:32307677

Mohapatra, L., Cheon, D., & Yoo, S. H. (2023). Carbon-Based Nanomaterials for Catalytic Wastewater Treatment: A Review. *Molecules (Basel, Switzerland)*, *28*(4), 1805. doi:10.3390/molecules28041805 PMID:36838793

Mondal, S., Manivasagan, P., Bharathiraja, S., Santha Moorthy, M., Kim, H. H., Seo, H., Lee, K. D., & Oh, J. (2017). Magnetic hydroxyapatite: A promising multifunctional platform for nanomedicine application. *International Journal of Nanomedicine*, *12*, 8389–8410. doi:10.2147/IJN.S147355 PMID:29200851

Morones, J. R., Elechiguerra, J. L., Camacho, A., Holt, K., Kouri, J. B., Ramírez, J. T., & Yacaman, M. J. (2005). The bactericidal effect of silver nanoparticles. *Nanotechnology*, *16*(10), 2346–2353. doi:10.1088/0957-4484/16/10/059 PMID:20818017

Morozovska, A. N., Eliseev, E. A., Svechnikov, G. S., & Kalinin, S. V. (2010). Pyroelectric response of ferroelectric nanowires: Size effect and electric energy harvesting. *Journal of Applied Physics*, *108*(4), 042009. doi:10.1063/1.3474964

Moustafa, E. B. (2021). Hybridization effect of BN and Al2O3 nanoparticles on the physical, wear, and electrical properties of aluminum AA1060 nanocomposites. *Applied Physics. A, Materials Science & Processing*, *127*(9), 724. doi:10.1007/s00339-021-04871-5

Mubeen, I., Tulaphol, S., Shengyong, L., Pan, D., Zhang, P., Sajid, M., Yan, M., & Stevens, W. R. (2021). Online measurement of 1, 2, 4-trichlorobenzene as dioxin indicator on multi-walled carbon nanotubes. *Environmental Pollution*, *268*, 115329. doi:10.1016/j.envpol.2020.115329 PMID:33137681

Muhamad Azim, M. K., Arifutzzaman, A., Saidur, R., Khandaker, M. U., & Bradley, D. A. (2022). Recent progress in emerging hybrid nanomaterials towards the energy storage and heat transfer applications: A review. *Journal of Molecular Liquids*, *360*, 119443. doi:10.1016/j.molliq.2022.119443

Mukhortova, Y. R., Pryadko, A. S., Chernozem, R. V., Pariy, I. O., Akoulina, E. A., Demianova, I. V., Zharkova, I. I., Ivanov, Y. F., Wagner, D. V., Bonartsev, A. P., Surmenev, R. A., & Surmeneva, M. A. (2022). Fabrication and characterization of a magnetic biocomposite of magnetite nanoparticles and reduced graphene oxide for biomedical applications. *Nano-Structures & Nano-Objects*, *29*, 100843. doi:10.1016/j.nanoso.2022.100843

Müller, R. H., & Peters, K. (1998). Nanosuspensions for the formulation of poorly soluble drugs: I. Preparation by a size-reduction technique. *International Journal of Pharmaceutics*, *160*(2), 229–237. doi:10.1016/S0378-5173(97)00311-6

Mutlu, Z., Debnath, B., Su, S., Li, C., Ozkan, M., Bozhilov, K. N., Lake, R. K., & Ozkan, C. S. (2018). Chemical vapor deposition and phase stability of pyrite on SiO_2. *Journal of Materials Chemistry. C, Materials for Optical and Electronic Devices*, *6*(17), 4753–4759. doi:10.1039/C8TC00584B

Naccache, R., Rodríguez, E. M., Bogdan, N., Sanz-Rodríguez, F., de la Cruz, M. D. C. I., de la Fuente, Á. J., ... Capobianco, J. A. (2012). High resolution fluorescence imaging of cancers using lanthanide ion-doped upconverting nanocrystals. *Cancers (Basel)*, *4*(4), 1067–1105. doi:10.3390/cancers4041067 PMID:24213500

Nadaf, S., Jena, G. K., Rarokar, N., Gurav, N., Ayyanar, M., Prasad, S., & Gurav, S. (2023). Biogenic and biomimetic functionalized magnetic nanosystem: Synthesis, properties, and biomedical applications. *Hybrid Advances*, *3*, 100038. doi:10.1016/j.hybadv.2023.100038

Nagtegaal, M., Küther, J., Ensling, J., Gütlich, P., & Tremel, W. (1999). Hydrothermal deposition of small α-Fe 2 O 3 (hematite) particles on ordered zirconium phosphonate multilayer SAMs on gold. *Journal of Materials Chemistry*, *9*(5), 1115–1120. doi:10.1039/a809319i

Nakamura, S., Ando, N., Sato, M., & Ishihara, M. (2020, April 30). Ultraviolet Irradiation Enhances the Microbicidal Activity of Silver Nanoparticles by Hydroxyl Radicals. *International Journal of Molecular Sciences*, *21*(9), 3204. doi:10.3390/ijms21093204 PMID:32366059

Nam, K. T., Kim, D. W., Yoo, P. J., Chiang, C. Y., Meethong, N., Hammond, P. T., Chiang, Y. M., & Belcher, A. M. (2006). Virus-Enabled synthesis and assembly of nanowires for lithium ion battery electrodes. *Science*, *312*(5775), 885–888. doi:10.1126/science.1122716 PMID:16601154

Nandi, D., Mohan, V. B., Bhowmick, A. K., & Bhattacharyya, D. (2020). Metal/metal oxide decorated graphene synthesis and application as supercapacitor: A review. *Journal of Materials Science*, *55*(15), 6375–6400. doi:10.1007/s10853-020-04475-z

Narang, A. S., & Mahato, R. I. (Eds.). (2010). *Targeted delivery of small and macromolecular drugs*. CRC Press. doi:10.1201/9781420087734

Narayanan, V., M A, G., & Rahman, A. (2019). How to 'train' your CVD to grow large-area 2D materials. *Materials Research Express*, *6*(12), 125002. doi:10.1088/2053-1591/ab5383

Naresh, V., & Lee, N. (2021). A review on biosensors and recent development of nanostructured materials-enabled biosensors. *Sensors (Basel)*, *21*(4), 1109. doi:10.3390/s21041109 PMID:33562639

Nathan, M., Golodnitsky, D., Yufit, V., Strauss, E., Ripenbein, T., Shechtman, I., Menkin, S., & Peled, E. (2005). Three-dimensional thin-film Li-ion microbatteries for autonomous MEMS. *Journal of Microelectromechanical Systems*, *14*(5), 879–885. doi:10.1109/JMEMS.2005.851860

Navabhatra, A., Maniratanachote, R., & Yingngam, B. (2022). Antiphotoaging properties of Zingiber montanum essential oil isolated by solvent-free microwave extraction against ultraviolet B-irradiated human dermal fibroblasts. *Toxicological Research*, *38*(2), 235–248. doi:10.1007/s43188-021-00107-z PMID:35419276

Navyatha, B., & Nara, S. (2021). Gold nanotheranostics: Future emblem of cancer nanomedicine. *Nanobiomedicine, 8*, 18495435211053945. doi:10.1177/18495435211053945 PMID:34737836

Nekoueian, K., Akhoundian, M., Wester, N., & Laurila, T. (2023). An ultrasensitive dopamine measurement platform based on molecularly imprinted polymer-carbon hybrid nanomaterials for *in vitro* use. *Electrochimica Acta, 445*, 142029. doi:10.1016/j.electacta.2023.142029

Nepal, J., Ahmad, W., Munsif, F., Khan, A., & Zou, Z. (2023). Advances and prospects of biochar in improving soil fertility, biochemical quality, and environmental applications. In *Frontiers in Environmental Science* (Vol. 11). Frontiers Media S.A. doi:10.3389/fenvs.2023.1114752

Nezhad, E. Z., Qu, X., Musharavati, F., Jaber, F., Appleford, M. R., Bae, S., Uzun, K., Struthers, M., Chowdhury, M. E. H., & Khandakar, A. (2021). Effects of titanium and carbon nanotubes on nano/micromechanical properties of HA/TNT/CNT nanocomposites. *Applied Surface Science, 538*, 148123. doi:10.1016/j.apsusc.2020.148123

Nguyen, K. T., & Zhao, Y. (2015). Engineered hybrid nanoparticles for on-demand diagnostics and therapeutics. *Accounts of Chemical Research, 48*(12), 3016–3025. doi:10.1021/acs.accounts.5b00316 PMID:26605438

Nickl, J. J., Schweitzer, K. K., & Luxenberg, P. (1972). Gasphasenabscheidung im system TiSiC. *Journal of the Less Common Metals, 26*(3), 335–353. doi:10.1016/0022-5088(72)90083-5

Nicks, J., Sasitharan, K., Prasad, R. R. R., Ashworth, D. J., & Foster, J. A. (2021). Metal–organic framework nanosheets: Programmable 2D materials for catalysis, sensing, electronics, and separation applications. *Advanced Functional Materials, 31*(42), 2103723. doi:10.1002/adfm.202103723

Nicole, L., Rozes, L., & Sánchez, C. (2010). Integrative approaches to hybrid multifunctional materials: From multidisciplinary research to applied technologies. *Advanced Materials, 22*(29), 3208–3214. doi:10.1002/adma.201000231 PMID:20552603

Nik Mohd Adnan, N. (2018). *Hybrid polymer-inorganic nanoparticles for biomedical applications.* UNSW Sydney.

Nishu, & Kumar, S. (2023). Smart and innovative nanotechnology applications for water purification. *Hybrid Advances, 3*, 100044. doi:10.1016/j.hybadv.2023.100044

Niu, B., Chen, Y., Zhang, L., & Tan, J. (2022). Organic–inorganic hybrid nanomaterials prepared via polymerization-induced self-assembly: Recent developments and future opportunities. *Polymer Chemistry, 13*(18), 2554–2569. doi:10.1039/D2PY00180B

Nordin, H., & Täljsten, B. (2004). Testing of hybrid FRP composite beams in bending. *Composites. Part B, Engineering, 35*(1), 27–33. doi:10.1016/j.compositesb.2003.08.010

Northfelt, D. W., Dezube, B. J., Thommes, J. A., Miller, B. J., Fischl, M. A., Friedman-Kien, A., Kaplan, L. D., Du Mond, C., Mamelok, R. D., & Henry, D. H. (1998). Pegylated-liposomal doxorubicin versus doxorubicin, bleomycin, and vincristine in the treatment of AIDS-related Kaposi's sarcoma: Results of a randomized phase III clinical trial. *Journal of Clinical Oncology*, *16*(7), 2445–2451. doi:10.1200/JCO.1998.16.7.2445 PMID:9667262

Novoselov, K. S., Geim, A. K., Morozov, S. V., Jiang, D. E., Zhang, Y., Dubonos, S. V., Grigorieva, I. V., & Firsov, A. A. (2004). Electric field effect in atomically thin carbon films. *Science*, *306*(5696), 666–669. doi:10.1126/science.1102896 PMID:15499015

Novoselov, K. S., Mishchenko, A., Carvalho, A., & Castro Neto, A. H. (2016). 2D materials and van der Waals heterostructures. *Science*, *353*(6298), aac9439. Advance online publication. doi:10.1126/science.aac9439 PMID:27471306

Nugba, B. E., Mousa, N. O., Osman, A., & El-Moneim, A. A. (2023). Nonenzymatic amperometric biosensor with anchored Ni nanoparticles for urinary glucose quantification. *Diamond and Related Materials*, *137*, 110171. doi:10.1016/j.diamond.2023.110171

Numan, A., Gill, A. A. S., Rafique, S., Guduri, M., Zhan, Y., Maddiboyina, B., Li, L., Singh, S., & Nguyen Dang, N. (2021). Rationally engineered nanosensors: A novel strategy for the detection of heavy metal ions in the environment. *Journal of Hazardous Materials*, *409*, 124493. doi:10.1016/j.jhazmat.2020.124493 PMID:33229259

O'Keeffe, C., Gannon, P., Gilson, P., Kafizas, A., Parkin, I. P., & Binions, R. (2013). Air purification by heterogeneous photocatalytic oxidation with multi-doped thin film titanium dioxide. *Thin Solid Films*, *537*, 131–136. doi:10.1016/j.tsf.2013.04.016

Obaidat, I. M., Narayanaswamy, V., Alaabed, S., Sambasivam, S., & Muralee Gopi, C. V. (2019). Principles of magnetic hyperthermia: A focus on using multifunctional hybrid magnetic nanoparticles. *Magnetochemistry (Basel, Switzerland)*, *5*(4), 67. doi:10.3390/magnetochemistry5040067

Okada, M., Okada, N., Chang, W. H., Endo, T., Ando, A., Shimizu, T., Kubo, T., Miyata, Y., & Irisawa, T. (2019). Gas-source CVD growth of atomic layered WS_2 from WF_6 and H_2S precursors with high grain size uniformity. *Scientific Reports*, *9*(1), 17678. doi:10.1038/s41598-019-54049-6 PMID:31776373

Oliveira, F. C., Carvalho, J. O., Magalhães, L. S. S. M., da Silva, J. M., Pereira, S. R., Gomes Júnior, A. L., Soares, L. M., Cariman, L. I. C., da Silva, R. I., Viana, B. C., Silva-Filho, E. C., Afewerki, S., da Cunha, H. N., Vega, M. L., Marciano, F. R., & Lobo, A. O. (2021). Biomineralization inspired engineering of nanobiomaterials promoting bone repair. *Materials Science and Engineering C*, *120*, 111776. doi:10.1016/j.msec.2020.111776 PMID:33545906

Oliveira, L. S., Avelino, K. Y. P. S., Oliveira, S. R. D. E., Lucena-Silva, N., de Oliveira, H. P., Andrade, C. A. S., & Oliveira, M. D. L. (2023). Flexible genosensors based on polypyrrole and graphene quantum dots for PML/RARα fusion gene detection: A study of acute promyelocytic leukemia in children. *Journal of Pharmaceutical and Biomedical Analysis*, *235*, 115606. doi:10.1016/j.jpba.2023.115606 PMID:37544275

Omar, S. H., Osman, R., Mamdouh, W., Abdel-Bar, H. M., & Awad, G. A. S. (2020). Bioinspired lipid-polysaccharide modified hybrid nanoparticles as a brain-targeted highly loaded carrier for a hydrophilic drug. *International Journal of Biological Macromolecules*, *165*, 483–494. doi:10.1016/j.ijbiomac.2020.09.170 PMID:32987085

Orillo, J. W., Dela Cruz, J., Agapito, L., Satimbre, P. J., & Valenzuela, I. (2014). Identification of diseases in rice plant (oryza sativa) using back propagation Artificial Neural Network. *2014 International Conference on Humanoid, Nanotechnology, Information Technology, Communication and Control, Environment and Management (HNICEM)*, 1–6. 10.1109/HNICEM.2014.7016248

ÖzkanÖzbek. (2023). Shear and Fracture Characteristics of Nano-silica and GNP Hybrid Nanoparticle Reinforced Single Lap Joints. *Journal*, *13*(3), 1970–1982.

Pal, A., & Kumar, V. (2023). AgriDet: Plant Leaf Disease severity classification using agriculture detection framework. *Engineering Applications of Artificial Intelligence*, *119*, 105754. doi:10.1016/j.engappai.2022.105754

Palacio-Vergara, M., Álvarez-Gómez, M., Gallego, J., & López, D. (2023). Biomass solvothermal treatment methodologies to obtain carbon quantum dots: A systematic review. *Talanta Open*, *8*, 100244. doi:10.1016/j.talo.2023.100244

Panda, P. K., Verma, A., Saraf, S., Raikwar, S., Da Bidla, P., & Jain, S. K. (2023). Poly(lactic acid) and poly(lactic-co-glycolic acid)-based nanocarrier systems for theranostic applications. In M. S. Hasnain, A. K. Nayak, & T. M. Aminabhavi (Eds.), *Polymeric Nanosystems* (pp. 13–36). Academic Press. doi:10.1016/B978-0-323-85656-0.00017-6

Pant, B., Pant, H. R., Barakat, N. A., Park, M., Jeon, K., Choi, Y., & Kim, H. Y. (2013). Carbon nanofibers decorated with binary semiconductor (TiO2/ZnO) nanocomposites for the effective removal of organic pollutants and the enhancement of antibacterial activities. *Ceramics International*, *39*(6), 7029–7035. doi:10.1016/j.ceramint.2013.02.041

Panyam, J., & Labhasetwar, V. (2003). Biodegradable nanoparticles for drug and gene delivery to cells and tissue. *Advanced Drug Delivery Reviews*, *55*(3), 329–347. doi:10.1016/S0169-409X(02)00228-4 PMID:12628320

Panzarini, E., Inguscio, V., Tenuzzo, B. A., Carata, E., & Dini, L. (2013). Nanomaterials and autophagy: New insights in cancer treatment. *Cancers (Basel)*, *5*(1), 296–319. doi:10.3390/cancers5010296 PMID:24216709

Park, J., Choudhary, N., Smith, J., Lee, G., Kim, M., & Choi, W. (2015). Thickness modulated MoS_2 grown by chemical vapor deposition for transparent and flexible electronic devices. *Applied Physics Letters*, *106*(1), 012104. Advance online publication. doi:10.1063/1.4905476

Park, S. S., & Ha, C. S. (2018). Hollow mesoporous functional hybrid materials: Fascinating platforms for advanced applications. *Advanced Functional Materials*, *28*(27), 1703814. doi:10.1002/adfm.201703814

Park, W., Shin, H., Choi, B., Rhim, W.-K., Na, K., & Han, D. K. (2020). Advanced hybrid nanomaterials for biomedical applications. *Progress in Materials Science*, *114*, 100686. doi:10.1016/j.pmatsci.2020.100686

Pasquini, L. (2022). Nanostructured Materials for Energy Storage and Conversion. *Nanomaterials (Basel, Switzerland)*, *12*(9), 1583. doi:10.3390/nano12091583 PMID:35564292

Pathak, R., Punetha, V. D., Bhatt, S., & Punetha, M. (2023). Carbon nanotube-based biocompatible polymer nanocomposites as an emerging tool for biomedical applications. *European Polymer Journal*, *196*, 112257. doi:10.1016/j.eurpolymj.2023.112257

Paul, A., & Srivastava, D. N. (2018). Amperometric glucose sensing at nanomolar level using MOF-encapsulated TiO2 platform. *ACS Omega*, *3*(11), 14634–14640. doi:10.1021/acsomega.8b01968 PMID:30555983

Paul, R., Chatterjee, D., Das Ghosh, L., Narayanswamy, V., Singh, M. P., Agarwal, M., Ghosh, D., Radhakrishna, M., Tiwary, C. S., Provazník, I., & Chattopadhyay, K. (2023). Synthesis, characterization and *In vitro* studies of CNT/Gd2O3 hybrid structure. *Carbon Trends*, *11*, 100272. doi:10.1016/j.cartre.2023.100272

Péchy, P., Rotzinger, F. P., Nazeeruddin, M. K., Kohle, O., Zakeeruddin, S. M., Humphry-Baker, R., & Grätzel, M. (1995). Preparation of phosphonated polypyridyl ligands to anchor transition-metal complexes on oxide surfaces: Application for the conversion of light to electricity with nanocrystalline TiO 2 films. *Journal of the Chemical Society. Chemical Communications*, *0*(1), 65–66. doi:10.1039/C39950000065

Pescaglini, A., & Iacopino, D. (2015). Metal nanoparticle–semiconductor nanowire hybrid nanostructures for plasmon-enhanced optoelectronics and sensing. *Journal of Materials Chemistry. C, Materials for Optical and Electronic Devices*, *3*(45), 11785–11800. doi:10.1039/C5TC01820J

Piumie, R. P., Power, A., Chandra, S., & Chapman, J. (2018). Graphene, electrospun membranes and granular activated carbon for eliminating heavy metals, pesticides and bacteria in water and wastewater treatment processes. *Analyst*, *143*(23), 5629–5645. doi:10.1039/C8AN00922H PMID:30357153

Pooja, P., & Chowdhury, P. (2021). Functionalized CdTe fluorescence nanosensor for the sensitive detection of water borne environmentally hazardous metal ions. *Optical Materials*, *111*, 110584. doi:10.1016/j.optmat.2020.110584

Popovtzer, R., Agrawal, A., Kotov, N. A., Popovtzer, A., Balter, J., Carey, T. E., & Kopelman, R. (2008). Targeted gold nanoparticles enable molecular CT imaging of cancer. *Nano Letters*, *8*(12), 4593–4596. doi:10.1021/nl8029114 PMID:19367807

Portet, C., Taberna, P., Simon, P., & Flahaut, E. (2006). Modification of al current Collector/Active Material Interface for power improvement of electrochemical capacitor electrodes. *Journal of the Electrochemical Society*, *153*(4), A649. doi:10.1149/1.2168298

Potara, M., Suarasan, S., Craciun, A.-M., Focsan, M., Hada, A.-M., & Astilean, S. (2022). Probing polyvinylpyrrolidone-passivated graphene oxide nanoflakes as contrast agents inside tissue-like phantoms via multimodal confocal microscopy. *Talanta*, *247*, 123581. doi:10.1016/j.talanta.2022.123581 PMID:35636370

Potdara, S. B. (2020). Sonochemical approach for synthesis of zinc oxide-poly methyl methacrylate hybrid nanoparticles and its application in corrosion inhibition. *Ultrasonics Sonochemistry*, *68*, 105200. doi:10.1016/j.ultsonch.2020.105200 PMID:32512431

Pötschke, K., Müller-Kirsch, L., Heitz, R., Sellin, R. L., Pohl, U. W., Bimberg, D., Zakharov, N., & Werner, P. (2004). Ripening of self-organized InAs quantum dots. *Physica E, Low-Dimensional Systems and Nanostructures*, *21*(2-4), 606–610. doi:10.1016/j.physe.2003.11.089

Poudel, B., Hao, Q., Ma, Y., Lan, Y., Minnich, A. J., Yu, B., Yan, X., Wang, D., Muto, A., Vashaee, D., Chen, X., Liu, J., Dresselhaus, M. S., Chen, G., & Ren, Z. (2008). High-Thermoelectric performance of nanostructured bismuth antimony telluride bulk alloys. *Science*, *320*(5876), 634–638. doi:10.1126/science.1156446 PMID:18356488

Pradhan, S., Patra, P., Das, S., Chandra, S., Mitra, S., Dey, K. K., Akbar, S., Palit, P., & Goswami, A. (2013). Photochemical modulation of biosafe manganese nanoparticles on Vigna radiata: A detailed molecular, biochemical, and biophysical study. *Environmental Science & Technology*, *47*(22), 13122–13131. doi:10.1021/es402659t PMID:24144189

Prasad, R. (2019). Nanotechnology in the life science. *Springer Nature*.

Prasad, T. N. V. K. V., Sudhakar, P., Sreenivasulu, Y., Latha, P., Munaswamy, V., Reddy, K. R., Sreeprasad, T. S., Sajanlal, P. R., & Pradeep, T. (2012). Effect of nanoscale zinc oxide particles on the germination, growth and yield of peanut. *Journal of Plant Nutrition*, *35*(6), 905–927. doi:10.1080/01904167.2012.663443

Price, L. S., Parkin, I. P., Hardy, A. M. E., Clark, R. J. H., Hibbert, T. G., & Molloy, K. C. (1999). Atmospheric Pressure Chemical Vapor Deposition of Tin Sulfides (SnS, Sn_2S_3, and SnS_2) on Glass. *Chemistry of Materials*, *11*(7), 1792–1799. doi:10.1021/cm990005z

Price, L. S., Parkin, I. P., Hibbert, T. G., & Molloy, K. C. (1998). Atmospheric Pressure CVD of SnS and SnS_2 on Glass. *Chemical Vapor Deposition*, *4*(6), 222–225. doi:10.1002/(SICI)1521-3862(199812)04:06<222::AID-CVDE222>3.0.CO;2-6

Priebe, A., & Michler, J. (2023). Review of Recent Advances in Gas-Assisted Focused Ion Beam Time-of-Flight Secondary Ion Mass Spectrometry (FIB-TOF-SIMS). *Materials (Basel)*, *16*(5), 2090. doi:10.3390/ma16052090 PMID:36903205

Qadri, S., Ganoe, A., & Haik, Y. (2009). Removal and recovery of acridine orange from solutions by use of magnetic nanoparticles. *Journal of Hazardous Materials*, *169*(1-3), 318–323. doi:10.1016/j.jhazmat.2009.03.103 PMID:19406571

Qian, L., Liu, S., Zhang, W., Chen, Y., Da Ouyang, L. H., Yan, J., & Chen, M. (2019). Enhanced reduction and adsorption of hexavalent chromium by palladium and silicon-rich biochar supported nanoscale zero-valent iron. *Journal of Colloid and Interface Science*, *533*, 428–436. doi:10.1016/j.jcis.2018.08.075 PMID:30172153

Qian, L., Shang, X., Zhang, B., Zhang, W., Su, A., Chen, Y., Da Ouyang, L. H., Yan, J., & Chen, M. (2019). Enhanced removal of Cr (VI) by silicon rich biochar-supported nanoscale zero-valent iron. *Chemosphere*, *215*, 739–745. doi:10.1016/j.chemosphere.2018.10.030 PMID:30347367

Qian, Y., Qin, C., Chen, M., & Lin, S. (2020). Nanotechnology in soil remediation - applications vs. implications. *Ecotoxicology and Environmental Safety*, *15*(201), 110815. doi:10.1016/j.ecoenv.2020.110815 PMID:32559688

Qi, K., Sun, B., Liu, S.-y., & Zhang, M. (2023). Research progress on carbon materials in tumor photothermal therapy. *Biomedicine and Pharmacotherapy*, *165*, 115070. doi:10.1016/j.biopha.2023.115070 PMID:37390711

Qin, B., Ma, H., Hossain, M., Zhong, M., Xia, Q., Li, B., & Duan, X. (2020). Substrates in the synthesis of two-dimensional materials via chemical vapor deposition. *Chemistry of Materials*, *32*(24), 10321–10347. doi:10.1021/acs.chemmater.0c03549

Qin, L., Huang, D., Xu, P., Zeng, G., Lai, C., Fu, Y., Yi, H., Li, B., Zhang, C., Cheng, M., Zhou, C., & Wen, X. (2019). In-situ deposition of gold nanoparticles onto polydopamine-decorated g-C3N4 for highly efficient reduction of nitroaromatics in environmental water purification. *Journal of Colloid and Interface Science*, *534*, 357–369. doi:10.1016/j.jcis.2018.09.051 PMID:30243177

Qiu, Y., Yu, J., Fang, G., Shi, H., Zhou, X., & Bai, X. (2008). Synthesis of Carbon/Carbon Core/Shell Nanotubes with a High Specific Surface Area. *The Journal of Physical Chemistry. C, Nanomaterials and Interfaces*, *113*(1), 61–68. doi:10.1021/jp806971e

Qi, X., Xiang, Y., Cai, E., Ge, X., Chen, X., Zhang, W., Li, Z., & Shen, J. (2023). Inorganic–organic hybrid nanomaterials for photothermal antibacterial therapy. *Coordination Chemistry Reviews*, *496*, 215426. doi:10.1016/j.ccr.2023.215426

Quadri, T. W., Fayemi, O. E., Olasunkanmi, L. O., & Ebenso, E. E. (2022). Survey of different electrochemical and analytical techniques for corrosion measurements. *Electrochemical and Analytical Techniques for Sustainable Corrosion Monitoring*, 293-323. doi:10.1016/B978-0-443-15783-7.00012-8

Qu, J., Luo, C., & Hou, J. (2011). Synthesis of ZnO nanoparticles from Zn-hyper- accumulator (Sedum alfredii Hance) plants. *Micro & Nano Letters*, *6*(3), 174–176. doi:10.1049/mnl.2011.0004

Qu, X., Alvarez, P. J., & Li, Q. (2013). Applications of nanotechnology in water and wastewater treatment. *Water Research*, *47*(12), 3931–3946. doi:10.1016/j.watres.2012.09.058 PMID:23571110

Rahimi, H., Salehiabar, M., Charmi, J., Barsbay, M., Ghaffarlou, M., Razlighi, M. R., Davaran, S., Khalilov, R., Sugiyama, M., & Nosrati, H. (2020). Harnessing nanoparticles for the efficient delivery of the CRISPR/Cas9 system. *Nano Today*, *34*, 100895. doi:10.1016/j.nantod.2020.100895

Rahman, A., Chowdhury, M. A., & Hossain, N. (2022). Green synthesis of hybrid nanoparticles for biomedical applications: A review. *Applied Surface Science Advances, 11*, 100296. https://doi.org/https://doi.org/10.1016/j.apsadv.2022.100296

Rahman, A., Chowdhury, M. A., & Hossain, N. (2022). Green synthesis of hybrid nanoparticles for biomedical applications: A review. *Applied Surface Science Advances*, *11*, 100296. doi:10.1016/j.apsadv.2022.100296

Rahn, R. O., & Landry, L. C. (1973). Ultraviolet irradiation of nucleic acids complexed with heavy atoms-II. Phosphorescence and photodimerization of DNA complexed with Ag. *Photochemistry and Photobiology*, *18*(1), 29–38. doi:10.1111/j.1751-1097.1973.tb06389.x PMID:4744137

Rajput, V., Minkina, T., Mazarji, M., Shende, S., Sushkova, S., Mandzhieva, S., Burachevskaya, M., Chaplygin, V., Singh, A., & Jatav, H. (2020). Accumulation of nanoparticles in the soil-plant systems and their effects on human health. *Annals of Agricultural Science*, *65*(2), 137–143. doi:10.1016/j.aoas.2020.08.001

Raliya, R., Biswas, P., & Tarafdar, J. C. (2015). TiO2 nanoparticle biosynthesis and its physiological effect on mung bean (Vigna radiata L.). *Biotechnology Reports (Amsterdam, Netherlands)*, *5*, 22–26. doi:10.1016/j.btre.2014.10.009 PMID:28626678

Raliya, R., & Tarafdar, J. C. (2014). Biosynthesis and characterization of zinc, magnesium and titanium nanoparticles: An eco-friendly approach. *International Nano Letters*, *4*(1), 93. doi:10.1007/s40089-014-0093-8

Ramachandran, R. P., Erkinbaev, C., & Vellaichamy, C. (2022). Smart nano-biosensors in sustainable agriculture and environmental applications. In Food, Medical, and Environmental Applications of Nanomaterials (pp. 527–542). Elsevier. doi:10.1016/B978-0-12-822858-6.00019-4

Ramadoss, J., Mariadass, M. C., Karuppasamy, J., Sonachalam, A., & Govindasamy, M. (2023). Enhancing Electrochemical Influence of Nanocomposite $Ti_3C_2T_x/MoS_2$ Layered Electrode for Hybrid Supercapacitor by Synergetic Effect, ACS. *The Journal of Physical Chemistry. C, Nanomaterials and Interfaces*, *127*(34), 16828–16840. doi:10.1021/acs.jpcc.3c04271

Ramakrishna Reddy, K. T., Koteswara Reddy, N., & Miles, R. W. (2006). Photovoltaic properties of SnS based solar cells. *Solar Energy Materials and Solar Cells*, *90*(18–19), 3041–3046. doi:10.1016/j.solmat.2006.06.012

Ramesh, M., Janani, R., Deepa, C., & Rajeshkumar, L. (2022). Nanotechnology-enabled biosensors: A review of fundamentals, design principles, materials, and applications. *Biosensors (Basel)*, *13*(1), 40. doi:10.3390/bios13010040 PMID:36671875

Rani, K. (2017). Biomedical applications of silver and gold nanoparticles: Effective and safe non-viral delivery vehicles. *Journal of Applied Biotechnology and Bioengineering*, *3*(2). Advance online publication. doi:10.15406/jabb.2017.03.00059

Ran, L., Duay, J., & Lee, S. B. (2010). Redox Exchange Induced MnO_2 Nanoparticle Enrichment in Poly(3,4-ethylenedioxythiophene) Nanowires for Electrochemical Energy Storage. *ACS Nano*, *4*(7), 4299–4307. doi:10.1021/nn1010182 PMID:20590128

Ran, L., & Lee, S. B. (2008). MnO_2/Poly(3,4-ethylenedioxythiophene) Coaxial Nanowires by One-Step Co electrodeposition for Electrochemical Energy Storage. *Journal of the American Chemical Society*, *130*(10), 2942–2943. doi:10.1021/ja7112382 PMID:18275200

Rastogi, A., Tripathi, D. K., Yadav, S., Chauhan, D. K., Živčák, M., Ghorbanpour, M., El-Sheery, N. I., & Brestic, M. (2019). Application of silicon nanoparticles in agriculture. In 3 Biotech (Vol. 9, Issue 3). Springer Verlag. doi:10.1007/s13205-019-1626-7

Ravindra, B. K., & Rajasab, A. H. (2014). A comparative study on biosynthesis of silver nanoparticles using four different fungal species. *International Journal of Pharmacy and Pharmaceutical Sciences*, *6*(1), 372–376.

Raychoudhury, T., & Scheytt, T. (2013). Potential of zerovalent iron nanoparticles for remediation of environmental organic contaminants in water: A review. *Water Sci Technol, 68*(7), 1425–1439. https://pubmed.ncbi.nlm.nih.gov/24135090/

Reddy, A. L. M., Shaijumon, M. M., Gowda, S. R., & Ajayan, P. M. (2010). Multisegmented Au-MnO 2 /Carbon Nanotube Hybrid Coaxial Arrays for High-Power Supercapacitor Applications. *The Journal of Physical Chemistry. C, Nanomaterials and Interfaces*, *114*(1), 658–663. doi:10.1021/jp908739q

Reid, S. D., Hector, A. L., Levason, W., Reid, G., Waller, B. J., & Webster, M. (2007). Thio- and seleno-ether complexes with Group 4 tetrahalides and tin tetrachloride: Preparation and use in CVD for metal chalcogenide films. *Dalton Transactions (Cambridge, England)*, *42*(42), 4769–4777. doi:10.1039/b708809d PMID:17955129

Rennie, C., Huang, Y., Siwakoti, P., Du, Z., Padula, M., Bao, G., Tuch, B. E., Xu, X., & McClements, L. (2023). *In vitro* evaluation of a hybrid drug-delivery nanosystem for fibrosis prevention in cell therapy for Type 1 diabetes. *Nanomedicine (London)*, *18*(1), 53–66. doi:10.2217/nnm-2022-0231 PMID:36938861

Revathy, R., Sajini, T., Augustine, C., & Joseph, N. (2023). Iron-based magnetic nanomaterials: Sustainable approaches of synthesis and applications. *Results in Engineering*, *18*, 101114. doi:10.1016/j.rineng.2023.101114

Reyngoud, B., Bishop, C., & Kral, M. (2015). Hybrid materials design to control creep in metallic pipes. *Materials & Design*, *84*, 25–35. doi:10.1016/j.matdes.2015.06.089

Rhodes, C. P., Long, J. W., Pettigrew, K. A., Stroud, R. M., & Rolison, D. R. (2011). Architectural integration of the components necessary for electrical energy storage on the nanoscale and in 3D. *Nanoscale, 3*(4), 1731. doi:10.1039/c0nr00731e PMID:21327256

Rho, Y. H., Kanamura, K., Fujisaki, M., Hamagami, J., Suda, S., & Umegaki, T. (2002). Preparation of Li4Ti5O12 and LiCoO2 thin film electrodes from precursors obtained by sol–gel method. *Solid State Ionics, 151*(1–4), 151–157. doi:10.1016/S0167-2738(02)00594-5

Rickerby, D. G., & Morrison, M. (2007). Nanotechnology and the environment: A European perspective. *Science and Technology of Advanced Materials, 8*(1-2), 19–24. doi:10.1016/j.stam.2006.10.002

Rico, C. M., Morales, M. I., Barrios, A. C., McCreary, R., Hong, J., Lee, W. Y., Nunez, J., Peralta-Videa, J. R., & Gardea-Torresdey, J. L. (2013). Effect of cerium oxide nanoparticles on the quality of rice (Oryza sativa L.) grains. *Journal of Agricultural and Food Chemistry, 61*(47), 11278–11285. doi:10.1021/jf404046v PMID:24188281

Ripenbein, T., Golodnitsky, D., Nathan, M., & Peled, E. (2010). Novel porous-silicon structures for 3D-interlaced microbatteries. *Electrochimica Acta, 56*(1), 37–41. doi:10.1016/j.electacta.2009.01.057

Rodríguez-Galván, A., & Contreras-Torres, F. F. (2022). Scanning tunneling microscopy of biological structures: An elusive goal for many years. *Nanomaterials (Basel, Switzerland), 12*(17), 3013. doi:10.3390/nano12173013 PMID:36080050

Rout, C. S., Late, D. J., & Morgan, H. (Eds.). (2019). *Fundamentals and Sensing Applications of 2D Materials*. Woodhead Publishing.

Roy, A. (2017). Silver Nanoparticles Synthesis from a Pharmaceutically Important Medicinal Plant Plumbago Zeylanica. *MOJ Bioequivalence & Bioavailability, 3*(5). Advance online publication. doi:10.15406/mojbb.2017.03.00046

Sagar, P., & Handa, A. (2020b). Prediction of wear resistance model for magnesium metal composite by response surface methodology using central composite design. *World Journal of Engineering*, ahead-of-print. doi:10.1108/WJE-08-2020-0379

Sagar, P., & Handa, A. (2020d). Selection of tool transverse speed considering trial run experimentations for AZ61/Tic composite developed via friction stir processing using triangular tool. *Materials Today: Proceedings, 38*, 198–203. doi:10.1016/j.matpr.2020.06.537

Sagar, P., & Handa, A. (2023). Exploring the mechanical, metallurgical, and fracture characteristics of hybrid-reinforced magnesium metal matrix composite synthesized via friction stir processing route. *Proceedings of the Institution of Mechanical Engineers, Part L: Journal of Materials: Design and Applications*. 10.1177/14644207231200640

Sagar, P., Kumar, G., & Handa, A. (2023). Progressive use of nanocomposite hydrogels materials for regeneration of damaged cartilage and their tribological mechanical properties. *Proceedings of the Institution of Mechanical Engineers, Part N: Journal of Nanomaterials, Nanoengineering and Nanosystems*. 10.1177/23977914231151487

Sagar, P., & Handa, A. (2020a). Role of tool rotational speed on the tribological characteristics of magnesium-based az61a/tic composite developed via friction stir processing route. *Journal of Achievements in Materials and Manufacturing Engineering*, *101*(2), 60–75. doi:10.5604/01.3001.0014.4921

Sagar, P., & Handa, A. (2020c). A comprehensive review of recent progress in fabrication of magnesium base composites by friction stir processing technique—A review. *AIMS Materials Science*, *7*(5), 684–704. doi:10.3934/matersci.2020.5.684

Sagar, P., Handa, A., & Kumar, G. (2021). Metallurgical, mechanical and tribological behaviour of Reinforced magnesium-based composite developed Via Friction stir processing. *Proceedings of the Institution of Mechanical Engineers. Part E, Journal of Process Mechanical Engineering*, *236*(4), 1440–1451. doi:10.1177/09544089211063099

Saha, S., & Bharadwaj, A. (2023). A step towards smart agriculture using metallic nanostructures. In *Plant Stress* (Vol. 10). Elsevier B.V. doi:10.1016/j.stress.2023.100216

Sahin, A., Ciki, B., & Karademir-Yilmaz, B. (2023). In vitro evaluation of biomaterials for neural tissue engineering. In O. Gunduz, C. B. Ustundag, & M. Sengor (Eds.), *Biomaterials for Neural Tissue Engineering* (pp. 367–415). Woodhead Publishing. doi:10.1016/B978-0-323-90554-1.00003-3

Sahoo, K., Gazi, T. R., Roy, S., & Chakraborty, I. (2023). Nanohybrids of atomically precise metal nanoclusters. *Communications Chemistry*, *6*(1), 157. doi:10.1038/s42004-023-00958-7 PMID:37495665

Saikia, K., Kakati, B. K., Boro, B., & Verma, A. (2018). Current advances and applications of fuel cell technologies. In Springer eBooks (pp. 303–337). doi:10.1007/978-981-13-1307-3_13

Salama, A., Abouzeid, R. E., Owda, M. E., Cruz-Maya, I., & Guarino, V. (2021). Cellulose–silver composites materials: Preparation and applications. In Biomolecules (Vol. 11, Issue 11). MDPI. doi:10.3390/biom11111684

Salamanca, M., Mar Peña, A. H., Prádanos, P., & Palacio, L. (2023). Forward Osmosis Application for the Removal of Emerging Contaminants from Municipal Wastewater: A Review. *Membranes*, *13*(7), 655. doi:10.3390/membranes13070655 PMID:37505021

Saleh, T. A., Shetti, N. P., Shanbhag, M. M., Raghava Reddy, K., & Aminabhavi, T. M. (2020). Recent trends in functionalized nanoparticles loaded polymeric composites: An energy application. *Materials Science for Energy Technologies*, *3*, 515–525. doi:10.1016/j.mset.2020.05.005

Samanta, A., Zhao, A., Shimizu, G. K. H., Sarkar, P., & Gupta, R. (2012). Post-Combustion CO 2 Capture Using Solid Sorbents. *RE:view*.

Sánchez, C., Ribot, F., & Lebeau, B. (1999). Molecular design of hybrid organic-inorganic nanocomposites synthesized via sol-gel chemistry. *Journal of Materials Chemistry*, *9*(1), 35–44. doi:10.1039/a805538f

Sanchis-Gual, R., Coronado-Puchau, M., Mallah, T., & Coronado, E. (2023). Hybrid nanostructures based on gold nanoparticles and functional coordination polymers: Chemistry, physics and applications in biomedicine, catalysis and magnetism. *Coordination Chemistry Reviews*, *480*, 215025. doi:10.1016/j.ccr.2023.215025

Sanità, G., Carrese, B., & Lamberti, A. (2020). Nanoparticle surface functionalization: How to improve biocompatibility and cellular internalization. *Frontiers in Molecular Biosciences*, *7*, 587012. doi:10.3389/fmolb.2020.587012 PMID:33324678

Sankaran, S., Mishra, A., Ehsani, R., & Davis, C. (2010). A review of advanced techniques for detecting plant diseases. *Computers and Electronics in Agriculture*, *72*(1), 1–13. doi:10.1016/j.compag.2010.02.007

Santhosh, C., Velmurugan, V., Jacob, G., Jeong, S. K., Grace, A. N., & Bhatnagar, A. (2016). Role of nanomaterials in water treatment applications: A review. *Chemical Engineering Journal*, *306*, 1116–1137. doi:10.1016/j.cej.2016.08.053

Santos, P., Valvez, S., Monjon, A., & Reis, P. N. B. (2020). The hybridisation effect on the viscoelastic properties of polymeric composites. *Procedia Structural Integrity*, *28*, 1816–1826. doi:10.1016/j.prostr.2020.11.004

Saravanan, R., Shankar, H., Prakash, T., Narayanan, V., & Stephen, A. (2011). ZnO/CdO composite nanorods for photocatalytic degradation of methylene blue under visible light. *Materials Chemistry and Physics*, *125*(1-2), 277–280. doi:10.1016/j.matchemphys.2010.09.030

Sargin, I., Baran, T., & Arslan, G. (2020). Environmental remediation by chitosan-carbon nanotube supported palladium nanoparticles: Conversion of toxic nitroarenes into aromatic amines, degradation of dye pollutants and green synthesis of biaryls. *Separation and Purification Technology*, *247*, 116987. doi:10.1016/j.seppur.2020.116987

Satishkumar, B. C., Vogl, E. M., Govindaraj, A., & Rao, C. N. R. (1996). The decoration of carbon nanotubes by metal nanoparticles. *Journal of Physics. D, Applied Physics*, *29*(12), 3173–3176. doi:10.1088/0022-3727/29/12/037

Savage, N., & Diallo, M. S. (2005). Nanomaterials and water purification: Opportunities and challenges. *Journal of Nanoparticle Research*, *7*(4-5), 331–342. doi:10.1007/s11051-005-7523-5

Sazonov, R., Kholodnaya, G., Ponomarev, D., Zhuravlev, M., Pyatkov, I., Konusov, F., Lapteva, O., & Gadirov, R. (2021). Physicochemical, optical, and photocatalytic properties of titanium dioxide modified with zinc oxide nanoparticles synthesized by a pulsed plasma chemical method. *Physica B, Condensed Matter*, *619*, 413208. doi:10.1016/j.physb.2021.413208

Schaibley, J. R., Yu, H., Clark, G., Rivera, P., Ross, J. S., Seyler, K. L., Yao, W., & Xu, X. (2016). Valleytronics in 2D materials. *Nature Reviews. Materials*, *1*(11), 1–15. doi:10.1038/natrevmats.2016.55

Schulman, D. S., Arnold, A. J., & Das, S. (2018). Contact engineering for 2D materials and devices. *Chemical Society Reviews*, *47*(9), 3037–3058. doi:10.1039/C7CS00828G PMID:29498729

Schwartz, D. A., Norberg, N. S., Nguyen, Q. P., Parker, J. M., & Gamelin, D. R. (2003). Magnetic quantum dots: Synthesis, spectroscopy, and magnetism of Co2+-and Ni2+-doped ZnO nanocrystals. *Journal of the American Chemical Society*, *125*(43), 13205–13218. doi:10.1021/ja036811v PMID:14570496

Seaberg, J., Montazerian, H., Hossen, M. N., Bhattacharya, R., Khademhosseini, A., & Mukherjee, P. (2021). Hybrid nanosystems for biomedical applications. *ACS Nano*, *15*(2), 2099–2142. doi:10.1021/acsnano.0c09382 PMID:33497197

Sears, J. (2005). TEX: the next generation of energy storage technology. *Telecommunications Energy Conference, INTELEC 2004*, 218–222. 10.1109/INTLEC.2004.1401469

Selvaraj, V., & Haribabu, K. (2020). Synthesis of graphene encased alumina and its application as nanofluid for cooling of heat-generating electronic devices. *Powder Technology*, *363*, 665–675. doi:10.1016/j.powtec.2020.01.050

Sen, K. (2021). Application of nanotechnology in pesticides adsorption with statistical optimization and modeling. In *Cognitive Data Models for Sustainable Environment* (pp. 65–99). Elsevier. doi:10.1016/B978-0-12-824038-0.00005-5

Shahinul Islam, M. (2023). Hybrid Nanostructured Materials as Electrodes in Energy Storage Devices. *Inorganics*, *11*(5), 183. doi:10.3390/inorganics11050183

Shaijumon, M. M., Ou, F. S., Ci, L., & Ajayan, P. M. (2008). Synthesis of hybrid nanowire arrays and their application as high power supercapacitor electrodes. *Chemical Communications*, *20*(20), 2373. doi:10.1039/b800866c PMID:18473074

Shamekhi, M. A., Mirzadeh, H., Mahdavi, H., Rabiee, A., Mohebbi-Kalhori, D., & Eslaminejad, M. B. (2019). Graphene oxide containing chitosan scaffolds for cartilage tissue engineering. *International Journal of Biological Macromolecules*, *127*, 396–405. doi:10.1016/j.ijbiomac.2019.01.020 PMID:30625354

Shan, Y., Cao, L., Muhammad, B., Xu, B., Zhao, P., Cao, C., & Huang, Q. (2020). Iron-based porous metal–organic frameworks with crop nutritional function as carriers for controlled fungicide release. *Journal of Colloid and Interface Science*, *566*, 383–393. doi:10.1016/j.jcis.2020.01.112 PMID:32018178

Shaqsi, A. Z. A., Sopian, K., & Al-Hinai, A. (2020). Review of energy storage services, applications, limitations, and benefits. *Energy Reports*, *6*, 288–306. doi:10.1016/j.egyr.2020.07.028

Sha, R., & Bhattacharyya, T. K. (2020). MoS2-based nanosensors in biomedical and environmental monitoring applications. *Electrochimica Acta*, *349*, 136370. doi:10.1016/j.electacta.2020.136370

Sharif, M., Tunio, S. A., & Bano, S. (2021). Synergistic effects of Zinc oxide nanoparticles and conventional antibiotics against methicillin resistant Staphylococcus aureus. *Advancements in Life Sciences*, *8*(2), 167–171.

Sharma, A., Chauhan, A. K., Kataria, N., & Garg, V. K. (2023). Hybrid nanomaterials for the removal of organic pollutants from wastewater. In *Hybrid Nanomaterials for Sustainable Applications* (pp. 71–102). Elsevier. doi:10.1016/B978-0-323-98371-6.00022-7

Sharma, S., & Dhadly, D. K. (2023). Nano-inspired biosensors and plant diseases: recent advances and challenges. In *Nanoparticles and Plant-Microbe Interactions: An Environmental Perspective* (pp. 135–162). Elsevier. doi:10.1016/B978-0-323-90619-7.00002-3

Sharma, S., Panwar, A. K., & Tripathi, M. (2020). Storage technologies for electric vehicles. [English Edition]. *Journal of Traffic and Transportation Engineering*, *7*(3), 340–361. doi:10.1016/j.jtte.2020.04.004

Sharmila, S., Gowri, S., Karthikeyan, C., & Faiyazuddin, M. (2023). Multifunctional hybrid nanomaterials and their antimicrobial activity. In C. M. Hussain, K. V. Anand, & S. Mallakpour (Eds.), *Antimicrobial Nanosystems* (pp. 213–225). Elsevier. doi:10.1016/B978-0-323-91156-6.00019-1

Sheberla, D., Bachman, J. C., Elias, J. S., Sun, C. J., Shao-Horn, Y., & Dincă, M. (2017). Conductive MOF electrodes for stable supercapacitors with high areal capacitance. *Nature Materials*, *16*(2), 220–224. doi:10.1038/nmat4766 PMID:27723738

Shen, Z., Wu, A., & Chen, X. (2017). Iron oxide nanoparticle based contrast agents for magnetic resonance imaging. *Molecular Pharmaceutics*, *14*(5), 1352–1364. doi:10.1021/acs.molpharmaceut.6b00839 PMID:27776215

Shephard, G. S., Stockenström, S., de Villiers, D., Engelbrecht, W. J., & Wessels, G. F. (2002). Degradation of microcystin toxins in a falling film photocatalytic reactor with immobilized titanium dioxide catalyst. *Water Research*, *36*(1), 140–146. doi:10.1016/S0043-1354(01)00213-5 PMID:11767740

Shevtsova, T., Cavallaro, G., Lazzara, G., Milioto, S., Donchak, V., Harhay, K., Korolko, S., Budkowski, A., & Stetsyshyn, Y. (2022). Temperature-responsive hybrid nanomaterials based on modified halloysite nanotubes uploaded with silver nanoparticles. *Colloids and Surfaces. A, Physicochemical and Engineering Aspects*, *641*, 128525. doi:10.1016/j.colsurfa.2022.128525

Shewale, M. V., & Daruwala, R. D. (2023). High performance deep learning architecture for early detection and classification of plant leaf disease. *Journal of Agriculture and Food Research*, *14*, 100675. Advance online publication. doi:10.1016/j.jafr.2023.100675

Sheykhbaglou, R., Sedghi, M., Shishevan, M. T., & Sharifi, R. S. (2010). Effects of nano-iron oxide particles on agronomic traits of soybean. *Notulae Scientia Biologicae*, *2*(2), 112–113. doi:10.15835/nsb224667

Shi, C., Zhu, N., Cao, Y., & Wu, P. (2015). Biosynthesis of gold nanoparticles assisted by the intracellular protein extract of Pycnoporus sanguineus and its catalysis in degradation of 4-nitroaniline. *Nanoscale Research Letters*, *10*(1), 1–8. doi:10.1186/s11671-015-0856-9 PMID:25852436

Shinde, P., & Rout, C. S. (2021). Advances in synthesis, properties and emerging applications of tin sulfides and its heterostructures. *Materials Chemistry Frontiers*, *5*(2), 516–556. doi:10.1039/D0QM00470G

Shirzadi-Ahodashti, M., Hashemi, Z., Mortazavi, Y., Khormali, K., Mortazavi-Derazkola, S., & Ebrahimzadeh, M. A. (2021). Discovery of high antibacterial and catalytic activities against multidrug resistant clinical bacteria and hazardous pollutants by biosynthesized of silver nanoparticles using Stachys inflata extract (AgNPs@SI). *Colloids and Surfaces. A, Physicochemical and Engineering Aspects*, *617*, 126383. doi:10.1016/j.colsurfa.2021.126383

Shi, Y., Fang, L., Han-Dong, P., Deng, X., & Sun, Z. (2022). Metal-Organic Framework-Derived NiSe Embedded into a Porous Multi-Heteroatom Self-Doped Carbon Matrix as a Promising Anode for Sodium-Ion Battery. *Nanomaterials (Basel, Switzerland)*, *12*(19), 3345. doi:10.3390/nano12193345 PMID:36234473

Shiyani, T., & Bagchi, T. (2020). Hybrid nanostructures for solar-energy-conversion applications. *Nanomaterials and Energy*, *9*(1), 39–46. doi:10.1680/jnaen.19.00029

Shnoudeh, A. J., Hamad, I., Abdo, R. W., Qadumii, L., Jaber, A. Y., Surchi, H. S., & Alkelany, S. Z. (2019). Synthesis, characterization, and applications of metal nanoparticles. In *Biomaterials and bionanotechnology* (pp. 527–612). Elsevier. doi:10.1016/B978-0-12-814427-5.00015-9

Silva, L. P., Silveira, A. P., Bonatto, C. C., Reis, I. G., & Milreu, P. V. (2017). Silver nanoparticles as antimicrobial agents: Past, present, and future. In *Nanostructures for antimicrobial therapy* (pp. 577–596). Elsevier. doi:10.1016/B978-0-323-46152-8.00026-3

Simon, P., & Gogotsi, Y. (2008). Materials for electrochemical capacitors. *Nature Materials*, *7*(11), 845–854. doi:10.1038/nmat2297 PMID:18956000

Singh, J., Singh, N., & Rathi, A. (2017). Facile approach to synthesize and characterization of silver nanoparticles by using mulberry leaves extract in aqueous medium and its application in antimicrobial activity. *J Nanostructures, 7*, 134–40. . doi:10.22052/jns.2017.02.007

Singh, V. V., & Malik, A. K. (2023). 3 - Approaches to produce hybrid nanomaterials for engineering applications. In *Micro and Nano Technologies* (pp. 55–69). Elsevier. https://doi.org/https://doi.org/10.1016/B978-0-323-98371-6.00013-6

Singhal, R., Echeverria, E., McIlroy, D. N., & Singh, R. N. (2021). Synthesis of hexagonal boron nitride films on silicon and sapphire substrates by low-pressure chemical vapor deposition. *Thin Solid Films*, *733*, 138812. doi:10.1016/j.tsf.2021.138812

Singh, G. P. (2018). Hydrothermal Synthesis of Hybrid Nanoparticles for Future Directions of Renewal Energy Applications, Exploring the Realms of Nature for Nanosynthesis. doi:10.1007/978-3-319-99570-0_14

Singh, J., Tripathi, N., & Mohapatra, S. (2019). Synthesis of Ag–TiO_2 hybrid nanoparticles with enhanced photocatalytic activity by a facile wet chemical method. *Nano-Structures and Nano-Objects, 18*, 100266. doi:10.1016/j.nanoso.2019.100266

Singh, M., Thakur, V., Kumar, V., Raj, M., Gupta, S., Devi, N., Upadhyay, S. K., Macho, M., Banerjee, A., Ewe, D., & Saurav, K. (2022). Silver nanoparticles and its mechanistic insight for chronic wound healing: Review on recent progress. *Molecules (Basel, Switzerland), 27*(17), 5587. doi:10.3390/molecules27175587 PMID:36080353

Singh, P. P. (2018). Environmental remediation by nanoadsorbents-based polymer nanocomposite. In C. M. Hussain & A. K. Mishra (Eds.), *New Polym. Nanocomposites Environ. Remediat* (pp. 223–241). Elsevier. doi:10.1016/B978-0-12-811033-1.00010-X

Singh, R., Altaee, A., & Gautam, S. (2020). Nanomaterials in the advancement of hydrogen energy storage. *Heliyon, 6*(7), e04487. doi:10.1016/j.heliyon.2020.e04487 PMID:32743097

Singla, M. K., Nijhawan, P., & Oberoi, A. S. (2021). Hydrogen fuel and fuel cell technology for cleaner future: A review. *Environmental Science and Pollution Research International, 28*(13), 15607–15626. doi:10.1007/s11356-020-12231-8 PMID:33538968

Singla, S., Sagar, P., & Handa, A. (2023a, March 28). Magnesium-based nanocomposites synthesized using friction stir processing: An experimental study. *Materials and Manufacturing Processes*, 1–18. doi:10.1080/10426914.2023.2195909

Singla, S., Sagar, P., Handa, A., & Kang, A. S. (2023b). Recent Advances in Magnesium-Based Metal Matrix Surface Composites Developed via Friction Stir Processing Route—An Overview. *Metallography, Microstructure, and Analysis, 0123456789*(3), 385–400. Advance online publication. doi:10.1007/s13632-023-00969-9

Sitia, L., Sevieri, M., Signati, L., Bonizzi, A., Chesi, A., Mainini, F., Corsi, F., & Mazzucchelli, S. (2022). HER-2-targeted nanoparticles for breast cancer diagnosis and treatment. *Cancers (Basel), 14*(10), 2424. Advance online publication. doi:10.3390/cancers14102424 PMID:35626028

Sivadasan, D., Sultan, M. H., Madkhali, O., Almoshari, Y., & Thangavel, N. (2021). Polymeric lipid hybrid nanoparticles (plns) as emerging drug delivery platform—A comprehensive review of their properties, preparation methods, and therapeutic applications. *Pharmaceutics, 13*(8), 1291. doi:10.3390/pharmaceutics13081291 PMID:34452251

Siwal, S. S., Zhang, Q., Devi, N., & Thakur, V. K. (2020). Carbon-Based Polymer Nanocomposite for High-Performance Energy Storage Applications. *Polymers, 12*(3), 505. doi:10.3390/polym12030505 PMID:32110927

Smith, A. J., Meek, P. E., & Liang, W. Y. (1977). Raman scattering studies of SnS2 and SnSe2. *Journal of Physics. Journal of Physics. C. Solid State Physics, 10*(8), 1321–1323. doi:10.1088/0022-3719/10/8/035

Solangi, N. H., Karri, R. R., Mubarak, N. M., & Mazari, S. A. (2023). *Mechanism of polymer composite-based nanomaterial for biomedical applications*. Advanced Industrial and Engineering Polymer Research. doi:10.1016/j.aiepr.2023.09.002

Solanki, P., Bhargava, A., Chhipa, H., Jain, N., & Panwar, J. (2015). Nano-fertilizers and their smart delivery system. In *Nanotechnologies in Food and Agriculture* (pp. 81–101). Springer International Publishing. doi:10.1007/978-3-319-14024-7_4

Soleymani, S., & Naghib, S. M. (2023). 3D and 4D printing hydroxyapatite-based scaffolds for bone tissue engineering and regeneration. *Heliyon, 9*(9), e19363. doi:10.1016/j.heliyon.2023.e19363 PMID:37662765

Song, J. G., Park, K., Park, J., & Kim, H. (2015). Vapor deposition techniques for synthesis of two-dimensional transition metal dichalcogenides. *Applied Microscopy, 45*(3), 119–125. doi:10.9729/AM.2015.45.3.119

Song, M., Liu, C., Chen, S., & Zhang, W. (2021). Nanocarrier-based drug delivery for melanoma therapeutics. *International Journal of Molecular Sciences, 22*(4), 1873. doi:10.3390/ijms22041873 PMID:33668591

Song, U., Shin, M., Lee, G., Roh, J., Kim, Y., & Lee, E. J. (2013). Functional analysis of TiO 2 nanoparticle toxicity in three plant species. *Biological Trace Element Research, 155*(1), 93–103. doi:10.1007/s12011-013-9765-x PMID:23900645

Song, X., Li, D., Li, R., Chen, Y., Huang, L., Wei, M., Pan, X., Zhou, H., & Tang, H. (2019). Ionic liquid modified inorganic nanoparticles for gaseous phenol adsorption. *Journal of Wuhan University of Technology. Materials Science Edition, 34*(4), 787–790. doi:10.1007/s11595-019-2118-7

Song, Y., Fang, G., Zhu, C., Zhu, F., Wu, S., Chen, N., Wu, T., Wang, Y., Gao, J., & Zhou, D. (2019). Zero-valent iron activated persulfate remediation of polycyclic aromatic hydrocarbon-contaminated soils: An in situ pilot-scale study. *Chemical Engineering Journal, 355*, 65–75. doi:10.1016/j.cej.2018.08.126

Sonika, S. K. V., Samanta, S., Srivastava, A. K., Biswas, S., Alsharabi, R. M., & Rajput, S. (2022). Conducting Polymer Nanocomposite for Energy Storage and Energy Harvesting Systems. *Advances in Materials Science and Engineering, 23*, 1–23. Advance online publication. doi:10.1155/2022/2266899

Soto, D., & Orozco, J. (2022). Hybrid Nanobioengineered nanomaterial-based electrochemical biosensors. *Molecules (Basel, Switzerland), 27*(12), 3841. doi:10.3390/molecules27123841 PMID:35744967

Sowmiya, P., Dhas, T. S., Inbakandan, D., Anandakumar, N., Nalini, S., Suganya, K. S. U., Remya, R. R., Karthick, V., & Kumar, C. M. V. (2023). Optically active organic and inorganic nanomaterials for biological imaging applications: A review. *Micron (Oxford, England)*, *172*, 103486. doi:10.1016/j.micron.2023.103486 PMID:37262930

Spadaro, J. A., Berger, T. J., Barranco, S. D., Chapin, S. E., & Becker, R. O. (1974). Antibacterial effects of silver electrodes with weak direct current. *Antimicrobial Agents and Chemotherapy*, *6*(5), 637–642. doi:10.1128/AAC.6.5.637 PMID:15825319

Spesivtseva, A., Gvarzhdets, N., Autlev, K., Kruchinin, E., & Kuznetsov, I. (2023). Diseases of the Neuroacanthocytosis Group: A Systematic Review of Clinical Cases and Difficulties in their Diagnosis. *Advancements in Life Sciences*, *10*(3), 335–340.

Srivatsan, T. S. (2009). A Review of: Hybrid Materials: Synthesis, Characterization, and Application, G. Kickelbick (Ed.). Materials and Manufacturing Processes, 24(10–11), 1231. https://doi.org/ doi:10.1080/10426910902984058

Staroń, A., Długosz, O., Pulit-Prociak, J., & Banach, M. (2020). Analysis of the Exposure of Organisms to the Action of Nanomaterials. *Materials (Basel)*, *13*(2), 349. doi:10.3390/ma13020349 PMID:31940903

Stefaniuk, M., Oleszczuk, P., & Ok, Y. S. (2016). Review on nano zerovalent iron (nZVI): From synthesis to environmental applications. *Chemical Engineering Journal*, *287*, 618–632. doi:10.1016/j.cej.2015.11.046

Su, Lu, & Hu. (2010). *Colloids and surfaces A: physicochemical and engineering aspects adsorption of benzene, toluene, ethylbenzene and p-xylene by NaOCloxidized carbon nanotubes.* https://doi.org/.colsurfa.2009.10.025 doi:10.1016/j

Suárez, P. L., García-Cortés, M., Fernández-Argüelles, M. T., Encinar, J. R., Valledor, M., Ferrero, F. J., Campo, J. C., & Costa-Fernández, J. M. (2019). Functionalized phosphorescent nanoparticles in (bio) chemical sensing and imaging–a review. *Analytica Chimica Acta*, *1046*, 16–31. doi:10.1016/j.aca.2018.08.018 PMID:30482294

Subrata Ghosh, S. R. (2022). Nanoparticle-enhanced multifunctional nanocarbons-recent advances on electrochemical energy storage applications. *Journal of Physics. D, Applied Physics*, *55*(41), 413001. doi:10.1088/1361-6463/ac7bb5

Su, C. (2017). Environmental implications and applications of engineered nanoscale magnetite and its hybrid nanocomposites: A review of recent literature. *Journal of Hazardous Materials*, *322*, 48–84. doi:10.1016/j.jhazmat.2016.06.060 PMID:27477792

Suh, H., Kim, G., Cho, S., Li, P., Son, D.-H., Koo, D., Lim, J., Choi, C.-S., Seok, S., & Bae, S. (2023). Comparative analysis of the synergistic effects of hybrid nanomaterial reinforcement in cementitious Composites: A perspective for pore refinement and thermal resistance. *Construction & Building Materials*, *401*, 132856. doi:10.1016/j.conbuildmat.2023.132856

Sun, L., Du, T., Hu, C., Chen, J., Lu, J., Lu, Z., & Han, H. (2017). Antibacterial activity of graphene oxide/g-C3N4 composite through photocatalytic disinfection under visible light. *ACS Sustainable Chemistry & Engineering*, *5*(10), 8693–8701. doi:10.1021/acssuschemeng.7b01431

Sun, S., Jin, C., He, W., Li, G., Zhu, H., & Huang, J. (2021). Management status of waste lithium-ion batteries in China and a complete closed-circuit recycling process. *The Science of the Total Environment*, *776*, 145913. doi:10.1016/j.scitotenv.2021.145913 PMID:33639457

Sun, S.-P., Zeng, X., Li, C., & Lemley, A. T. (2014). Enhanced heterogeneous and homogeneous Fenton-like degradation of carbamazepine by nano-Fe3O4/H2O2 with nitrilotriacetic acid. *Chemical Engineering Journal*, *244*, 44–49. doi:10.1016/j.cej.2014.01.039

Sun, T., Li, Y., Liu, Y., Deng, B., Liao, C., & Zhu, Y. (2023). Advanced scanning electron microscopy and microanalysis: Applications to nanomaterials. In Y. Yin, Y. Lu, & Y. Xia (Eds.), *Encyclopedia of Nanomaterials* (1st ed., pp. 183–209). Elsevier. doi:10.1016/B978-0-12-822425-0.00104-4

Sun, X., Cai, W., & Chen, X. (2015). Positron emission tomography imaging using radiolabeled inorganic nanomaterials. *Accounts of Chemical Research*, *48*(2), 286–294. doi:10.1021/ar500362y PMID:25635467

Sun, X.-F., Liu, B., Jing, Z., & Wang, H. (2015). Preparation and adsorption property of xylan/poly (acrylic acid) magnetic nanocomposite hydrogel adsorbent. *Carbohydrate Polymers*, *118*, 16–23. doi:10.1016/j.carbpol.2014.11.013 PMID:25542101

Taberna, P., Mitra, S., Poizot, P., Simon, P., & Tarascon, J. (2006). High rate capabilities Fe3O4-based Cu nano-architectured electrodes for lithium-ion battery applications. *Nature Materials*, *5*(7), 567–573. doi:10.1038/nmat1672 PMID:16783360

Taffelli, A., Dirè, S., Quaranta, A., & Pancheri, L. (2021). MoS2 based photodetectors: A review. *Sensors (Basel)*, *21*(8), 2758. doi:10.3390/s21082758 PMID:33919731

Tahir, M. N., Eberhardt, M., Theato, P., Faiß, S., Janshoff, A., Gorelik, T., Kolb, U., & Tremel, W. (2006b). Reactive polymers: A versatile toolbox for the immobilization of functional molecules on TiO2 nanoparticles. *Angewandte Chemie International Edition*, *45*(6), 908–912. doi:10.1002/anie.200502517 PMID:16381051

Tahir, M. N., Eberhardt, M., Therese, H. A., Kolb, U., Theato, P., Müller, W. E., Schröder, H.-C., & Tremel, W. (2006a). From Single Molecules to Nanoscopically Structured Functional Materials: Au Nanocrystal Growth on TiO2 Nanowires Controlled by Surface-Bound Silicatein. *Angewandte Chemie International Edition*, *45*(29), 4803–4809. doi:10.1002/anie.200503770 PMID:16795093

Tahir, M., Zink, N., Eberhardt, M., Therese, H. A., Faiss, S., Janshoff, A., Kolb, U., Theato, P., & Tremel, W. (2007). Hierarchical assembly of TiO2 nanoparticles on WS2 nanotubes achieved through multifunctional polymeric ligands. *Small*, *3*(5), 829–834. doi:10.1002/smll.200600663 PMID:17407103

Tamayo, L., Palza, H., Bejarano, J., & Zapata, P. A. (2018). *Polymer Composites With Metal Nanoparticles: Synthesis, Properties, and Applications. In Polymer Composites with Functionalized Nanoparticles: Synthesis, Properties, and Applications*. Elsevier Inc.

Tang, L., Tan, J., Nong, H., Liu, B., & Cheng, H. M. (2021). Chemical vapor deposition growth of two-dimensional compound materials: Controllability, material quality, and growth mechanism. *Accounts of Materials Research*, *2*(1), 36–47. doi:10.1021/accountsmr.0c00063

Tanks, J., Sharp, S., & Harris, D. (2016). Charpy Impact Testing to Assess the Quality and Durability of Unidirectional CFRP Rods. *Polymer Testing*, *51*, 63–68. Advance online publication. doi:10.1016/j.polymertesting.2016.02.009

Tan, Q., Zhao, S., Xu, T., Wang, Q., Zhang, M., Yan, L., Chen, X., & Lan, M. (2023). Inorganic nanodrug delivery systems for crossing the blood–brain barrier: Advances and challenges. *Coordination Chemistry Reviews*, *494*, 215344. doi:10.1016/j.ccr.2023.215344

Tan, X., Wang, X., Chen, C., & Sun, A. (2007). Effect of soil humic and fulvic acids, pH and ionic strength on Th(IV) sorption to TiO2 nanoparticles. *Applied Radiation and Isotopes*, *65*(4), 375–381. doi:10.1016/j.apradiso.2006.10.014 PMID:17157512

Tarasenka, N., Kornev, V., Cai, X., Li, R., Dudchik, N., Emeliyanova, O., & Tarasenko, N. (2023). Nanoceria and hybrid silver–ceria nanoparticles fabricated by liquid-mediated laser ablation as antimicrobial agents. *Nano-Structures & Nano-Objects*, *34*, 100971. doi:10.1016/j.nanoso.2023.100971

Tarkistani, M. A. M., Komalla, V., & Kayser, V. (2021). Recent advances in the use of iron–gold hybrid nanoparticles for biomedical applications. *Nanomaterials (Basel, Switzerland)*, *11*(5), 1227. doi:10.3390/nano11051227 PMID:34066549

Taylor, M. L., Wilson, R. E. Jr, Amrhein, K. D., & Huang, X. (2022). Gold Nanorod-Assisted Photothermal Therapy and Improvement Strategies. *Bioengineering (Basel, Switzerland)*, *9*(5), 200. Advance online publication. doi:10.3390/bioengineering9050200 PMID:35621478

Taymaz, H., Bircan, M. D., Kamış, H., Orhan, H., Aydoğan, Z., & Akıllı, A. (2022). Facile and green synthesis of ZnO nanoparticles for effective photocatalytic degradation of organic dyes and real textile wastewater. *International Journal of Phytoremediation*, 1–12. PMID:36437748

Tehrani, Z., Thomas, D. J., Korochkina, T., Phillips, C., Lupo, D., Lehtimäki, S., O'Mahony, J., & Gethin, D. T. (2017). Large-area printed supercapacitor technology for low-cost domestic green energy storage. *Energy*, *118*, 1313–1321. doi:10.1016/j.energy.2016.11.019

Teja, A. S., & Koh, P. Y. (2009). Synthesis, properties, and applications of magnetic iron oxide nanoparticles. *Progress in Crystal Growth and Characterization of Materials*, *55*(1-2), 22–45. doi:10.1016/j.pcrysgrow.2008.08.003

Teper, P., Sotirova, A., Mitova, V., Oleszko-Torbus, N., Utrata-Wesołek, A., Koseva, N., Kowalczuk, A., & Mendrek, B. (2020). Antimicrobial activity of hybrid nanomaterials based on star and linear polymers of N, N′-dimethylaminoethyl methacrylate with in situ produced silver nanoparticles. *Materials (Basel)*, *13*(13), 3037. doi:10.3390/ma13133037 PMID:32645957

Tewari, C., Tatrari, G., Kumar, S., Pathak, M., Rawat, K. S., Kim, Y. N., Saha, B., Jung, Y. C., Mukhopadhyay, P., & Sahoo, N. G. (2023). Can graphene-based composites and membranes solve current water purification challenges - a comprehensive review. In *Desalination* (Vol. 567). Elsevier B.V. doi:10.1016/j.desal.2023.116952

Thakur, S., Kaur, M., Lim, W. F., & Lal, M. (2020). Fabrication and characterization of electrospun ZnO nanofibers; antimicrobial assessement. *Materials Letters*, *264*, 127279. doi:10.1016/j.matlet.2019.127279

Thakur, S., Thakur, S., Shandilya, M., Lal, M., & Rai, R. (2017). Biosynthesis of nanoparticles using plant extracts. In R. Rai (Ed.), *Smart Materials for Smart Living* (pp. 365–382). Nova Scientific Publishers.

The State of Food Security and Nutrition in the World. (2022). 2022. In *The State of Food Security and Nutrition in the World 2022*. FAO. doi:10.4060/cc0639en

Tian, L., Zhang, Y., Wang, L., Geng, Q., Liu, D., Duan, L., Wang, Y., & Cui, J. (2020). Ratiometric dual signal-enhancing-based electrochemical biosensor for ultrasensitive kanamycin detection. *ACS Applied Materials & Interfaces*, *12*(47), 52713–52720. doi:10.1021/acsami.0c15898 PMID:33170623

Tohamy, H. A. S., El-Sakhawy, M., & Kamel, S. (2022). Development of Magnetite/Graphene Oxide Hydrogels from Agricultural Wastes for Water Treatment. *Journal of Renewable Materials*, *10*(7), 1889–1909. doi:10.32604/jrm.2022.019211

Tong, Y., Boldoo, T., Mr, J. H., & Cho, H. (2020). Improvement of photo-thermal energy conversion performance of MWCNT/Fe3O4 hybrid nanofluid compared to Fe_3O_4 nanofluid. *Energy*, *196*, 117086. doi:10.1016/j.energy.2020.117086

Torchilin, V. (2011). Tumor delivery of macromolecular drugs based on the EPR effect. *Advanced Drug Delivery Reviews*, *63*(3), 131–135. doi:10.1016/j.addr.2010.03.011 PMID:20304019

Townsend, A., & Gouws, R. (2022). A Comparative Review of Lead-Acid, Lithium-Ion and Ultra-Capacitor Technologies and Their Degradation Mechanisms. *Energies*, *15*(13), 4930. doi:10.3390/en15134930

Tri, Ouellet-Plamondon, Rtimi, Assadi, & Nguyen. (2019). *Methods for Synthesis of Hybrid Nanoparticles, Micro and Nano Technologies.* doi:10.1016/B978-0-12-814134-2.00003-6

Tripathi, S., Mahra, S., Tiwari, K., Rana, S., Tripathi, D. K., Sharma, S., & Sahi, S. (2023). Recent Advances and Perspectives of Nanomaterials in Agricultural Management and Associated Environmental Risk: A Review. *Nanomaterials (Basel, Switzerland)*, *13*(10), 1604. doi:10.3390/nano13101604 PMID:37242021

Truong, H. T. X., Martinez, M. J., Ochoa, O. O., & Lagoudas, D. C. (2020). Mode I fracture toughness of hybrid co-cured Al-CFRP and NiTi-CFRP interfaces: An experimental and computational study. *Composites. Part A, Applied Science and Manufacturing*, *135*, 105925. doi:10.1016/j.compositesa.2020.105925

Tseng, W. J., Chuang, Y. C., & Chen, Y. A. (2018). Mesoporous Fe3O4@ Ag@ TiO2 nanocomposite particles for magnetically recyclable photocatalysis and bactericide. *Advanced Powder Technology*, *29*(3), 664–671. doi:10.1016/j.apt.2017.12.008

Tyagi, D., Wang, H., Huang, W., Hu, L., Tang, Y., Guo, Z., Ouyang, Z., & Zhang, H. (2020). Recent advances in two-dimensional-material-based sensing technology toward health and environmental monitoring applications. *Nanoscale*, *12*(6), 3535–3559. doi:10.1039/C9NR10178K PMID:32003390

Umair Wani, T., Hamid Rather, A., Saleem Khan, R., Macossay, J., Jadhav, A. H., Srinivasappa, P. M., Abdal-hay, A., Rather, S., & Sheikh, F. A. (2023). Titanium dioxide functionalized multi-walled carbon nanotubes, and silver nanoparticles reinforced polyurethane nanofibers as a novel scaffold for tissue engineering applications. *Journal of Industrial and Engineering Chemistry*, *121*, 200–214. doi:10.1016/j.jiec.2023.01.024

Usman, M., Farooq, M., Wakeel, A., Nawaz, A., Cheema, S. A., Rehman, H., Ashraf, I., & Sanaullah, M. (2020). Nanotechnology in agriculture: Current status, challenges and future opportunities. *The Science of the Total Environment*, *721*, 137778. doi:10.1016/j.scitotenv.2020.137778 PMID:32179352

Usman, M., Faure, P., Ruby, C., & Hanna, K. (2012). Remediation of PAH-contaminated soils by magnetite catalyzed Fenton-like oxidation. *Applied Catalysis B: Environmental*, *117*, 10–17. doi:10.1016/j.apcatb.2012.01.007

Vangijzegem, T., Stanicki, D., & Laurent, S. (2019). Magnetic iron oxide nanoparticles for drug delivery: Applications and characteristics. *Expert Opinion on Drug Delivery*, *16*(1), 69–78. doi:10.1080/17425247.2019.1554647 PMID:30496697

Varghese, S. S., Varghese, S. H., Swaminathan, S., Singh, K. K., & Mittal, V. (2015). Two-dimensional materials for sensing: Graphene and beyond. *Electronics (Basel)*, *4*(3), 651–687. doi:10.3390/electronics4030651

Vaseashta, A., Vaclavikova, M., Vaseashta, S., Gallios, G., Roy, P., & Pummakarnchana, O. (2007). Nanostructures in environmental pollution detection, monitoring, and remediation. *Science and Technology of Advanced Materials*, *8*(1-2), 47–59. doi:10.1016/j.stam.2006.11.003

Vassileva, E. D., & Koseva, N. S. (2010). Sonochemically born proteinaceous micro-and nanocapsules. *Advances in Protein Chemistry and Structural Biology*, *80*, 205–252. doi:10.1016/B978-0-12-381264-3.00006-0 PMID:21109221

Vecitis, C. D., Zodrow, K. R., Kang, S., & Elimelech, M. (2010). Electronic-structure-dependent bacterial cytotoxicity of single-walled carbon nanotubes. *ACS Nano*, *4*(9), 5471–5479. doi:10.1021/nn101558x PMID:20812689

Vendrame, L. F. O., Zuchetto, T., Fagan, S. B., & Zanella, I. (2019). Nanofilter based on functionalized carbon nanostructures for the adsorption of pentachlorophenol molecules. *Computational & Theoretical Chemistry*, *1165*, 112561. doi:10.1016/j.comptc.2019.112561

Vernickaite, E., Bubniene, U., Cesiulis, H., Ramanavicius, A., & Podlaha, E. J. (2016). A Hybrid Approach to Fabricated Nanowire-Nanoparticle Composites of a Co-W Alloy and Au Nanoparticles. *Journal of the Electrochemical Society*, *163*(7), 344–348. doi:10.1149/2.1401607jes

Vivero-Escoto, J. L., & Huang, Y.-T. (2011). Inorganic-organic hybrid nanomaterials for therapeutic and diagnostic imaging applications. *International Journal of Molecular Sciences*, *12*(6), 3888–3927. doi:10.3390/ijms12063888 PMID:21747714

Walkey, C., Das, S., Seal, S., Erlichman, J., Heckman, K., Ghibelli, L., Traversa, E., McGinnis, J. F., & Self, W. T. (2015). Catalytic properties and biomedical applications of cerium oxide nanoparticles. *Environmental Science. Nano*, *2*(1), 33–53. doi:10.1039/C4EN00138A PMID:26207185

Wang, B., Qin, X., Meng, K., Zhu, L., & Li, Z. (2022). Classification of Amino Acids Using Hybrid Terahertz Spectrum and an Efficient Channel Attention Convolutional Neural Network. *Nanomaterials (Basel, Switzerland)*, *12*(12), 2114. doi:10.3390/nano12122114 PMID:35745458

Wang, D., Zhou, J., Shi, R., Wu, H., Chen, R., Duan, B., Xia, G., Xu, P., Wang, H., Zhou, S., Wang, C., Wang, H., Guo, Z., & Chen, Q. (2017). Biodegradable core-shell dual-metal-organic-frameworks nanotheranostic agent for multiple imaging guided combination cancer therapy. *Theranostics*, *7*(18), 4605–4617. doi:10.7150/thno.20363 PMID:29158848

Wang, H., & Dai, H. (2013). strongly coupled inorganic–nano-carbon hybrid materials for energy storage. *Chemical Society Reviews*, *42*(7), 3088–3113. doi:10.1039/c2cs35307e PMID:23361617

Wang, J., & Lin, Y. (2008). Functionalized carbon nanotubes and nanofibers for biosensing applications. *Trends in Analytical Chemistry*, *27*(7), 619–626. doi:10.1016/j.trac.2008.05.009 PMID:19122842

Wang, J., Ni, Q., Wang, Y., Zhang, Y., He, H., Gao, D., Ma, X., & Liang, X.-J. (2021). Nanoscale drug delivery systems for controllable drug behaviors by multi-stage barrier penetration. *Journal of Controlled Release*, *331*, 282–295. doi:10.1016/j.jconrel.2020.08.045 PMID:32866590

Wang, J., Shao, X., Zhang, Q., Tian, G., Ji, X., & Bao, W. (2017). Preparation of mesoporous magnetic Fe2O3 nanoparticle and its application for organic dyes removal. *Journal of Molecular Liquids*, *248*, 13–18. doi:10.1016/j.molliq.2017.10.026

Wang, J., Wu, X., Shen, P., Wang, J., Shen, Y., Shen, Y., Webster, T. J., & Deng, J. (2020). Applications of inorganic nanomaterials in photothermal therapy based on combinational cancer treatment. *International Journal of Nanomedicine*, *15*, 1903–1914. doi:10.2147/IJN.S239751 PMID:32256067

Wang, P., Li, N., Li, S., & Zhang, Y. (2022). Strategies for preparing hybrid nanomaterials via polymerization-induced self-assembly. *European Polymer Journal*, *172*, 111234. doi:10.1016/j.eurpolymj.2022.111234

Wang, W., Liu, M., Shafiq, M., Li, H., Hashim, R., El-Newehy, M., EL-Hamshary, H., Morsi, Y., & Mo, X. (2023). Synthesis of oxidized sodium alginate and its electrospun bio-hybrids with zinc oxide nanoparticles to promote wound healing. *International Journal of Biological Macromolecules*, *232*, 123480. doi:10.1016/j.ijbiomac.2023.123480 PMID:36720331

Wang, X., Kong, L., Zhou, S., Ma, C., Lin, W., Sun, X., Kirsanov, D., Legin, A., Wan, H., & Wang, P. (2021). Development of QDs-based nanosensors for heavy metal detection: A review on transducer principles and in-situ detection. *Talanta*, *122903*. Advance online publication. doi:10.1016/j.talanta.2021.122903 PMID:34857381

Wang, X., Zheng, K., Cheng, W., Li, J., Liang, X., Shen, J., Dou, D., Yin, M., & Yan, S. (2021). Field application of star polymer-delivered chitosan to amplify plant defense against potato late blight. *Chemical Engineering Journal*, *417*, 129327. doi:10.1016/j.cej.2021.129327

Wang, Y., & Herron, N. (1991). Nanometer-sized semiconductor clusters: Materials synthesis, quantum size effects, and photophysical properties. *Journal of Physical Chemistry*, *95*(2), 525–532. doi:10.1021/j100155a009

Wang, Y., Qi, W., Mao, Z., Wang, J., Zhao, R. C., & Chen, H. (2023). rPDAs doped antibacterial MOF-hydrogel: Bio-inspired synergistic whole-process wound healing. *Materials Today Nano*, *23*, 100363. doi:10.1016/j.mtnano.2023.100363

Wang, Y., Wang, X., Ma, X., Chen, Q., He, H., Nau, W. M., & Huang, F. (2019). Coassembly of gold nanoclusters with nucleic acids: Sensing, bioimaging, and gene transfection. *Particle & Particle Systems Characterization*, *36*(10), 1900281. doi:10.1002/ppsc.201900281

Wang, Y., Wang, Y., Hosono, E., Wang, K., & Zhou, H. (2008). The Design of a $LiFePO_4$/Carbon Nanocomposite With a Core–Shell Structure and Its Synthesis by an In Situ Polymerization Restriction Method. *Angewandte Chemie International Edition*, *47*(39), 7461–7465. doi:10.1002/anie.200802539 PMID:18720357

Wang, Y., Zhang, S. T., Li, L., Yang, X., Lu, L., & Li, D. (2023). Dopant-free passivating contacts for crystalline silicon solar cells: Progress and prospects. *EcoMat*, *5*(2), e12292. doi:10.1002/eom2.12292

Wang, Y., Zhou, Z., Yang, Z., Chen, X., Xu, D., & Zhang, Y. (2009). Gas sensors based on deposited single-walled carbon nanotube networks for DMMP detection. *Nanotechnology*, *20*(34), 345502. doi:10.1088/0957-4484/20/34/345502 PMID:19652278

Weber, W. J. Jr. (2002). Distributed optimal technology networks: A concept and strategy for potable water sustainability. *Water Science and Technology*, *46*(6-7), 241–246. doi:10.2166/wst.2002.0685 PMID:12380997

Wei, C., Lin, W. Y., Zainal, Z., Williams, N. E., Zhu, K., Kruzic, A. P., Smith, R. L., & Rajeshwar, K. (1994). Bactericidal activity of TiO2 photocatalyst in aqueous media: Toward a solar-assisted water disinfection system. *Environmental Science & Technology*, *28*(5), 934–938. doi:10.1021/es00054a027 PMID:22191837

Wei, R., Hu, J., Zhou, T., Zhou, X., Liu, J., & Li, J. (2014). Ultrathin SnS_2 nanosheets with exposed {001} facets and enhanced photocatalytic properties. *Acta Materialia*, *66*, 163–171. doi:10.1016/j.actamat.2013.11.076

Weng, R., Chen, L., Lin, S., Zhang, H., Wu, H., Liu, K., Cao, S., & Huang, L. (2017). Preparation and Characterization of Antibacterial Cellulose/chitosan Nanofiltration Membranes. *Polymers*, *9*(12), 1–13. doi:10.3390/polym9040116 PMID:30970796

White, J. M. L., Powell, A., Brady, K., & Russell-Jones, R. (2003). Severe generalized argyria secondary to ingestion of colloidal silver protein. *Clinical and Experimental Dermatology*, *28*(3), 254–256. doi:10.1046/j.1365-2230.2003.01214.x PMID:12780705

Whittingham, M. S. (2004). Lithium batteries and cathode materials. *Chemical Reviews*, *104*(10), 4271–4302. doi:10.1021/cr020731c PMID:15669156

Wibowo, Y. G., Ramadan, B. S., Taher, T., & Khairurrijal, K. (2023). Advancements of Nanotechnology and Nanomaterials in Environmental and Human Protection for Combatting the COVID-19 During and Post-pandemic Era: A Comprehensive Scientific Review. *Biomedical Materials & Devices*, 1–24. doi:10.1007/s44174-023-00086-9 PMID:37363141

Wick, P., Manser, P., Limbach, L. K., Dettlaff-Weglikowska, U., Krumeich, F., Roth, S., Stark, W., & Bruinink, A. (2007). The degree and kind of agglomeration affect carbon nanotube cytotoxicity. *Toxicology Letters*, *168*(2), 121–131. doi:10.1016/j.toxlet.2006.08.019 PMID:17169512

Wickramaratne & Jaroniec. (2013). *Activated Carbon Spheres for CO 2 Adsorption*. Academic Press.

Wiggers, H., Sehlleier, Y. H., Kunze, F., Xiao, L., Schnurre, S. M., & Schulz, C. (2020). Self-assembled nano-silicon/graphite hybrid embedded in a conductive polyaniline matrix for the performance enhancement of industrial applicable lithium-ion battery anodes. *Solid State Ionics*, *344*, 115117. doi:10.1016/j.ssi.2019.115117

William, W. Y., Chang, E., Drezek, R., & Colvin, V. L. (2006). Water-soluble quantum dots for biomedical applications. *Biochemical and Biophysical Research Communications*, *348*(3), 781–786. doi:10.1016/j.bbrc.2006.07.160 PMID:16904647

Willis, A. L., Turro, N. J., & O'Brien, S. (2005). Spectroscopic characterization of the surface of iron oxide nanocrystals. *Chemistry of Materials*, *17*(24), 5970–5975. doi:10.1021/cm051370v

Wright, C., Powell, L., Johnson, D., & Hilal, N. (2013). MICROSCOPY | Atomic Force Microscopy. Encyclopedia of Food Microbiology (Second Edition), 666-675. doi:10.1016/B978-0-12-384730-0.00217-2

Wu, L. J., You, Q. S., Duan, J. L., Luo, Y. X., Liu, L. J., Li, X., Gao, Q., Zhu, H. P., He, Y., Xu, L., Jonas, J. B., Wang, W., & Guo, X. H. (2015). Prevalence and associated factors of myopia in high-school students in Beijing. *PLoS One*, *10*(3), e0120764. doi:10.1371/journal.pone.0120764 PMID:25803875

Wu, W., He, Q., & Jiang, C. (2008). Magnetic iron oxide nanoparticles: Synthesis and surface functionalization strategies. *Nanoscale Research Letters*, *3*(11), 397–415. doi:10.1007/s11671-008-9174-9 PMID:21749733

Wu, W., Wang, L., Li, Y., Zhang, F., Lin, L., Niu, S., Chenet, D., Zhang, X., Hao, Y., Heinz, T. F., Hone, J., & Wang, Z. L. (2014). Piezoelectricity of single-atomic-layer MoS_2 for energy conversion and piezotronics. *Nature*, *514*(7523), 470–474. doi:10.1038/nature13792 PMID:25317560

Xia, F., Wang, H., Xiao, D., Dubey, M., & Ramasubramaniam, A. (2014). Two-dimensional material nanophotonics. *Nature Photonics*, *8*(12), 899–907. doi:10.1038/nphoton.2014.271

Xia, H., Li, L., & Ceder, G. (2006). Li diffusion in LiCoO2 thin films prepared by pulsed laser deposition. *Journal of Power Sources*, *159*(2), 1422–1427. doi:10.1016/j.jpowsour.2005.12.012

Xiang, Y., Zhang, G., Chen, C., Liu, B., Cai, D., & Wu, Z. (2018). Fabrication of a pH-responsively controlled-release pesticide using an attapulgite-based hydrogel. *ACS Sustainable Chemistry & Engineering*, *6*(1), 1192–1201. doi:10.1021/acssuschemeng.7b03469

Xia, T., Kovochich, M., Liong, M., Mädler, L., Gilbert, B., Shi, H., Yeh, J. I., Zink, J. I., & Nel, A. E. (2008). Comparison of the mechanism of toxicity of zinc oxide and cerium oxide nanoparticles based on dissolution and oxidative stress properties. *ACS Nano*, *2*(10), 2121–2134. doi:10.1021/nn800511k PMID:19206459

Xia, Y. (2014). Are we entering the nano era? *Angewandte Chemie International Edition*, *53*(46), 12268–12271. doi:10.1002/anie.201409223 PMID:25256914

Xiu, Z. M., Ma, J., & Alvarez, P. J. (2011). Differential effect of common ligands and molecular oxygen on antimicrobial activity of silver nanoparticles versus silver ions. *Environmental Science & Technology*, *45*(20), 9003–9008. doi:10.1021/es201918f PMID:21950450

Xiu, Z. M., Zhang, Q. B., Puppala, H. L., Colvin, V. L., & Alvarez, P. J. (2012). Negligible particle-specific antibacterial activity of silver nanoparticles. *Nano Letters*, *12*(8), 4271–4275. doi:10.1021/nl301934w PMID:22765771

Xu, Niu, Li, Yang, Gao, Ding, Ni, Zhu, Liu, Tang, Lv, Peng, Hu, Zhou, & Xu. (2022). Recent progress of quantum dots for energy storage applications. *Carbon Neutrality*, 1-13. doi:10.1007/s43979-022-00002-y

Xue, X., Cheng, K. W. E., & Sutanto, D. (2006). A study of the status and future of superconducting magnetic energy storage in power systems. *Superconductor Science and Technology*, *19*(6), R31–R39. doi:10.1088/0953-2048/19/6/R01

Xu, M., Mao, C., Chen, H., Liu, L., Wang, Y., Hussain, A., Li, S., Zhang, X., Tuguntaev, R. G., Liang, X.-J., Guo, W., & Cao, F. (2022). Osteopontin targeted theranostic nanoprobes for laser-induced synergistic regression of vulnerable atherosclerotic plaques. *Acta Pharmaceutica Sinica. B*, *12*(4), 2014–2028. doi:10.1016/j.apsb.2021.12.020 PMID:35847489

Xu, P., Zeng, G. M., Huang, D. L., Feng, C. L., Hu, S., Zhao, M. H., Lai, C., Wei, Z., Huang, C., Xie, G. X., & Liu, Z. F. (2012). Use of iron oxide nanomaterials in wastewater treatment: A review. *The Science of the Total Environment*, *424*, 1–10. doi:10.1016/j.scitotenv.2012.02.023 PMID:22391097

Xu, S., Qin, Y., Xu, C., Wei, Y., Yang, R., & Wang, Z. L. (2010). Self-powered nanowire devices. *Nature Nanotechnology*, *5*(5), 366–373. doi:10.1038/nnano.2010.46 PMID:20348913

Xu, T., Wang, Y., Xue, Y., Li, J., & Wan, Y. (2023). MXenes@metal-organic framework hybrids for energy storage and electrocatalytic application: Insights into recent advances. *Chemical Engineering Journal*, *470*, 144247. doi:10.1016/j.cej.2023.144247

Xu, W., Gao, Y., Ming, W., He, F., Li, J., Zhu, X., Kang, F., Li, J., & Wei, G. (2020). Suppressing Defects-Induced Nonradiative Recombination for Efficient Perovskite Solar Cells through Green Antisolvent Engineering. *Advanced Materials*, *32*(38), 2003965. Advance online publication. doi:10.1002/adma.202003965 PMID:32767422

Xu, Y., Xu, C., Huang, Q., Cao, L., Teng, F., Zhao, P., & Jia, M. (2021). Size effect of mesoporous silica nanoparticles on pesticide loading, release, and delivery in cucumber plants. *Applied Sciences (Basel, Switzerland)*, *11*(2), 575. doi:10.3390/app11020575

Xu, Y., Xu, H., Wang, L., Yan, J., Li, H., Song, Y., Huang, L., & Cai, G. (2013). The CNT modified white C 3 N 4 composite photocatalyst with enhanced visible-light response photoactivity. *Dalton Transactions (Cambridge, England)*, *42*(21), 7604–7613. doi:10.1039/c3dt32871f PMID:23535886

Xu, Z., Xu, Z., Gu, J., Zhou, J., Sha, G., Huang, Y., Wang, T., Fan, L., Zhang, Y., & Xi, J. (2023). In situ formation of ferrous sulfide in glycyrrhizic acid hydrogels to promote healing of multi-drug resistant Staphylococcus aureus-infected diabetic wounds. *Journal of Colloid and Interface Science*, *650*, 1918–1929. doi:10.1016/j.jcis.2023.07.141 PMID:37517191

Yadav, K., Sahu, K. K., Sucheta, Gnanakani, S. P. E., Sure, P., Vijayalakshmi, R., Sundar, V. D., Sharma, V., Antil, R., Jha, M., Minz, S., Bagchi, A., & Pradhan, M. (2023). Biomedical applications of nanomaterials in the advancement of nucleic acid therapy: Mechanistic challenges, delivery strategies, and therapeutic applications. *International Journal of Biological Macromolecules*, *241*, 124582. doi:10.1016/j.ijbiomac.2023.124582 PMID:37116843

Yamada, H. (2009). Secondary Batteries – Lithium Rechargeable Systems – Lithium-Ion | Lithium Vanadium Oxide/Niobium Oxide Batteries. In Elsevier eBooks (pp. 368–374). doi:10.1016/B978-044452745-5.00220-3

Yamanaka, M., Hara, K., & Kudo, J. (2005). Bactericidal actions of a silver ion solution on Escherichia coli, studied by energy-filtering transmission electron microscopy and proteomic analysis. *Applied and Environmental Microbiology*, *71*(11), 7589–7593. doi:10.1128/AEM.71.11.7589-7593.2005 PMID:16269810

Yang, C., Florent, M., de Falco, G., Fan, H., & Bandosz, T. J. (2020). ZnFe2O4/activated carbon as a regenerable adsorbent for catalytic removal of H2S from air at room temperature. *Chemical Engineering Journal*, *394*, 124906. doi:10.1016/j.cej.2020.124906

Yang, D., Cui, B., Wang, C., Zhao, X., Zeng, Z., Wang, Y., Sun, C., Liu, G., & Cui, H. (2017). Preparation and characterization of emamectin benzoate solid nanodispersion. *Journal of Nanomaterials*, *2017*, 2017. doi:10.1155/2017/6560780

Yang, G., Liu, F., Zhao, J., Fu, L., Gu, Y., Qu, L., Zhu, C., Zhu, J.-J., & Lin, Y. (2023). MXenes-based nanomaterials for biosensing and biomedicine. *Coordination Chemistry Reviews*, *479*, 215002. doi:10.1016/j.ccr.2022.215002

Yang, H., Huang, Y., Song, B., Kainer, K. U., & Dieringa, H. (2019). Enhancing the creep resistance of AlN/Al nanoparticles reinforced Mg-2.85Nd-0.92Gd-0.41Zr-0.29Zn alloy by a high shear dispersion technique. *Materials Science and Engineering A*, *755*, 18–27. doi:10.1016/j.msea.2019.03.131

Yang, J., Grzech, A., Mulder, F. M., & Dingemans, T. J. (2011). Methyl modified MOF-5: A water stable hydrogen storage material. *Chemical Communications*, *47*(18), 5244. doi:10.1039/c1cc11054c PMID:21451855

Yang, Y., Guo, W., Pradel, K. C., Zhu, G., Zhou, Y., Zhang, Y., Hu, Y., Lin, L., & Wang, Z. L. (2012). Pyroelectric nanogenerators for harvesting thermoelectric energy. *Nano Letters*, *12*(6), 2833–2838. doi:10.1021/nl3003039 PMID:22545631

Yang, Z., Zhang, J., Kintner-Meyer, M. C., Lu, X., Choi, D., Lemmon, J. P., & Li, J. (2011b). Electrochemical energy storage for green grid. *Chemical Reviews*, *111*(5), 3577–3613. doi:10.1021/cr100290v PMID:21375330

Yan, J., Qiu, W., Wu, G., Heremans, P., & Chen, H. (2018). Recent progress in 2D/quasi-2D layered metal halide perovskites for solar cells. *Journal of Materials Chemistry. A, Materials for Energy and Sustainability*, *6*(24), 11063–11077. doi:10.1039/C8TA02288G

Yan, S., Hu, Q., Li, J., Chao, Z., Cai, C., Yin, M., Du, X., & Shen, J. (2019). A star polycation acts as a drug nanocarrier to improve the toxicity and persistence of botanical pesticides. *ACS Sustainable Chemistry & Engineering*, *7*(20), 17406–17413. doi:10.1021/acssuschemeng.9b04567

Yan, S., Hu, Q., Wei, Y., Jiang, Q., Yin, M., Dong, M., Shen, J., & Du, X. (2022). Calcium nutrition nanoagent rescues tomatoes from mosaic virus disease by accelerating calcium transport and activating antiviral immunity. *Frontiers in Plant Science*, *13*, 1092774. doi:10.3389/fpls.2022.1092774 PMID:36561462

Yan, S., Qian, J., Cai, C., Ma, Z., Li, J., Yin, M., Ren, B., & Shen, J. (2020a). Spray method application of transdermal dsRNA delivery system for efficient gene silencing and pest control on soybean aphid Aphis glycines. *Journal of Pest Science*, *93*(1), 449–459. doi:10.1007/s10340-019-01157-x

Yan, S., Ren, B. Y., & Shen, J. (2021). Nanoparticle-mediated double-stranded RNA delivery system: A promising approach for sustainable pest management. *Insect Science*, *28*(1), 21–34. doi:10.1111/1744-7917.12822 PMID:32478473

Yan, S., Ren, B., Zeng, B., & Shen, J. (2020b). Improving RNAi efficiency for pest control in crop species. *BioTechniques*, *68*(5), 283–290. doi:10.2144/btn-2019-0171 PMID:32202134

Yao, L., Yang, B., Cui, H., Zhuang, J., Ye, J., & Xue, J. (2016). Challenges and progresses of energy storage technology and its application in power systems. *Journal of Modern Power Systems and Clean Energy*, *4*(4), 519–528. doi:10.1007/s40565-016-0248-x

Yao, Y., Zhou, Y., Liu, L., Xu, Y., Chen, Q., Wang, Y., Wu, S., Deng, Y., Zhang, J., & Shao, A. (2020). Nanoparticle-based drug delivery in cancer therapy and its role in overcoming drug resistance. *Frontiers in Molecular Biosciences*, *7*, 193. doi:10.3389/fmolb.2020.00193 PMID:32974385

Yasmeen, F., Raja, N. I., Razzaq, A., & Komatsu, S. (2017). Proteomic and physiological analyses of wheat seeds exposed to copper and iron nanoparticles. *Biochimica et Biophysica Acta (BBA)-. Biochimica et Biophysica Acta. Proteins and Proteomics*, *1865*(1), 28–42. doi:10.1016/j.bbapap.2016.10.001 PMID:27717896

Yayehrad, A. T., Siraj, E. A., Matsabisa, M., & Birhanu, G. (2023). 3D printed drug loaded nanomaterials for wound healing applications. *Regenerative Therapy*, *24*, 361–376. doi:10.1016/j.reth.2023.08.007 PMID:37692197

Ye, G., Gong, Y., Lei, S., He, Y., Li, B., Zhang, X., Jin, Z., Dong, L., Lou, J., Vajtai, R., Zhou, W., & Ajayan, P. M. (2017). Synthesis of large-scale atomic-layer SnS_2 through chemical vapor deposition. *Nano Research*, *10*(7), 2386–2394. doi:10.1007/s12274-017-1436-3

Ye, Z., Jiang, Y., Li, L., Wu, F., & Chen, R. (2021). Self-Assembly of 0D–2D Heterostructure Electrocatalyst from MOF and MXene for Boosted Lithium Polysulfide Conversion Reaction. *Advanced Materials*, *33*(33), 2101204. Advance online publication. doi:10.1002/adma.202101204 PMID:34245063

Yezhelyev, M. V., Gao, X., Xing, Y., Al-Hajj, A., Nie, S., & O'Regan, R. M. (2006). Emerging use of nanoparticles in diagnosis and treatment of breast cancer. *The Lancet. Oncology*, *7*(8), 657–667. doi:10.1016/S1470-2045(06)70793-8 PMID:16887483

Yguerabide, J., & Yguerabide, E. E. (1998). Light-scattering submicroscopic particles as highly fluorescent analogs and their use as tracer labels in clinical and biological applications: II. Experimental characterization. *Analytical Biochemistry*, *262*(2), 157–176. doi:10.1006/abio.1998.2760 PMID:9750129

Yingngam, B., Chiangsom, A., Pharikarn, P., Vonganakasame, K., Kanoknitthiran, V., Rungseevijitprapa, W., & Prasitpuriprecha, C. (2019). Optimization of menthol-loaded nanocapsules for skin application using the response surface methodology. *Journal of Drug Delivery Science and Technology*, *53*, 101138. doi:10.1016/j.jddst.2019.101138

Yingngam, B., Kacha, W., Rungseevijitprapa, W., Sudta, P., Prasitpuriprecha, C., & Brantner, A. (2019). Response surface optimization of spray-dried citronella oil microcapsules with reduced volatility and irritation for cosmetic textile uses. *Powder Technology*, *355*, 372–385. doi:10.1016/j.powtec.2019.07.065

Yin, J. M., Quan, M. P., Wang, Z., Wang, J., Yang, Z. K., Duan, L. S., Li, Z., Li, Q. X., Wang, H., & Tan, W. M. (2022). Cationic modified lignin: Regulation of synthetic microspheres for achieving anti-photolysis and sustained release of the abscisic acid. *Industrial Crops and Products*, *177*, 114573. doi:10.1016/j.indcrop.2022.114573

Yin, J. M., Wang, H. L., Yang, Z. K., Wang, J., Wang, Z., Duan, L. S., Li, Z., & Tan, W. M. (2020). Engineering lignin nanomicroparticles for the antiphotolysis and controlled release of the plant growth regulator abscisic acid. *Journal of Agricultural and Food Chemistry*, *68*(28), 7360–7368. doi:10.1021/acs.jafc.0c02835 PMID:32627551

Yin, Z., Cui, C., Chen, H., Duoni, Yu, X., & Qian, W. (2020). The Application of Carbon Nanotube/Graphene-Based Nanomaterials in Wastewater Treatment. *Small*, *16*(15), 1902301. doi:10.1002/smll.201902301

Yoo, Y. K., Varela-Guerrero, V., & Jeong, H. (2011). Isoreticular Metal−Organic Frameworks and Their Membranes with Enhanced Crack Resistance and Moisture Stability by Surfactant-Assisted Drying. *Langmuir*, *27*(6), 2652–2657. doi:10.1021/la104775d PMID:21299194

You, J., Dou, L., Yoshimura, K., Kato, T., Ohya, K., Moriarty, T., Emery, K., Chen, C. C., Gao, J., Li, G., & Yang, Y. (2013). A polymer tandem solar cell with 10.6% power conversion efficiency. *Nature Communications*, *4*(1), 1446. Advance online publication. doi:10.1038/ncomms2411 PMID:23385590

Younis, M. A., Tawfeek, H. M., Abdellatif, A. A., Abdel-Aleem, J. A., & Harashima, H. (2022). Clinical translation of nanomedicines: Challenges, opportunities, and keys. *Advanced Drug Delivery Reviews*, *181*, 114083. doi:10.1016/j.addr.2021.114083 PMID:34929251

Yuan, Q., & Williams, R. A. (2007). Large scale manufacture of magnetic polymer particles using membranes and microfluidic devices. *China Particuology*, *5*(1-2), 26–42. doi:10.1016/j.cpart.2007.02.001

Yu, H., Zhang, J., Shaheen, I., Ahmad, M., Chen, X., Akkinepally, B., & Hussain, I. (2023). Nature-resembled nanostructures for energy storage/conversion applications. *Journal of Industrial and Engineering Chemistry*. Advance online publication. doi:10.1016/j.jiec.2023.08.041

Yu, J. C., Ho, W., Yu, J., Yip, H., Wong, P. K., & Zhao, J. (2005). Efficient visible-light-induced photocatalytic disinfection on sulfur-doped nanocrystalline titania. *Environmental Science & Technology*, *39*(4), 1175–1179. doi:10.1021/es035374h PMID:15773492

Yu, J., Li, J., Zhang, W., & Chang, H. (2015). Synthesis of high quality two-dimensional materials via chemical vapor deposition. *Chemical Science (Cambridge)*, *6*(12), 6705–6716. doi:10.1039/C5SC01941A PMID:29861920

Yun, S., Zhang, Y., Xu, Q., Liu, J., & Qin, Y. (2019). Recent advance in new-generation integrated devices for energy harvesting and storage. *Nano Energy*, *60*, 600–619. doi:10.1016/j.nanoen.2019.03.074

Yu, W., Liu, C., Hou, P., Zhang, L., Shan, X., Li, F., & Cheng, H. (2015). Lithiation of silicon nanoparticles confined in carbon nanotubes. *ACS Nano*, *9*(5), 5063–5071. doi:10.1021/acsnano.5b00157 PMID:25869474

Yu, Y., Ouyang, C., Gao, Y., Si, Z., Chen, W., Wang, Z., & Xue, G. (2005). Synthesis and characterization of carbon nanotube/polypyrrole core–shell nanocomposites via *in situ* inverse microemulsion. *Journal of Polymer Science. Part A, Polymer Chemistry*, *43*(23), 6105–6115. doi:10.1002/pola.21114

Zan, L., Fa, W., Peng, T., & Gong, Z. K. (2007). Photocatalysis effect of nanometer TiO2 and TiO2-coated ceramic plate on Hepatitis B virus. *Journal of Photochemistry and Photobiology. B, Biology*, *86*(2), 165–169. doi:10.1016/j.jphotobiol.2006.09.002 PMID:17055286

Zhang, P., Guo, Z., Ullah, S., Melagraki, G., Afantitis, A., & Lynch, I. (2021). Nanotechnology and artificial intelligence to enable sustainable and precision agriculture. In Nature Plants (Vol. 7, Issue 7, pp. 864–876). Nature Research. doi:10.1038/s41477-021-00946-6

Zhang, F., Shi, Y., Sun, X., Zhao, D., & Stucky, G. D. (2009). Formation of hollow upconversion rare-earth fluoride nanospheres: Nanoscale kirkendall effect during ion exchange. *Chemistry of Materials*, *21*(21), 5237–5243. doi:10.1021/cm902231s

Zhang, G., Cui, X., Qin, Z., Wang, Z., Lu, Y., Xu, Y., Xu, S., Tang, L., Zhang, L., Liu, G., Wang, X., Zhang, J., & Tang, J. (2023). Atherosclerotic plaque vulnerability quantification system for clinical and biological interpretability. *iScience*, *26*(9), 107587. doi:10.1016/j.isci.2023.107587 PMID:37664595

Zhang, H., Cao, G., Wang, Z., Yang, Y., Shi, Z., & Gu, Z. (2008). Growth of manganese oxide nanoflowers on Vertically-Aligned carbon nanotube arrays for High-Rate Electrochemical capacitive Energy Storage. *Nano Letters*, *8*(9), 2664–2668. doi:10.1021/nl800925j PMID:18715042

Zhang, H., Chen, F., Zhai, Y., Jiang, K., Li, Q., & Fan, S. (2009). Cross-Stacked Carbon Nanotube Sheets Uniformly Loaded with SnO_2 Nanoparticles: A Novel Binder-Free and High-Capacity Anode Material for Lithium-Ion Batteries. *Advanced Materials*, *21*(22), 2299–2304. doi:10.1002/adma.200802290

Zhang, H., Feng, J., Fei, T., Liu, S., & Zhang, T. (2014). SnO2 nanoparticles-reduced graphene oxide nanocomposites for NO2 sensing at low operating temperature. *Sensors and Actuators. B, Chemical*, *190*, 472–478. doi:10.1016/j.snb.2013.08.067

Zhang, J., Huang, H., Song, G., Huang, K., Luo, Y., Liu, Q., He, X., & Cheng, N. (2022). Intelligent biosensing strategies for rapid detection in food safety: A review. *Biosensors & Bioelectronics*, *202*, 114003. Advance online publication. doi:10.1016/j.bios.2022.114003 PMID:35065479

Zhang, J., Hu, K., Di, L., Wang, P., Liu, Z., Zhang, J., ... Chen, T. (2021). Traditional herbal medicine and nanomedicine: Converging disciplines to improve therapeutic efficacy and human health. *Advanced Drug Delivery Reviews*, *178*, 113964. doi:10.1016/j.addr.2021.113964 PMID:34499982

Zhang, J., Wang, F., Shenoy, V. B., Tang, M., & Lou, J. (2020). Towards controlled synthesis of 2D crystals by chemical vapor deposition (CVD). *Materials Today*, *40*, 132–139. doi:10.1016/j.mattod.2020.06.012

Zhang, L., Chen, Q., Ma, Y., & Sun, J. (2019). Microfluidic methods for fabrication and engineering of nanoparticle drug delivery systems. *ACS Applied Bio Materials*, *3*(1), 107–120. doi:10.1021/acsabm.9b00853 PMID:35019430

Zhang, M., He, F., Zhao, D., & Hao, X. (2011). Degradation of soil-sorbed trichloroethylene by stabilized zero valent iron nanoparticles: Effects of sorption, surfactants, and natural organic matter. *Water Research*, *45*(7), 2401–2414. doi:10.1016/j.watres.2011.01.028 PMID:21376362

Zhang, M., Shao, S., Yue, H., Wang, X., Zhang, W., Chen, F., Zheng, L., Xing, J., & Qin, Y. (2021). High stability au nps: From design to application in nanomedicine. *International Journal of Nanomedicine*, *16*, 6067–6094. doi:10.2147/IJN.S322900 PMID:34511906

Zhang, S., Gao, H., & Bao, G. (2015). Physical principles of nanoparticle cellular endocytosis. *ACS Nano*, *9*(9), 8655–8671. doi:10.1021/acsnano.5b03184 PMID:26256227

Zhang, T., Lowry, G., Capiro, N. L., Chen, W., Chen, Y., Dionysiou, D. D., Elliott, D., Choshal, S., Hofmann, T., Hsu-Kim, H., Hughes, J., Jiang, C., Jiang, G., Jing, C., Kavanaugh, M., Li, Q., Liu, S., Ma, J., Pan, B., ... Alvarez, P. J. J. (2019). In situ remediation of subsurface contamination: Opportunities and challenges for nanotechnology and advances materials. *Environmental Science. Nano*, *6*(5), 1283–1302. doi:10.1039/C9EN00143C

Zhang, W., Zhang, Y., Qiu, J., Zhao, Z., & Liu, N. (2021). Topological structures of transition metal dichalcogenides: A review on fabrication, effects, applications, and potential. *InfoMat*, *3*(2), 133–154. doi:10.1002/inf2.12156

Zhang, Y. I., Zhang, L., & Zhou, C. (2013). Review of chemical vapor deposition of graphene and related applications. *Accounts of Chemical Research*, *46*(10), 2329–2339. doi:10.1021/ar300203n PMID:23480816

Zhang, Y., Li, Z., Zhao, Z., Li, Y., Liu, Z., & Sun, S. (2023). Mesoporous carbon in biomedicine: Modification strategies and biocompatibility. *Carbon*, *212*, 118121. doi:10.1016/j.carbon.2023.118121

Zhang, Y., Tang, T. T., Girit, C., Hao, Z., Martin, M. C., Zettl, A., Crommie, M. F., Shen, Y. R., & Wang, F. (2009). Direct observation of a widely tunable bandgap in bilayer graphene. *Nature*, *459*(7248), 820–823. doi:10.1038/nature08105 PMID:19516337

Zhang, Y., Wu, B., Xu, H., Liu, H., Wang, M., He, Y., & Pan, B. (2016). Nanomaterials-enabled water and wastewater treatment. *NanoImpact*, *3*, 22–39. doi:10.1016/j.impact.2016.09.004

Zhan, Y., Liu, Z., Najmaei, S., Ajayan, P. M., & Lou, J. (2012). Large-area vapor-phase growth and characterization of MoS_2 atomic layers on a SiO_2 substrate. *Small*, *8*(7), 966–971. doi:10.1002/smll.201102654 PMID:22334392

Zhao, L., Sun, Y., Hernandez-Viezcas, J. A., Servin, A. D., Hong, J., Niu, G., Peralta-Videa, J. R., Duarte-Gardea, M., & Gardea-Torresdey, J. L. (2013). Influence of CeO2 and ZnO nanoparticles on cucumber physiological markers and bioaccumulation of Ce and Zn: A life cycle study. *Journal of Agricultural and Food Chemistry*, *61*(49), 11945–11951. doi:10.1021/jf404328e PMID:24245665

Zhao, W., Liu, Y., Zhang, P., Zhou, P., Wu, Z., Lou, B., Jiang, Y., Shakoor, N., Li, M., Li, Y., Lynch, I., Rui, Y., & Tan, Z. (2022). Engineered Zn-based nano-pesticides as an opportunity for treatment of phytopathogens in agriculture. *NanoImpact*, *28*, 100420. Advance online publication. doi:10.1016/j.impact.2022.100420 PMID:36038133

Zhao, X., Liu, W., Cai, Z., Han, B., Qian, T., & Zhao, D. (2016). An overview of preparation and applications of stabilized zero-valent iron nanoparticles for soil and groundwater remediation. *Water Research*, *100*, 245–266. doi:10.1016/j.watres.2016.05.019 PMID:27206054

Zheng, K., Wu, J., Li, W., Dippold, D., Wan, Y., & Boccaccini, A. R. (2018). Incorporation of Cu-containing bioactive glass nanoparticles in gelatin-coated scaffolds enhances bioactivity and osteogenic activity. *ACS Biomaterials Science & Engineering*, *4*(5), 1546–1557. doi:10.1021/acsbiomaterials.8b00051 PMID:33445312

Zheng, M., Pan, M., Zhang, W., Lin, H., Wu, S., Lu, C., Tang, S., Liu, D., & Cai, J. (2021). Poly (α-l-lysine)-based nanomaterials for versatile biomedical applications: Current advances and perspectives. *Bioactive Materials*, *6*(7), 1878–1909. doi:10.1016/j.bioactmat.2020.12.001 PMID:33364529

Zheng, Y., Liu, Y., Guo, X., Chen, Z., Zhang, W., Wang, Y., Tang, X., Zhang, Y., & Zhao, Y. (2020). Sulfur-doped g-C3N4/rGO porous nanosheets for highly efficient photocatalytic degradation of refractory contaminants. *Journal of Materials Science and Technology*, *41*, 117–126. doi:10.1016/j.jmst.2019.09.018

Zhou, Y., Jing, X., & Chen, Y. (2017). Material chemistry of graphene oxide-based nanocomposites for theranostic nanomedicine. *Journal of Materials Chemistry. B*, *5*(32), 6451–6470. doi:10.1039/C7TB00680B PMID:32264411

Zhu, D., Kong, H., Sun, Z., Xu, Y., Han, P., Xi, Y., & Wei, G. (2023). Recent advance in tailoring the structure and functions of self-assembled peptide nanomaterials for biomedical applications. *Coordination Chemistry Reviews*, *494*, 215374. Advance online publication. doi:10.1016/j.ccr.2023.215374

Zhu, G., Chen, J., Tie-Jun, Z., Jing, Q., & Wang, Z. L. (2014). Radial-arrayed rotary electrification for high performance triboelectric generator. *Nature Communications*, *5*(1), 3426. doi:10.1038/ncomms4426 PMID:24594501

Zhu, Y., Liu, Y., Khan, K., Arkin, G., Tareen, A. K., Xie, Z., He, T., Su, L., Guo, F., Lai, X., Xu, J., & Zhang, J. (2023). Ultrasound combined with nanomaterials for cancer therapy. *Materials Today. Advances*, *17*, 100330. doi:10.1016/j.mtadv.2022.100330

About the Contributors

Virat Khanna is an Associate Professor and Head in the Department of Mechanical Engineering at Maharaja Agrasen University, HP, India and has 12 years of teaching experience. He is Ph.D. in Mechanical Engineering from Maharaja Agrasen University, HP, India. He Completed his M Tech in Mechanical Engineering in 2013 and B Tech in Mechanical Engineering in 2008 from I. K. Gujral Punjab Technical University, Jalandhar, India. His areas of research are Aluminium Metal matrix nanocomposite (Graphene, CNT, Processing technique, optimization and mathematical modelling techniques). He has published and presented more than 20 papers in Journals and conferences of national and international repute. He is also the editor of several books and conference proceedings with various international publishers like Elsevier, CRC Press, IGI Global, Bentham books etc.

Priyanka Mahajan is currently working as an Assistant Professor (Botany) in the School of Basic and Applied Sciences, Maharaja Agrasen University, HP, India. She has 2 years of teaching experience and 5 years of Post-Doctoral research experience in the field of plant sciences. She has completed her graduation and post-graduation degree from Panjab University, Chandigarh in Botany (Hons.). She has acquired her Ph.D. degree from Department of Botany, Panjab University, Chandigarh in the field of Botany (Eco- physiology) in 2015. The potential of monoterpene (Beta-pinene)- a natural plant product was explored to ameliorate the phytotoxic effects of hexavalent chromium in the cereal crops during her doctorate and post- doctorate research. Her academic and research achievements include Gold-Medal in M.sc, 8 research and review articles in the international journal of high impact factors and 4 chapters in the books of international publishers. She has participated and presented her research work in number of national and international conferences.

* * *

Madhu Bala is working as Assistant Professor in the Department of Computer Science and Engineering, Maharaja Agrasen University, Solan, H.P., India. She has more than 10 years of teaching experience. Her research area is plant leaf disease identification using deep learning techniques. She is highly motivated and accomplished individual with a passion for using cutting-edge technology to address pressing challenges in the field of agriculture. She did her M. Tech in Computer Science and Engineering from CDL University, Sirsa, India and awarded as gold medalist for doing academically best. Her areas of interest are image processing, computer vision, artificial intelligence and machine learning. She has several National and International Publications to her credits and published two patents. She has mentored various technical events like Hackathons and has won two awards at National Level. She has also received recognition from the Ministry of Human Resource and Development, India for her excellent efforts as a convener and an Innovation Ambassador for Institution Innovation Council.

Sisira Donsamak, a lecturer in the Department of Pharmacy Practice at Ubon Ratchathani University in Thailand, earned her bachelor's degree with first-class honors and a Master of Pharmacy in Clinical Pharmacy from Khon Kaen University, Thailand. She then went on to complete her Ph.D. in Pharmacy at Cardiff University in the United Kingdom, funded by a scholarship from the Office of the Civil Service Commission of the Thai Government. Dr. Donsamak is dedicated to advancing community pharmacy services and promoting the rational use of medications, a commitment she fulfills through both her role as an educator and her research endeavors. https://orcid.org/0000-0003-4031-4760.ccbaeb6c-7b63-4844-9c12-e2e3be0f0769

Shilpa Gupta is working as an Assistant Professor in Department of Electrical and Electronics Engineering, Maharaja Agrasen University, H.P., India. She has 15+ years of teaching experience with 1.5 year experience of working in industry including patent analyst in 3aip solutions. She is guiding various Ph.D Scholars in the field of soft robotics and speech recognition. She did her Ph.D in Electrical Engineering from National Institute of Technology (NIT), Kurukshetra, India, in the field of Evaluation Of Reliability Metrics Of Shuffle Exchange And Gamma Interconnection. She did her M.Tech from NIT, Kurukshetra in VLSI Design. Her areas of interest are Intellectual Property right (IPR), Reliability, Fault Tolerance, Quality of service, Computer networks and crossbars, Soft Robotics. She has 16 publications in National and International Journals with various indexings including scopus and web of science, 13 international conferences with best paper award to her credits. She has published three patents till date. She has mentored various technical events like Hackathons and has won two awards.

Jeerisuda Khumsikiew serves in the Department of Pharmacy Practice at Ubon Ratchathani University in Thailand. She holds a Bachelor's, Master's, and Ph.D. in pharmaceutical sciences and clinical pharmacy from Ubon Ratchathani and Khon Kaen Universities. She further honed her skills through a six-month research stint at the University of Nottingham, UK, funded by a prestigious Royal Golden Jubilee scholarship. Committed to advancing clinical pharmacy and primary care, her research focuses on bridging theory with practice, particularly in patient-centered pharmaceutical care. Through her work, she aims to elevate the standards of evidence-based pharmaceutical care. https://orcid.org/0000-0002-1930-5977.11fc2f3c-ed38-4f2c-81c6-b10e36bf630b

Abhiruj Navabhatra, an assistant professor at Rangsit University in Thailand, specializes in pharmacology and toxicology. He earned his pharmacy degree in 2005 and a Ph.D. from Chulalongkorn University in 2016. During a 2014-2016 fellowship at Nagoya University, Japan, he focused on enzyme mechanisms, specifically cytochrome P450, and drug interactions, earning a Scopus H-index of 5. He has published over ten scholarly works, mainly in pharmacology, toxicology, and machine learning applications in pharmaceutical sciences. His research has appeared in esteemed journals, and he continues to make significant contributions to his field. https://orcid.org/0000-0003-4129-1302.1067fc47-c129-44be-940d-b5c6594facd0

Rojjares Netthong, affiliated with Ubon Ratchathani University's Department of Pharmacy Practice, holds a PhD in Pharmacy and Pharmaceutical Sciences. With a nine-year background in the field, she has diversified her expertise through an MBA focused on Appreciative Inquiry theory. She has served on the university faculty, managed its pharmacy store, and initiated community pharmacy projects. Recently, completing her PhD from the University of Lincoln in 2023, her research aims to advance the smart use of antibiotics in Thailand. Dr. Netthong is well-versed in international healthcare policies, particularly those of the UK. She presented her work in various academic settings, including a national forum on Antimicrobial Resistance in Thailand in 2022. Additionally, she has contributed to the journal "Antibiotics" and maintains an active academic network through platforms such as LinkedIn, ResearchGate, and ORCID. https://orcid.org/0000-0001-9751-7564.0c370a23-3cbb-4952-9c63-22e95e31565d

Benjabhorn Sethabouppha, Ph.D., earned her Doctorate in Pharmaceutical Sciences from the University of Manchester, UK, and currently serves in the Department of Pharmaceutical Chemistry and Technology at Ubon Ratchathani University, Thailand. Known for her multidisciplinary expertise, she specializes in areas such as industrial pharmacy, good manufacturing practices, production process optimization, quality assurance and control, formulation sciences, and physiochemical characterization of solid-state materials, among others. Her scholarly work serves as a vital bridge between academia and industry, significantly contributing to the advancement of pharmaceutical sciences and the education of emerging experts in the field.

Ritika Sharma is an Assistant Professor in Computer Science and Engineering Department at Maharaja Agrasen Institute of Technology, Maharaja Agrasen University, H.P, India and has 6 years of teaching experience. She has completed her B.Tech. (Honors) in Computer Science and Engineering from Himachal Pradesh University, Shimla, India in 2013, M.Tech. (Honors) from Department of Computer Science and Applications, Kurukshetra University, Kurukshetra, India in 2016 and pursuing her Ph.D. from MAIT, Maharaja Agrasen University, H.P, India. She has presented and published various research papers in national and international conferences and also published 4 patents. Under her guidance students participates in various project competitions and has won two prizes in National Level Smart India Hackathons. Her research interest is in Image Processing, Neural Networks, Artificial Intelligence, Machine Learning, Deep Learning, Data Science.

Archna Talwar has experience of 36 years in research and teaching. She has published a number of papers in international and national repute journals. She had also received three national awards.

Bancha Yingngam, an Associate Professor at Ubon Ratchathani University, Thailand, specializes in Pharmaceutical Sciences and Technology. He earned his pharmacy degree in 2005 and a Ph.D. in 2011 from the same institution. During a postdoctoral fellowship at Karl-Franzens-Universität Graz, Austria, from 2013 to 2021, he expanded his research to include optimization and machine learning algorithms, earning a Scopus H-index of 10. He has published over 20 scholarly works and currently focuses on artificial intelligence applications in green chemistry and pharmacy. Dr. Yingngam has received the "Outstanding Research Award in Health Science" twice and has mentored 11 graduate students in Austria. He has also reviewed over 100 Scopus-listed articles, underscoring his commitment to advancing his field. https://orcid.org/0000-0001-7215-9123.028abe51-615e-466f-a319-307c6044b3a5

Krit Yingngam is a part-time research assistant in the Integrated Biopharmaceutical Research Cluster at the Faculty of Pharmaceutical Sciences, Ubon Ratchathani University. His core duties are with the Advanced Program at Benchama Maharat School in Ubon Ratchathani Province, Thailand. https://orcid.org/0009-0008-8919-6607.152cb349-89b8-4b0f-a1b5-0813648905a9

Index

A

Artificial Intelligence 1-3, 7, 11, 16, 18, 20, 22

B

Biomedical 1-10, 12-24, 26-27, 29-33, 41-43, 45, 47, 49-53, 55-56
Biomedical Applications 1-9, 12-15, 17-19, 21-24, 26-27, 31-33, 42, 45, 47, 50, 52-53, 55-56
Biosensors 1-4, 7, 9, 11-15, 18, 20-22, 32-35, 43

C

Cancerous 28-29
Carbon Nanotubes 3-6, 8-14, 16-17, 19-26, 28, 30, 32-33, 47, 49, 51, 54
Carbon-Based Nanomaterials 1, 3, 6, 8-9, 14, 17, 23, 27-28
Catalytic Activity 1-2, 13, 19
Chemical Vapor Deposition 1-4, 9, 11, 18-25

D

Drug Delivery 1-4, 7-23, 25, 29, 32, 40-47, 50, 55

E

Electronics 1-2, 4-7, 13-15, 20-22, 24
Energy Harvesting 1-3, 7, 9, 13-16, 18-19, 23, 26
Energy Storage 1-10, 12-21, 23-25, 27-29, 31-32
Environmental Sustainability 1-2, 8, 15

G

Green Synthesis 7, 16-17, 19, 21, 24, 27, 31-32, 48, 52

H

Heterostructures 3-4, 10, 18-23
Hybrid Nano Composite Materials 1
Hybrid Nanomaterial 1, 4, 7, 9-11, 13, 16-18, 48, 54
Hybridization 1-2, 4, 8, 17, 23-24

I

Imaging Probes 1, 13, 27

M

Material Efficiency 19
Mechanical testing 6, 16
Medicine 1-3, 5-6, 10-17, 19, 22-28, 31, 34, 41-42, 44
Microstructural characteristics 7
Multifunctional 1-2, 5, 8-10, 13-24, 26, 28, 31, 46, 53

N

Nanobots 1, 3
Nanocarriers 2-4, 6, 8, 10-11, 13-14, 16, 20, 45

Nanoclusters 5-6, 19, 21-22
Nanofertilizers 13
Nanomaterials 1-36, 40-56
Nanomedicine 2-3, 13-16, 18-19, 21-24, 27-28, 50, 52
Nanometer 1-2, 10, 12, 16, 21, 35-36
Nanosensors 11, 13-15, 20, 29, 31, 33
Nanotechnology 1-7, 10-24, 26-34

P

Photocatalysis 3, 6-7, 12, 17, 21, 24-25, 32, 36
Phototherapy 1, 8
Precision Farming 1, 3, 8, 10

S

Semiconductors 2, 5-6, 8, 12, 14-15, 34-35
Sensor Design 1
Smart Irrigation 14
Soil Health 4, 8, 10
Soil Remediation 16, 29
Solar Cells 1-2, 5-8, 10, 13-17, 20, 23-24, 31
Surface Morphology 6, 10

T

Theragnostic 15, 31-32, 43
Theranostics 1, 11-17, 19, 21, 46
Theratonics 1, 12
Tissue Engineering 1-3, 7-13, 16-17, 19-20, 22-25, 41, 43-45, 49, 52-54
TMCs 1-2
TMDs 1-3, 6-7, 9, 20

W

Waste Water Management 14, 17

www.ingramcontent.com/pod-product-compliance
Lightning Source LLC
LaVergne TN
LVHW080340011026
849161LV00009B/2138